The Periodic Table

THE PERIODIC TABLE

Its Story and Its Significance

ERIC R. SCERRI

OXFORD
UNIVERSITY PRESS
2007

OXFORD
UNIVERSITY PRESS

Oxford University Press, Inc., publishes works that further
Oxford University's objective of excellence
in research, scholarship, and education.

Oxford   New York
Auckland   Cape Town   Dar es Salaam   Hong Kong   Karachi
Kuala Lumpur   Madrid   Melbourne   Mexico City   Nairobi
New Delhi   Shanghai   Taipei   Toronto

With offices in
Argentina   Austria   Brazil   Chile   Czech Republic   France   Greece
Guatemala   Hungary   Italy   Japan   Poland   Portugal   Singapore
South Korea   Switzerland   Thailand   Turkey   Ukraine   Vietnam

Published by Oxford University Press, Inc.
198 Madison Avenue, New York, New York 10016

www.oup.com

Oxford is a registered trademark of Oxford University Press

Library of Congress Cataloging-in-Publication Data
Scerri, Eric R.
The periodic table : its story and its significance / Eric R. Scerri.
p. cm.
Includes bibliographical references.
ISBN-13 978-0-19-530573-9

1. Periodic law—Tables. 2. Chemical elements. I. Title.
QD467.S345 2006
546'.8—dc22   2005037784

3 5 7 9 8 6 4

Printed in the United States of America
on acid-free paper

I dedicate this book to my mother, Ines,
and my late father, Edward Scerri,
for steering me toward the scholarly life.

I also dedicate this book to the 100th
anniversary of the death of Dimitri
Mendeleev (1834–1907).

ACKNOWLEDGMENTS

This book has been in the making for about six years, although perhaps I should say about 20 years since it was that long ago that I undertook my Ph.D. at what was then Chelsea College, University of London, under the excellent supervision of the late Heinz Post. Of course, I could go back even further and mention that my love affair with the periodic table began when I was still in my teens and attending Walpole grammar school in the West London borough of Ealing.

Now that this book is completed, I have the opportunity to thank all those who contributed to it either directly or indirectly as colleagues or mentors at various stages of my own development. At Walpole grammar school, Mrs. Davis was the chemistry teacher who noticed that I was fooling around at the back of the class and ordered me to sit in the front row. At this point, I had no choice but to listen to the lesson and soon discovered that chemistry was rather interesting.

Moving on to Westfield College, which was part of the University of London, I had many wonderful professors, among them John Throssell and Bernard Aylett, a theoretician and an inorganic chemist, respectively. This was followed by a year of theoretical work at Cambridge under the great David Buckingham, who despaired of my asking too many philosophical questions. Then I moved to Southampton University, where I obtained a Master of Philosophy degree in Physical Chemistry with the inimitable Pat Hendra. At this point, I began teaching chemistry in high schools and tutorial colleges. I eventually went back to research and wrote my Ph.D. in history and philosophy of science on the question of the reduction of chemistry to quantum mechanics. I cannot overestimate the debt that I owe to Heinz Post, and as all who knew him recall, he was perhaps the nearest thing to the archcritic Wolfgang Pauli that ever graced the philosophy of science scene in the United Kingdom. Not that I ever witnessed Pauli, however.

It was Heinz Post who encouraged me to try to develop the philosophy of chemistry, which I have sought to do ever since. I think it was also Heinz who first planted the idea of my going to the United States to teach and carry out research. But before moving on to my story in the United States, let me pause to mention a few other folks in London who have been influential and helpful: Mike Melrose, a theoretical chemist from King's College, London, and John Worrall from the London School of Economics. It has been a great privilege to have subsequently co-authored an article with each of them.[1]

I went to the United States as a postdoctoral fellow at Caltech. Here I must thank my colleagues Diana Kormos-Buchwald, Fiona Cowie, Alan Hayek, and James Woodward in the Humanities Division. I subsequently went for a year to Bradley University in the heart of Illinois, where I was warmly received by Don Glover and Kurt Field, among others, in the chemistry department. Then followed another visiting professorship at Purdue University, where I interacted mainly with George Bodner and historian-chemist-educator Derek Davenport. In the year 2000, I moved to the chemistry department at UCLA, where I am blessed with numerous great colleagues, among others, Miguel Garcia-Garibay, Robin Garrell, Steve Hardinger, Ken Houk, Herb Kaesz, Richard Kaner, Laurence Lavelle, Tom Mason, Craig Merlic and Harold Martinson.

In addition, I am grateful to all the members of the International Philosophy of Chemistry Society, which a small group of us founded in the early 1990s after we realized that there were a sufficient number of people with an interest in this field.[2] My thanks to Michael Akeroyd, Davis Baird, Nalini Bhushan, Paul Boogard, Joseph Earley, Rom Harré, Robin Hendry, David Knight, Mark Leach, Paul Needham, Mary Jo Nye, Jeff Ramsay, Joachim Schummer, Jaap van Brakel, Krishna Vemulapalli, Stephen Weininger, Michael Weisberg, and many others.

Perhaps the largest group to acknowledge consists of the many scholars of the periodic table from diverse fields, who include Peter Atkins, Henry Bent, Bernadette Bensaude, Nathan Brooks, Fernando Dufour, John Emsley, Michael Gordin, Ray Hefferlin, Bill Jensen, Masanori Kaji, Maurice Kibler, Bruce King, Mike Laing, Dennis Rouvray, Oliver Sacks, Mark Winters, and others.

I thank my various co-editors at *Foundations of Chemistry*, both past and present, including John Bloor, Carmen Giunta, Jeffrey Kovac, and Lee McIntyre. I thank my UCLA colleagues in the Department of Philosophy, including Calvin Normore, Sheldon Smith, and Chris Smeenk and in the Department of History, Ted Porter and Norton Wise.

My thanks to members of various online discussion lists, including Chemed, History of Chemistry, Philchem, Hopos, and CCL (Computational Chemistry Listserver), with whom various points were ironed out, sometimes amidst heated debate.

Last but not least, there are a number of people who helped me specifically with the compilation of this book, especially with collecting photos and images. They include Ted Benfey, Gordon Woods, Ernst Homberg, and Frenando Dufour and Susan Zoske; George Helfand and Andreana Adler from the photographic unit at UCLA who scanned the diagrams; and Marion Peters in the chemistry department library. Special thanks to Daniel Contreras, who was always patient in helping me to unearth those obscure early sources—I am sure he grew quite tired of ordering volumes of *Science News* for me from the vaults on the other side of campus. Special thanks also go out to Geoffrey Rayner Canham and William Brock for their detailed comments on the entire manuscript and to Jan Van Spronsen, the doyen of the periodic table, for his comments on some early chapters of the book. Thanks to Amy Bianco. At OUP I thank my editor Jeremy Lewis, as well as Abby Russell, Michael Seiden, Laura Ikwild, and Lisa Stallings, all of whom made the publishing process a pleasurable experience.

PHOTO CREDITS

The Edgar Fahs Smith Collection at the University of Pennsylvania for providing photos and permission for use of photos of Cannizzaro, Dalton, Lavoisier, Lewis, and Ramsay, and for permission to use a photo of Lothar Meyer.

The Emilio Segrè collection at the American Institute of Physics for photos and permission for use of photos of Bohr; the Burbidges, Fowler, and Hoyle; Curie; G.N. Lewis; Mendeleev; and Seaborg, and for permission to use photos of Moseley and Pauli.

Gordon Woods for providing the photo and permission for "The Consolidators of the Periodic Law."

Fernando Dufour for providing the photo of his own 3-D periodic system.

ARTICLE CREDITS

I have drawn from my own previous articles and particularly from four publications:

British Journal for the Philosophy of Science, 42, 309–325, 1991 (published by Oxford University Press, UK).
Annals of Science, 51, 137–150, 1994 (published by Taylor & Francis).
Studies in History and Philosophy of Science, 32, 47–452, 2001 (published by Elsevier).
Foundations of Chemistry, 6, 93–116, 2004 (published by Springer).

All these articles were used by permission from the publishers.

CONTENTS

Introduction   xiii

CHAPTER 1
The Periodic System: An Overview   3

CHAPTER 2
Quantitative Relationships among the Elements
and the Origins of the Periodic Table   29

CHAPTER 3
Discoverers of the Periodic System   63

CHAPTER 4
Mendeleev   101

CHAPTER 5
Prediction and Accommodation: The Acceptance
of Mendeleev's Periodic System   123

CHAPTER 6
The Nucleus and the Periodic Table: Radioactivity,
Atomic Number, and Isotopy   159

CHAPTER 7
The Electron and Chemical Periodicity   183

CHAPTER 8
Electronic Explanations of the Periodic System
Developed by Chemists   205

CHAPTER 9
Quantum Mechanics and the Periodic Table   227

CHAPTER 10
Astrophysics, Nucleosynthesis, and More Chemistry   249

Notes   287

Index   329

INTRODUCTION

> As long as chemistry is studied there will be a periodic table.
> And even if someday we communicate with another part of
> the universe, we can be sure that one thing that both cul-
> tures will have in common is an ordered system of the ele-
> ments that will be instantly recognizable by both intelligent
> life forms.
>
> J. Emsley, *The Elements*

The periodic table of the elements is one of the most powerful icons in science:
a single document that captures the essence of chemistry in an elegant pattern.
Indeed, nothing quite like it exists in biology or physics, or any other branch of
science, for that matter. One sees periodic tables everywhere: in industrial labs,
workshops, academic labs, and of course, lecture halls.

THE PERIODIC SYSTEM OF THE ELEMENTS

It is sometimes said that chemistry has no deep ideas, unlike physics, which can
boast quantum mechanics and relativity, and biology, which has produced the theory of evolution. This view is mistaken, however, since there are in fact two big
ideas in chemistry. They are chemical periodicity and chemical bonding, and they
are deeply interconnected.

The observation that certain elements prefer to combine with specific kinds
of elements prompted early chemists to classify the elements in tables of chemical
affinity. Later these tables would lead, somewhat indirectly, to the discovery of the
periodic system, perhaps the biggest idea in the whole of chemistry. Indeed, periodic tables arose partly through the attempts by Dimitri Mendeleev and numerous
others to make sense of the way in which particular elements enter into chemical
bonding.

The periodic table of the elements is a wonderful mnemonic and a tool that
serves to organize the whole of chemistry. All of the various periodic tables that
have been produced are attempts to depict the periodic system. The periodic system is so fundamental and all pervasive in the study of chemistry, as well as in pro-

fessional research, that it is often taken for granted, as very familiar things in life so frequently are.

In spite of the central, or some might say homely, role of the periodic table, very few authors have felt drawn to write books on its evolution. There is no book that deals adequately with the historical, and especially the conceptual, aspects of the periodic system or its significance in chemistry and science generally.[1] It is with the aim of injecting a more philosophical treatment to understanding the periodic system that the present work has been undertaken. I make no apologies for this approach, which I believe is long overdue and can perhaps be understood in the context of the almost complete neglect of the study of the philosophy of chemistry until its recent resurgence in the mid-1990s.

Only two major books on the periodic system have appeared in the English language, one of these being a translation from the Dutch original. The more contemporary of these books, published in 1969 and authored by J. van Spronsen, is an excellent and detailed exposition of the history of the periodic system. One of the few omissions from van Spronsen's book is a discussion of the way in which modern physics is generally claimed to have explained the periodic system. Van Spronsen at times accepts the usual unspoken, or sometimes explicit, claim that the periodic system has been "reduced" to quantum mechanics, to use a phrase popular in philosophy of science.[2] On my own view, the extent to which quantum mechanics reduces the periodic system is frequently overemphasized. Of course, quantum mechanics provides a better explanation than was available from the classical theories of physics, but in some crucial respects the modern explanation is still lacking, as I hope to explain.

The only other serious treatise on the periodic system, written in English, is a masterly and detailed exposition, published in 1896, by F.P. Venable of the University of North Carolina–Chapel Hill.[3] It goes without saying that, for all its strengths, this book is severely limited, as it covers a period that ended more than 100 years ago, before modern physics began to exercise a major influence on the way the periodic system is understood.

There is also a compilation of more than 700 representations of the periodic system in a book by E. Mazurs, who devoted a lifetime of study to the topic. However, this book is neither a history nor a philosophy of the periodic system but a rather idiosyncratic attempt to develop a system of classification for periodic classifications themselves. It serves as a repository of the huge variety of forms in which the periodic system has been represented and is a testament to how expansive and energetic the quest for the ultimate form of the periodic system has been.[4] This quest appears to be with us to this day, an issue that will be taken up in later chapters. Another virtue of the Mazurs book is that it provides a wealth of references to the primary and secondary literature on the periodic system, although this, too, is now some 25 or so years out of date.[5]

The textbook author Peter Atkins has published a short popular book on the periodic system.[6] There are also a number of books, including those by Puddephatt and Monaghan,[7] as well as by Cooper,[8] Pode,[9] and Sanderson,[10] which use the periodic system as a means of presenting the chemistry of the elements but make little attempt to evaluate critically the foundational basis of the system. The continuing interest in the periodic system is further exemplified by the appearance of three recent books aimed at the nonspecialist by Strathern, Sacks, and Morris.[11] Although the focus of these books is on chemistry generally, they contain sections on the development of the periodic system. Very recently, M. Gordin has published a biography of Mendeleev, which is historically sensitive as well as scientifically accurate, and benefits from the author's first-hand knowledge of the original Russian documents.[12]

THE ELEMENTS

In this book I examine the concept of an element in some detail, starting from the views expressed by the ancient Greek philosophers and bringing us right up to modern times. Although this topic has seldom been discussed in the context of the evolution of the periodic system, it is difficult to fully understand the classification of the elements without first attempting to understand what an element is and how such a concept has changed over time. There is a sense in which ancient views on the nature of the elements have not been entirely rejected, although they have been changed considerably.

The study of the nature of elements and compounds is at the heart of much of Aristotle's philosophy of substance and matter and even his most general views of "being" and "becoming." This was also true of many of the pre-Socratic philosophers, who were the first to discuss and theorize about the elements. About 20 centuries later, the nature of the elements was a major issue in the revolution of chemistry. Antoine Lavoisier seems to have been one of the first chemists to renounce the metaphysical view of the elements, which he replaced with a form of empiricism, which considered only substances that could actually be isolated as elements. Elements in this latter sense of the term are often called "simple substances."

This essentially philosophical question regarding the nature of elements returned and profoundly shaped the views of Mendeleev, who is arguably the leading discoverer of the periodic system. Indeed, it appears that Mendeleev may have been able to make more progress than some of his contemporaries, who were also developing periodic systems, because of his philosophical ideas about the nature of the elements. Even in the twentieth century, following the discovery of isotopes, fierce debates were waged on the nature and correct definition of the term "element."

Mendeleev held a dual view on the nature of elements, whereby they could be regarded as unobservable basic substances and also as Lavoisier's simple substances at the same time. Mendeleev thus acknowledged one of the central mysteries running throughout the long history of chemistry, which is the question of how, if at all, the elements survive in the compounds they form when they are combined together. For example, how can it be claimed that a poisonous gray metal like sodium is still present when it combines with a green poisonous gas chlorine, given that the compound formed, sodium chloride, or common table salt, is white and not only nonpoisonous but also essential for life? These are the kinds of questions the ancient Greek philosophers wrestled with while trying to understand the nature of matter and change. As I will show, such questions are still with us today, although some aspects of them have been explained by modern physical theory and the theories of chemical bonding.

Alchemy

Although in this book I briefly examine the nature of the elements, and of atomism from their earliest origins, not too much time is devoted to issues surrounding alchemy, for various reasons. First, the study of alchemy has been fraught with the obvious difficulties of trying to understand a complex set of practices spanning a number of areas, including what today would be considered religion, psychology, numerology, metallurgy, and so on. In addition, alchemical texts were frequently shrouded in deliberate mystery and obfuscation to protect the practitioners, who were regularly accused of being charlatans. Such mystery also served to restrict alchemical knowledge to a few initiates belonging to particular secret cults.

The question of whether modern chemistry is a direct outgrowth of alchemy, or whether alchemy's fundamental tenets had to be rejected in order for chemistry to get started, has been the source of much debate and continues to be disputed by scholars. All I do here is refer the reader to a few detailed treatments containing more serious discussion of this vast field of study.

One interesting aspect of this issue that has emerged in recent years is a questioning of the notion that the giants of modern science, such as Newton and Boyle, turned their backs on alchemy. Starting about 30 years ago, historians of science, and Betty Jo Dobbs in particular, have argued rather persuasively that Newton was a dedicated alchemist and that he might even have devoted more time to this field than to his work in theoretical physics, for which he is now universally revered. More recently, Lawrence Principe has re-alchemized Boyle in a similar way that Dobbs had re-alchemized Newton.[13] Through painstaking analysis of Boyle's writing, Principe argues that, contrary to the accepted view, Boyle did not reject alchemical ways in his seminal book, *The Sceptical Chymist*. In fact, Principe writes:

We now see that Boyle himself in no way rejected transmutational alchemy but rather pursued it avidly and appropriated several of its theoretical principles. . . . Boyle was not as "modern" as we thought, nor alchemy as "ancient." What we are witnessing, then, is a rapprochement between what have been previously seen as two separate and irreconcilable halves of the history of chemistry.[14]

A Philosophical Approach

As I have already suggested, the study of the periodic system is philosophically important in several ways. Let me be a little more specific. For some time now, philosophers of science have realized that they have placed too much emphasis on the study of scientific theories and not enough on other important aspects of science, such as experimental work and scientific practice in general.[15] This has led many researchers to initiate the study of the philosophy of experimentation. But even within the philosophical investigation of theoretical work, there has been a growing sense that there is much more to scientific theorizing than just appealing to high-level theories.

In many cases, the theory in question is too difficult to apply, and so scientists tend to base their work on models and approximations. The full acceptance of this fact has produced a subdiscipline that studies the nature of scientific models.[16] And yet, as I argue in this book, the periodic table of chemistry is neither a theory nor a model but more akin to an "organizing principle," for want of a better term. This book is partly an attempt to encourage philosophers of science to study the periodic table as an example of yet another scientific entity that does a lot of useful scientific work without being a theory.[17]

Another reason why the periodic table is philosophically important is that it provides an excellent testing ground for the question of whether chemistry is nothing but physics deep down or, as philosophers like to say, whether chemistry reduces to physics. But even asking such a question has become controversial in modern scholarship. The view that physics is the most fundamental of the sciences or, indeed, the very notion of one field being more fundamental than another one is under severe threat from disciplines such a literary criticism, cultural anthropology, and postmodern critiques of science.[18] Such issues have become highly controversial in recent times, producing what is perhaps the major debate in today's academic world, namely, the "science wars." Many scholars, scientists, and intellectuals find themselves pitted against each other over the question of whether science provides a form of objective truth or whether it is no more than a social construction not necessarily governed by the way the world actually is. The traditional view of scientific objectivity is increasingly regarded as a thing of the past, and some scholars are even prepared to embrace a form of relativism, or the view that all forms of knowledge are equally valid.[19]

But many others believe the question of fundamentalism and reduction can still be studied within the context of science. One can still consider the more modest question of whether chemistry reduces to its sister science of physics. This question can be approached in a scientific manner by examining the extent to which chemical models or, indeed, the periodic system, can be explained by the most basic theory of physics, namely, quantum mechanics. It is this question that forms the underlying theme for this entire book, and it is a question that is addressed more and more explicitly in later chapters as the story reaches the impact of modern physical theories on our understanding of the periodic system.

The Evolution of the Periodic System

As I try to show in this book, several intermediate and anticipatory steps preceded every important stage in the development of the periodic system. Of all the major developments in the history of science, there may be no better example than that of the periodic system to argue against Thomas Kuhn's thesis that scientific progress occurs through a series of sharp revolutionary stages.[20] Indeed, Kuhn's insistence on the centrality of revolutions in the development of science and his efforts to single out revolutionary contributors has probably unwittingly contributed to the retention of a Whiggish history of science, whereby only the heroes count while blind alleys and failed attempts are written out of the story.[21]

Science is, above all, a collective endeavor involving a large variety of people working sometimes in teams, sometimes in isolation, sometimes aware of the work of their contemporaries, and sometimes not. When trying to examine the development of a system of knowledge such as the periodic system, it may be more important to look at the overall picture complete with wrinkles than to concentrate on who came first or whether a certain development really is an anticipation of a later one. Nevertheless, since priority issues are part of this fascinating story, in this book I try to give an account of some of the most important ones without claiming to provide the final word on any of the long-standing disputes.

Perhaps a further word on a different sense of the term "Whiggism" is appropriate. Since this book is not intended as a work of historical scholarship, there will be many instances in which the story will be driven by what eventually took place in the history of science. I make no apology for this approach since part of the interest is in trying to trace the development of the modern periodic system. For example, when discussing triads of elements, which were based on atomic weights, I will not avoid looking ahead to the use of atomic numbers to see what effects this change might have on the validity or otherwise of triads.

So without further delay, what follows is a brief synopsis of the chapters of the present book. I adopt a historical approach in order to convey the gradual evolution that has taken place around the chemical icon that is the periodic system. However, my primary concern is the evolution of concepts and ideas rather than trying to produce a detailed historical account.[22] At times, I even use strictly ahistorical examples to illustrate particular points.

The book takes the reader on an interdisciplinary tour of the many areas of science that are connected with the periodic system, including physics, mathematics, computational methods, history and philosophy of science, and of course, chemistry. The story begins with the pre-Socratic philosophers in ancient Greece and progresses through the birth of atomism and on to Aristotle's four elements of earth, water, fire, and air. By the Middle Ages, when the full impact of alchemy was reached, a few other elements, such as sulfur and mercury, were added to the list. But this book does not explore the state of chemical knowledge of the elements in the Middle Ages, early medicine, or Arabic chemistry, although these are important preliminaries to modern chemistry. Nor does it visit the theory of phlogiston, which was deposed by the chemical revolution; it merely examines Lavoisier's famous list of 37 fundamental substances.[23]

Instead, the story of the periodic system will take the plunge with the work of William Prout, Johann Döbereiner, Leopold Gmelin, and others who began to explore numerical relationships among the elements in addition to the previously known chemical analogies between them. We encounter the first true periodic system, which was the helical periodic system of Alexandre Emile Béguyer De Chancourtois, as well as the early periodic systems of William Odling, Gustav Hinrichs, Jean Baptiste André Dumas, Max Pettenkofer, John Newlands, and Julius Lothar Meyer, culminating with Mendeleev's tables and his deductions concerning existing as well as completely new elements. In each case, we look into some of the historical background involved as well as specific aspects of the periodic system proposed.

The discovery in the 1890s of the noble gases, a group of elements that did not initially appear to fit into the periodic system, is analyzed, as is the eventual resolution of this problem. The turn of the twentieth century saw the discovery of radioactivity, which led to new ideas about the structure of the atom from J.J. Thomson and Ernest Rutherford. Very soon, isotopes of many of the elements were discovered, and this produced a major challenge to the periodic system. Niels Bohr, Wolfgang Pauli, Erwin Schrödinger, and Werner Heisenberg, who provided the modern explanation of the periodic system in terms of orbiting electrons and quantum numbers, continued the invasion of physics into the understanding of the periodic table.

Whenever scientists are presented with a useful pattern or system of classification, it is only a matter of time before they begin to ask whether there may be some underlying explanation for the pattern. The periodic system is no exception. Attempts to produce explanations of the periodic system have led to major advances in areas of science other than chemistry, especially theoretical physics. The notion that the atom consists of a nucleus with electrons in orbit around it, which is taken for granted in modern science, originated when British physicist

J.J. Thomson tried to explain the order of the elements displayed in the periodic table. Similarly, when Bohr, one of the founders of quantum mechanics, applied new ideas about the quantum of energy to the atom, he was specifically trying to obtain a deeper understanding of the periodic system of the elements.[24]

A few years later, Pauli produced his celebrated Exclusion Principle, which is now known to govern the behavior of all matter from materials used to make transistors to the matter in neutron stars. Pauli's original research, carried out in atomic physics, was initially an attempt to explain the form of the periodic system and why the various electron shells of the atom can contain only specific numbers of electrons. In the process, Pauli produced one of the most general principles known to science. His Exclusion Principle tells us, in simple terms, that an electronic orbital can contain only two electrons, which must be spinning in opposite directions. A careful analysis of this, and other general principles of quantum mechanics, has produced the discipline of quantum chemistry, which nowadays is exploited in the development of new materials from superconductors to pharmaceutical drugs.

Now a word on the subject of chemical education. Two of the leading discoverers of the periodic system, Lothar Meyer and Mendeleev, were outstanding chemical educators who developed their versions of the periodic system while writing chemistry textbooks. One of the principal roles of the periodic table is as a teaching tool, given that it unifies so much chemical information and establishes unity amidst the diversity of chemical phenomena. In recent years, there has been a growing awareness that chemistry is being taught as though it were a subdiscipline of physics. This tendency has occurred because physics, in the form of quantum mechanics, has been successful in explaining many aspects of chemistry. But this success is frequently overemphasized.

Chemistry students are increasingly fed a diet of orbitals, electronic configurations, and other theoretical concepts instead of being exposed to the more tangible colors, smells, and even explosions of "real chemistry." Some authors advocate making chemical education more "chemical" while at the same time introducing students to the necessary concepts in modern physics. In such an endeavor, the periodic table can serve as an excellent link between macroscopic chemical properties and the underlying quantum mechanical explanations.

But in addition to any pedagogical implications, the relationship between chemistry and physics has become increasingly important in philosophy of science. In particular, the recent growth of the philosophy of chemistry as a distinct subdiscipline has been based to some extent on examining the question of the reduction of chemical laws, chemical models, and representations, such as the periodic system, to fundamental physics.

But even before the advent of philosophy of chemistry, the question of the reduction of scientific theories to successor theories has been an important concern, as has the question of whether any of the special sciences reduce to basic physics. Broadly speaking, as the logical positivist approach to philosophy has been superseded, claims for the reduction of theories and fields of science have been increasingly challenged.[25] The failure to establish the full reduction of theories and the special sciences has been one of the reasons for the demise of logical positivism in philosophy. But this failure of reduction in the manner prescribed by logical positivism has not led to the abandonment of another central tenet of logical positivism, namely, a belief in the unity of the sciences.[26]

In contemporary philosophy of science, the question of reduction is no longer approached in an axiomatic manner. It is rather pursued in a more naturalistic manner by examining the extent to which the periodic system, for example, can be deduced from the first principles of quantum mechanics. While this approach is still rigorous, it is not rigorous in the sense of using formal logic in order to establish the required connection.[27] It is rather by examining the extent to which the facts in the secondary science, if one must use such terms, can be deduced in an *ab initio* manner, to use a contemporary phrase, from computational chemistry. One needs to examine the way in which the Schrödinger Equation explains the structure of the periodic system, a topic that is specifically addressed in chapter 9. But leading up to these more contemporary developments, there were already claims made by Bohr, and on his behalf, that he had given a reduction of the periodic system using just the old quantum theory.[28] The story of the periodic system is inextricably linked with the increasing influence of modern physics upon chemistry. The question of reduction in many forms thus underlies the developments discussed in this book.

And even further back in the story of the periodic system, one can see the influence of numerical approaches dating back to Prout's hypothesis and Döbereiner's triads, both of which predate the discovery of the periodic system. Hence, it is not just chemistry that enables one to classify the elements but a combination of chemistry with the urge to reduce, in the most general Pythagorean sense of describing facts mathematically. The story of the periodic system is the story of the blending of chemistry, Pythagoreanism, and most recently, quantum physics.

If one takes a realistic view concerning the periodic law, one might claim that there is a definite fact of the matter concerning the point at which approximate repetition occurs among the elements as the atomic number sequence increases. For example, the position of the element helium has led to a certain amount of debate. While most chemists insist that the element is a noble gas, an appeal to the electronic configuration of its atoms suggests that it might be placed among the alkaline earths. A chemist having an antirealist disposition on these issues might consider that the representation of the elements is a matter of convention and that there is no real fact of the matter concerning where helium and other troublesome elements should be placed. These issues are discussed in chapter 10, which also considers the astrophysical origin of the elements as well as some unusual chemical regularities embodied in the periodic table.

The question of reduction raises another interesting issue concerning the reduction of chemistry to quantum mechanics. It appears that most chemists are quite willing to accept the reductive claims from physics insofar as it bestows greater theoretical underpinning to chemistry. Nevertheless, in cases such as the positioning of helium, chemists retain the right to classify the element in chemical terms even at the risk of overruling the findings of the reducing science.

Along with the realist view of the periodic system, as referred to above, comes the question of whether to regard the elements as "natural kinds," meaning realistic scientific entities that are differentiated by nature itself rather than by our human attempts at classification. This in turn opens up further dialogue with mainstream philosophy of science, which concerns itself with the question of natural kinds. In philosophy of biology, species have been deemed not to be natural kinds since biological species evolve over time. Many philosophers have sought to locate natural kinds at the chemical level.[29] Elements, in particular, are regarded by many as the quintessential natural kind term. To be gold is to possess atomic number 79 and vice versa. Natural kinds have been regularly invoked in the debates among philosophers of language concerning how linguistic terms such as "gold" or "water" refer to objects in the world. According to the widely held Kripke-Putnam view, we are urged to take a scientific view of natural kind terms. The term "water," for example, is to be taken as denoting just what modern science stipulates water to be, usually taken to be molecules of H_2O. This approach raises many issues that continue to exercise contemporary philosophers of science. Water is not simply H_2O since it may contain impurities or may be present in ionized form, to cite just two of many objections that have been raised.[30] Even the notion that elements may be natural kinds has been criticized on the basis of the existence of isotopes of many elements. Not all atoms of gold have the same mass, and so it has been claimed that gold is not a unique natural kind.[31]

It appears that one of the best ways to explore the relationship between chemistry and modern physics is to consider the status of the periodic system.[32] Given the renewed interest in the philosophy of chemistry and in the periodic system itself,[33] a reassessment of these basic issues is now required, and this is attempted in the chapters of this book.

The Periodic Table

THE PERIODIC SYSTEM
An Overview

THE ELEMENTS

In ancient Greek times philosophers recognized just four elements, earth, water, air, and fire, all of which survive in the astrological classification of the 12 signs of the zodiac. At least some of these philosophers believed that these different elements consisted of microscopic components with differing shapes and that this explained the various properties of the elements. These shapes or structures were believed to be in the form of Platonic solids (figure 1.1) made up entirely of the same two-dimensional shape. The Greeks believed that earth consisted of microscopic cubic particles, which explained why it was difficult to move earth. Meanwhile, the liquidity of water was explained by an appeal to the smoother shape possessed by the icosahedron, while fire was said to painful to the touch because it consisted of the sharp particles in the form of tetrahedra. Air was thought to consist of octahedra since that was the only remaining Platonic solid. A little later, a fifth Platonic solid, the dodecahedron, was discovered, and this led to the proposal that there might be a fifth element or "quintessence," which also became known as ether.

Although the notion that elements are made up of Platonic solids is regarded as incorrect from a modern point of view, it is the origin of the very fruitful notion that macroscopic properties of substances are governed by the structures of the microscopic components of which they are comprised. These "elements" survived well into the Middle Ages and beyond, augmented with a few others discovered by the alchemists, the precursors of modern-day chemists. One of the many goals of the alchemists seems to have been the transmutation of elements. Not surprisingly, perhaps, the particular transmutation that most enticed them was the attempt to change the base metal lead into the noble metal gold, whose unusual color, rarity, and chemical inertness have made it one of the most treasured substances since the dawn of civilization.

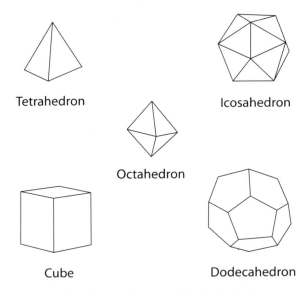

Tetrahedron   Icosahedron

Octahedron

Cube   Dodecahedron

FIGURE 1.1 The five Platonic solids. O. Benfey, Precursors and Cocursors of the Mendeleev Table: The Pythagorean Spirit in Element Classification, *Bulletin for the History of Chemistry*, 13–14, 60–66, 1992–1993, figure on p. 60 (by permission).

The earliest understanding of the term "element" among the Greek philosophers was of a "tendency" or "potentiality" that gave rise to the observable properties of the element. This rather subtle distinction between the abstract form of an element and its observable form has been all but forgotten in modern times. It has nonetheless served as a fundamental guiding principle to such noted contributors to the periodic system as Dimitri Mendeleev, its major discoverer.

According to most textbook accounts, chemistry began in earnest only when it turned its back on alchemy and on this seemingly mystical understanding of the nature of elements. The triumph of modern science is generally regarded as resting on direct experimentation and the adoption of an empiricist outlook, which holds that only that which can be observed should count. Not surprisingly, therefore, the more subtle and perhaps more fundamental sense of the concept of elements was rejected. For example, Robert Boyle and Antoine Lavoisier both took the view that an element should be defined by an appeal to empirical observations, thus denying the role of abstract elements. They recommended that an element should be defined as a material substance that has yet to be broken down into any more fundamental components by chemical means. In 1789, Lavoisier published a list of 33 simple substances or elements according to this empiricist criterion (figure 1.2).[1] Gone were the ancient elements of earth, water, air, and fire, which had by now been shown to consist of simpler substances.[2]

	Noms nouveaux.	Noms anciens correfpondans.
Subftances fimples qui appartiennent aux trois règnes, & qu'on peut regarder comme les élémens des corps.	Lumière	Lumière.
	Calorique	Chaleur. / Principe de la chaleur. / Fluide igné. / Feu. / Matière du feu & de la chaleur.
	Oxygène	Air déphlogiftiqué. / Air empiréal. / Air vital. / Bafe de l'air vital.
	Azote	Gaz phlogiftiqué. / Mofète. / Bafe de la mofète.
	Hydrogène	Gaz inflammable. / Bafe du gaz inflammable.
Subftances fimples non métalliques oxidables & acidifiables.	Soufre	Soufre.
	Phofphore	Phofphore.
	Carbone	Charbon pur.
	Radical muriatique	Inconnu.
	Radical fluorique	Inconnu.
	Radical boracique	Inconnu.
Subftances fimples métalliques oxidables & acidifiables.	Antimoine	Antimoine.
	Argent	Argent.
	Arfenic	Arfenic.
	Bifmuth	Bifmuth.
	Cobalt	Cobalt.
	Cuivre	Cuivre.
	Etain	Etain.
	Fer	Fer.
	Manganèfe	Manganèfe.
	Mercure	Mercure.
	Molybdène	Molybdène.
	Nickel	Nickel.
	Or	Or.
	Platine	Platine.
	Plomb	Plomb.
	Tungftène	Tungftène.
	Zinc	Zinc.
Subftances fimples falifiables terreufes.	Chaux	Terre calcaire, chaux.
	Magnéfie	Magnéfie, bafe du fel d'epfom.
	Baryte	Barote, terre pefante.
	Alumine	Argile, terre de l'alun, bafe de l'alun.
	Silice	Terre filiceufe, terre vitrifiable.

FIGURE 1.2
List of 33 simple substances compiled by Lavoisier. *Traité Elémentaire de Chimie*, Cuchet, Paris, 1789, p. 192

Many of these substances would qualify as elements by modern standards, while others, such as *lumière* (light) and *calorique* (heat), are certainly no longer regarded as elements.[3] Rapid advances in techniques of separation and characterization of chemical substances over the forthcoming years would help chemists expand and refine this list. The important technique of spectroscopy, which measures the emission and absorption spectra of various kinds of radiation, would eventually yield a very accurate means by which each element could be identified through its unique "fingerprint." In modern times, we recognize 91 naturally occurring elements, and it has even been possible to extend the range of the elements beyond those that occur naturally.[4]

THE DISCOVERY OF THE ELEMENTS

The story of the discovery of the elements is a fascinating one and has been the subject of at least one classic account.[5] A time line for the discoveries is given in table 1.1. This story is not systematically addressed in the present book, although references to predictions and discovery of elements are made throughout.

There have been a number of major episodes in the history of chemistry when half a dozen or so elements were discovered almost at once, or within a period of a few years. Of course, some elements, such as iron, copper, gold, and other metals, have been known since antiquity. In fact, historians and archeologists refer to certain epochs in human history as the Iron Age or the Copper Age. The alchemists added several more elements to the list, including sulfur, mercury, and phosphorus. In relatively modern times, the discovery of electricity enabled chemists to isolate many of the more reactive elements that, unlike copper and iron, could not be obtained by heating their ores with carbon. The English chemist Humphry Davy seized upon the use of electricity or, more specifically, electrolysis to isolate as many as 10 elements, including calcium, barium, magnesium, sodium, and chlorine.

Following the discovery of radioactivity and nuclear fission, and the development of techniques in radiochemistry, it became possible to fill the remaining few gaps in the periodic table. The last gap to be filled was that corresponding to element 43, which became known as technetium from the Greek *techne*, meaning artificial or manufactured. It was "manufactured" in the course of some radiochemical reactions that would not have been feasible before the advent of nuclear physics. Until recently, it was believed that this element did not occur naturally, but a reexamination of old evidence has now suggested that it does in fact occur naturally and that early reports of its discovery made in 1925 may have been unjustly discredited.[6]

The most recent spate of elemental discoveries is also based on technological developments, involving the production and harnessing of beams of pure atoms or pure elementary particles such as neutrons. These particles can be fired at each other with great precision to achieve nuclear fusion reactions and to thereby create new elements with extremely high atomic numbers. The initiator of this field was the American chemist Glenn Seaborg, who first synthesized plutonium in 1943 and went on to head research teams that were responsible for the synthesis of many more trans-uranium elements.

NAMES AND SYMBOLS OF THE MODERN ELEMENTS

Part of the appeal of the periodic table derives from the individual nature of the elements and from their names.[7] The chemist and concentration camp survivor Primo Levi began each chapter of his much-acclaimed book *The Periodic Table*[8]

TABLE 1.1

Discovery Time Line for the Elements and Approximate Dates of Contributions
from Major Chemists and Physicists Connected with the Periodic System

Antiquity	Au, Ag, Cu, Fe, Sn, Pb, Sb, Hg, S, C	
Middle Ages	As, Bi, Zn, P, Pt	
1700		
1710		
1720		
1730	Co	
1740		
1750	Ni, Mg	
1760	H	
1770	N, O, Cl, Mn, Ba	
1780	Mo, W, Te, Zr, U	Lavoisier
1790	Ti Y, Be	
1800	V, Nb, Ta, Rh, Pa, Os, Ir, Ce	Dalton, Avogadro
	K, Na, B, Ca, Sr, Ru, Ba	Davy
1810	I, Th, Li, Se, Cd	
1820	Si, Al, Br	Döbereiner
1830	La	
1840	Er	Gmelin
1850		Cannizzaro
1860	Cs, Rb, Tl, In, He	Mendeleev, Lothar Meyer
1870	Ga, Ho, Yb, Sc, Tm	
1880	Gd, Pr, Nd, Ge, F, Dy	
1890	Ar, He, Kr, Ne, Xe, Po, Ra, Ac	Ramsay, Rayleigh
1900	Rn, Eu, Lu	Thomson
1910	Pa	Lewis, van den Broek, Moseley
1920	Hf, Re, Tc	Bohr, Pauli, Schrödinger
1930	Fr	
1940	Np, At, Pu, Cm, Am, Pm, Bk	Seaborg
1950	Cf, Es, Fm, Md, No	
1960	Lr, Rf, Db	
1970	Sg	
1980	Bh, Mt, Hs	
1990	Ds, Rg, 112, 114, 116	
2000		

Compiled by the author

with a vivid description of an element such as gold, lead, or oxygen. The book itself is about his relations and acquaintances, but each anecdote is motivated by Levi's love of a particular element.[9] More recently, the well-known neurologist and author Oliver Sacks has written a book called *Uncle Tungsten*, in which he tells of his boyhood fascination with chemistry and in particular the periodic table.[10]

During the many centuries over which the elements have been discovered, many different themes have been used to select their names.[11] Just reading a list of the names of elements can conjure up episodes from Greek mythology. Promethium, element 61, takes its name from Prometheus, the god who stole fire from heaven and gave it to human beings only to be punished for this act by Zeus.[12] The connection of this tale to element 61 seems to be the extreme effort that was needed to isolate it, just as the task performed by Prometheus was difficult and dangerous. Promethium is one of the very few elements that do not occur naturally on the earth. It was obtained as a decay product from the fission of another element, uranium.

Planets and other celestial bodies have been used to name some elements. For example, palladium, which was discovered in 1803, is named after Pallas, or the second asteroid that was itself discovered just one year earlier in 1802. Helium is named after *helios*, the Greek name for the sun. It was first observed in the spectrum of the sun in 1868, and it was not until 1895 that it was first identified in terrestrial samples.

Many elements derive their names from colors. Cesium is named after the Latin color *caesium*, which means gray-blue, because it has prominent gray-blue lines in its spectrum. The yellow-green gas chlorine comes from the Greek word *khloros*, which denotes the color yellow-green.[13] The salts of the element rhodium often have a pink color, and this explains why the name of the element was chosen from *rhodon*, the Greek for rose. In cases of more recently synthesized elements, their names come from those of the discoverer or a person that the discoverers wish to honor. This is why we have bohrium, curium, einsteinium, fermium, gadolinium, lawrencium, meitnerium, mendeleevium, nobelium, roentgenium, rutherfordium, and seaborgium.[14]

A large number of elements—15, to be precise—have come from the place where their discoverer lived, or wished to honor: americium, berkelium, californium, darmstadtium, europium, francium, germanium, hassium, polonium, gallium, hafnium, lutetium, rhenium, ruthenium, and scandium. Yet other element names are derived from geographical locations connected with minerals in which they were found. This category includes the remarkable case of four elements named after the Swedish village of Ytterby, which lies close to Stockholm. Erbium, terbium, ytterbium, and yttrium were all found in ores located around this village, while a fifth element, holmium, was named after the Latin for Stockholm.

The naming of the later trans-uranium elements is a separate story in itself, complete with nationalistic controversies and, in some cases, acrimonious disputes over who first synthesized the element and should therefore be accorded the honor of selecting a name for it. In an attempt to resolve such disputes, the International Union of Pure and Applied Chemistry (IUPAC) decreed that the elements should be named impartially and systematically with the Latin numerals for the atomic number of the element in each case. Element 105, for example, would be known as un-nil-pentium, while element 106 would be un-nil-hexium. But more recently, after much deliberation over the true discoverers of some of these later superheavy elements, IUPAC has returned the naming rights to the discoverers or synthesizers who were judged to have established priority in each case. Instead of their impersonal Latin names, elements 105 and 106 are now called bohrium and seaborgium, respectively.[15]

Seaborgium is a particularly interesting case, since for many years the committee did not approve of the choice of the American chemist Glenn Seaborg's name even though he had been responsible for the synthesis of about 10 new elements, including number 106. Their official reason seems to have been an old rule that required that no element could be named after a person still living.[16] Following much campaigning by chemists in the United States and other parts of the world, Seaborg was finally granted his element while he was still alive.

Another curious case concerns the German chemist Otto Hahn, whose name was unofficially given to the element hahnium only to be removed later and changed to the name dubnium after the place where several trans-uranium elements were synthesized. Meanwhile, an element has been named after Hahn's one-time colleague Lise Meitner. To many observers, this is a just move since Hahn had been awarded the Nobel Prize for the discovery of nuclear fission while Meitner, who had participated in many of the crucial steps in the work, was denied the prize.[17] To others, it represents an excess of political correctness.

The symbols that are used to depict each element in the periodic table also have a rich and interesting story. In alchemical times, the symbols for the elements often coincided with those of the planets from which they were named or with which they were associated (figure 1.3). The element mercury, for example, shared the same symbol as that of Mercury, the innermost planetary body. Copper was associated with the planet Venus, and both the element and the planet shared the same symbol.

When Dalton published his atomic theory in 1805, he retained several of the alchemical symbols for the elements. These were rather cumbersome, however, and did not lend themselves easily to reproduction in articles and books. The modern use of simple letter symbols was introduced by the Swedish chemist Jacob Berzelius a little later in 1813.

In the modern periodic table, a small minority of elements are represented by a single letter of the alphabet. These include hydrogen, carbon, oxygen, nitrogen, sulfur, and fluorine, which appear as H, C, O, N, S, and F.[18] Most elements are depicted by two letters, the first of which is a capital letter and the second a

Metal							
	gold	silver	iron	mercury	tin	copper	lead

Symbol							
	○	☾	♂	☿	♃	♀	♄

Celestial Body							
	Sun	Moon	Mars	Mercury	Jupiter	Venus	Saturn

Day							
Lat. (dies)	Solis	Lunae	Martis	Mercurii	Jovis (pater)	Veneris	Saturni
Fr.	dimanche	lundi	mardi	mercredi	jeudi	vendredi	samdi
Eng.	Sunday	Monday	Tuesday	Wednesday	Thursday	Friday	Saturday

FIGURE 1.3 Names and symbols of the ancient metals compared to names of celestial bodies and days of the week. V. Rignes, *Journal of Chemical Education*, 66, 731–738, 1989, p. 731 (by permission).

lowercase letter. This gives rise to such element symbols as Li, Be, Ne, Ca, and Sc, for lithium, beryllium, neon, calcium, and scandium, respectively. Some of these two-letter symbols are by no means intuitively obvious, such as Cu, Na, Fe, Pb, Hg, Ag, and Au, which are derived from the Latin names for the elements copper, sodium, iron, lead, mercury, silver, and gold.[19] Tungsten is represented by a W after the German name for the element, which is *wolfram*. For a brief period of time, some heavy elements were depicted by three letters, as mentioned above, but many have now been given particular names, such as seaborgium for element 106. But the very recently discovered elements, such as 112, 114, and 116, whose official discoverers are still under discussion, continue to be depicted by three letters, namely, Uub, Uuq, and Uuh.[20]

THE MODERN PERIODIC TABLE

The manner in which the elements are arranged in rows and columns in the modern periodic table, also called the medium-long form (figure 1.4), reveals many relationships among them. Some of these relationships are very well known, while others still await discovery. To take just one example, in the 1990s scientists discovered that the property of high temperature superconductivity, the flow of current with zero resistance, could be observed at relatively high temperatures of about 100 Kelvin. This discovery was partly serendipitous. When the elements lanthanum, copper, oxygen, and barium were combined together in a particular manner, the resulting compound happened to display high-temperature superconductivity. There followed a flurry of worldwide activity in an effort to raise the temperature at which the effect could be maintained. The ultimate goal was to achieve room-temperature superconductivity, which would allow technological breakthroughs such as levitating trains gliding effortlessly along superconducting rails. One of the main guiding principles used in this quest was the periodic table of the elements. The table allowed researchers to replace some of the elements in the compound with others that are known to behave in a similar manner and then examine the effect on superconducting behavior. This is how the element yttrium was incorporated into a new set of superconducting compounds, to produce the compound $YBa_2Cu_3O_7$ with a superconducting temperature of 93K.[21] This knowledge, and undoubtedly much more, lies dormant within the periodic system waiting to be discovered and put to good use.

The conventional periodic table consists of rows and columns. Trends can be observed among the elements going across and down the table. Each horizontal row represents a single period of the table. On crossing a period, one passes from metals such as potassium and calcium on the left, through transition metals such as iron, cobalt, and nickel, then through some semimetallic elements such as germanium, and on to some nonmetals such as arsenic, selenium, and bromine on the right side of the table. In general, there is a smooth gradation in chemical and physical properties as a period is crossed, but exceptions to this general rule abound and make the study of chemistry a fascinating and unpredictably complex field.

Metals themselves can vary from soft dull solids such as sodium or potassium to hard shiny substances such as chromium, platinum, and gold. Nonmetals, on the other hand, tend to be solids or gases, such as carbon and oxygen, respectively. In terms of their appearance, it is sometimes difficult to distinguish between solid metals and solid nonmetals. To the layperson, a hard and shiny metal may seem to be more metallic than a soft metal such as sodium. But in a chemical sense, elements that have the greater ability to lose electrons (lower ionization energies) are regarded as being the more metallic. Sodium is therefore regarded by chemists as being more metallic than such elements as iron or copper. The periodic trend from metals to nonmetals is repeated with each period, such that when the rows are stacked they form columns, or groups, of similar elements. Elements within a single group tend to share many important physical and chemical properties, although there are many exceptions.

The manner in which the groups in the modern periodic table are labeled is complicated and controversial. The groups, or columns, of main-group elements, which are also referred to as representative elements, lie on the extreme left and right of the modern periodic table. In the United States, these groups are generally labeled with Roman numerals from I to VIII, with the letter A sometimes added to differentiate them from transition metals or groups IB to VIIIB, which lie in the central portion of the table. However, in Europe the convention is different in that all groups are sequentially labeled from left to right as IA to VIIIA until one reaches the group headed by copper, where the labeling becomes IB up to the noble gases, which are said to be in group VIIIB (figure 1.5).[22] Both of these sys-

H																	He
Li	Be											B	C	N	O	F	Ne
Na	Mg											Al	Si	P	S	Cl	Ar
K	Ca	Sc	Ti	V	Cr	Mn	Fe	Co	Ni	Cu	Zn	Ga	Ge	As	Se	Br	Kr
Rb	Sr	Y	Zr	Nb	Mo	Tc	Ru	Rh	Pd	Ag	Cd	In	Sn	Sb	Te	I	Xe
Cs	Ba	Lu	Hf	Ta	W	Re	Os	Ir	Pt	Au	Hg	Tl	Pb	Bi	Po	At	Rn
Fr	Ra	Lr	Rf	Db	Sg	Bh	Hs	Mt	Ds	Rg							

La	Ce	Pr	Nd	Pm	Sm	Eu	Gd	Tb	Dy	Ho	Er	Tm	Yb
Ac	Th	Pa	U	Np	Pu	Am	Cm	Bk	Cf	Es	Fm	Md	No

FIGURE 1.4 The modern or medium-long form table.

FIGURE 1.5 — Diagram of conventional periodic table format with alternative numbering systems for groups.

Group numbering legend (IUPAC / U.S. system / European system):

IUPAC	U.S. system	European system
1	IA	IA
2	IIA	IIA
3	IIIB	IIIA
4	IVB	IVA
5	VB	VA
6	VIB	VIA
7	VIIB	VIIA
8	—	—
9	—	—
10	—	—
11	IB	IB
12	IIB	IIB
13	IIIA	IIIB
14	IVA	IVB
15	VA	VB
16	VIA	VIB
17	VIIA	VIIB
18	VIIIA	VIIIB

Main table:

1	2	3	4	5	6	7	8	9	10	11	12	13	14	15	16	17	18
H																	He
Li	Be											B	C	N	O	F	Ne
Na	Mg											Al	Si	P	S	Cl	Ar
K	Ca	Sc	Ti	V	Cr	Mn	Fe	Co	Ni	Cu	Zn	Ga	Ge	As	Se	Br	Kr
Rb	Sr	Y	Zr	Nb	Mo	Tc	Ru	Rh	Pd	Ag	Cd	In	Sn	Sb	Te	I	Xe
Cs	Ba	Lu	Hf	Ta	W	Re	Os	Ir	Pt	Au	Hg	Tl	Pb	Bi	Po	At	Rn
Fr	Ra	Lr	Rf	Db	Sg	Bh	Hs	Mt	Ds	Rg							

f-block:

La	Ce	Pr	Nd	Pm	Sm	Eu	Gd	Tb	Dy	Ho	Er	Tm	Yb
Ac	Th	Pa	U	Np	Pu	Am	Cm	Bk	Cf	Es	Fm	Md	No

FIGURE 1.5 Diagram of conventional periodic table format with alternative numbering systems for groups: the more recent IUPAC system (top line), U.S. system (second line), and European system (third line). Note that three columns are labeled VIII in the U.S./European systems but that each column has a distinct number in the IUPAC system.

tems use the same Roman numeral for each column, which, in the case of main-group elements, also denotes the number of outer-shell electrons.

Given the confusion that these conventions have caused, there has been much attention directed at obtaining a unified system. Recently, IUPAC recommended that groups should be sequentially numbered with Arabic numerals from left to right, as groups 1 to 18, without the use of the letters A or B. The unfortunate result of this proposal is that the direct correlation between the number of outer-shell electrons in the atoms of main-group elements and the group labels in the old U.S. and European systems is lost. For example, the atom of oxygen has six outer-shell electrons and is said to be in group VI (followed by an A or B) in the older systems, whereas in the IUPAC system it is considered to be in group 16-. As a result, although many textbooks display the IUPAC recommendation on periodic tables, they generally fail to adhere to it when discussing the properties of the elements.[23]

This book mainly uses Roman numerals for the representative, or main-group elements, and refers to transition metal groups by the name of their first element. For example, group IVA in the U.S. system (carbon, silicon, germanium, tin, and lead) is referred to as simply group IV. Meanwhile, group IVB in the U.S. system (chromium, molybdenum, tungsten) is referred to as the chromium group.[24] Nevertheless, in chapter 10 the IUPAC system of numbering groups is used to avoid any possible confusion.

And so with this proviso, on the extreme left of the table, group I contains such elements as the metals sodium, potassium, and rubidium. These are unusually soft and reactive substances, quite unlike what are normally considered metals, such as iron, chromium, gold, and silver. The metals of group I are so reactive that merely placing a small piece of one of them into pure water gives rise to a vigorous reaction that produces hydrogen gas and leaves behind a colorless alkaline solution.[25] The elements in group II include magnesium, calcium, and barium and tend to be less reactive than those of group I in most respects.

Moving to the right, one encounters a central rectangular block of elements collectively known as the transition metals, which includes such examples as iron, copper, and zinc. In early periodic tables, known as short-form tables (figure 1.6), these elements were placed among the groups of what are now called the main-group elements. Several valuable features of the chemistry of these elements are lost in the modern table because of the manner in which they have been separated from the main body of the table, although the advantages of this later organization outweigh these losses.[26] To the right of the transition metals lies another block of representative elements starting with group III and ending with group VIII, the noble gases on the extreme right of the table.

Sometimes the properties a group shares are not immediately obvious. This is the case with group IV, which consists of carbon, silicon, germanium, tin, and lead. Here one notices a great diversity on progressing down the group. Carbon, at the

преимущественно найдти общую систему элементовъ. Вотъ этотъ опытъ:

			Ti=50	Zr=90	?=180.
			V=51	Nb=94	Ta=182.
			Cr=52	Mo=96	W=186.
			Mn=55	Rh=104,4	Pt=197,4
			Fe=56	Ru=104,4	Ir=198.
			Ni=Co=59	Pl=106,6	Os=199.
H=1			Cu=63,4	Ag=108	Hg=200.
	Be=9,4	Mg=24	Zn=65,2	Cd=112	
	B=11	Al=27,4	?=68	Ur=116	Au=197?
	C=12	Si=28	?=70	Sn=118	
	N=14	P=31	As=75	Sb=122	Bi=210
	O=16	S=32	Se=79,4	Te=128?	
	F=19	Cl=35,5	Br=80	I=127	
Li=7	Na=23	K=39	Rb=85,4	Cs=133	Tl=204
		Ca=40	Sr=87,6	Ba=137	Pb=207.
		?=45	Ce=92		
		?Er=56	La=94		
		?Yt=60	Di=95		
		?In=75,6	Th=118?		

FIGURE 1.6 Short-form table: the original Mendeleev table published in 1869. D.I. Mendeleev, Sootnoshenie svoistv s atomnym vesom elementov, *Zhurnal Russkeo Fiziko-Khimicheskoe Obshchestv,* 1, 60–77, 1869, table on p. 70.

head of the group, is a nonmetal solid that occurs in three completely different structural forms (diamond, graphite, and fullerenes)[27] and forms the basis of all living systems. The next element below, silicon, is a semimetal that, interestingly, may form the basis of artificial life, or at least "artificial intelligence", since it lies at the heart of all computers. The next element down, germanium, is a more recently discovered semimetal that was predicted by Mendeleev and later found to have many of the properties he foresaw. On moving down to tin and lead, one arrives at two metals known since antiquity. In spite of this wide variation among them, in terms of metal–nonmetal behavior, the elements of group IV nevertheless are similar in an important chemical sense in that they all display a maximum combining power, or valence, of 4.[28]

The apparent diversity of the elements in group VII is even more pronounced. The elements fluorine and chlorine, which head the group, are both poisonous gases. The next member, bromine, is one of the only two known elements that exist as a liquid at room temperature, the other one being the metal mercury.[29]

Moving further down the group, one then encounters iodine, a violet-black solid element.[30] If a novice chemist were asked to group these elements according to their appearances, it is inconceivable that he or she would consider classifying together fluorine, chlorine, bromine, and iodine. This is one instance where the subtle distinction between the observable and the abstract sense of the concept of an element can be helpful. The similarity between them lies primarily in the nature of the abstract elements and not the elements as substances that can be isolated and observed.[31]

On moving all the way to the right, a remarkable group of elements, the noble gases, is encountered, all of which were first isolated just before, or at, the turn of the twentieth century.[32] Their main property, rather paradoxically, at least when they were first isolated, was that they lacked chemical properties.[33] These elements, such as helium, neon, argon, and krypton, were not even included in early periodic tables, since they were unknown and completely unanticipated. When they were discovered, their existence posed a formidable challenge to the periodic system, but one that was eventually successfully accommodated by the extension of the table to include a new group, labeled group VIII, or group 18 in the IUPAC system.

Another block of elements, found at the foot of the modern table, consists of the rare earths that are commonly depicted as being literally disconnected. But this is just an apparent feature of this generally used display of the periodic system. Just as the transition metals are generally inserted as a block into the main body of the table, it is quite possible to do the same with the rare earths. Indeed, many such long-form displays have been published. While the long-form tables (figure 1.7) give the rare earths a more natural place among the rest of the elements, they are rather cumbersome and do not readily lend themselves to conveniently shaped wall charts of the periodic system. Although there are a number of different forms of the periodic table, what underlies the entire edifice, no matter the form of its representation, is the periodic law.

THE PERIODIC LAW

The periodic law states that after certain regular but varying intervals the chemical elements show an approximate repetition in their properties. For example, fluorine, chlorine, and bromine, which all fall into group VII, share the property of forming white crystalline salts of general formula NaX with the metal sodium. This periodic repetition of properties is the essential fact that underlies all aspects of the periodic system.

This talk of the periodic law raises some interesting philosophical issues. First of all, periodicity among the elements is neither constant nor exact. In the generally used medium–long form of the periodic table, the first row has two elements,

FIGURE 1.7 Long–form periodic table.

H																														He
Li	Be																								B	C	N	O	F	Ne
Na	Mg																								Al	Si	P	S	Cl	Ar
K	Ca												Sc	Ti	V	Cr	Mn	Fe	Co	Ni	Cu	Zn	Ga	Ge	As	Se	Br	Kr
Rb	Sr												Y	Zr	Nb	Mo	Tc	Ru	Rh	Pd	Ag	Cd	In	Sn	Sb	Te	I	Xe
Cs	Ba	La	Ce	Pr	Nd	Sm	Eu	Gd	Tb	Dy	Ho	Er	Tm	Yb	Lu	Hf	Ta	W	Re	Os	Ir	Pt	Au	Hg	Tl	Pb	Bi	Po	At	Rn
Fr	Ra	Ac	Th	Pa	U	Pu	Am	Cm	Bk	Cf	Es	Fm	Md	No	Lr	Rf	Db	Sg	Bh	Hs	Mt	Ds	Rg							

the second and third each contains eight, the fourth and fifth contain 18, and so on. This implies a varying periodicity consisting of 3, 9, 9, 19, and so on,[34] quite unlike the kind of periodicity one finds in the days of the week or notes in a musical scale. In these latter cases, the period length is constant, such as eight for the days of the week as well as the number of notes on a Western musical scale.

Among the elements, however, not only does the period length vary, but also the periodicity is not exact. The elements within any column of the periodic table are not exact recurrences of each other. In this respect, their periodicity is not unlike the musical scale, in which one returns to a note denoted by the same letter, which sounds like the original note but is definitely not identical to it, being an octave higher.

The varying length of the periods of elements and the approximate nature of the repetition have caused some chemists to abandon the term "law" in connection with chemical periodicity. Chemical periodicity may not seem as lawlike as the laws of physics, but whether this fact is of great importance is a matter of debate.[35] It can be argued that chemical periodicity offers an example of a typically chemical law, approximate and complex, but still fundamentally displaying lawlike behavior.[36]

Perhaps this is a good place to discuss some other points of terminology. How is a periodic table different from a periodic system? The term "periodic system" is the more general of the two. The periodic system is the more abstract notion that holds that there is a fundamental relationship among the elements. Once it becomes a matter of displaying the periodic system, one can choose a three-dimensional arrangement, a circular shape, or any number of different two-dimensional tables. Of course, the term "table" strictly implies a two-dimensional dimensional representation.[37] So although the term "periodic table" is by far the best known of the three terms law, system, and table, it is actually the most restricted.

REACTING ELEMENTS AND ORDERING THE ELEMENTS

Much of what is known about the elements has been learned from the way they react with other elements and from their bonding properties. The metals on the left-hand side of the conventional periodic table are the complementary opposites of the nonmetals, which tend to lie toward the right-hand side. This is so because, in modern terms, metals form positive ions by the loss of electrons, while nonmetals gain electrons to form negative ions. Such oppositely charged ions combine together to form neutrally charged salts such as sodium chloride or calcium bromide. There are further complementary aspects of metals and nonmetals. Metal oxides or hydroxides dissolve in water to form bases while nonmetal oxides or hydroxides dissolve in water to form acids. An acid and a base react together in a "neutralization" reaction to form a salt and water. Bases and acids, just like metals and nonmetals from which they are formed, are also opposite but complementary.[38]

Acids and bases have a connection with the origins of the periodic system since they featured prominently in the concept of equivalent weights, which was first used to order the elements. The equivalent weight of any particular metal, for example, was originally obtained from the amount of metal that reacts with a certain amount of a chosen standard acid. The term "equivalent weight" was subsequently generalized to denote the amount of an element that reacts with a standard amount of oxygen. Historically, the ordering of the elements across periods was determined by equivalent weight, then later by atomic weight, and eventually by atomic number.[39]

Chemists first began to make quantitative comparisons among the amounts of acids and bases that reacted together. This procedure was then extended to reactions between acids and metals. This allowed chemists to order the metals on a numerical scale according to their equivalent weight, which, as mentioned, is just the amount of the metal that combines with a fixed amount of acid. The concept of equivalent weights is, at least in principle, an empirical one since it seems not to rest on the theoretical assumption that the elements are ultimately composed of atoms.[40]

Atomic weights, as distinct from equivalent weights, were first obtained in the early 1800s by John Dalton, who indirectly inferred them from measurements on the masses of the relevant elements combined together. But there were complications in this apparently simple method that forced Dalton to make assumptions about the chemical formulas of the compounds in question. The key to this question is the valence, or combining power, of an element. For example, a univalent atom combines with hydrogen atoms in a ratio of 1:1; divalent atoms, such as oxygen, combine in a ratio of 2:1; and so on.

Equivalent weight, as mentioned above, is sometimes regarded as a purely empirical concept since it does not seem to depend upon whether one believes in the existence of atoms. Following the introduction of atomic weights, many chemists who felt uneasy about the notion of atoms attempted to revert to the older concept of equivalent weights. They believed that equivalent weights would be purely empirical and therefore more reliable. But as many authors have argued, most recently Alan Rocke, such hopes were an illusion since equivalent weights also rested on the assumption of particular formulas for compounds, and formulas are theoretical notions.

For many years, there was a great deal of confusion created by the alternative use of equivalent weight and atomic weight. Dalton himself assumed that water consisted of one atom of hydrogen combined with one atom of oxygen, which would make its atomic weight and equivalent weight the same, but his guess at the valence of oxygen turned out to be incorrect. Many authors used the terms "equivalent weight" and "atomic weight" interchangeably, thus further adding to the confusion. The true relationship between equivalent weight, atomic weight, and valency was clearly established only in 1860 at the first major scientific conference, which was held in Karlsruhe, Germany.[41] This clarification and the general adoption of consistent atomic weights cleared the path for the independent discovery of the periodic system by as many as six individuals in various countries, who each proposed forms of the periodic table that were successful to varying degrees. Each placed the elements generally in order of increasing atomic weight.[42]

The third, and most modern, of the ordering concepts mentioned above is atomic number. Once atomic number was understood, it displaced atomic weight as the ordering principle for the elements. No longer dependent on combining weights in any way, atomic number can be given a simple microscopic interpretation in terms of the structure of the atoms of any element. The atomic number of an element is given by the number of protons, or units of positive charge, in the nucleus of any of its atoms. Thus, each element on the periodic table has one more proton than the element preceding it. Since the number of neutrons in the nucleus also tends to increase as one moves through the periodic table, this makes atomic number and atomic weights roughly correspondent, but it is the atomic number that identifies any element. This is to say that atoms of any particular element always have the same number of protons, although they can differ in the number of neutrons they contain.[43]

DIFFERENT REPRESENTATIONS OF THE PERIODIC SYSTEM

The modern periodic system succeeds remarkably well in ordering the elements by atomic number in such a way that they fall into natural groups, but this system can be represented in more than one way. Thus, there are many forms of the periodic table, some designed for different uses. Whereas a chemist might favor a form that highlights the reactivity of the elements, an electrical engineer might wish to focus on similarities and patterns in electrical conductivities.[44]

The way in which the periodic system is displayed is a fascinating issue, and one that especially appeals to the popular imagination. Since the time of the early periodic tables of John Newlands, Julius Lothar Meyer, and Dimitri Mendeleev, there have been many attempts to obtain the "ultimate" periodic table. Indeed, it has been estimated that within 100 years of the introduction of Mendeleev's famous table of 1869, approximately 700 different versions of the periodic table were published. These include all kinds of alternatives, including three-dimensional tables, helices, concentric circles, spirals, zigzags, step tables, and mirror image tables. Even today, articles are regularly published in the *Journal of Chemi-*

cal Education, for example, purporting to show new and improved versions of the periodic system.[45]

What is fundamental to all these attempts is the periodic *law* itself, which exists in only one form. None of the multitude of displays changes this aspect of the periodic system. Many chemists stress that it does not matter how this law is physically represented, provided that certain basic requirements are met. Nevertheless, from a philosophical point of view, it may still be relevant to consider the most fundamental representation of the elements, or the ultimate form of the periodic system, especially as this relates to the question of whether the periodic law should be regarded in a realistic manner or as a matter of convention.[46] The usual response that representation is only a matter of convention would seem to clash with the realist notion that there may be a fact of the matter concerning the points at which the repetitions in properties occur.

RECENT CHANGES IN THE PERIODIC TABLE

In 1945, Glenn Seaborg (figure 1.8) suggested that the elements beginning with actinium, number 89, should be considered a rare earth series, whereas it had previously been supposed that the new series of rare earths would begin after element 92, or uranium (figure 1.9). Seaborg's new periodic table revealed an analogy between europium (63) and gadolinium (64) and the as yet undiscovered elements 95 and 96, respectively. On the basis of these analogies, Seaborg succeeded in synthesizing and identifying the two new elements, which were subsequently named americium and curium. A number of further trans-uranium elements have subsequently been synthesized.[47]

The standard form of the periodic table has also undergone some minor changes regarding the elements that mark the beginning of the third and fourth rows of the transition elements. Whereas older periodic tables show these elements to be lanthanum (57) and actinium (89), more recent experimental evidence and analysis have put lutetium (71) and lawrencium (103) in their former places.[48] It is also interesting to note that some even older periodic tables based on macroscopic properties had anticipated these changes.

These are examples of ambiguities in what may be termed secondary classification, which is not as unequivocal as primary classification, or the sequential ordering of the elements. In classical chemical terms, secondary classification corresponds to the chemical similarities between the various elements in a group. Meanwhile, in modern terms, secondary classification is explained by recourse to the concept of electronic configurations. Regardless of whether one takes a classical qualitative chemical approach or a more physical approach based on electronic configurations, secondary classification of this type is more tenuous than primary

FIGURE 1.8
Glen Seaborg. Photo from Emilio
Segrè Collection, by permission.

classification and cannot be established as categorically.[49] The way in which secondary classification, as defined here, is established is a modern example of the tension between using chemical properties or physical properties for classification. The precise placement of an element within groups of the periodic table can vary depending on whether one puts more emphasis on electronic configuration (a physical property) or its chemical properties. In fact, many recent debates on the placement of helium in the periodic system revolve around the relative importance that should be assigned to these two approaches.[50]

In recent years, the number of elements has increased well beyond 100 as the result of the synthesis of artificial elements. At the time of writing, conclusive evidence has been reported for element 111.[51] Such elements are typically very unstable, and only a few atoms are produced at any time. However, ingenious chemical techniques have been devised that permit the chemical properties of these so-called superheavy elements to be examined and allow one to check whether extrapolations of chemical properties are maintained for such highly massive atoms. On a more philosophical note, the production of these elements allows us to examine whether the periodic law is an exceptionless law, of the same kind as Newton's law of gravitation, or whether deviations to the expected recurrences in chemical properties might take place once a sufficiently high atomic number is reached. No surprises have been found so far, but the question of whether some of these superheavy elements have the expected chemical properties is far from being fully resolved. One important complication that arises in

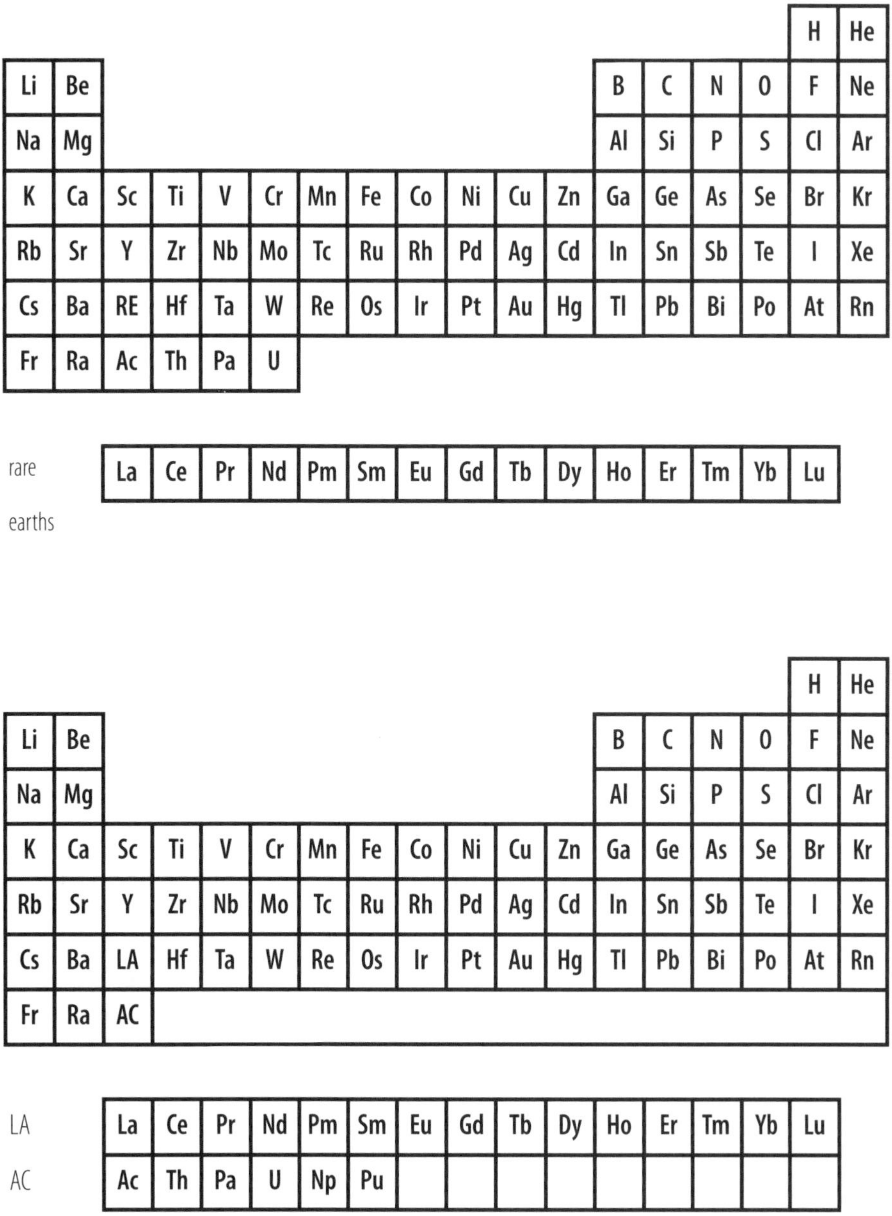

FIGURE 1.9 Pre-Seaborg (a) and post-Seaborg (b) periodic tables. RE denotes rare earth elements from 57–71 inclusive; LA, lanthanides ($Z = 57$–71); AC, actinides beginning with $Z = 89$, where Z is atomic number.

this region of the periodic table is the increasing significance of relativistic effects due to very rapidly moving electrons.[52] These effects cause the adoption of unexpected electronic configurations in some atoms and may result in equally unexpected chemical properties.

UNDERSTANDING THE PERIODIC SYSTEM

Developments in physics have had a profound influence on the manner in which the periodic system is now understood. The two important theories in modern physics are Einstein's theory of relativity and quantum mechanics.

The first of these has had a limited impact on our understanding of the periodic system but is becoming increasingly important in accurate calculations carried out on atoms and molecules. The need to take account of relativity arises whenever objects move at speeds close to that of light. Inner electrons, especially those in the heavier atoms in the periodic system, can readily attain such relativistic velocities. It would be impossible to carry out an accurate calculation, especially on a heavy atom, without applying the necessary relativistic corrections. In addition, many seemingly mundane properties of elements such as the characteristic color of gold or the liquidity of mercury can best be explained as relativistic effects due to fast-moving inner-shell electrons.[53]

But it is the second theory of modern physics that has exerted by far the more important influence in attempts to understand the periodic system theoretically. Quantum theory was actually born in the year 1900, some 14 years before the discovery of atomic number. It was first applied to atoms by Niels Bohr, who pursued the notion that the similarities between the elements in any group of the periodic table could be explained by their having equal numbers of outer-shell electrons.[54] The very notion of a particular number of electrons in an electron shell is an essentially quantumlike concept. Electrons are assumed to possess only certain quanta, or packets, of energy, and depending on how many such quanta they possess, they lie in one or another shell around the nucleus of the atom.

Soon after Bohr had introduced the concept of the quantum to the understanding of the atom, many others developed his theory until the old quantum theory gave rise to quantum mechanics. Under the new description, electrons are regarded as much as waves as they are as particles. Even stranger is the notion that electrons no longer follow definite trajectories or orbits around the nucleus. Instead, the description changes to talk of smeared-out electron clouds, which occupy so-called orbitals.[55] The most recent explanation of the periodic system is given in terms of how many such orbitals are populated by electrons. The explanation depends on the electron arrangement or "configuration" of an atom, which is spelled out in terms of the occupation of its orbitals.[56]

The interesting question raised here is the relationship between chemistry and modern atomic physics and, in particular, quantum mechanics. The popular view reinforced in most textbooks is that chemistry is nothing but physics "deep down" and that all chemical phenomena, and especially the periodic system, can be developed on the basis of quantum mechanics. There are some problems with this view, however, which are considered in this book.

For example, in chapter 9 it is suggested that the quantum mechanical explanation for the periodic system is still far from perfect. This is important because chemistry books, especially textbooks aimed at teaching, tend to give the impression that our current explanation of the periodic system is essentially complete. This is just not the case, or so it will be argued.[57]

MOLECULAR TABLES

Another recent departure has been the invention of periodic tables designed to summarize the properties of compounds rather than elements. In 1980, Ray Hefferlin[58] produced a periodic system for all the conceivable diatomic molecules that can be formed between the first 118 elements.[59] In order to represent this vast number of entries, Hefferlin used four three-dimensional blocks of varying sizes. His representation reveals that interatomic distances, spectroscopic frequencies, and molecular ionization energies are periodic properties. It also provided successful predictions regarding the properties of diatomic molecules.

Jerry Dias, a chemist at the University of Missouri–Kansas City, has devised a periodic classification of a class of organic molecules called benzenoid aromatic hydrocarbons, of which naphthalene, $C_{10}H_8$, is the simplest example (figure 1.10). By analogy with Johann Döbereiner's triads of elements, described in chapter 2, these molecules can be sorted into groups of three in which the central molecule has a total number of carbon and hydrogen atoms that is the mean of the flanking entries, both downward and across the table. This periodic scheme has been applied to making a systematic study of the properties of benzenoid aromatic hydrocarbons, which has led to the predictions of the stability and reactivity of many of their isomers.

However, it is the periodic table of elements that has had the widest and most enduring influence. The periodic table ranks as one of the most fruitful and unifying ideas in the whole of modern science, comparable perhaps with Darwin's theory of evolution by natural selection. Unlike such theories as Newtonian mechanics, the periodic table has not been falsified by developments in modern physics but has evolved while remaining essentially unchanged. After evolving for nearly 150 years through the work of numerous individuals, the periodic table remains at the heart of the study of chemistry. This is mainly because it is of immense practical

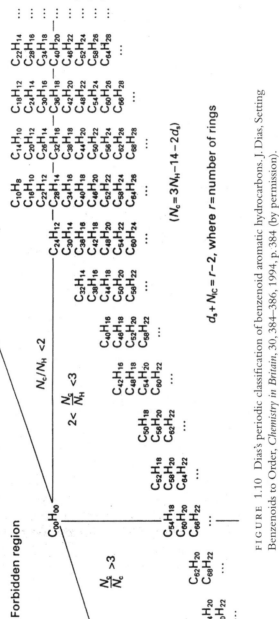

FIGURE 1.10 Dias's periodic classification of benzenoid aromatic hydrocarbons. J. Dias, Setting Benzenoids to Order, *Chemistry in Britain*, 30, 384–386, 1994, p. 384 (by permission).

benefit for making predictions about all manner of chemical and physical properties of the elements and possibilities for bond formation. Instead of having to learn the properties of the more than 100 elements, the modern chemist, or the student of chemistry, can make effective predictions from knowing the properties of typical members of each of the eight main groups and those of the transition metals and rare earth elements.

Having laid some thematic foundations and defined some key terms, in chapter 1 I begin the story of the development of the modern periodic system, starting with its birth in the eighteenth and nineteenth centuries.

QUANTITATIVE RELATIONSHIPS
AMONG THE ELEMENTS AND THE
ORIGINS OF THE PERIODIC TABLE

Elements within a vertical group on the periodic table share certain chemical similarities, but the modern periodic system is not derived purely from descriptive characteristics. If chemical similarities were the sole basis for their classification, there would be many cases where the order and placement of the elements would be ambiguous. The development of the modern periodic system began when it was recognized that there are precise numerical relationships among the elements. Its subsequent evolution has also involved contributions from physics, as described in subsequent chapters. But whereas the latter contributions drew on fundamental physical theories, the ones that are examined in this chapter do not share this aspect. Instead, they involved looking for patterns among the numerical properties, such as equivalent weight or atomic weight, associated with each element.[1]

Throughout its history, the development of the periodic table has involved a delicate interplay between two contrasting approaches: discerning quantitative physical data, on one hand, and observing qualitative similarities among the elements as a form of natural history, on the other. Both approaches are essential, and the balance that has been struck between them has been of crucial importance at various stages in our story.

QUANTITATIVE ANALYSIS

Whereas attention to qualitative aspects has always been an essential part of chemistry, the use of quantitative data has been a relatively new addition. The time when chemists began to pay attention to quantitative aspects of chemical reactions and chemical substances has been the source of much debate among historians. The traditional view has been that this step was taken by Antoine Lavoisier (figure 2.1), who is regarded as the founder of modern chemistry. The

Antoine Lavoisier,
né à Paris, le 26 Aout 1745, mort le 6 Avril 1794

FIGURE 2.1
Antoine Lavoisier. Photo
from Edgar Fahs Smith
Collection by permission.

more recent historical opinion is that Lavoisier made few original contributions and that much of his fame lay in his abilities as an organizer and presenter of chemical knowledge.[2]

Nevertheless, Lavoisier was able to dispel some of the vagueness and confusion that dogged the field of chemistry as he found it. The confusion included the chaotic way in which substances were named as well as the uncertain knowledge of weight changes accompanying chemical reactions. Prior to Lavoisier and his contemporaries, it was believed that when substances burned they would release a substance called phlogiston. Although some substances do appear to lose weight when they are burned, many others show a gain in weight. Lavoisier used his considerable personal wealth to commission the making of the finest balances of his day, some of which could measure changes as accurately as one part in 600,000. As a result of his weighing experiments, Lavoisier succeeded in showing that substances that burned did not in fact give off phlogiston and that the notion of phlogiston was redundant.[3] He also showed that what is essential for burning is the element oxygen, a substance that had previously been discovered by Swedish chemist

Carl Scheele and had been subjected to several earlier studies by the Englishman Joseph Priestley.[4]

Moreover, by accurately weighing reacting substances, Lavoisier was able to announce the law of conservation of matter, which states:

> *In every chemical operation, an equal quantity of matter*
> *exists before and after the operation.*

Lavoisier's emphasis on the quantification of chemistry also paved the way for the laws of chemical combination, which soon prompted John Dalton to develop his atomic theory.

Returning to the revisionary accounts of Lavoisier, it has been argued that a more significant development from the dismissal of phlogiston was the question of composition. What Lavoisier achieved was a reversal of the compositional order that had been held by earlier chemists starting with Georg Stahl. In Lavoisier's chemistry, sulfur and phosphorus were simpler than their acids, thus displaying the opposite order than in the old chemistry. Contrary to the view of the old chemistry, metals were simpler than their calxes (oxides) according to Lavoisier. Likewise, hydrogen and oxygen were regarded as simpler than water in Lavoisier's compositional order, once again quite opposite the view held in the old chemistry.[5] Some historians even regard Lavoisier's work as a culmination of the tradition begun a good deal earlier by the likes of Stahl on the question of chemical composition, rather than the start of a new tradition in chemistry.[6]

But perhaps Lavoisier's greatest contribution, particularly for our story, was one already mentioned briefly in chapter 1: Lavoisier was highly critical of the classical abstract element scheme of the Greeks and subsequent chemists. By adopting an empiricist approach, he attempted to eradicate any talk of abstract elements or principles in favor of elements as simple substances, which could be isolated and which could not be further decomposed. This anti-metaphysical departure may have been just what was needed in chemistry at the time,[7] although Lavoisier did not succeed in completely dispensing with the need for elements as principles, as many authors have pointed out.[8]

EQUIVALENT WEIGHTS

One of the next major developments, following along quantitative lines, was due to Jeremias Benjamin Richter, who between 1792 and 1794 published a set of quantities that later became known as equivalent weights (table 2.1). He first measured amounts of acids that combined with certain amounts of bases. He then extended this procedure to measuring the amount of some metals that combine with a certain fixed amount of acid, and thus obtained an indirect measure of the relative amounts in which elements can combine together. This perhaps marked the first

TABLE 2.1

Richter's table of equivalent weights,
as modified by E. Fischer in 1802

Base		Acid	
Alumina	525	Hydrofluoric	427
Magnesia	615	Carbonic	577
Ammonia	672	Sebaic	706
Lime	793	Muriatic	712
Soda	859	Oxalic	755
Strontia	1,329	Phosphoric	979
Potash	1,605	Sulfuric	1,000
Baryta	2,222	Succinic	1,209
		Nitric	1,405
		Acetic	1,480
		Citric	1,583
		Tartaric	1,694

E.G. Fischer, *Claude Louis Bethollet über die Gesetze der Verwandtschaft,*
Berlin, 1802. Table on p. 232

time that the properties of the elements could be compared to each other on a simple numerical scale. This irresistible urge to find numerical patterns in nature would prove to be a powerful force in the development of the periodic table.

A SHORT DIGRESSION ON GREEK ATOMISM

The ancient Greek philosophers introduced atomism partly as a response to what they considered as the awkward notion of infinity.[9] Zeno had introduced a famous paradox whose effect depended on the existence of infinity. According to the paradox, if a person needs to cover a certain distance between points A and B, he or she may do so by a series of steps. In the first step, the person covers half the distance. The second step involves covering half of the remaining distance, and so on. Clearly, this process will continue ad infinitum since each time a step is taken it takes the person closer to the destination but never allows arrival. This paradox and many others like it depend on taking an infinite number of steps between points A and B.

If infinity is deemed to be unreal or unphysical, however, the problem appears to evaporate. One does indeed reach a destination because one cannot take an infinite number of steps. If distances are not infinitely subdivisible, there is no longer any paradox. But the Greek philosophers did not stop at denying infinite subdivisibility of distances. They applied the same denial to matter. They reasoned that a chunk of matter could likewise not be infinitely subdivided, that there would come a point in the subdivision when one had reached the smallest possible chunk of matter or *atomos*, meaning indivisible matter. Atoms of distance and atoms of matter were thus born of a philosophical desire to banish infinities from distances and from matter. But atomism of this kind remained a purely philosophical idea, and the Greek philosophers showed little inclination to perform experiments to support their notion.

DALTON'S ATOMIC THEORY

In 1801, Dalton (figure 2.2), a Manchester school teacher, published an article on meteorology, which was one of his main scientific interests. This work was to be the beginning of his reintroduction of atomic theory into science. Atomism had been proposed in ancient Greece, but it had subsequently been abandoned for about 2,000 years, although the mention of atoms or small "particles" had not been entirely forgotten in scientific circles.[10] For example, Isaac Newton referred frequently to atoms, although not by name, including in the following passage, which Dalton knew well:

> It seems probable to me, that God in the beginning form'd matter in solid, massy, hard, impenetrable, moveable particles of such sizes and figures, and with such other properties, and in such proportion to space, as most conduced to the end for which he form'd them; and that these primitive particles being

FIGURE 2.2
John Dalton. Photo from Edgar Fahs
Smith Collection by permission.

solids, are incomparably harder than any porous bodies compounded of them, even so very hard, as never to wear or break in pieces; no ordinary power being able to divide what God himself made one in the first creation.[11]

But in spite of Newton's notions on the quantification of chemistry, expressed in other passages, little more had been achieved in the field apart from the work of Lavoisier and others using the balance to try to understand chemical reactions, and Richter's little known work on equivalent weights. Lavoisier had taken a firmly empiricist[12] approach of ignoring the possible existence of unobservable atoms and by focusing on elements as the final stages of chemical analysis of substances. Dalton rejected this position and embraced a realistic conception in which atoms actually exist and have particular sizes and weights.

His ideas on atomism can be summarized under three main points. First, Dalton assumed that all matter was composed of atoms that were indestructible and nonchangeable, thus denying the possibility of the transmutation of elements. Dalton thus belonged to the new chemical tradition that consciously distanced itself from the alchemical doctrine of transmutation. Dalton was not the only one to do so, but he provides an interesting contrast to Robert Boyle, for example, who 150 years earlier also did important quantitative work in chemistry while at the same time being steeped in alchemy, as recent scholarly work has shown.

Second, and contrary to many of his contemporaries, who believed strongly in the unity of all matter, Dalton believed that there were as many different kinds of atoms as there were elements.[13] Finally, Dalton suggested that the weights of atoms would serve as a kind of bridge between the realm of microscopic unobservable atoms and the world of observable properties. But atomic weight is not necessarily the same as an equivalent weight, an issue raised in chapter 1 that will be revisited here.

The precise origin of Dalton's ideas has been traced by historians to his research into the nature of the air, which had been found to consist of a mixture of gases. At the time, it was not understood why the various component gases of air did not separate out according to their different densities. In broad terms, Dalton reasoned that if gases consist of tiny particles, or atoms, they would be more likely to form a mixture than if they consisted of continuous fluids. This argument is plausible if one accepts that continuous fluids cannot intermingle to the extent that tiny isolated particles can. It is also clear that part of Dalton's motivation for supporting an atomic theory lay in Newton's view that like particles should repel each other. According to this view, the different gases in the air intermingled rather than forming separate strata, because particles of each gas would move away from each other to fill the available space, ignoring the particles of the other gases in the space.[14]

It is interesting to examine briefly how Dalton arrived at the values of atomic weights of the elements shown in table 2.2.[15] For example, he referred

TABLE 2.2

Part of an early table of atomic and
molecular weights published by Dalton

Element	Weight
Hydrogen	1
Azot	4.2
Carbon (charcoal)	4.3
Ammonia	5.2
Oxygen	5.5
Water	6.5
Phosphorus	7.2
Nitrous gas	9.3
Ether	9.6
Nitrous oxide	13.7
Sulphur	14.4
Nitric acid	15.2

Adapted from J. Dalton, *Memoirs of the Literary
and Philosophical Society of Manchester,* 2(1), 207,
1805, table on p. 287. (Azot is the old name for
nitrogen.)

to Lavoisier's data on the composition of water, namely, 85% oxygen and 15% hydrogen. Assuming a formula of HO, Dalton calculated the weight of an oxygen atom to be 85/15 = approximately 5.5, by taking the weight of a hydrogen atom as unity. Similarly, Dalton obtained a value for the nitrogen atom by drawing on William Austin's data showing that ammonia consists of 80% nitrogen and 20% hydrogen. Again, Dalton assumed the formula of this compound to be of the binary form NH.[16]

Dalton also suggested that the gases in the air diffused into one another because the particles were of different sizes, but he soon realized that it was the different weights of these particles, and not their sizes, which was the key feature in determining how gases would combine. In a paper published in 1803, he estimated the relative atomic weights of a number of different elements. The publication associated with this work represents the first ever list of atomic weights. Eventually, systems of classification would begin to be developed in which atomic weight would replace equivalent weight as the chief criterion by which elements were arranged, but this process was to take a period of about 60 years. Table 2.2 shows an early set of atomic and molecular weights, in modern terms, published by Dalton in 1805.

Another important consequence of Dalton's hypothesis that matter consists of atoms was that it provided an explanation of the long recognized law of constant proportion. As Richter had pointed out, when any two elements combine together, for example, hydrogen and oxygen, they always do so in a constant ratio of their masses. This fact can be understood if one assumes that a certain precise number of atoms of one element combine with a particular number of atoms of the other element. According to this view, macroscopic observations summarized in the law of constant proportion represent a scaled up version of millions of such atomic combinations. If, on the other hand, matter did not consist of atoms but could be infinitely subdivided, it is not clear why oxygen and hydrogen, or any other elements, should always react together in the same particular ratio of masses.[17]

Meanwhile, others had made observations concerning the combination of masses of any two elements in more than one compound. It had been realized that if A reacts with B to form more than one compound, the various amounts of B that react with a fixed amount of A bear a simple whole number ratio to each other. Dalton carried out further experiments on this relationship, and as a result of his work, this, too, became regarded as a law of chemical combination (the law of multiple proportions), and one that his atomic hypothesis could readily explain. On the atomic hypothesis, the law of multiple proportions results from the fact that, for example, one oxygen atom can combine with one atom of carbon to form a compound, and in addition, two atoms of oxygen can combine with one atom of carbon to form a different compound. The ratio of amounts of oxygen combining with a fixed amount of carbon is therefore the simple ratio of 2 to 1. These two compounds are carbon monoxide and carbon dioxide, respectively.

At first, Dalton's concept of atomic weight did not improve the prospects of classifying the elements, since there were problems involved in calculating this quantity. While the equivalent weights introduced by Richter[18] at least appeared to have a clear experimental basis, Dalton's atomic weights, and those published by several of his contemporaries, seemed to be more theoretical, although this difference later turned out to be an illusion. The determination of atomic weights depended on assuming a particular formula for a compound since formulas could not yet be verified experimentally. The case of water provides a good example. One gram of hydrogen always reacts with approximately 8 grams of oxygen,[19] and thus the equivalent weight of oxygen is given as 8 relative to that of hydrogen. What Dalton did was assume that hydrogen and oxygen occur as individual atoms and that they combine at the atomic level, thereby accounting for the macroscopic facts about the combination of specific volumes of hydrogen and oxygen. The problem is that unless the formula of water is known, this assumption can tell us nothing about the relative weights of the atoms of hydrogen and oxygen since it is not known how many hydrogen atoms are combining with each oxygen atom. At this point, Dalton was forced to make a guess as to whether one atom of each element had combined together or whether it was two of hydrogen and one of oxygen, or perhaps vice versa, or any other ratio of atoms. Dalton based such choices on what he called the rule of simplicity, meaning that in the absence of additional information, he would assume the simplest possible ratio of 1:1.[20] Accordingly he assumed the formula of water to be HO and determined the atomic weight of oxygen to be eight, just like its equivalent weight.

The question of finding the correct formulas for compounds was only conclusively resolved a good deal later when the concept of valency, the combining power of a particular element, was clarified by the chemists Edward Frankland and Auguste Kekulé working separately. Hydrogen, for example, has a valence of 1, while the value for oxygen is 2. It follows that two atoms of hydrogen combine with one of oxygen. With this new knowledge, the relationship between atomic weight and equivalent weight could be stated simply:

$$\text{atomic weight} = \text{valence} \times \text{equivalent weight}$$

Since oxygen has a valence of 2 and its equivalent weight is 8, as many early chemists had determined, its correct atomic weight is therefore twice that number, or 16. The correct formula for water is H_2O and not HO, as Dalton had assumed.

LAW OF DEFINITE PROPORTIONS BY VOLUMES

Soon after Dalton began to publish articles on his atomic theory, experiments were performed by Alexander von Humbolt and Joseph Louis Gay-Lussac leading to what was termed a law of definite proportions by volumes.[21] These scientists experimented on forming water vapor by passing electric sparks through a mixture of oxygen and hydrogen. They found that whatever volume of oxygen reacted, it was necessary to use almost exactly twice the volume of hydrogen to within ± 0.19%. They also noted that the volume of water vapor formed was almost identical to the volume of hydrogen initially used. Thus:

2 volumes of hydrogen + 1 volume of oxygen → 2 volumes of water vapor

They were able to extend this finding of whole number ratios to several other reactions involving gases. For example:

3 volumes of hydrogen + 1 volume of nitrogen → 2 volumes of ammonia

2 volumes of carbon monoxide + 1 volume of oxygen →
2 volumes of carbon dioxide

Gay-Lussac summarized these results in a new law announced in 1809, which stated:

> *The volumes of gases entering into chemical reaction and the*
> *gaseous products are in a ratio of small integers.*

Recall that Dalton's main idea was that matter was composed of indivisible elementary atoms of fixed characteristic weight. Their combination in simple numbers gave rise to compound atoms and accounted for chemical laws such as the law of constant composition and the law of multiple proportions. However, when it came to Gay-Lussac's law, Dalton's original idea was unable to explain the observations summarized in equations such as those above. Consider the first equation again:

2 volumes of hydrogen + 1 volume of oxygen → 2 volumes of water[22]

Using the benefit of hindsight, the cause of the problem can easily be appreciated. Since one volume of oxygen was combining with two volumes of hydrogen, this implied that particles of oxygen must have been dividing.

But this idea contradicts the very heart of Dalton's notion that the smallest particles of any element are supposed to be indivisible. Dalton's own reaction to Gay-Lussac's law was to question the data and to repeat the experiments. This led to his claiming that the ratios were not in fact as simple as reported by von Humbolt and Gay-Lussac. Nevertheless, the simple ratio continued to be reproduced by others and has passed the test of time. Dalton could have accepted the existence of molecules of elements composed of two or more atoms of an element while still holding that such a body represented the simplest chemical unit that retains the properties of the element in question, but he failed to do so.

Meanwhile, Gay-Lussac suggested the very plausible notion that the appearance of small integers relating volumes in the reactions implied that *equal volumes* of gases contained *equal numbers* of particles, EVEN.[23] The step of reconciling Dalton's ideas on the existence of atoms with Gay-Lussac's law was taken by the Italian scientist Amadeo Avogadro in 1811. The crucial new ingredient introduced by Avogadro was that, contrary to what had previously been believed, the ultimate particles of a gas were not necessarily composed of single atoms but could equally well be assemblies of two or more atoms. Such assemblies or molecules could thus form the ultimate particles of any gaseous element.

Avogadro's idea was to provide the solution to Gay-Lussac's law while still maintaining the existence of Dalton's ultimate particles, although such particles could now be diatomic molecules, which were capable of subdivision. Unfortunately, Avogadro's resolution was understood by very few chemists at the time and had to wait a further 50 years before it became firmly established at the hands of fellow Italian chemist Stanislao Cannizzaro. It was Cannizzaro who helped to bring order to the prevailing confusion regarding atomic weights, as described in chapter 3.

PROUT'S HYPOTHESIS

A rather remarkable fact began to emerge after values of equivalent weights and atomic weights had been published by various people. Apart from some exceptions, many of the equivalent weights and atomic weights appeared to be approximate whole number multiples of the weight of hydrogen. To some chemists, this stubborn fact pointed against Dalton's idea of distinct elements and represented support for the essential unity of all the elements. More specifically, it suggested that all the elements, or their atoms, might be multiples of atoms of hydrogen. This would mean that there was really only one kind of matter, which could occur in different states of combination.

The first person to articulate this view clearly was the Scottish physician William Prout. Since the equivalent weights and atomic weights of the elements were nowhere near being exact multiples of each other, Prout rounded them off to the nearest whole number and gave hydrogen a value of 1. Such was the seductive lure of the Pythagorean tradition for seeking simple ratios, and it was this conviction that allowed Prout to ignore the apparent discrepancies as seen in the nonintegral values of the weights of some elements.

Prout's first articles on this subject appeared anonymously in Thomas Thomson's *Annals of Philosophy* along with a rather modest disclaimer:

> The author of the following essay submits it to the public with the greatest diffidence; for though he has taken the utmost pains to arrive at the truth, he has not such confidence in his abilities as an experimentalist as to induce him to dictate to others far superior to himself in chemical acquirements and fame.[24]

In his second paper on this hypothesis, Prout adds,

> If the views we have ventured to advance be correct, we may almost consider the *protyle* of the ancients to be realized in hydrogen; an opinion by-the-bye, not altogether new.[25]

The term "protyle" refers to an underlying primary matter, which was believed by some Greek philosophers to be the basis of all matter. The word itself is derived from *proto-hyle*, or "first stuff." Whereas for Dalton there were numerous distinct kinds of basic substances or elements, those with a more unitary view of matter, such as Prout, could not accept such a notion.[26] Prout's mention of the fact that his idea was not entirely new in the above quotation has given rise to much speculation by subsequent commentators. It seems to be agreed that he may well have obtained at least part of his idea from the writings of the English chemist Humphry Davy, who believed that many elements literally contained hydrogen. In 1808, Davy had written,

> The existence of hydrogen in sulphur is fully proved, and we have no right to consider a substance, which can be produced from it in such large quantities, merely as an accidental ingredient.[27]

Experiments inspired by Prout's hypothesis provided an increasingly accurate set of atomic weights, which could then be used to try to order the elements in the periodic system. Many of the pioneers of the periodic system, including Wolfgang Döbereiner, Leopold Gmelin, Max Pettenkofer, Jean Baptiste André Dumas, and Alexandre De Chancourtois, were very interested in Prout's hypothesis, and it figured prominently in their ideas regarding the classification of the elements.

Though Prout's hypothesis fared well initially, at least in England, where it was supported by Thomas Thomson, who had first published the work, it would fall in and out of favor for years to come. In 1825, Jons Jacob Berzelius, who would become one of the most influential chemists of his era, compiled a set of improved atomic weights that refuted Prout's hypothesis.[28] For example, the values in table 2.3 are the atomic weights of some selected elements according to Dalton and Berzelius, respectively.[29]

Berzelius objected to the practice of rounding off atomic weights to obtain whole numbers, which was common among supporters of the hypothesis, and had the following rather harsh words to say about Thomson, who as mentioned above was a supporter of Prout:

> This investigator belongs to that very small class from which science can derive no advantage whatever . . . and the greatest consideration which contemporaries can show to the author is to treat this work as if it had never happened.[30]

In 1827, the German chemist Gmelin, undaunted by such warnings, proceeded to round off even Berzelius's values, such as the ones shown in table 2.3, and to thereby reassert support for Prout's hypothesis:

> It is surprising that in the case of many substances the combining [equivalent] weight is an integral multiple of that of hydrogen, and it may be a law of nature that the combining weights of all other substances can be evenly divided by that of the smallest of them all.[31]

As other chemists continued to improve the accuracy of the atomic weights for existing elements and determined values for new elements, much of their data seemed to point away from the hypothesis. At the same time, however, new coincidences emerged where the more accurate atomic weights of some key elements were found to be very close to exact ratios.[32] This in part inspired Dumas, another

TABLE 2.3
Comparison of some of Dalton's and Berzelius's atomic weights

	H	N	Mg	Na
Dalton (1810)	1	5	10	21
Berzelius (1827)	1.06	14.2	25.3	46.5

Based on J. Dalton, *A New System of Chemical Philosophy*, R. Bickerstaff, Manchester & London, part II, 1810, p. 248; J.J. Berzelius, *Lehrbuch der Chemie*, 2nd ed., Arnold, Dresden, vol. 3, part I, 1827, p. 112.

influential French chemist, to revive Prout's hypothesis once again in 1857. These ratios included those for carbon, oxygen, and nitrogen thus:

$$C:H \sim 12:1$$
$$O:H \sim 16:1$$
$$N:H \sim 14:1$$

Throughout this period, however, there was one unavoidable obstacle to the ready acceptance of Prout's hypothesis, which no amount of rounding or remeasuring would redress. This was the fact that the atomic weight of chlorine stubbornly refused to change from its measured value of about 35.5, thus providing an apparently clear contradiction to the hypothesis.

Then, in 1844 the French chemist Charles Marignac made the ingenious suggestion of considering the basic unit of measurement as half the mass of the hydrogen atom, thus making chlorine almost exactly 71 times the weight of this unit. In 1858, Dumas took a further step and suggested a quarter of the weight of hydrogen as the basic unit, which made even more elements fall into line with the revised form of Prout's hypothesis. Of course there is no limit to how small one might make the basic unit, but the smaller it became, the more the strength of Prout's original hypothesis appeared to be weakened.

The person who did the most to refute Prout's hypothesis was the Belgian Jean Servais Stas, who began his researches into atomic weight determination in 1841 by writing, "I will say it loudly. When I undertook my researches I had an almost absolute faith in the correctness of Prout's hypothesis."[33]

Almost 25 years later, however, after measuring the atomic weights of numerous elements with as yet unheard of precision, Stas drastically changed his opinion and declared, "One must consider Prout's hypothesis as pure illusion."[34] Whereas he had originally thought that an increased precision in atomic weight determination would reveal integral multiples of the value for hydrogen, it served only to show the opposite. Far too many elements showed weights that were clearly not whole number multiples of the weight of hydrogen.

Despite the apparent problems with Prout's hypothesis, it remained true that many elements, far more than chance seemed to allow, had atomic weight values that are almost integral. As R.J. Strutt (later Lord Rayleigh) wrote in 1901, long after Prout's hypothesis seemed to have fallen by the wayside,

> The atomic weights tend to approximate to whole numbers far more closely than can reasonably be accounted for by any accidental coincidence. . . . [T]he chance of any such coincidence being the explanation is not more than one in one thousand.[35]

The explanation for why some elements did not show integral atomic weights had to await the discovery of isotopes. Eventually, it would be understood that the elements whose values on a hydrogen scale are close to being whole numbers are those that exist only in one form or whose other forms, or isotopes, occur only in very small amounts.[36] By contrast, many elements showed values that differed markedly from whole numbers, such as chlorine (35.46), copper (63.57), zinc (65.38), and mercury (200.6). Their atomic weights were not exact or even close to exact multiples of hydrogen's because they occur as mixtures of several isotopes that are present in comparable amounts.

So Prout's hypothesis turned out to be incorrect in that the elements are not composites of hydrogen according to their atomic weights, and yet there is a sense in which his idea can be said to have now been vindicated by modern physics. In terms of numbers of protons, the nuclei of all the elements are indeed composites of the nucleus of the hydrogen atom, which contains just a single proton. But, even at the time it was first proposed, Prout's hypothesis proved to be very fruitful, because it encouraged the determination of accurate atomic weights by numerous chemists who were trying to either confirm or refute it.[37] From the point of view of Karl Popper's philosophy of science, this all makes perfect sense. A useful scientific idea need not necessarily be correct, but it is essential that it should be refutable in the light of experimental evidence.

DÖBEREINER DISCOVERS TRIADS

Just as the examination of atomic weight data led to Prout's hypothesis, so it was to produce another fruitful philosophical principle, that of triads. This development originated with the German chemist Döbereiner, who was active in the city of Jena in the early 1800s. Döbereiner became interested in the emerging study of stoichiometry, the study of proportions in chemical reactions, and became an early adherent of the newly developed theory of chemical atomism. He was the first to notice the existence of various groups of three elements, subsequently called triads, which showed chemical similarities and which displayed an important numerical relationship, namely, that the equivalent weight, or atomic weight, of the middle element is the approximate mean of the values of the two flanking elements in the triad.

In 1817, Döbereiner found that if certain elements were combined with oxygen in binary compounds, a numerical relationship could be discerned among the equivalent weights of these compounds. Thus, when oxides of calcium, strontium, and barium were considered, the equivalent weight of strontium oxide was approximately the mean of those of calcium oxide and barium oxide.[38] The three elements in question, strontium, calcium, and barium, were said to form a triad:[39]

$$SrO = (CaO + BaO)/2 = 107$$
$$= (59 + 155)/2$$

Though Döbereiner was working with weights that had been deduced with the relatively crude experimental methods of the time, his values compare rather well with current values for the triad:[40]

$$104.75 = (56 + 153.5)/2$$

Döbereiner's observation had little impact on the chemical world at first but later became very influential. He is now regarded as one of the earliest pioneers of the development of the periodic system. What is not often reported in modern accounts is that Döbereiner considered the possibility that the middle element of his triad might actually be a mixture of the other two elements in question and that his observations might support the notion of transmutation among the three elements.[41]

Very little happened regarding triads until 12 years later, in 1829, when Döbereiner added three new triads. The first involved the element bromine, which had been isolated in the previous year. He compared bromine to chlorine and iodine, using the atomic weights obtained earlier by Berzelius:

$$Br = (Cl + I)/2 = (35.470 + 126.470)/2 = 80.470^{[42]}$$

The mean value for this triad is reasonably close to Berzelius's value for bromine of 78.383. Döbereiner also obtained a triad involving some alkali metals, sodium, lithium, and potassium, which were known to share many chemical properties:

$$Na = (Li + K)/2 = (15.25 + 78.39)/2 = 46.82^{[43]}$$

In addition, he produced a fourth triad:

$$Se = (S + Te)/2 = (39.239 + 129.243)/2 = 80.741^{[44]}$$

Once again, the mean of the flanking elements, sulfur and tellurium, compares well with Berzelius's value of 79.5 for selenium.

Döbereiner also required that, in order to be meaningful, his triads should reveal chemical relationships among the elements as well as numerical relationships. On the other hand, he refused to group fluorine, a halogen, together with chlorine, bromine, and iodine, as he might have done on chemical grounds, because he failed to find a triadic relationship among the atomic weights of fluorine and those of these other halogens. He was also reluctant to take the occurrence of triads among dissimilar elements, such as nitrogen, carbon, and oxygen, as being in any sense significant even though they did display the triadic numerical relationship.

Suffice it to say that Döbereiner's research established the notion of triads as a powerful concept, which several other chemists were soon to take up with much effect. Indeed, Döbereiner's triads, which would appear on the periodic table grouped in vertical columns, represented the first step in fitting the elements into a system that would account for their chemical properties and would reveal their physical relationships.[45]

Before the correct relationship between atomic and equivalent weights had been discovered, some chemists regularly referred to atomic weights as equivalents and vice versa. To make matters worse, even when the same terminology was used by any given two chemists, there was still disagreement as to actual values, since various standards were used by different workers. In addition, the methods for obtaining atomic weights were applicable only to gases. Initially, it was not possible to estimate the atomic weights of liquids and solids, and this made it difficult to recognize periodic relationships, since on crossing a period one typically moves from solids to gases.

It is therefore not surprising that groups of similar elements in the periodic table were discovered long before periods involving dissimilar elements or, in other words, that vertical relationships were discovered before horizontal ones, in modern terms. Of course, there is a more immediate reason why groups were discovered long before periods: elements within groups share chemical properties, thus rendering their grouping intuitively obvious. Although this is quite true, what is being addressed here is the separate issue of the recognition of numerical relationships among elements in groups. The existence of the periodic table depends not only on chemical properties but also almost as much on numerical aspects and on physical principles, although the latter in particular raises certain philosophical questions concerning the reduction of chemistry.

GMELIN'S REMARKABLE SYSTEM

In 1843, a full 26 years before the publication of Dimitri Mendeleev's famous system of 1869, a much neglected and underrated periodic system was published (figure 2.3).[46] This was the work of Gmelin, the author of the rather voluminous *Handbuch der Chemie* and one of the most influential chemical writers of this time.[47]

Although Döbereiner is rightly regarded as the originator of the notion of triads, Gmelin also did much useful work in this area, and it was he who coined the term "triad." Like Döbereiner, Gmelin considered both chemical and numerical relationships when looking for triads, and he was able to extend his predecessor's work using improved atomic weights that had been unavailable to Döbereiner. For example, whereas Döbereiner had grouped magnesium together with the alkaline earths based on their chemical similarities, he was unable to find a triad relationship involving it and other alkaline earth elements. Gmelin, on the other hand, was able to discern the following relationship among magnesium, barium, and calcium, using his own newly obtained values for atomic weights, which he published in the same book in 1827:

$$(Mg + Ba)/4 = Ca$$
$$(12 + 68.6)/4 = 20.15 \ (Ca = 20.5)^{48}$$

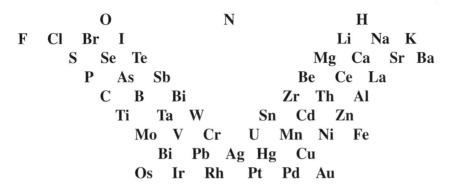

FIGURE 2.3 System of 1843. L. Gmelin, *Handbuch der anorganischen chemie* 4th ed., Heidelberg, 1843, vol. 1, p. 52.

But let us now turn to the more remarkable aspects of Gmelin's system of 1843. From the existence of four unconnected triads discovered by Döbereiner, Gmelin was able to make a huge leap forward in obtaining a system based on triads consisting of as many as 55 elements. In addition, his system as a whole was essentially ordered according to increasing atomic weight. With this work, Gmelin succeeded in capturing the correct grouping of most of the then known main-group elements. Gmelin arranged his triads horizontally in the V-shaped schematic shown in figure 2.3.

Suppose we take the right arm of Gmelin's V-shape and make it point downward, and then consider the arrangement obtained (figure 2.4). It is important to appreciate that this change does not alter Gmelin's table in any fundamental way but merely re-presents its contents. If atomic weights are introduced explicitly into the table, something that Gmelin did not do, a general increase in this quantity in both wings of main-group elements is seen.

On removing all the elements in the central portion of figure 2.4, the entire table can be rotated by 90°, and all the columns can be stacked together, as shown in figure 2.5. What Gmelin's table achieves, granted this artistic license, is an essentially correct grouping of many of the representative, or main-group, elements. Though he failed to arrange the transition metals and inner transition metals correctly, this can hardly be taken as a reason for thinking any less of his system, since problems with transition metals were common even in later, more mature periodic systems.

The fact that Gmelin could have produced such an arrangement of the main-group elements so early in the evolution of the periodic system, as shown in figure 2.5, is rather remarkable. In the case of groups I, II, IV, V, VI, and VII, *all* the elements included are shown in the correct order of increasing atomic weight going left to right. Boron and bismuth, which would have been placed incorrectly in

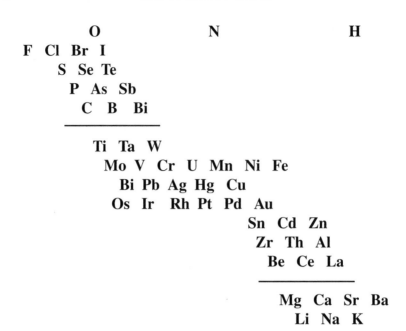

FIGURE 2.4 Flattened version of Gmelin's system.

group IV on moving from figure 2.4 to 2.5, have been omitted from figure 2.5. The only main-group misplacements appear to be nitrogen and oxygen, but Gmelin clearly recognizes that oxygen belongs with sulfur, selenium, and tellurium when he points out the following relationship, which also includes antimony:

$$O = 8, S = 16, Se = 40, Te = 64, Sb = 129 = 1:2:5:8:16$$

Perhaps Gmelin's table cannot be properly called a periodic system, since it does not depict the well-known tendency of the elements to recur, that is, to show periodicity after certain regular intervals. Moreover, Gmelin's system does not *explicitly* arrange the elements in increasing order of atomic weight. But there may have been an implicit use of atomic weight ordering in Gmelin's system, since his placing of many of the triads side by side produces the correct order as found in subsequent more mature periodic systems.[49] It is also known that Gmelin was very interested in values of atomic weights, since in 1827 he produced an early list of atomic weights for as many as 45 elements.[50]

About 25 years before Mendeleev, Gmelin used his own rudimentary system of the elements to give an overall structure and direction to his chemical textbook. He was thus possibly the first chemistry textbook author to do so. Although Mendeleev is usually credited with basing a textbook around the periodic system of the elements, he used an inductive approach, not presenting his system until the final chapter of the first volume of his textbook, even in later editions.[51] Gmelin,

Li		C			F
Na	Mg		P	S	Cl
K	Ca		As	Se	Br
	Sr		Sb	Te	I
	Ba				

FIGURE 2.5 Gmelin's table rotated, after removing the 30 elements
in the central position, as well as B, Bi, H, O, and N.

on the other hand, gives the system immediately on the very first page, at the start
of volume 2 of his series, and the remainder of this volume is a detailed 500 or so
page survey of the chemistry of 12 nonmetallic elements.[52]

Moreover, the order that Gmelin elects to follow in his presentation is dictated by the system itself. He begins with oxygen and hydrogen, two of the three
elements at the head of his table. His following chapters discuss the chemistry of
carbon and boron, which Gmelin has placed together in the same group. He then
discusses the chemistry of phosphorus, the only nonmetal in group V of the modern table apart from nitrogen. This is followed by chapters on sulfur and selenium,
which are the nonmetals in what became group VI. After that comes a survey of
the chemistry of all four of the then-known nonmetals in group VII of the modern periodic table.

Finally, the volume closes with the chemistry of nitrogen, the remaining one of
the three elements that Gmelin placed at the head of his system of elements. Apart
from the misplacement of boron along with carbon,[53] in what would correspond
to the modern group IV, Gmelin has given a systematic survey of most of the
important nonmetals in the order of groups IV, V, VI, and VII, from the perspective
of the mature periodic system, which emerged only with the work of Mendeleev
in 1869. This, I submit, suggests remarkable foresight and intuition on the part of
Gmelin, as does the way in which he uses his system to ground the presentation
of the chemistry of these elements. Yet Gmelin's contribution to the classification
of the elements has not been sufficiently appreciated by historians of chemistry,
or even historians of the periodic system.[54] Johannes van Spronsen, the author of
the only scholarly book on the history of the periodic system, mentions Gmelin's
remarkable table of 1843, but does so somewhat dismissively:

> In 1843 Gmelin also tried to find a relationship existing between all elements.
> This however, meant demoting the atomic weight. . . . The elements oxygen,
> nitrogen and hydrogen, for which he apparently could find no homologues,
> form the basis for his classification.[55]

But van Spronsen appears not to notice that Gmelin did in fact correctly classify
at least one of these elements, oxygen, correctly with sulfur, selenium, and tellu-

rium.[56] Perhaps Gmelin's system should no longer be regarded as something of a footnote to Döbereiner's discovery of the existence of triads but as an important discovery of almost equal stature.

A QUALITATIVE INTERMISSION

In order to appreciate the advances that were made by considering quantitative properties of the elements, it is necessary to also consider what was known in qualitative terms at the time that the periodic system was beginning to ferment. One obvious way to do this is to consult some chemistry textbooks of the period, since they serve as the repository of the chemical knowledge known at the time. For the sake of brevity, I describe the set of textbooks written by Gmelin, whose successive editions would eventually give rise to a very influential set of books on inorganic chemistry.[57] It is not possible to divorce quantitative aspects altogether, given that, as noted above, Gmelin recognized triads based on atomic weights and even appears to have ordered the elements to some implicit extent on their atomic weights. Nevertheless, the influential series of books by Gmelin was primarily a summary of all the qualitative knowledge on the elements available at the time.

In volume 2 of his series, Gmelin gives a classification system for the elements that had been identified at the time (figure 2.3). As he puts it, these are 61 "undecomposable ponderable bodies," of which 12 are nonmetallic and 49 are metallic.

These elements are arranged in groups horizontally according to their chemical and physical properties, with the more electronegative elements on the left and the more electropositive ones on the right side. Gmelin does not specify more precisely what determines the order of the elements from left to right. One can only suppose that it might be degrees of electronegativity.[58] For example, the halogens are arranged in a row starting with fluorine, which is the most electronegative of them, followed, in decreasing order of electronegativity, by chlorine, bromine, and iodine.

The halogens represent a family of elements whose group similarities became apparent almost as soon as they were isolated and, in the case of fluorine, even before its isolation. Considering these elements in historical order, chlorine was first discovered by Scheele in 1774, although he believed that it contained oxygen. It was first isolated in 1810 by Davy, who was the first to recognize that it was an element. That same year saw the first isolation of iodine by Bernard Courtois, followed by bromine in 1826 by Antoine Balard. Fluorine had not yet been isolated at the time that Gmelin was writing, although it had been recognized as an element. It was finally isolated in 1886, sometime after Gmelin devised his system. However, it had been studied in compound form by many chemists, including Davy, Gay-Lussac, and Lavoisier, all of whom experimented on one of its most common compounds, hydrofluoric acid.

Writing in 1843, Gmelin devoted a total of 123 pages to these four elements alone. Starting with the least electronegative of them, he discusses the reactions of iodine with water and oxygen and the existence of various oxides of iodine, such as IO_5 and IO_7. This is followed by discussion of the reactions of iodine with a number of other elements, namely, hydrogen, boron, phosphorus, sulfur, selenium, and the other halogens. He proceeds to discuss the chemistry of bromine and chlorine, which give analogous reactions with all the same elements as iodine and which form analogous acids with a few minor exceptions.[59]

In the case of fluorine, the same pattern of reactivity is described with a few more exceptions. These more noticeable differences might be explained by the fact that the element had not yet been isolated, so it was not as easy to examine its reactions with all the elements as had been possible in the case of iodine, bromine, and chlorine.[60] Overall, the well-established similarity among chlorine, bromine, and iodine would explain why it was one of the first set of elements grouped together as a chemically significant triad by Döbereiner. The differences with fluorine might explain why the latter element was frequently omitted from triads or extended triads, called tetrads.

Another nonmetallic group of elements given by Gmelin in his classification consists of sulfur, selenium, and tellurium. As in the case of the halogens, Gmelin does not specifically discuss the analogies among these elements, but the reader is left in no doubt by a study of the detailed chemistry of the elements in Gmelin's text. Sulfur and selenium are discussed, one after the other, over a total of 94 pages, only 20 of which are devoted to selenium. As in the case of the halogens, the chemical similarities, at least for sulfur and selenium, are abundantly obvious. They include reaction with oxygen, acid formation, and reaction with hydrogen, phosphorus, sulfur, bromine, chlorine, and a number of metals. Although the element tellurium is clearly grouped with sulfur and selenium, Gmelin does not discuss its chemistry along with these latter two elements. This would seem to indicate some ambiguity in the classification of the element tellurium. In fact, the chemistry of tellurium is delayed until volume 4, when it finally makes an appearance following descriptions of the reactions of the group consisting of phosphorus, antimony, and bismuth. It would appear that Gmelin is ambivalent about the classification of tellurium. Whereas in the system shown in figure 2.2 it is included among the elements of what would become group VI in the modern periodic table, in discussing the detailed chemistry of the element Gmelin appears to contradict his earlier choice. The chemical and physical properties of tellurium are discussed in the context of the elements of the modern group V, along with phosphorus, arsenic, and antimony.[61]

The element nitrogen, which heads the modern group V, is separated from all these elements, however, and is discussed at a very early stage in volume 2. Recall that nitrogen, along with hydrogen and oxygen, is shown at the head of the system with no particular group membership.[62]

Turning to Gmelin's grouping of metals, we encounter a group containing lithium, sodium, and potassium. As well as being one of the triads discovered by Döbereiner, the chemical grouping of the elements is virtually inescapable. The elements are all soft, have a low density, and react with water to form alkaline solutions.[63] The analogies among them are remarkable given that all of them react, without fail, with oxygen, boron, carbon, phosphorus, sulfur, selenium, and the halogens, although to varying degrees.

Similarly, Gmelin summarizes the chemical similarities among another group of metals consisting of magnesium, calcium, strontium, and barium,[64] the last three of which had been recognized as the very first triad of elements by Döbereiner.[65] These elements are physically harder and less reactive than those of Gmelin's group consisting of lithium, sodium, and potassium. In modern terms, the major difference between these two groups of elements is that lithium, sodium, and potassium show a valence of 1 while magnesium, calcium, strontium, and barium show a valence of 2. Gmelin appears not to have recognized this feature however, since the formulas he gives for all their oxides, for example, consist of one atom of the metal combined to one atom of oxygen.[66] It would seem that the qualitative differences alone between the members of the two groups sufficed to convince Gmelin and others that the elements concerned belonged in two different groups.

PETTENKOFER'S DIFFERENCE RELATIONSHIPS

In 1850, Pettenkofer, at the University of Munich, another supporter of Prout's hypothesis, published an article dealing with numerical relationships among the equivalent weights of the elements. But unlike his predecessors, he did not focus on triads, believing that the findings by Döbereiner and others were due to mere chance. As an example, Pettenkofer pointed out that while the atomic weight of the middle member of the chlorine, bromine, iodine triad was indeed the mean of the flanking elements, this was not the case in the chemically analogous triad of fluorine, chlorine, and bromine.

Nevertheless, Pettenkofer, too, created what essentially amounted to triads, and even larger groups of elements, although by a quite different approach. In the course of an examination of the data on the known elements, Pettenkofer realized that some series of chemically similar elements tended to show constant differences among their equivalent weights. He noted, for example, that the weights of lithium, sodium, and potassium differed by gaps of 16 units (table 2.4). As some authors remark, Pettenkofer thus failed to notice that this was tantamount to Döbereiner's recognition that the middle element has an equivalent weight that is the mean of the two flanking elements.

TABLE 2.4

Pettenkofer's atomic weight gaps

	Li	Na	K
Equivalent weights	7	23	39
Differences		16	16

Based on M. Pettenkofer, Ueber die regelmassigen Abstande der Aequivalentzahalen der sogennanten einfachen Radicale, *Annalen der Chemie und
Pharmazie*, 105, 187-202, 1858.

In some other series of elements, Pettenkofer pointed out that the differences
in equivalent weights were a multiple of a certain number, such as 8 in the case of
the alkaline earths and the oxygen group (table 2.5).

In taking these steps, Pettenkofer was already going beyond triads to considering larger series of elements. In the case of carbon, boron, and silicon, which were
often grouped together in early classifications of the elements, as well as in the case
of the halogens, the differences were factors of 5. In addition, there was another
series, containing nitrogen, phosphorus, arsenic, and antimony, in which the differences involved factors of both 5 and 8, as shown in table 2.6.

On the basis of this theory of constant differences, and multiples of constant
differences, Pettenkofer proposed the idea of calculating the equivalent weights
for elements whose values would be difficult to measure otherwise. It is perhaps
significant that Mendeleev some time later mentioned the name of Pettenkofer
in his own articles as one of the few who had influenced his work on the periodic system. As is well known, Mendeleev made much use of predictions based on
interpolations among atomic weights, and also used interpolations to correct the

TABLE 2.5

Pettenkofer's differences in atomic weights for alkaline earths and oxygen group

	Mg	Ca	Sr	Ba
Equivalent weight	12	20	44	68
Differences		8	3(8)	3(8)

	O	S	Se	Te
Equivalent weights	8	16	40	64
Differences		8	3(8)	3(8)

Based on M. Pettenkofer, Ueber die regelmassigen Abstande der Aequivalentzahlen der sogennanten
einfachen Radicale, *Annalen der Chemie und Pharmazie*, 105, 187-202, 1858.

TABLE 2.6
Pettenkofer's atomic weight differences for the "nitrogen series"

	N	P	As	Sb
Equivalent weights	14	32	75	129
Differences		18	43	54
		2(5)+8	7(5)+8	9(5)+8

Based on M. Pettenkofer, Ueber die regelmassigen Abstande der Aequivalentzahalen der sogennanten einfachen Radicale, *Annalen der Chemie und Pharmazie*, 105, 187-202, 1858.

atomic weights of already known elements. Given the work of such chemists as Pettenkofer, it is clear that the idea of making such predictions did not originate with Mendeleev.[67]

DUMAS'S CONTRIBUTIONS AND HIS REVIVAL OF TRANSMUTATION

The year 1851 was a rather busy one for the famous French chemist Dumas, who published two important papers and delivered an influential public lecture in Ipswich, England, on the classification of the elements.[68] Throughout this work, Dumas drew attention to four triads—(S, Se, Te), (Cl, Br, I), (Li, Na, K), and (Ca, Sr, Ba)—but without mentioning Döbereiner as their original discoverer. Whereas the German chemist had suggested that the middle member of each triad might be a mixture of the two extreme elements, Dumas thought that the middle member was a compound of its flanking partners and offered this idea as support for Prout's hypothesis.

Dumas went as far as to suggest that transmutation might be possible among the elements in each triad and that research should be carried out to discover the mechanism of these possible transformations. He also took the fact that elements such as cobalt and nickel are often found associated together in nature as further evidence of the possibility of elemental transmutation. Interestingly, the English scientist Michael Faraday praised Dumas's public lecture and agreed that some form of transmutation was suggested by these findings. Faraday said:

> Thus we have here one of the many scientific developments of late origin, which tend to lead us back into speculations analogous with those of the alchemists . . . and now we find, after our attention has been led in that direction, that the triad of chlorine, bromine and iodine not only offers well-marked progression of certain chemical manifestations, but that the same progression is accordant with the numerical exponents of their combining weights. We seem

here to have a dawning of a new light, indicative of the mutual convertibility of certain groups of elements, although under conditions which as yet are hidden from scrutiny.[69]

I draw attention to this passage because it reveals that at least some leading chemists appeared to continue to believe in this central alchemical doctrine long after alchemy is supposed to have been abandoned.

KREMERS GOES HORIZONTAL

Peter Kremers, working in Koln, Germany, was one of the earliest chemists to begin to consider what would eventually form horizontal series of elements in the mature periodic systems of the future. He did this by examining the numerical relationships among the atomic weights of elements with little in common. For example, he noted the regularity among a short series of elements that included oxygen, sulfur, titanium, phosphorus, and selenium shown in table 2.7.[70] Kremers also discovered some new triads, such as

$$Mg = \frac{O + S}{2}, \quad Ca = \frac{S + Ti}{2}, \quad Fe = \frac{Ti + P}{2}$$

From a modern standpoint, these triads may not seem to be chemically significant. There are two reasons for this. The modern medium-long form of the periodic table fails to display secondary kinships among some elements. Sulfur and titanium both show a valence of 4, for example, though they do not appear in the same group in the medium-long form of the periodic table.[71] But it is not so far fetched to consider them as being chemically analogous. Given the fact that both titanium and phosphorus commonly display valences of 3, this grouping, too, is not as incorrect as a modern reader may think. The second reason why one should not be too surprised by some of the less plausible triad groupings made by Kremers is that the notion of a triad had begun to take on a life of its own as distinct from chemical analogies. The aim had become one of finding triad relationships

TABLE 2.7
Kremers's atomic weight differences for the oxygen series

	O	*S*	*Ti*	*P*	*Se*
Equivalent wt.	8	16	24.12	32	39.62
Difference		8	$\cong 8$	$\cong 8$	$\cong 8$

Based on P. Kremers, *Annalen der Physik und Chemie (Poggendorff)*, 85, 37, 246, 1852.

among the weights of the elements irrespective of whether or not this had a chemical significance. Mendeleev later described such activity among his colleagues as an obsession with triads, which he believed to have delayed the discovery of the mature periodic system.

But to return to Kremers, perhaps his most incisive contribution lay in the suggestion of a bidirectional scheme of what he termed "conjugated triads." Here, certain elements would serve as members of two distinct triads lying perpendicularly to each other,

Li 6.5	Na 23	K 39.2
Mg 12	Zn 32.6	Cd 56
Ca 20	Sr 43.8	Ba 68.5

Thus, in a more profound way than any of his predecessors, Kremers was comparing chemically dissimilar elements, a practice that would reach full maturity only with the tables of Julius Lothar Meyer and Mendeleev.

In 1856, Kremers mistakenly claimed that the divergences from exact values in the relationships among triads were caused by changes in temperature and that each triad would show an exact numerical relationship at a particular temperature. In the course of this research, he produced the table shown in table 2.8. One suspects that most people studying this table would have been seduced into thinking that the divergences were indeed very small, and might well be spurious, thus further strengthening the notion of triads.

TABLE 2.8
Kremers's differences between calculated and observed
atomic weight triads

Triad (T)	Middle element (M)	$(T - M)/T$
K + Li/2	Na	−0.007
Ba + Ca/2	Sr	+0.010
Ag + Hg/2	Pb	+0.003
J + Cl/2	Br	+0.016
S + Se/2	Cr	+0.038
S + Te/2	Se	+0.017
Cr + Va/2	Mo	+0.035
P + Sb/2	As	+0.009

The headings for each column have been adapted from the original table, which uses different symbols. The modern symbol for vanadium is V rather than Va as in this table. J denotes iodine, or *jod* in German.

P. Kremers, *Annalen der Physik und Chemie (Poggendorff)*, 99, 58–63, 1856.

SUPERTRIADS

The most ambitious scheme of all involving triads was created by the 20-year-old Ernst Lenssen while he was working at an agricultural institute in Wiesbaden. In 1857, Lenssen published an article in which virtually all of the 58 known elements were arranged into a total of 20 triads, with the exception of niobium, which he could not fit into any triad (table 2.9).[72] Ten of his triads consisted of nonmetals and acid-forming metals, and the remaining 10 of just metals.

Lenssen also suggested further relationships involving groups of triads. Using the 20 triads in table 2.9, he was able to identify a total of seven enneads, or supertriads, in which the mean equivalent weight of each middle triad lies approximately midway between the mean weights of the other triads in a group of three triads (table 2.10). However, as the table shows, he was forced to combine his tri-

TABLE 2.9
The 20 triads of Lenssen

		Calculated atomic weight			*Determined atomic weights*		
1	(K + Li)/2	= Na	= 23.03	39.11	23.00	6.95
2	(Ba + Ca)/2	= Sr	= 44.29	68.59	47.63	20
3	(Mg + Cd)/2	= Zn	= 33.8	12	32.5	55.7
4	(Mn + Co)/2	= Fe	= 28.5	27.5	28	29.5
5	(La + Di)/2	= Ce	= 48.3	47.3	47	49.6
6	Yt Er Tb			32	?	?
7	Th Norium Al			59.5	?	13.7
8	(Be + Ur)/2	= Zr	= 33.5	7	33.6	60
9	(Cr + Cu)/2	= Ni	= 29.3	26.8	29.6	31.7
10	(Ag + Hg)/2	= Pb	= 104	108	103.6	100
11	(O + C)/2	= N	= 7	8	7	6
12	(Si + Fl)/2	= Bo	= 12.2	15	11	9.5
13	(Cl + J)/2	= Br	= 40.6	17.7	40	63.5
14	(S + Te)/2	= Se	= 40.1	16	39.7	64.2
15	(P + Sb)/2	= As	= 38	16	37.5	60
16	(Ta + Ti)/2	= Sn	= 58.7	92.3	59	25
17	(W + Mo)/2	=V	= 69	92	68.5	46
18	(Pa + Rh)2	= Ru	= 52.5	53.2	52.1	51.2
19	(Os + Ir)/2	= Pt	= 98.9	99.4	99	98.5
20	(Bi + Au)/2	= Hg	= 101.2	104	100★	98.4★

★In the original version, these two atomic weights have been inadvertently interchanged.
Norium was an element that was first reported in 1845 but was not later comfirmed.

E. Lenssen, Uber die gruppirung der elemente nach ihrem chemisch-physikalischen charackter, *Annalen der Chemie und Pharmazie*, 103, 121–131, 1857.

TABLE 2.10

Lenssen's supertriads

Triad	Mean equivalent weight	
1	23	
3	33	$(23 + 44)/2 = 33.5$
2	44	
4	28	
6	?	
5	47	
9	29.5	
8	33.5	$(29.5 + 37)/2 = 33.3$
7	37	
H	1	
11	7	$(1 + 12)/2 = 6.5$
12	12	
15	38	
14	40	$(38 + 40)/2 = 38$
13	40	
18	52.1	
16	61	$(99 + 104)/2 = 101.5$
17	69	
19	99	
20	101	
10	104	

Triads are numbered as in table 2.9.

E. Lenssen, Uber die gruppirung der elemente nach ihrem
chemisch-physikalischen charackter, *Annalen der Chemie und
Pharmazie*, 103, 121–131, 1857.

ads in a somewhat arbitrary order to achieve this goal. In addition, one supertriad
involves just one element, hydrogen, rather than a triad of elements.

Lenssen was another early pioneer who was prepared to make predictions on
the basis of his system. For example, he predicted the atomic weights of erbium
and terbium, neither of which had yet been isolated. This is mentioned just to
emphasize again that Mendeleev did not "invent" the idea of making predictions
using classification systems for the elements, as seems to be popularly believed. The
fact remains, however, that Lenssen's predictions were found to be incorrect.

Many philosophical issues regarding the prediction of new elements and properties of elements are examined at various points in this book. One of these issues
concerns the importance that is attributed to predictions while a theory is in the
process of being accepted. This is a theme that has been actively debated by recent
historians and philosophers of science, and it has some implications for the story of

the periodic table, the development of which is usually presented as relying heavily on the prediction of new elements.[73]

But at the close of this chapter, it should be appreciated how far chemistry had advanced from the introduction of numerical aspects by Lavoisier and others at the time of the chemical revolution to the growing realization that such numerical aspects of chemistry were also essential to the classification of the elements. Indeed, it appears that real progress was achieved only when the debate over Prout's hypothesis and the hunt for triads served to focus attention on the role of numerical values, whereas the previous attempts to sort the elements according to chemical similarities had failed to produce any coherent scheme. And all of this occurred before it was discovered that simply ordering the elements according to increasing values of atomic weight would reveal a periodicity in their properties.

The story of the early development of the periodic table demonstrates very effectively how scientific ideas can progress in spite of what later appear to be mistakes. For example, many of the triads that were identified have turned out to be incorrect in that they made no chemical sense, and yet the general project of examining triad relationships was still rather fruitful. Local mistakes do not seem to matter too much. One might consider the case of Dalton, for example. Three of his main ideas—the importance of repulsions between like particles; the existence of a caloric, or heat envelope, around atoms; and the formula of OH he assumed for water—have all turned out to be incorrect. Nevertheless, Dalton's general project has been tremendously influential. His atomic theory represents one of the pillars of modern science, providing a theoretical understanding of the observed laws of chemical combination, among other chemical facts. It is as if scientific evolution somehow transcends any logical stepwise progression yet still displays a form of organic growth within which any "mistakes" are subsumed. This overall evolution seems to have a life of its own, which overrides mistakes on the part of individual scientists, or even collective errors, in the light of subsequent knowledge.[74]

POSTSCRIPT ON TRIADS

As I have shown, the recognition of triads represented the first important step toward the eventual construction of the modern periodic system. The limitations of the concept of triads had more to do with the data that were being used by the early pioneers than with the concept itself. It is interesting to consider how triads fare in the modern periodic system.

It is now realized that atomic weight is not the fundamental property determining the placement of each element in the periodic table. The elements can indeed for the most part be ordered by increasing weight, but the atomic weight of

TABLE 2.11
Döbereiner's triads computed using atomic numbers

Element	Atomic weight	Mean	Atomic number	Mean
Chlorine	35.457		17	
Bromine	79.916	81.19	35	35
Iodine	126.932		53	
Sulfur	32.064		16	
Selenium	79.2	79.78	34	34
Tellurium	127.5		52	
Calcium	40.07		20	
Strontium	87.63	88.72	38	38
Barium	157.37		56	
Phosphorus	31.027		15	
Arsenic	74.96	76.40	33	33
Antimony	121.77		51	

Compiled by the author.

any particular element depends upon the contingencies of terrestrial abundances of all the isotopes of that particular element. When they measured the atomic weight of an element, nineteenth-century chemists were unwittingly measuring the average weight of a mixture of isotopes (except for those elements that occur as a single isotope in nature.[75]

As was discovered well after the turn of the twentieth century, when the structure of the atom was discerned, the order of the elements is determined unambiguously by the property of atomic number, which corresponds to the number of protons in the nucleus of the atoms of any particular element. It emerges that in certain parts of the modern periodic table the triad relationship turns out to be exact if atomic numbers are used instead of atomic weights (table 2.11). For example, a number of the triads discovered by Döbereiner behave in this manner.

I postpone the full explanation for why perfect triads occur in parts of the periodic system until chapter 6, where I give an account of the discovery of isotopes and atomic number. From the perspective of the modern periodic table, about 50% of all possible vertical triads, using atomic numbers, are in fact exact.

ATOMIC WEIGHT DETERMINATION

In this final section I take up the question of the determination of atomic weights that was begun by Dalton at the beginning of the nineteenth century. As noted above, Dalton adopted the rule of maximum simplicity regarding formulas, but unfortunately, this rule is broken in the vast majority of compounds such as water, ammonia, and all the oxides of nitrogen apart from NO. Moreover, whereas Dalton began by using comparative densities of gases to estimate atomic weights along with the notion of equal volumes equal numbers of particles (EVEN), he later turned against this hypothesis.

The next major contributor to the project of determining atomic weights was Berzelius, who published tables of atomic weights in 1814 and 1818, which he greatly extended and revised in 1826. Among the many problems encountered by chemists working to determine atomic weights was the question of selection of a standard. Whereas Dalton had quite reasonably chosen H = 1, not all elements combine with hydrogen. To determine the atomic weight of an element that does not react directly with hydrogen required using an intermediate element, thus increasing the sources of errors.[76]

Since oxygen forms compounds with most elements, it was adopted as the standard but was confusingly given different values by different chemists.[77] Berzelius was one of the few who tried to go beyond equivalent weights in order to determine atomic weights. As mentioned above, the determination of atomic weights depended on recognizing the correct formula for a compound. Using Gay-Lussac's law of definite proportions by volumes, Berzelius arrived at the correct formulas for water, ammonia, hydrogen chloride, and hydrogen sulfide: H_2O, NH_3, HCl, and H_2S. But Gay-Lussac's law was limited to combining gases, although Berzelius devised a method for compounds such as $PbSO_4$ and also ventured to estimate the atomic weights of a number of metals by similar approaches. In his earliest tables of 1814 and 1818, Berzelius regarded metals as dioxides, giving them formulas such as AgO_2 and FeO_2. In 1826, he changed to regarding them as monoxides, thus making the values of alkali metals twice what they should have been but obtaining correct values for the alkaline earths.

Two major developments then followed that permitted better atomic weight determinations. They were the law of Pierre-Louis Dulong and Alexis-Thérèse Petit and the law of isomorphism.

Dulong and Petit discovered that the specific heat of any solid element multiplied by its atomic weight is approximately equal to a constant (table 2.12). In fact, they adjusted the atomic weights of many elements with uncertain weights so that their behavior would fit their new law. There was little justification for taking such an action except that it seemed to preserve the putative law. Although they had originally hoped to use the law in order to determine unknown atomic weights, its approximate nature meant that this was not possible. Nevertheless, the law of Dulong and Petit proved to have another important use. It could be used to check possible atomic weights and to settle cases in which there was some doubt as to whether the supposed value needed to be halved, doubled, or kept intact. For example, from Dulong and Petit's law, it quickly became clear that Berzelius's val-

TABLE 2.12
Table from Dulong and Petit's article of 1819

Element	Specific heat	Atomic weight O = 1	Product of atomic weight and specific heat
Bismuth	0.0288	13.30	0.3830
Lead	0.0293	12.95	0.3794
Gold	0.0298	12.43	0.3704
Platinum	0.014	11.16	0.3740
Tin	0.0514	7.35	0.3779
Silver	0.0557	6.75	0.3759
Zinc	0.0927	4.03	0.3736
Tellurium	0.0012	4.03	0.3675
Copper	0.0949	3.957	0.3755
Nickel	0.1035	3.69	0.3819
Iron	0.1100	3.392	0.3731
Cobalt	0.1498	2.46	0.3685
Sulfur	0.1880	2.11	0.3780

As the attentive reader may note, Dulong and Petit are not consistent in quoting values in the fourth column to the correct number of significant figures.

P.L. Dulong, A.T. Petit, Recherches sur quelques points importants de la Théorie de la Chaleur, *Annales de Chimie Physique*, 10, 395-413, 1819.

ues for the alkali metals were in error by a factor of 2 and that his formulas needed adjustment to M_2O instead of MO.[78]

In 1819, Eilhard Mitscherlich began to publish articles in Berlin on what would become a law named after him. He found that certain elements could substitute for each other to produce analogous or, as he termed them, isomorphic structures having the same crystalline form apart from some minor variations in the angles between crystal planes.[79] He suggested that the crystal form was determined uniquely by the number of atoms in the chemical compound in question. One could therefore deduce the atomic weight of one element in a compound from the known atomic weight of another element that could act as a substitute for the first element. Recall that the problem had been to assign a particular number of atoms of any particular element to a compound, such as the classic case of Dalton's HO as opposed to Berzelius's H_2O. Here was a new way of settling such questions in the case of solid crystalline compounds. For example, by considering the two isomorphous compounds of potassium sulfate (K_2SO_4) and potassium selenate (K_2SeO_4), and from the known atomic weight of sulfur, which was 32, Mitscherlich was able to deduce the correct atomic weight of selenium to be approximately 79.[80]

Another chemist to make important contributions to the determination of atomic weights was Dumas, as mentioned above in connection with his work on triads. In 1826, he devised a new method that permitted atomic weights to be determined for any liquid or solid substance that could be vaporized.[81] The density of the vapor produced could be compared with that of hydrogen, and on the basis of the EVEN hypothesis, the atomic weight of the vaporized element could be determined. Dumas was thus one of the few chemists making use of the hypothesis of Avogadro and André Ampère. But later, Dumas found a number of clearly anomalous atomic weights for elements, including sulfur, phosphorus, and arsenic. Faced with these problems, Dumas turned against the EVEN hypothesis, believing it to be defective in the case of elements in the gaseous state. In 1836, Dumas became even more pessimistic and now directed his criticisms against the use of atoms in chemistry. He wrote,

> What remains of the ambitious excursion we allowed ourselves into the domain of atoms? Nothing, at least nothing necessary. What remains is the conviction that chemistry lost its way, as usual when, abandoning experiment, it tried to find its way through the mists without a guide. . . . If I were master I would erase the word "atom" from the science, persuaded that it goes beyond experiment; and in chemistry we should never go beyond experiment.[82]

CONCLUSION

It is rather surprising that both Prout's hypothesis and the notion of triads are essentially correct and appeared problematic only because the early researchers were working with the wrong data. Prout was essentially correct when he asserted that the elements could be ordered in such a way that they represented whole multiples of hydrogen. It is now known that each element contains one more proton than the last as one moves across the periodic table, and this is what determines atomic number. Thus, in a sense all the elements are indeed composites of the hydrogen atom, since the number of protons in any element is an exact multiple of the single proton found in the nucleus of the hydrogen atom. The problem had been that chemists had focused on trying to make atomic weights exact multiples of each other, not realizing that atomic weights included contributions from neutrons. But the numbers of neutrons vary among each isotope, thus upsetting the simple ratios expected from Prout's hypothesis. Eventually, the switch from atomic weight to atomic number would remove the inexactness that caused chemists to abandon Prout's hypothesis as a useful tool in the classification of elements.

Similarly, the notion of triads was essentially correct but did not always work perfectly because early chemists were using the wrong data. Indeed, from the per-

FIGURE 2.6
Stanislao Cannizzaro. Photo from Edgar
Fahs Smith Collection by permission.

spective of the modern table and using atomic numbers, about 50% of all possible triads are exact.

Finally, it should be borne in mind that much of this work was carried out using equivalent weights or incorrect atomic weights. It became possible to develop successful periodic systems, which would accommodate all the elements into coherent systems, only when a set of consistent atomic weight could be obtained.

Atomic weights as distinct from equivalent weights were determined starting with Dalton's researches. However, these weights depended on knowing how many atoms of a particular element were present in a compound, something that was not well understood and over which Dalton made many errors because of his assumption of maximum simplicity. Atomic weight measurements were originally restricted to gaseous elements such as oxygen or nitrogen. Soon new methods were developed by Dulong and Petit as well as Mitscherlich and Dumas, which could be applied to elements in other states of aggregation.

Nevertheless, much confusion still remained. For example, very few chemists accepted the EVEN hypothesis as first announced by Avogadro and Ampère. Dumas, one of the few who did, lost his nerve when he ran across elements that yielded what seemed to be highly anomalous atomic weights. The problems would be resolved only when Cannizzaro (figure 2.6) insisted on the correctness of Avogadro's hypothesis and elaborated a method that finally gave a set of correct and consistent atomic weights.[83] It is to Cannizzaro's work that we turn our attention to in the next chapter.

DISCOVERERS OF THE PERIODIC SYSTEM

The periodic system was not discovered by Dimitri Mendeleev alone, as is commonly thought, or even just by Mendeleev and Julius Lothar Meyer. It was discovered by as many as five or six individuals at about the same time, in the decade of the 1860s, following the rationalization of atomic weights at the Karlsruhe conference.[1]

It became apparent by the middle of the nineteenth century that something needed to be done to resolve the widespread confusion over equivalent and atomic weights. Amedeo Avogadro had already proposed a solution to Joseph Louis Gay-Lussac's law that preserved John Dalton's indivisible elemental particles. Recall that Gay-Lussac had observed that volumes of gases entering into chemical combination and their gaseous products are in a ratio of small integers. Dalton had refused to accept this because it implied that atoms appeared to divide in some instances, such as the combination of hydrogen and oxygen to create steam. Avogadro had suggested that such "atoms" must be diatomic; that is, in their most elemental form they must be double. Thus, the oxygen atom was not dividing; rather, it was an oxygen molecule, which consisted of two oxygen atoms, that was coming apart.

Unfortunately, the terms in which Avogadro expressed his views were rather obscure and failed to make much impression on the chemists of the day. Two exceptions were the French physicist and chemist André Ampère[2] and the Alsatian chemist Charles Gerhardt,[3] both of whom adopted the view that elemental gases were composed of diatomic molecules.

One of the consequences of the general refusal to recognize the existence of diatomic molecules as the ultimate "atoms" of gaseous elements was that, as mentioned in chapter 2, the confusion between equivalent weights and atomic weights continued to reign. Although the relative weights of oxygen to hydrogen in water are approximately 8 to 1, the relative weight of the oxygen atom to the hydrogen

atom takes on values of 8 or 16 depending on what one considers the correct formula for water to be. Dalton opted for a formula of HO for water, which meant
that he was forced to assume an atomic weight of 8 for oxygen. Dalton allowed his
insistence on the indivisibility of atoms to obscure the possibility that in some substances the smallest atom, in the chemical sense, consisted of two atoms combined
together. Since this unit is still the smallest possible unit of the element in question,
Dalton need not have been concerned.

But although Avogadro had proposed the solution as early as 1811, its acceptance would have to wait until 1860, by which time great confusion had developed among different chemists regarding the atomic weights of many elements
and, consequently, the formulas of many compounds. The rapid growth of
organic chemistry during this period and the proliferation of varying formulas
for the same compounds added to the need to find a solution. When the chemist August Kekulé prepared a textbook of chemistry in the early 1860s, he listed
as many as 19 different formulas for acetic acid, all of which had been used in
the literature.

It was against this background that the Karlsruhe conference was convened. Its
aims were to clarify the notions of "atom" and "molecule" and the related issues
of equivalent weight and atomic weight. It fell to Stanislao Cannizzaro to resuscitate the work of his countryman Avogadro and to make it more palatable to the
chemists who attended the conference. No new science was required, just a careful
analysis of the problems at hand and a desire to bring order to the chaos surrounding the conflicting use of different atomic weights and the consequent different
formulas. Once the notion that elemental gases consisted of diatomic molecules
was accepted, the whole substructure of chemistry was corrected. At last, chemists
interested in classifying the elements had a firm foundation on which they could
build with confidence.[4]

Cannizzaro accepted Avogadro's hypothesis, namely, that equal volumes of
all gases, at the same temperature and pressure, contain the same number of
particles.[5] As a result, he argued that the relative density of a gas would provide
a measure of its relative mass. This much was not new and had been assumed
by others. What Cannizzaro did was to pursue Avogadro's hypothesis in a comprehensive manner in such a way as to finally usher in the acceptance of the
hypothesis and to break the deadlock over the measurement of atomic weights.
He began with the elementary assumption that if the molecular mass of hydrogen were M and if that of an element is found to be N times that of hydrogen,
and then the molecular mass of the unknown element would be NM. But the
aim is to obtain the atomic mass (a) of any element A; Cannizzaro recognized
that one could analyze a large number of compounds of the element. If it
turned out, as it always did, that all the intramolecular masses of A were whole
number multiples of 1 and the same mass, then that mass had the right to be
called the atomic weight of A.[6]

TABLE 3.1
Cannizzaro's method applied to finding the atomic weight of carbon

Species	Mass of carbon within molecule of the species
Carbonic acid	12
Carbonic oxide	12
Sulfide of carbon	12
Marsh gas	12
Ethylene	24
Propylene	36
Ether	48

The highest common factor of 12 from the column on the right, on this evidence, is the atomic weight of carbon.

Based on S. Cannizzaro, Sketch of a Course of Chemical Philosophy, *Il Nuovo Cimento*, 7, 321–366, 1858, Table from p. 335, translated in Alembic Club Reprints No. 18, Alembic Club Publications, Edinburgh, 1910.

For example, in the case of carbon, Cannizzaro published a table, on which table 3.1 is based, and proceeded to lend his full support to the idea of regarding atoms realistically:

> Compare, I say . . . the various quantities of the same element contained in the molecule of the free substance and in those of all its different compounds, and you will not be able to escape the following law: The different quantities of the same element contained in different molecules are all whole multiples of one and the same quantity, which always being entire, has the right to be called an atom. . . .[7]

ANOTHER BRIEF INTERLUDE ON QUALITATIVE CHEMISTRY

Let us consider another textbook, written in 1867 by the French chemist Alfred Naquet, to see how much was known about the chemistry of the elements at the time when the mature periodic system was being discovered. Table 3.2, of families of the elements, has been constructed on the basis of the groups, or families, listed by Naquet.[8]

Several improvements can be seen when this table is compared with the grouping by Leopold Gmelin given in chapter 2. Naquet's 2° *famille* (family) shows that oxygen has now been correctly included among elements such as sulfur and selenium. In addition, Naquet correctly includes nitrogen among the group containing phosphorus, arsenic, and antimony and also adds bismuth.[9] Whereas Gmelin had

TABLE 3.2
Families of elements according to Naquet's textbook of 1864

Metalloids 1° famille	2° famille	3° famille	4° famille	5° famille	
Cl	O	B	Si	N	
Br	S		C	P	
I	Se		Sn	As	
H	Te		Zr	Sb	
		Ti	Bi		
		Tl	U		

Metals 1° famille	2° famille	3° famille	4° famille	5° famille	6° famille
K	Ca	Au	Al		Mo
Na	Sr		Mn		W
Li	Ba		Fe		Ir
Rb	Mg		Cr		Rh
Cs	Zn		Co		Ru
Ag	Cd		No		
	Cu		Pb		
	Hg		Pt		

Based on A. Naquet, *Principes de Chimie*, F. Savy, Paris, 1867. Compiled by the author

grouped only carbon and silicon together in his own system, Naquet includes tin, in addition to three transition metals.[10] The final improvement is seen in Naquet's 1° family of metals, in which he has included the newly discovered rubidium and cesium. Different textbook authors of this period listed similar families of elements, and this kind of information would have been available to all discoverers of the periodic system, at least in principle. It is in relation to such qualitatively based systems that one must consider the discovery of the quantitatively based periodic systems that follow.

THE RAPID APPEARANCE
OF SEVERAL PERIODIC SYSTEMS

What permitted the rapid progress toward the development of the periodic system in several different countries during the 1860s was the publication, between 1858 and 1860, of a set of consistent atomic weights by Cannizzaro, based on the above method, which he compiled in preparation for the Karlsruhe conference. Once Cannizzaro had clearly established the distinction between molecular and atomic weights, the relative weights of the known elements could be compared in a reliable manner, although a number of these values were still incorrect and would be corrected only by the discoverers of the periodic system.

Despite the pivotal role played by the rationalization of atomic weights in preparing the way for the successful sorting of the elements, it is debatable whether the fairly rapid and independent discovery of the mature periodic system over the following decade represented a scientific revolution in the Kuhnian sense. Indeed, as remarked in the introduction, the history of the periodic system appears to be the supreme counterexample to Thomas Kuhn's thesis, whereby scientific developments proceed in a sudden, revolutionary fashion. The more one examines the development of the periodic system, the more one sees continuity rather than sudden breaks in understanding. Looking at the events leading up to the introduction of Mendeleev's periodic system in 1869, the concept of periodicity can be seen as evolving in distinct stages through the work of other chemists. Thus, rather than six actual discoveries of the system, it may be more correct to see it as an evolution through several systems, discovered within a period of less than 10 years.[11] The final one of these was by Mendeleev, who also worked harder than anyone else to establish the validity of the fully mature system. There are good reasons for singling out Mendeleev (and Lothar Meyer) in this story, and further grounds for making Mendeleev the one leading discoverer of the periodic system. But as I argue in this chapter, the *idea* of periodicity, which is central to the periodic system, did not originate with Mendeleev.

Other factors hastened the sudden explosion of periodic systems published in the 1860s. One of these was the discovery of new elements as a result of the development of the novel technique of spectroscopy. Having more elements to work with meant that there would be fewer gaps among them, making periodicity easier to discern. And spectroscopy itself, which permitted the characterization of each element by its unique spectral fingerprint, would in turn allow a much greater understanding of the chemical nature of the elements.

Another important change that occurred around this time and helped to make the discovery of the periodic system possible was the increased questioning of William Prout's hypothesis (that all elements are composites of hydrogen), which had figured rather prominently in the previous wave of discoveries leading up to the periodic system. In fact, such was the decline in support for Prout's hypothesis in the 1860s that chemists who still harbored such thoughts felt compelled to conceal their names. This was the case with one "Studiosus" who published an article in 1864, in response to John Newlands's periodic system, claiming that the atomic weights of the elements were multiples of 8. Meanwhile another pro-Proutian author, who called himself "Inquirer," attempted to mediate the controversy that ensued between Newlands and Studiosus.[12]

With the decline of the Prout's hypothesis, chemists became less concerned with finding neat, integral relationships among the elements. At the same time, other kinds of numerical regularities that had held the fascination of noted chemists, such as Alexandre Dumas and Max Pettenkofer, also began to subside. While the craze for searching for numerical regularities started to fade away, the work of chemists began to show a different aim and method. Instead of trying to find isolated triads or unconnected groups, researchers were now free to focus on seeking an integrated system that would include all the known elements in a meaningful way.

In examining the work of the six discoverers of the periodic system, it is important to consider their published articles in some detail. In trying to portray an overall picture of the evolution of the periodic system, I do not concentrate exclusively on the final published tables given by these authors. In the case of Newlands and John Odling, as well as some of the others, I examine several subsidiary tables, sometimes dealing with specific comparisons between the elements. This approach reveals important aspects in the evolution of their ideas that are missed by concentrating only on the finished work of any of the discoverers.

ALEXANDRE EMILE BÉGUYER DE CHANCOURTOIS

There are valid reasons for declaring that the periodic system was essentially discovered in 1862 by De Chancourtois, a French geologist. De Chancourtois appears to have taken not just an important step in the story of the periodic system but, in many ways, the single most important step. It was he who first recognized that the properties of the elements are a periodic function of their atomic weights, a full seven years before Mendeleev arrived at the same conclusion.

Although he hit upon this crucial notion underlying the entire edifice of the periodic system, De Chancourtois is not generally accorded very much credit, partly because his publication did not appear in a chemistry journal, and because he did not develop his insight any further over subsequent years. Indeed, it was only about 30 years after his paper appeared that De Chancourtois's claim to priority came to light through the efforts of Philip Hartog in England and Paul Emile Le Coq De Boisbaudran and Albert Auguste De Lapparent in France.[13]

De Chancourtois became professor of subterranean topography at the Ecole de Mines in Paris in 1848 and then in 1856 assumed a professorship in geology at the same institution. He attempted to systematize many different areas, including knowledge of minerals, geology, and geography, and even produced a form of universal alphabet. De Chancourtois presented his system of the chemical elements to the Académie des Sciences and also published it in its journal, the *Comptes Rendus*.[14] He proposed a three-dimensional representation of the periodicity of properties as a function of atomic weight (figure 3.1). De Chancourtois used equivalent weights for the elements, although he divided many values by 2, as a result of which most of his values agreed approximately with the new atomic weights of Cannizzaro. De Chancourtois also consistently rounded off the weights to produce whole number values. Although he did not commit himself specifically to the Proutian idea of atoms being composites of hydrogen, De Chancourtois did express his support for what he called "Prout's law," whereby the values for all elements should be whole number multiples of the value for the element hydrogen.

In 1862, De Chancourtois arranged the elements according to what he termed increasing "numbers" along a spiral. These numbers were written along a vertical line that served to generate a vertical cylinder. The circular base of the cylinder was divided into 16 equal parts. The helix was traced at an angle of 45° to its vertical axis, and its screw thread was similarly divided, at each of its turns, into 16 portions. Thus, the seventeenth point along the thread was directly above the first, the eighteenth above the second, and so on. As a result of this representation, elements whose characteristic numbers differed by 16 units were aligned in vertical columns. Sodium, for example, with a weight of 23, appeared one complete turn above lithium, whose value was taken as 7. The next column contained the elements magnesium, calcium, iron, strontium, uranium, and barium. One may begin to see the modern alkaline earth group emerging, the only difference being that several transition elements have also been included along the same vertical alignment. But this feature is not surprising since De Chancourtois's table is a short-form table that does not separate main-group elements from transition metals.

The first full turn of the spiral ended with the element oxygen, and the second full turn was completed at sulfur. Periodic relationships, or chemical groupings, could be seen in De Chancourtois's system, although only approximately, by moving vertically downward along the surface of the cylinder. The eighth such turn, and coincidentally the halfway point down the cylinder, occurred at tellurium. This rather arbitrary feature provided De Chancourtois with the name of *vis tellurique*, or telluric screw, for his system. This name may also have been chosen by De Chancourtois from *tellos*, Greek for earth, given that as a geologist, he was primarily interested in classifying the elements of the earth.

De Chancourtois's system did not create much impression on chemists for a number of reasons. The original published article failed to include a diagram, mainly because of the complexity faced by the publisher in trying to reproduce it, with the result that its visual force was lost. Another problem was that the system did not convey chemical similarities convincingly, as a result of the style of representation (the spiral) adopted by its author. While some of the intended chemical groupings, such as the alkali metals, the alkaline earths, and the halogens, did indeed fall into vertical columns, many others did not, thus making it a less successful system than it might have been. Yet another drawback to the system the inclusion of radicals such as NH_4^+ and CH_3, as well as such compounds as cyanogen, some oxides and acids, and even some alloys.

7 Avril 1862

VIS TELLURIQUE

CLASSEMENT NATUREL DES CORPS SIMPLES OU RADICAUX

obtenu au moyen d'un

SYSTÈME DE CLASSIFICATION HÉLICOIDAL ET NUMÉRIQUE

par A.E. BÉGUYER DE CHANCOURTOIS

Tableau des
Caractères Géometriques

0 1 2 3 4 5 6 7 8 9 10 11 12 13 14 15 16

Hydrogène	H	1
Lithium	Li	7
Glucyum	Gl	9
Bore	Bo	11
Carbone	C	12
Azote	Az	14
Oxygène	O	16
Fluor	Fl	19
Sodium	Na	23
Magnesium	Mg	24
Aluminium	Al	27
Silicium	Si	28
Phosphore	Ph	31
Soufre	S	32
Chlore	Cl	35
Potassium	K	39
Calcium	Ca	40
Titane	Ti	48
Chrome	Cr	53
Manganèse	Mn	55
Fer	Fe	56
Nikel	Ni	59
Cobalt	Co	60
Cuivre	Cu	63
Yttrium	Yt	64
Zinc	Zn	65
Zirconium	Zr	67
Arsenic	As	75
Brome	Br	79
Selenium	Se	80
Rubidium	Rb	87
Strontium	Sr	88
Lanthane	La	91
Cerium	Ce	92
Molybdène	Mo	96
Didyme	Di	99
Yttrium	Yt	100
Thallium	Tl	103
Rhodium	Rh	104
Palladium	Pd	107
Argent	Ag	108
Cadmium	Cd	111
Etain	Sn	115
Thorium	Th	119
Urane	Ur	120
Antimone	Sb	121
Caesium	Cs	124
Iode	Io	127
Tellure	Te	128
Tantale	Ta	184
Tungstène	W	185
Iridium	Ir	197
Platine	Pt	199
Or	Au	200
Mercure	Hg	204
Ruthenium	Ru	205
Osmium	Os	208
Bismuth	Bi	209

FIGURE 3.1
Telluric Screw of 1862.
A.E. Béguyer De Chancourtois, Vis Tellurique: Classement naturel des corps simples ou radicaux, obtenu au moyen d'un système de classification hélicoïdale et numérique, *Comptes Rendus de L'Academie*, 54, 757–761, 840–843, 967–971, 1862. Redrawn in J. van Spronsen, *The Periodic System of the Chemical Elements, the First One Hundred Years*, Elsevier, Amsterdam, 1969, p. 99. Reproduced by permission of Elsevier.

Frustrated that the journal *Comptes Rendus* failed to include a diagram, De Chancourtois had his system republished in 1863. But, because it was published privately, this further article received even less notice from other scientists than did the original one.[15] Still, it cannot be denied that De Chancourtois was the first to show that the properties of the elements are a periodic function of their atomic weights, or as he said himself, "*Les proprietées des corps sont les proprietées des nombres.* [The properties of bodies are the properties of numbers]."[16]

Of course, De Chancourtois intended the term "numbers" to mean the values of atomic weights, and even the improved atomic weights did not always yield clean intervals among the elements or line them up in what would appear to be the right order. However, following the eventual discovery of atomic numbers, De Chancourtois turned out to be even more correct than he might have himself imagined, as the properties of the elements are indeed a periodic function of their atomic numbers. De Chancourtois was also inadvertently prophetic in that he used whole number atomic weights, thus in effect creating an ordinal series of elements. To regard this as an anticipation of atomic number is not altogether implausible, although unlike Newlands (described below), De Chancourtois did not have a complete sequence of whole numbers in his system.

De Chancourtois's system was later criticized by Mendeleev, who in his Faraday lecture rather unfairly stated that De Chancourtois himself had not regarded his work as being a "natural system" of the elements. This lecture, given in London in 1889, seems to have provoked the English chemist Hartog, who had studied extensively in France, into making a belated priority claim on behalf of De Chancourtois. A couple of years later, De Chancourtois's cause was taken up by the French chemists Paul Emile Le Coq De Boisbaudran and Albert Auguste De Laparrent, who attempted to make their fellow Frenchmen more aware of the neglected work on the telluric screw.

It is also interesting to note another remark De Chancourtois made in his article:

> Will not my series, for instance, essentially chromatic as they are, be a guide in researches on the spectrum? Will not the relations of the different rays of the spectrum prove to be derived directly from the law of numerical characteristics, or vice versa?[17]

The periodic table did indeed turn out to be a very powerful guide to the study of atomic spectra, and vice versa, as shown when the influence of quantum theory is considered in later chapters. In many instances, the periodic system reveals periodicity in physical as well as chemical properties. The way in which spectral lines are split by a magnetic field, for example, be it into doublets, triplets, or quartets, is something that shows periodicity, just as do chemical properties such as reactivities toward particular elements.

One final comment should perhaps be made about De Chancourtois. His lack of chemical knowledge may have been a hindrance in some cases, and conversely, his emphasis on geological factors may have misled him in the development of the periodic system. For example, he stated that the isomorphism between feldspars and pyroxenes had been the starting point of his system. The element aluminum appears to function analogously to the alkali metals, a fact that does not necessarily indicate that aluminum should be grouped together with the alkali metals such as sodium and potassium. But this is precisely what De Chancourtois did in his system. In fact, he even changed the atomic weight, or characteristic weight, as he termed it, in the case of aluminum to make it fall neatly into line with the alkali metals. Had he known more chemistry, he might not have taken this unjustified step.

JOHN NEWLANDS

Newlands was born in 1837 in Southwark, a suburb of London, which by a coincidence was also the birthplace of Odling, another pioneer of the periodic system. After studying at the Royal College of Chemistry in London, Newlands became the assistant to the chief chemist of the Royal Agricultural Society of Great Britain. In 1860, he served briefly as an army volunteer with Giuseppe Garibaldi, who was fighting the revolutionary war in Italy. The reason for Newlands's sortie seems to have been connected to the fact that his mother was of Italian descent. It also meant that Newlands was not able to attend the Karlsruhe conference of the same year, although since he was not a major chemist at the time he would probably not have been invited. After returning to London, Newlands began working as a sugar chemist, while also supplementing his income by teaching chemistry privately, but he was never to hold an academic position.

Newlands's first attempt at classification concerned a system for organic compounds that he published in 1862 along with proposals for a new system of nomenclature.[18] In the following year, he published the first of what would be many classification systems for the elements. Although the year was 1863, Newlands developed his first system without the benefit of the atomic weight values that had been issued following the Karlsruhe conference of 1860, as he was unaware of them. Nevertheless, he did use the atomic weight values favored by Gerhardt, who had begun to revise atomic weights even before the Karlsruhe conference. Thus, Newlands was able to produce a table consisting of 11 groups of elements with analogous properties whose weights differed by a factor of 8 or some multiple of 8. Because it was unfashionably Proutian, Newlands published his first article on classification of the elements anonymously, although he revealed his identity soon afterward in response to criticisms by the equally anonymous "Studiosus" (figure 3.2).

Group I. Metals of the alkalies :—Lithium, 7 ; sodium, 23 ; potassium, 39 ; rubidium, 85 ; cæsium, 123 ; thallium, 204.

The relation among the equivalents of this group (see CHEMICAL NEWS, January 10, 1863) may, perhaps, be most simply stated as follows :—

$$1 \text{ of lithium} + 1 \text{ of potassium} = 2 \text{ of sodium.}$$

$$1 \quad ,, \quad + 2 \quad ,, \quad = 1 \text{ of rubidium.}$$
$$1 \quad ,, \quad + 3 \quad ,, \quad = 1 \text{ of cæsium.}$$
$$1 \quad ,, \quad + 4 \quad ,, \quad = 163, \text{ the equivalent of a metal not yet discovered.}$$
$$1 \quad ,, \quad + 5 \quad ,, \quad = 1 \text{ of thallium.}$$

Group II. Metals of the alkaline earths :—Magnesium, 12 ; calcium, 20 ; strontium, 43·8 ; barium, 68·5.

In this group, strontium is the mean of calcium and barium.

Group III. Metals of the earths :—Beryllium, 6·9 ; aluminium, 13·7 ; zirconium, 33·6 ; cerium, 47 ; lanthanium, 47 ; didymium, 48 ; thorium, 59·6.

Aluminium equals two of beryllium, or one-third of the sum of beryllium and zirconium. (Aluminium also is one-half of manganese, which, with iron and chromium, forms sesquioxides, isomorphous, with alumina.)

$$1 \text{ of zirconium} + 1 \text{ of aluminium} = 1 \text{ of cerium.}$$
$$1 \quad ,, \quad + 2 \quad ,, \quad = 1 \text{ of thorium.}$$

Lanthanium and didymium are identical with cerium, or nearly so.

Group IV. Metals whose protoxides are isomorphous with magnesia :—Magnesium, 12 ; chromium, 26·7 ; manganese, 27·6 ; iron, 28 ; cobalt, 29·5 ; nickel, 29·5 ; copper, 31·7 ; zinc, 32·6 ; cadmium, 56.

Between magnesium and cadmium, the extremities of this group, zinc is the mean. Cobalt and nickel are identical. Between cobalt and zinc, copper is the mean. Iron is one-half of cadmium. Between iron and chromium, manganese is the mean.

Group V.—Fluorine, 19 ; chlorine, 35·5 ; bromine, 80 ; iodine, 127.

In this group bromine is the mean between chlorine and iodine.

Group VI.—Oxygen, 8 ; sulphur, 16 ; selenium, 39·5 ; tellurium, 64·2.

In this group selenium is the mean between sulphur and tellurium.

Group VII.—Nitrogen, 14 ; phosphorus, 31 ; arsenic, 75 ; osmium, 99·6 ; antimony, 120·3 ; bismuth, 213.

FIGURE 3.2
Newlands's groups of elements. J.A.R. Newlands, On Relations among the Equivalents, *Chemical News*, 7, 70–72, 1863, table on p. 71

Newlands's grouping of elements in 1863 is surprisingly suggestive, especially bearing in mind that it utilizes pre-Karlsruhe atomic weights. Ever since Prout, investigators had struggled with the fact that arithmetic intervals in atomic weights among the elements are not as exact or as regular as it seems they should be. The fact that atomic weights depend upon the vagaries of isotopic mixtures for any particular element was not, of course, suspected at the time. In addition to the isotope issue, the atomic weights of many elements had not been correctly determined. Nevertheless, one cannot fail to be struck by the good fortune that Newlands and the other pioneers of the periodic system experienced in that the ordering of the elements according to atomic weight, despite their irregular intervals, corresponds almost exactly to that based on atomic number. It is almost as though nature's mixtures of isotopes had conspired together to announce the ordering that would later be discovered in terms of atomic numbers.

In his 1863 article, Newlands described a relationship among atomic weights of the alkali metals and used it to predict the existence of a new element of weight 163, as well as a new element that would be placed between iridium and rhodium. Unfortunately for Newlands, neither of these elements ever materialized. However, as has recently been pointed out, Mendeleev also made similar predictions that failed to materialize among elements with high atomic weights.[19] These failures can be attributed to the existence of the lanthanide elements, which occur between the second and third transition series of elements in modern terms. The lanthanides would be a problem for all the discoverers of the periodic system, as only 6 of the 14 lanthanides had been discovered prior to the 1860s, when these early periodic systems were being developed.[20]

In 1864, Newlands published his second article on the classification of the elements (table 3.3, figure 3.3). This time he drew on the more correct, post-Karlsruhe atomic weights, a version of which had been published in England by Alexander Williamson. Newlands now found a difference of 16, or very close to this value, instead of 8, between the weights of six sets of first and second members among groups of similar elements. Again, this finding seems unexpectedly accurate given that he was working with atomic weights and not atomic numbers. A very similar table comparing differences in atomic weights between first and second members of groups of analogous elements was discovered independently and published in the very same year by Odling (described below). Indeed, Odling outdid Newlands in recognizing 10 such relationships, to Newlands's six. This fact has not been given any exposure in histories of the periodic system, which sometimes fail to even mention Odling as one of the discoverers.

Less than a month after his second system appeared in 1864, Newlands published a third system that same year (figure 3.4), but in this table he included fewer

TABLE 3.3
Newlands's first table of 1864

Member of a group having lowest equivalent		One element immediately above the preceding one		Difference $H = 1 \ O = 16$	
Magnesium	24	Calcium	40	16	1
Oxygen	16	Sulphur	32	16	1
Lithium	7	Sodium	23	16	1
Carbon	12	Silicon	28	16	1
Fluorine	19	Chlorine	35.5	16.5	1.031
Nitrogen	14	Phosphorus	31	17	1.062

Remade from J.A.R. Newlands, Relations Between Equivalents, *Chemical News*, 10, 59–60, 1864, table on p. 59.

Triad

No.	Element	Lowest Term	Mean	Highest Term	
I.	Li 7	+17 = Mg 24	Zn 65	Cd 112	Au 196
II.	B 11				
III.	C 12	+16 = Si 28	As 75	Sn 118	
IV.	N 14	+17 = P 31	Se 79.5	Sb 122	+88 = Bi 210
V.	O 16	+16 = S 32	Er 80	Te 129	+70 = Os 199
VI.	F 19	+16.5 = Cl 35.5		I 127	
VII.	Li 7 +16 = Na 23	+16 = K 39	Rb 85	Cs 133	+70 = Tl 203
VIII.	Li 7 +17 = Mg 24	+16 = Ca 40	Sr 87.5	Ba 137	+70 = Pb 207
IX.		Mo 96	V 137	W 184	
X.		Pd 106.5		Pt 197	

FIGURE 3.3 System of 1864. J.A.R. Newlands, Relations between Equivalents, *Chemical News*, 10, 59–60, 1864, p. 59.

		No.		No.		No.		No.		No.
Group	*a*	N 6	P 13	As 26	Sb 40	Bi 54
„	*b*	O 7	S 14	Se 27	Te 42	Os 50
„	*c*	Fl 8	Cl 15	Br 28	I 41	— —
„	*d*	Na 9	K 16	Rb 29	Cs 43	Tl 52
„	*e*	Mg 10	Ca 17	Sr 30	Ba 44	Pb 53

FIGURE 3.4 Newlands's later system of 1864. J.A.R. Newlands, On Relations among Equivalents, *Chemical News*, 10, 94–95, 1864, p. 94.

elements (24, plus a space for a new element) and made no mention of atomic weights. The article is nevertheless of considerable merit since Newlands assigned an ordinal number to each of the elements, thus in a sense anticipating the modern notion of atomic number. Abandoning the arithmetic progressions in atomic weights that had bedeviled earlier investigators, Newlands simply lined the elements up in order of increasing atomic weight without concern for the values of those weights. Nevertheless, any anticipation of the modern concept of atomic weight is marred by the several cases where the sequence of elements does not strictly follow Newlands's ordinal numbers. The modern ordering based on atomic number does not show any such exceptions.

The most important thing Newland did in his third publication on the classification of the elements was to present a periodic *system*; that is, he revealed a pattern of repetition in the properties of the elements after certain regular intervals. This, of course, is the essence of the periodic law, and Newlands deserves credit for having recognized this fact so early, along with De Chancourtois. Another innovation of Newlands's later system of 1864, which is almost universally attributed to Mendeleev, although it was also carried out by Odling, was the way in which Newlands reversed the positions of the elements iodine and tellurium in order to give precedence to chemical properties over the apparent atomic weight ordering.[21] Newlands thus holds the distinction of having been the first of these three discoverers to make a so-called pair reversal.[22] It is somewhat surprising, however, especially given his emphasis on chemical properties, that Newlands failed to display analogies between several other obviously related elements, such as lithium and sodium.[23]

The Law of Octaves

In 1865, Newlands developed yet another system, which was a vast improvement on that of the previous year in that he now included 65 elements, in increasing order of atomic weight, while once again using ordinal numbers rather than actual values of atomic weight. This system was built upon his famous "law of octaves," whereby the elements showed a repetition in their chemical properties after intervals of eight elements.[24] Newlands went so far as to draw an analogy between a period of elements and a musical octave, in which the tones display a repetition involving an interval of eight notes (counting from one note of C, e.g., to the next note C inclusive). In the words of Newlands himself:

> If the elements are arranged in the order of their equivalents with a few slight transpositions, as in the accompanying table, it will be observed that elements belonging to the same group usually appear on the same horizontal line. It will also be seen that the numbers of analogous elements differ either by 7 or by some multiple of seven; in other words, members of the same group stand to each other in the same relation as the extremities of one or more octaves in music. . . . The eighth element starting from a given one is a kind of repetition of the first. This particular relationship I propose to term the *Law of Octaves*.[25]

This statement marks a rather important step in the evolution of the periodic system since it represents the first clear announcement of a new law of nature relating to the repetition of the properties of the elements after certain intervals in their sequence. As mentioned before, the periodic *law*, though not a fashionable term nowadays, is perhaps the most important aspect of the periodic table. The periodic table in all its many forms is, after all, just an attempt to represent this law graphically.

There remains the question raised earlier as to whether De Chancourtois might have been the first to recognize the existence of the periodic law. As Wendell Taylor has suggested, Newlands was far more explicit about the existence of a periodic law than was De Chancourtois, who merely mentioned it as a possibility.[26] There is little doubt that Odling also failed to recognize the existence of a fundamental law, though he did recognize the existence of a periodic system. Odling specifically claimed that, after a detailed examination of the numerical differences between the atomic weights of analogous elements, he had decided that these relations were "too numerous to depend upon some hitherto unrecognized law."[27]

Returning to Newlands's system of 1865, even though it is a genuine *periodic* system, compared with his earlier lists or groups of elements, Newlands did not see the need to separate the elements into subgroups as Mendeleev later did by offsetting certain elements within main groups. In modern terms, he did not see the need to separate out the transition metals, as is now carried out in the modern medium-long form of the periodic table. (See chapter 1 for diagrams of the short and medium-long forms of the periodic table.) The law of octaves applies perfectly to the first two periods, excluding the noble gases, which had not yet been discovered. Beyond that, Newlands's periodicities were bound to run into difficulties since the inclusion of the transition metals makes the later periods much longer than 8. Only his fellow London chemist, Odling, anticipated this problem, as described below.

Newlands first announced his law of octaves in a paper delivered to the London Chemical Society in 1866, but to his great misfortune, his insight was badly received. This event is perhaps Newlands's best-known legacy to the history of the periodic system and is repeated *ad nauseam* in textbooks and popular accounts. What Newlands presented to the society was an improved version of his 1865 system, in which more elements were arranged strictly according to his ordinal numbers. In his earlier published table of 1865, this had not been the case, especially for the elements with ordinal numbers beyond 50. The new table presented to the society (table 3.4) also shows some chemical improvements in that the element lead is now placed in the same group as carbon, silicon, and tin, whereas it had not appeared in the table of 1865.

As the popular story goes, Newlands included the mention of a law of octaves in his presentation and proceeded to draw an analogy with the musical scale. Whether he seriously intended to suggest a connection between chemistry and music is not clear. In any case, his fanciful analogy was probably not the reason why the chemists in attendance were quick to dismiss Newlands's scheme. Their hostility is perhaps better attributed to the British tendency, of the time, to be suspicious of theoretical ideas in general. The best-known response to Newlands is the much-quoted one of George Carey Foster, who suggested that Newlands might well have obtained a superior classification scheme if he had merely ordered the elements alphabetically according to the first letter of each of their names.

Some modern commentators have tried to exonerate Newlands by saying that he was unlucky to have been working at the time before the noble gases had been discovered. They suggest that if he had known of this additional group, he would have realized that chemical repetition follows a "nonet law," not an octet rule. In that case, he might not have been tempted to make analogies with the musical scale and thus might not have fallen prey to the assembled scientists at the London Chemical Society. These attempts to exonerate Newlands are in fact rather unnecessary, since he specifically anticipated the possibility of the repeat period being greater than 8, as discussed below.

Another aspect of the Newlands mythology concerns the fact that the chemists gathered at the London Chemical Society meeting decided not to permit publication of Newlands's article in the society's proceedings. Although this is quite true, it should not be taken to imply that Newlands was prevented from publishing his ideas on the classification of the elements. In fact, he had already published several articles in the highly respected journal *Chemical News* and would succeed in publishing the contents of his presentation to the London Chemical Society a few months later in this same journal. The reason why Newlands's ill-fated talk had been denied publication by the London Chemical Society only emerged seven years after the event, when Odling, who had chaired the meeting, wrote that the society had made it a rule not to publish papers of a purely theoretical nature, since it was likely to lead to controversy. One cannot rule out the possibility that there may have been a certain amount of rivalry between Odling and Newlands regarding the construction of periodic systems and that this may have influenced

TABLE 3.4

Newlands's table illustrating the law of octaves as presented to the Chemical Society in 1866

	No.		No.		No.		No.		No.		No.		No.		No.
H	1	F	8	Cl	15	Co & Ni	22	Br & Ni	29	Pd	36	I	42	Pt & Ir	50
Li	2	Na	9	K	16	Cu	23	Rb	30	Ag	37	Cs	44	Os	51
G	3	Mg	10	Ca	17	Zn	24	Sr	31	Cd	38	Ba & V	45	Hg	52
Bo	4	Al	11	Cr	19	Y	25	Ce & La	33	U	40	Ta	46	Tl	53
C	5	Si	12	Ti	18	In	26	Zr	32	Sn	39	W	47	Pb	54
N	6	P	13	Mn	20	As	27	Di & Mo	34	Sb	41	Nb	48	Bi	55
O	7	S	14	Fe	21	Se	28	Ro & Ru	35	Te	43	Au	49	Th	56

Note the inclusion of a number of nonconventional symbols, from the contemporary point of view. These are G for glucinium, subsequently called beryllium; Bo for boron; Di for didymium, which later turned out to be a mixture of two rare earth elements; and Ro for rhodium.

J.A.R. Newlands, as reported in Proceedings of Chemical Societies: *Chemical News*, 13, 113–114, 1866, table on p. 113.

Odling's view. Nevertheless, this seems rather tenuous since Newlands was something of an outsider among academic chemists, and it is unlikely that Odling would have regarded him as a threat. Odling was a more rounded chemist whose main interests lay in the wider question of the relationship between atomic weight and equivalent weight and the related question of the difference between atoms and molecules. Unlike in the case of Newlands, the classification of the elements was only a sideline for Odling.

In an article published in 1866, Newlands tried to answer the criticisms that had been leveled at him in the course of his fateful presentation to the London Chemical Society. The accompanying table published by Newlands represents the first time that he arranged chemical groups arranged in vertical columns, and once again, the ordering of the elements follows a numerical order with the exception of three reversals (Ce and La with Zr, U with Sn, and Te with I). Newlands responded to a criticism that he had not left any gaps and that this would be a problem when future elements were discovered:

> The fact that such a simple relation [the law of octaves] exists now, affords a strong presumptive proof that it will always continue to exist, even should hundreds of new elements be discovered. For, although the difference in the number of analogous elements might, in that case, be altered from 7, to a multiple of 7, of 8, 9, 10, 20, or any conceivable figure, the existence of a simple relation among the elements would be none the less evident.[28]

Newlands is, of course, correct.[29] In fact he was vindicated by the subsequent discovery of the noble gases, which, instead of disrupting the repeating pattern, simply increased the repeat distance between successive periods to eight rather than seven.[30] In a later system, of 1878, Newlands would carry out just such an expansion by establishing periods with 10 elements, although the net result of this was to create far too many empty spaces that were not subsequently filled by new elements (figure 3.5).

Following the publication of Mendeleev's periodic system in 1869, Newlands began to publish a series of letters in which he attempted to establish his priority in arriving at the first successful periodic system. Meanwhile, much to his chagrin, the Davy medal was awarded jointly to Mendeleev and Lothar Meyer in 1882 for their discovery of the periodic system. Newlands renewed his efforts, publishing a further summary of his own achievements in 1882 and again in 1884 in book form.[31] His tenacity was at least partly rewarded when the Davy medal was finally awarded to him in 1887. As late as 1890, Newlands published a rejoinder to a critique that had been expressed by Mendeleev in his Faraday lecture of two years before. It is also worth noting that, despite this critique, Mendeleev regarded the work of Newlands more highly that he did that of Lothar Meyer.

During the period 1863–1890, Newlands published a total of 16 articles, in which he tried many different schemes on the classification of the elements. These

TABLE II.

HORIZONTAL ARRANGEMENT IN SEVENS. AT. WT. ÷ 23, OR Na = 10·00.

No.			No.			No.			No.			No.			No.			No.		
1. H	0·435																			
2.	Li	3·04	9.	Na	10·00	16.	K	17·00	23.	—	—	30.	Cu	27·57	37.	Rb	37·13	44.	—	—
3.	Be	4·09	10.	Mg	10·43	17.	Ca	17·39	24.	—	—	31.	Zn	28·35	38.	Sr	38·09	45.	—	—
4.	B	4·80	11.	Al	11·91	18.	—	—	25.	Fe	24·35	32.	Ga	30·39	39.	Y	38·26	46.	—	—
5.	C	5·22	12.	Si	12·20	19.	Ti	21·74	26.	—	—	33.	—	—	40.	Zr	38·96	47.	Rh	45·39
6.	N	6·09	13.	P	13·48	20.	V	22·26	27.	—	—	34.	As	32·61	41.	Nb	40·87	48.	Ru	45·39
7.	O	6·96	14.	S	13·91	21.	Cr	22·70	28.	Ni	25·57	35.	Se	34·52	42.	Mo	41·74	49.	Pd	46·35
8.	F	8·26	15.	Cl	15·43	22.	Mn	23·91	29.	Co	25·57	36.	Br	34·78	43.	—	—	50.	—	—

TABLE II. (continued.)

No.			No.			No.			No.		
51.	Ag	46·96	58.	Cs	57·83	65.	—	—	72.	—	—
52.	Cd	48·70	59.	Ba	59·57	66.	—	—	73.	—	—
53.	In	49·30	60.	Di	60·00	67.	—	—	74.	Er	77·39
54.	Sn	51·30	61.	Ce	60·87	68.	—	—	75.	La	78·26
55.	Sb	53·04	62.	—	—	69.	—	—	76.	Ta	79·13
56.	Te	55·65	63.	—	—	70.	—	—	77.	W	80·00
57.	I	55·22	64.	—	—	71.	—	—	78.	—	—

No.			No.			No.			No.		
79.	—	—	86.	Au	85·65	93.	—	—	100.	—	—
80.	—	—	87.	Hg	86·96	94.	—	—	101.	—	—
81.	Pt	85·83	88.	Tl	88·52	95.	—	—	102.	—	—
82.	Ir	86·09	89.	Pb	90·00	96.	—	—	103.	Th	102·17
83.	Os	86·61	90.	Bi	91·30	97.	—	—	104.	—	—
84.	—	—	91.	—	—	98.	—	—	105.	U	104·35
85.	—	—	92.	—	—	99.	—	—	106.	—	—

FIGURE 3.5 Table of 1878. J.A.R. Newlands, On Relations among the Atomic Weights of the Elements, *Chemical News*, 37, 255–258, 1878, pp. 256–257.

met with varying degrees of success both in scientific terms and in terms of recognition. There can be no doubt, however, that Newlands ranks among the true
pioneers of the modern periodic system, in particular for being the first to recognize explicitly the existence of the periodic law, which in many ways is the real
crux of the matter.

WILLIAM ODLING

Unlike many of the discoverers of the periodic system, who were otherwise marginal figures in the history of chemistry, Odling was a distinguished chemist and
scientist who held some very important positions in the course of his career. Most
notably, he succeeded Michael Faraday as director of the Royal Institution in London. Odling also had the advantage of having attended the Karlsruhe conference,
where he had given a lecture on the need to adopt a unified system of atomic
weights. Unlike Newlands, whose first attempts at a periodic system were carried
out in ignorance of Cannizzaro's recommended values of atomic weights, Odling
was able to avail himself of these values from the beginning of his attempts at producing a table of the elements. In fact, after Karlsruhe, Odling rapidly became the
leading champion of the views of Cannizzaro and Avogadro in England. Odling
above all others would therefore have recognized the significance of the new
atomic weight values.

Odling's main article on the periodic system appeared in 1864, while he was a
reader in chemistry at St. Bartholomew's Hospital in London. Whereas Newlands's
system of the same year had only included 24 of the 60 known elements, Odling
succeeded in including 57 of them (figure 3.6). Furthermore, Odling's paper preceded Newlands's announcement of periodicity to the London Chemical Society,
which was made shortly afterward in 1865. Nevertheless, it appears that the two
chemists worked quite independently of each other.

Odling begins his article by stating, "Upon arranging the atomic weights or proportional numbers of the sixty or so recognized elements in order of their several
magnitudes, we observe a marked continuity in the resulting arithmetical series,"[32]
and goes on to point out a few exceptions to this regularity. Then he makes an
observation that amounts to an independent discovery of the periodic system:

> With what ease this purely arithmetical seriation may be made to accord with
> a horizontal arrangement of the elements according to their usually received
> groupings, is shown in the following table, in the first three columns of which
> the numerical sequence is perfect, while in the other two the irregularities are
> but few and trivial.[33]

That Odling had recognized the periodicity in chemical properties is clearly seen
in the horizontal groupings that he organizes in this table.

			Ro 104	Pt 197
			Ru 104	Ir 197
			Pd 106·5	Os 199
H 1	„	„	Ag 108	Au 196·5
„	„	Zn 65	Cd 112	Hg 200
L 7	„	„	„	Tl 203
G 9	„	„	„	Pb 207
B 11	Al 27·5	„	U 120	„
C 12	Si 28	„	Sn 118	„
N 14	P 31	As 75	Sb 122	Bi 210
O 16	S 32	Se 79·5	Te 129	„
F 19	Cl 35·5	Br 80	I 127	„
Na 23	K 39	Rb 85	Cs 133	„
Mg 24	Ca 40	Sr 87·5	Ba 137	„
	Ti 50	Zr 89·5	Ta 138	Th 231·5
	„	Ce 92	„	
	Cr 52·5	Mo 96	V 137	
	Mn 55		W 184	
	Fe 56			
	Co 59			
	Ni 59			
	Cu 63·5			

FIGURE 3.6   Table of 1864. W. Odling, On the Proportional Numbers of the Elements, *Quarterly Journal of Science*, 1, 642–648, 1864, p. 643.

Odling notes that there are a considerable number of pairs of chemically analogous elements, indeed, half of all the known elements, whose difference in atomic weights lies between the values of 84.5 and 97. Some of these pairs are shown in table 3.5. He then notices that about half of these cases include the first and third members of previously known triads. He suggests that a middle member might be found for the other half, stating that, "the discovery of intermediate elements in the case of some or all of the other pairs is not altogether improbable."[34] This is clearly an example of a prediction made on the basis of a periodic system, although admittedly a rather tentative one that was not further developed with specific examples.

TABLE 3.5
Odling's first table of differences

I	–	Cl	or	127	–	35.5	=	91.5
Au	–	Ag		296.5	–	108	=	88.5
Ag	–	Na		108	–	23	=	85
Cs	–	K		133	–	39	=	97

Based on W. Odling, On the Proportional Numbers of the Elements
Quarterly Journal of Science, 1, 642–648, 1864, table on p. 644.

A table is then given of as many as 17 pairs of elements whose members differ by atomic weights of 40–48 (table 3.6). This is followed by yet a third set of pairs of elements, 10 instances in all, of "more or less analogous elements" that differ in atomic weight by 16 units, or something close to this amount. It is worth noting perhaps, that in 7 out of these 10 instances, the element with the lower atomic weight of the pair is the first member of the group of similar chemical elements to which they both belong.

It would appear that, in identifying these gaps, Odling was making a rather remarkable observation that seemed to go beyond the earlier triadic relationships, since the 16-unit gap appears with approximate consistency in so many of Odling's three sets of element pairs. One might claim that just this observation constitutes the recognition of periodicity. What Odling appears to have realized, particularly in the case of the last set of elements, is that, in as many as 10 important cases, there is an approximate repetition in the properties of any of these elements following

TABLE 3.6
Odling's third table of differences

Cl	–	F	or	35.5	–	19	=	16.5
K	–	Na		39	–	23	=	16
Na	–	Li		23	–	7	=	16
Mo	–	Se		96	–	80	=	16
S	–	O		32	–	16	=	16
Ca	–	Mg		40	–	24	=	16
Mg	–	G		24	–	9	=	15
P	–	N		31	–	14	=	17
Al	–	B		27.5	–	11	=	16.5
Si	–	C		28	–	12	=	16

W. Odling, On the Proportional Numbers of the Elements, *Quarterly Journal of Science*, 1, 642–648, 1864, table on p. 645.

a difference in atomic weight of 16, or very close to this value. Bearing in mind that he used atomic weights, which are approximately double the values of atomic number, this is also very close to being the recognition of the law of octaves. In other words, Odling appears to have realized that the repetition occurs after a difference in atomic number of eight units, which corresponds to an atomic weight difference of 16.

Odling makes the further claim that the chemical similarities between elements separated by differences of about 48 in atomic weight, such as cadmium and zinc, are greater than those between pairs of elements, such as zinc and magnesium, that are separated by other intervals, such as 41 in this case. Thus, it would appear that he recognized the need to separate certain elements (those that would eventually become known as the transition metals) from the main body of the table. In this way, periodicity could be retained in the properties of the majority of the elements, as is done in the modern medium-long form of the table.[35] If the transition metals are separated out of the short-form table, the primary periodic relationship between main-group elements is emphasized and the fact that period lengths vary is accommodated in a natural manner.

It so happens that Odling was correct in this case. From the perspective of the modern periodic table, cadmium and zinc are both transition metals that show a primary kinship, whereas zinc and magnesium belong to the transition metals and main-group elements, respectively, and show only secondary kinships. Odling may have anticipated the modern trend to separate zinc and magnesium into different groups and, indeed, different blocks of the periodic table.

Any claim that Odling is making a significant anticipation here is vitiated, however, by the fact that in the same paragraph he goes on to give what he considers to be other examples of this behavior, all of which are incorrect. He claims that there is a greater chemical similarity among cesium, rubidium, and potassium, as well as among barium, strontium, and calcium, both of which sets show common differences of about 48 between closest members, than between potassium and sodium, where the difference is only 18. This is simply not the case.

If we are to judge these suggestions from the perspective of the modern table, we see that Odling is correct in drawing the first distinction, given that magnesium does not belong with the transition elements zinc and cadmium. However, in the second example, it has turned out that no comparable difference exists between potassium and sodium, both of which are now classified as main-group elements of group I. In any case, separating the transition metals from the main body of the table would not affect any of these groupings, as they are all composed of main-group elements.

What is confusing the issue, as far as the numerical relationships are concerned, is precisely the fact that successive periods in the mature periodic table do not all have the same length. Odling does not appear to have realized that different periods have different lengths, even though he has deliberately separated some elements out from the main body of the table. The suggestion that Odling anticipated the existence of transition metal groups in the periodic table to preserve periodicity is thus somewhat debatable.[36]

GUSTAVUS HINRICHS

The case of Hinrichs is rather unusual among the discoverers of the periodic system. This is because his scientific interests were so far ranging, and the evidence he brought to bear on producing a classification of the elements was so diverse as to lead some commentators to regard him as a mere crank. Although he held a number of academic appointments, first at University of Iowa and later at the University of Missouri–St. Louis, Hinrichs seemed to go out of his way to cultivate eccentricity. In addition, he seldom gave references to other authors in his numerous publications, thus making a balanced assessment of his contributions rather more difficult.

Hinrichs was born in 1836 in Holstein, which was then a part of Denmark but later became a German province. Hinrichs published his first book at the age of 20, while attending the University of Copenhagen. He immigrated to the United States in 1861 to escape political persecution and, after a year of teaching high school, was appointed head of Modern Languages at the University of Iowa. A mere one year later, he became Professor of Natural Philosophy, Chemistry, and Modern Languages. He is also credited with founding the first meteorological station in the United States in 1875, acting as its director for 14 years.

Hinrichs was a prolific author who published some 3,000 articles in Danish, French, and German, as well as in English, in addition to about 25 books of varying lengths in English and German. These books include the highly eccentric *Atomechanik* of 1867, in German, in which Hinrichs gives his definitive views on the classification of the elements. It is interesting to note that the majority of Hinrichs's articles were published in languages other than English. He seems to have disliked American journals, complaining that their insistence on correcting his work caused unacceptable delays in publication. Karl Zapffe, the author of a detailed analysis of Hinrichs's work, has suggested that Hinrichs's disaffection with American journals may have been part of his distaste for all things American. This may have included his American colleagues and may have led to his eventual dismissal from the University of Iowa in 1885.[37]

As Zapffe writes:

> It is not necessary to read far into Hinrichs' numerous publications to recognize the marks of an egocentric zeal which defaced many of his contributions with an untrustworthy eccentricity. Only at this late date does it become possible to separate those inspirations which were real—and which swept him off

his feet—from background material which was captured in the course of his own learning. Whatever the source, Hinrichs usually dressed it with multilingual ostentation, and to such a point of disguise that he even came to regard Greek philosophy as his own.[38]

The jury is still out on Hinrichs. While Jan van Spronsen includes him in his list of six genuine discoverers of the periodic system, William Jensen, a chemist and chemical educator at the University of Cincinnati, is among those who regard Hinrichs as a scientific maverick and a crank.[39] This also seems to be the conclusion of Cassebaum and George Kauffman, who include just six lines on Hinrichs in an article on the codiscoverers of the periodic system and who devote considerable space to a footnote pointing out his unconventional scientific attitudes.[40] But careful consideration of Hinrichs's work shows that there was much useful science, if one is prepared to take time to examine the various strands of his research.

Hinrichs took a rather Pythagorean approach to science in that he was captivated by numerical relationships, even those involving very diverse phenomena. Whereas Pythagoreanism had already figured in the early research on triads and Prout's hypothesis, Hinrichs's own brand of Pythagoreanism was far more extreme. By an ingenious argument (examined shortly below), he was led to postulate the notion that atomic spectra can provide information on the dimensions of atoms, an idea that is essentially correct from the modern perspective.[41] Since Hinrichs's idea has not been clearly described in previous accounts of the evolution of the periodic system, or at least the few accounts that even mention his work, I attempt to describe it here.

Hinrichs's wide range of interests extended to astronomy. Like many authors before him, as far back as Plato, Hinrichs noticed some numerical regularities regarding the sizes of the planetary orbits. In an article published in 1864, Hinrichs showed a table (table 3.7) that he proceeded to interpret. Hinrichs expressed the differences in these distances by the formula $2^x \times n$, in which n is the difference in the distances of Venus and Mercury from the sun, or 20 units. Depending on the value of x, the formula therefore gives the following distances:[42]

$$2^0 \times 20 = 20$$
$$2^1 \times 20 = 40$$
$$2^2 \times 20 = 80$$
$$2^3 \times 20 = 160$$
$$2^4 \times 20 = 320$$

etc.

A few years previously, in 1859, the Germans Gustav Robert Kirchoff and Robert Bunsen had discovered that each element could be made to emit light, which could then be dispersed with a glass prism and analyzed quantitatively.[43] What they also discovered was that every single element gave a unique spectrum

TABLE 3.7

Hinrichs's 1864 table of planetary distances

	Distance to the Sun
Mercury	60
Venus	80
Earth	120
Mars	200
Asteroid	360
Jupiter	680
Saturn	1,320
Uranus	2,600
Neptune	5,160

G.D. Hinrichs, The Density, Rotation and Relative Age of the Planets, *American Journal of Science and Arts*, 2(37), 36–56, 1864, table on p. 43.

consisting of a set of specific spectral lines, which they set about measuring and publishing in elaborate tables. Some authors suggested that these spectral lines might provide information about the various elements that had produced them, but these suggestions met with strenuous criticism from one of their discoverers, Bunsen. Indeed, Bunsen remained quite opposed to the idea of studying spectra in order to study atoms or to classify them in some way.[44]

Hinrichs, however, had no hesitation in connecting spectra with the atoms of the elements. In particular, he became interested in the fact that, with any particular element, the frequencies of its spectral lines always seemed to be whole number multiples of the smallest difference. For example, in the case of calcium, a ratio of 1:2:4 had been observed among its spectral frequencies. Hinrichs's interpretation of this fact was bold and elegant: If the sizes of planetary orbits produce a regular series of whole numbers, and if the ratios among spectral line differences also produce whole number ratios, the cause of the latter might lie in the size ratios among the atomic dimensions of the various elements (table 3.8). This is Pythagoreanism with a vengeance, but it proved to be fruitful in that it led Hinrichs to a successful, and highly novel, means of classifying the elements into a periodic system.

By closely studying the work of Kirchoff and Bunsen, Hinrichs found that some of the spectral line frequencies, those referred to as "dark lines," could be related to the chemistry of the elements through their atomic weights, as well as to their postulated atomic dimensions. The difference between the spectral line frequencies seemed to be inversely proportional to the atomic weights of the elements in question. Hinrichs quoted the values of calcium, where the frequency difference is 4.8 units, and barium, which is chemically similar but has a higher atomic weight and shows a frequency difference of 4.4 units.[45]

TABLE 3.8
Schematic form of Hinrichs's argument

From astronomy	Size ratios among orbits	$\rightarrow$	Whole number ratios
From spectra	Observation of whole		Size ratios among
	number ratios	$\rightarrow$	atomic dimensions

Hinrichs then proposed the following formula to connect the atomic weight of any element with its atomic dimensions:

$$A = a \times b \times c$$

A is the atomic weight and a, b, and c are the respective lengths of the sides of a prism denoting the shape of the atom. The base of the prism, which is taken as the dimension a, would be of the same size for all the elements belonging to a particular chemical group. If a particular group contained square prisms, their formula would reduce to

$$A = a^2 \times b$$

In other cases, where the base of the prism took on a triangular shape, the formula would be expressed as

$$A = (a \times b \times c) + k,$$

where k is a constant. Given how improbable this whole approach might seem, it is quite remarkable how useful it turned out to be when Hinrichs applied it to rationalizing the atomic weights of the elements.[46] For example, it served quite successfully as a basis for deciding which elements should be grouped together in his periodic system. Table 3.9 shows some of his groups and demonstrates how in each case one of the formulas given above is able to accommodate, rather accurately, the atomic weight of each element in the proposed groups. Of course, this was not the only reason Hinrichs grouped elements together. Many of the groupings suggested themselves primarily on the basis of chemical similarities, with which Hinrichs would have been well acquainted through his knowledge of chemistry.

In the course of this work, Hinrichs expressed his support for the notion of primary matter, which had been the basis of Prout's hypothesis of half a century earlier. Hinrichs was convinced that the atomic weights of the elements were whole numbers. Because the value of chlorine was 35.5 according to Cannizzaro's atomic weights, Hinrichs concluded that the primary atom had a weight of half the value of hydrogen and so took H/2 to be the basic unit for expressing all the other weights. The weight of chlorine therefore assumed a value of 71, and the Cannizzaro atomic weights of all the other elements were similarly doubled. These

TABLE 3.9

Hinrichs's table of atomic weights and atomic dimensions
for several groups of elements

	n	A	Calculated	Determined	Difference
Oxygen group: quadratic formula			$A = n \times 4^2$		
Oxygen	1	1×4^2	$= 16$	16	0.0
Sulfur	2	2×4^2	$= 32$	32	0.0
Selenium	5	5×4^2	$= 80$	80	0.0
Tellurium	8	8×4^2	$= 128$	128	0.0
Alkali metal group: quadratic with pyramid			$A = 7 + (n \times 4^2)$		
Lithium	0	7	7		0.0
Sodium	1	$7 + (1 \times 4^2)$	$= 23$	23	0.0
Potassium	2	$7 + (2 \times 4^2)$	$= 39$	39	0.0
Rubidium	5	$7 + (5 \times 4^2)$	$= 87$	85.4	$- 1.6$
Cesium	8	$7 + (8 \times 4^2)$	$= 135$	133	$- 2.0$
Chlorine group: quadratic formula			$A = (n \times 3^2) \pm 1$		
Fluorine	2	$(2 \times 3^2) + 1$	$= 19$	19	0.0
Chlorine	4	$(4 \times 3^2) - 1$	$= 35$	35.5	$+ 0.5$
Bromine	9	$(9 \times 3^2) - 1$	$= 80$	80	0.0
Iodine	14	$(14 \times 3^2) + 1$	$= 127$	127	0.0
Alkaline earth group: quadratic formula			$A = n \times 2^2$		
Magnesium	3	3×2^2	$= 12$	12	0.0
Calcium	5	5×2^2	$= 20$	20	0.0
Strontium	11	11×2^2	$= 44$	43.8	$- 0.2$
Barium	17	17×2^2	$= 68$	68.5	$+ 0.5$

Remade from G.D. Hinrichs, On the Spectra and Composition of the Elements, *American Journal of Science and Arts*, 92, 350–368, 1866, table on p. 365.

are the values that are seen in the culmination of Hinrichs's work on the classification of the elements, his spiral periodic system, as shown in figure 3.7.

The 11 "spokes" radiating from the center of this wheel-like system consist of three predominantly nonmetal groups and eight groups containing metals. From a modern perspective, the nonmetal groups appear to be incorrectly ordered, in that the sequence is groups VI, V, and then VII when proceeding from left to right at the top of the spiral. The group containing carbon and silicon is classed with the metallic groups by Hinrichs, presumably because it also includes the metals nickel, palladium, and platinum. In the modern table, these three metals are indeed grouped together, but not in the same group as carbon and silicon, which belong with germanium, tin, and lead in group IV.

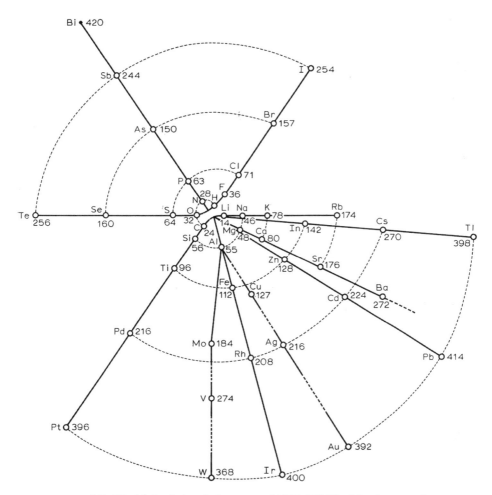

FIGURE 3.7 Hinrichs's spiral periodic system of 1867. G.D. Hinrichs, *Programm der Atomechanik oder die Chemie eine Mechanik de Pantome*, Augustus Hageboek, Iowa City, IA, 1867. As simplified by J. van Spronsen, *The Periodic System of the Chemical Elements, the First One Hundred Years*, Elsevier, Amsterdam, 1969 (by permission from Elsevier).

Overall, however, Hinrichs's periodic system is rather successful in grouping together many important elements. One of its main advantages is the clarity of its groupings, compared, say, with the more elaborate but less successful periodic systems of Newlands in 1864 and 1865. For example, Hinrichs groups together oxygen, sulfur, selenium, and tellurium. Newlands also groups these elements together but includes osmium (Os) with them. Hinrichs groups together nitrogen, phosphorus, arsenic, tin, and bismuth. So does Newlands, but he incorrectly includes manganese, as well as didymium[47] and molybdenum in one space. Hinrichs groups together lithium, sodium, potassium, and rubidium. Newlands also groups these elements together but also incorrectly includes copper, silver, cesium, gold, and tellurium.[48]

Although it is not arranged as a long-form table, Hinrichs's classification seems to capture many of the primary periodicity relationships seen in the modern periodic table, and unlike many of Newlands's tables, it is not cluttered by attempts to show secondary kinship relationships. Hinrichs, for example, groups together copper, silver, and gold. In the case of Newlands, these elements are grouped separately, with the exception of one table in 1865, which classifies the three elements together and also intersperses them with such other elements as potassium, rubidium, and cesium.[49]

It is clear from his books that Hinrichs possessed a deep knowledge of chemistry, as well as a proficiency in mineralogy.[50] Yet his approach to the classification of the elements was only partly chemical. He was perhaps the most interdisciplinary of all the discoverers of the periodic system. Indeed, the fact that Hinrichs arrived at his system from such a different direction as the others might be taken to lend the periodic system itself independent support, just as Lothar Meyer's studies of physical periodicity (described below) also do.

In an article published in *The Pharmacists* in 1869,[51] Hinrichs discusses previous unsuccessful attempts to classify the elements, but in doing so fails to mention any of his codiscoverers, such as De Chancourtois, Newlands, Odling, Lothar Meyer, and Mendeleev. Hinrichs characteristically appears to have completely ignored all other attempts to base the classification of the elements directly on atomic weights, though one can assume that he was aware of them given his knowledge of foreign languages. This is not to say that his classification is unconnected with atomic weights, only that the connection is rather indirect in view of the astronomical argument that seems to be the basis of the approach.

Finally, it should be stressed that Hinrichs appeared to be ahead of his time in assigning great importance to the analysis of the spectra of the elements and in trying to relate these facts to the periodic classification. However, his spectral studies are by no means universally accepted. Some contemporary historians, including Klaus Hentschel, have criticized Hinrichs's work, claiming that he was somewhat selective in what data he admitted into his calculations.[52]

More than that of any other scientist discussed in this book, the work of Hinrichs is so idiosyncratic and labyrinthine that a more complete study will be required before anyone can venture to pronounce on its real value.

JULIUS LOTHAR MEYER

Lothar Meyer (figure 3.8) was born in 1830 in Heilbronn, Germany. He was the fourth of seven children of a physician father and a mother whose own father was also a local physician. Julius and one of his brothers, Oskar, began their studies with

FIGURE 3.8
Julius Lothar Meyer. Photo from
author's collection, permission from
Edgar Fahs Smith Collection.

the intention of continuing this family medical tradition, but it was not long before both of them had turned to other fields. Oskar became a physicist, while Julius became one of the most influential chemists of his time.

Lothar Meyer is best remembered for his independent discovery of the periodic system, although more credit is invariably accorded to Mendeleev. The two chemists eventually became engaged in a rather bitter priority dispute, which Mendeleev apparently won, although how much of that was due to Mendeleev's more forceful personality is difficult to ascertain fully. Certainly Mendeleev had a more complete system and went on to make predictions on the basis of his system. He was also to champion the cause of the periodic law to a far greater extent than was Lothar Meyer. But if one asks the question of who arrived at the mature periodic system first, a strong case can be made for saying that in many crucial details the system of Lothar Meyer was not only first but also more correct.

Lothar Meyer attended the Karlsruhe conference in 1860 and learned firsthand of Cannizzaro's groundbreaking work on the atomic weights of the elements.[53] He then edited a version of Cannizzaro's article that appeared in Germany in Wilhelm Ostwald's series under the title *Klassiker der Wissenschaften*. Lothar Meyer later described the effect that Cannizzaro's article had on him by saying, "[T]he scales fell from my eyes and my doubts disappeared and were replaced by a feeling of quiet certainty."[54] In 1864, Lothar Meyer published the first edition of a chemistry textbook, *Die Modernen Theorie der Chemie*, which was deeply influenced by the work of Cannizzaro. The book appeared in five editions and was translated into English, French, and Russian, eventually becoming one of the most authoritative treatments on the theoretical principles of chemistry before the advent of physical chemistry in the late 1800s.

By the time Lothar Meyer had written the manuscript for his book in 1862, he had produced a table of 28 elements arranged in order of increasing atomic weight. An adjacent table containing a further 22 elements also appeared in the book, although these were not arranged according to atomic weight order. All this took place only two years after the Karlsruhe conference. It should perhaps be noted in passing that it took Mendeleev something like nine years from the time of his attending the same conference before he, too, produced a table of elements arranged in order of increasing atomic weights.

Lothar Meyer was also deeply influenced by the work of Johann Döbereiner and Pettenkofer, both of whom, as described in chapter 2, had published articles on the existence of triads of elements, where the weight of the middle member was the approximate mean of that of the flanking elements. Going further, Pettenkofer had pointed to an analogy between the regular increase in the weights of successive members of any homologous series in organic chemistry and the almost regular increase in the atomic weights of similar elements within any triad, something that had also been noticed by Dumas.

Most organic compounds can be classified according to the homologous series to which they belong. Such series are created in an iterative fashion with the repeated addition of a chemical unit, such as CH_2 in the case of the alkanes.[55] This regularity suggested to Pettenkofer and Dumas that the molecules of such series must be composed of regular units.

If the analogy between these organic compounds were also applied to inorganic atoms, it would suggest that atoms are likewise composed of parts. In other words, just as the regularity in the increasing molecular weights in a homologous series suggests that its members contain some sort of building block universal to that series, so the regularity seen in the intervals between atomic weights of members of a triad would suggest that the atoms of those members are somehow modular. Lothar Meyer did indeed regard such evidence as pointing to the composite nature of inorganic atoms,[56] something that Mendeleev never accepted throughout his life.[57]

Lothar Meyer published his table of 28 elements for the first time in 1864 (figure 3.9). His arrangement of elements in order of increasing atomic weights and the clear establishment of horizontal relationships among these elements are other instances in which Lothar Meyer anticipated Mendeleev by several years.[58] As described in chapter 4, where we encounter the details of Mendeleev's work, this recognition of the need to order the elements in terms of increasing atomic weight, and especially the recognition of horizontal relationships, has wrongly been regarded as a first by Mendeleev. Yet here in 1864,

	4 werthig	3 werthig	2 werthig	1 werthig	1 werthig	2 werthig
	—	—	—	—	Li = 7.03	(Be = 9.3?)
Differenz =					16.02	(14.7)
	C = 12.0	N = 14.04	O = 16.00	Fl = 19.0	Na = 23.05	Mg = 24.0
Differenz =	16.5	16.96	16.07	16.46	16.08	16.0
	Si = 28.5	P = 31.0	S = 32.07	Cl = 35.46	K = 39.13	Ca = 40.0
Differenz =	$\frac{89.1}{2} = 44.55$	44.0	46.7	44.51	46.3	47.6
	—	As = 75.0	Se = 78.8	Br = 79.97	Rb = 85.4	Sr = 87.6
Differenz =	$\frac{89.1}{2} = 44.55$	45.6	49.5	46.8	47.6	49.5
	Sn = 117.6	Sb = 120.6	Te = 128.3	I = 126.8	Cs = 133.0	Ba = 137.1
Differenz =	$89.4 = 2 \times 44.7$	$87.4 = 2 \times 43.7$	—	—	$(71 = 2 \times 35.5)$	—
	Pb = 207.0	Bi = 208.0	—	—	(Tl = 204?)	—

FIGURE 3.9 Table of 1864. J. Lothar Meyer, *Die modernen thoerien und ihre Bedeutung für die chemische Statisik*, Breslau (Wroclaw), 1864, p. 135.

Lothar Meyer is publishing both ideas simultaneously, without, in most cases, receiving due recognition for these advances from contemporary, or later, commentators.

Lothar Meyer's 1864 table also showed clearly for the first time a regular variation in valency of the elements, from 4 to 1 on moving from left to right across the table, followed by a repetition of valence 1 and a further increase to elements with valence 2.[59] This table suggests that Lothar Meyer struggled to arrange elements in terms of atomic weight as well as chemical properties. He seems to have decided to let chemical properties outweigh strict atomic weight ordering in some cases. An example of this is in his grouping of tellurium with elements such as oxygen and sulfur, while iodine is grouped with the halogens, in spite of their ordering according to atomic weight. Lothar Meyer also separated the elements into two tables in a manner corresponding to the separation of our modern main-group elements from the modern transition elements. As mentioned above in the case of Odling, such a separation has become a feature of the modern medium-long–form and long-form tables.

Another very noteworthy feature of Lothar Meyer's table of 1862 (published in 1864) is the presence of many gaps to denote unknown elements. Once again, it appears that the leaving of gaps did not originate with Mendeleev, who was to wait a further five years before even venturing to publish a periodic system and eventually making the detailed predictions for which he subsequently became so well known. Lothar Meyer's table contains interpolations between neighboring elements. In the space below the element silicon, for example, he indicates that there should be an element whose atomic weight would be greater than silicon's by a difference of 44.55. This implies an atomic weight of 73.1 for this unknown element, which when discovered was found to have an atomic weight of 72.3. This prediction of the element germanium, which was first isolated in 1886, is usually attributed to Mendeleev even though it was clearly anticipated by Lothar Meyer in this early table of 1864.

The criticism has been made that Lothar Meyer did not explicitly refer to atomic weight in his 1864 table.[60] This objection seems a little excessive, however, since with regard to the 28-element table, the arrangement is clearly based on increasing atomic weight, such that Lothar Meyer may not have felt the need to comment on this rather obvious feature. Of course, the same cannot be said for the smaller table consisting of 22 elements. But the fact that these elements have been separated from the other 28 may indicate that Lothar Meyer realized that in these cases the concept of increasing atomic weight did not apply strictly to the classification he chose to adopt.[61] Nevertheless, atomic weight increases vertically down each column, and there are only six inconsistencies in the increase in atomic weight going across the table. Given that Lothar Meyer had classified a total of 50 elements while showing only six mistaken reversals

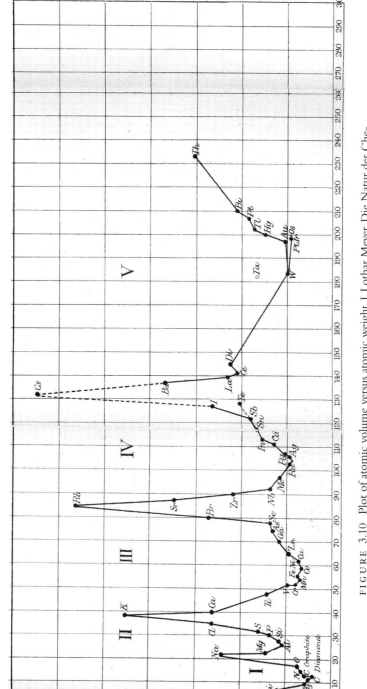

FIGURE 3.10 Plot of atomic volume versus atomic weight. J. Lothar Meyer, Die Natur der Chemischen Elemente als Function ihrer Atomgewichte, *Annalen der Chemie, Supplementband*, 7, 354–364, 1870. Redrawn by T. Bayley, *Philosophical Magazine*, 13, 26–37, 1882, p. 26.

in atomic weights, all of which occur among the problematic transition metals (in the modern usage of the term),[62] this cannot be considered a significant
failing on his part. The only serious misplacements he made in terms of atomic
weight increase concern just two elements, molybdenum and vanadium. All of
his other reversals are quite within the possible bounds of error in measured
atomic weights.

But perhaps Lothar Meyer's greatest strength lay in his additional knowledge
of physical properties and his use of them in constructing representations of the
periodic system. He paid close attention to atomic volumes, densities, and fusibilities of the elements, for example. His published diagram showing the periodicity
among atomic volumes of the elements (i.e., atomic weight divided by specific
gravity), in particular, is generally considered to have contributed favorably to the
general acceptance of the periodic system (figure 3.10). Indeed, one can see the
periodicity among the elements almost at a glance from this diagram. Mendeleev,
too, was aware of the importance of atomic volume. In fact, he made predictions
on atomic volumes beginning in his first article of 1869. But he did not emphasize
the periodicity in this physical property of atoms, nor did he display such suggestive diagrams of its trend.

THE REMELÉ–SEUBERT EPISODE:
THE UNPUBLISHED TABLE OF 1868

In the course of the controversy between Mendeleev and Lothar Meyer, which
followed the publication of their respective periodic systems, it seems fair to say
that Mendeleev was the victor at least as far as the scientific public was concerned.
However, there is a rather intriguing episode that did not come to light until much
later, and that might have made a significant difference in this controversy had it
become known earlier. In 1868, when Lothar Meyer was preparing the second
edition of his book, he produced a vastly expanded periodic system that included
a further 24 elements and nine new vertical families of elements (figure 3.11). This
system preceded Mendeleev's famous table of 1869 that subsequently claimed all
the glory. Moreover, Lothar Meyer's system was more accurate than Mendeleev's.
For example, Lothar Meyer correctly placed mercury with cadmium, lead with
tin, and thallium with boron. In all three of these cases, Mendeleev's table failed to
make this connection.[63]

It appears that for some reason Lothar Meyer's 1868 table was not published.
A full 25 years later, Adolf Remelé, a German chemist who succeeded Lothar
Meyer as professor of chemistry in Eberswalde, showed the table to Lothar
Meyer, who in the meantime seemed to have forgotten all about its existence. In
1895, after Lothar Meyer's death, Carl Seubert, one of his colleagues, finally pub-

MEYER'S TABLE OF 1868.

1	2	3	4	5	6	7	8
		Al=27.3 2⁄₃l=14.8	Al.=27.3			.	C=12.00 16.5 Si=28.5 ²⁄₃l=44.5
Cr=52.6	Mn=55.1 49.2 Ru=104.3 92.8=2.46.4 Pt=197.1	Fe=56.0 48.9 Rh=103.4 92.8=2.46.4 Ir=197.1	Co=58.7 47.8 Pd=106.0 93=2.465 Os=199.	Ni=58.7	Cu=63.5 44.4 Ag=107.9 88.8=2.44.4 Au=196.7	Zn=65.0 46.9 Cd=111.9 88.3=2.44.5 Hg=200.2	²⁄₃l=44.5 Sn=117.6 89.4=2.41.7 Pb=207.0

9	10	11	12	13	14	15
N=14.4 16.96 P=31.0 44.0 As=75.0 45.6 Sb=120.6 87.4=2.43.7 Bi=208.0	O=16.00 16.07 S=32.07 46.7 Se=78.8 49.5 Te=128.3	F=19.0 16.46 Cl=35.46 44.5 Br=79.9 46.8 I=126.8	Li=7.03 16.02 Na=23.05 16.08 K=39.13 46.3 Rb=85.4 47.6 Cs=133.0 71=2.35.5 Te=204.0	Be=9.3 14.7 Mg=24.0 16.0 Ca=40.0 47.6 Sr=87.6 49.5 Ba=137.1	Ti=48 42.0 Zr=90.0 47.6 Ta=137.6	Mo.=92.0 45.0 Vd=137.0 47.0 W=184.0

FIGURE 3.11 Unpublished system of 1868. J. Lothar Meyer.

lished the forgotten table. Unfortunately, this attempt to restore some semblance of priority to Lothar Meyer, after this almost comical time delay, fell largely on deaf ears.

CONCLUSION

As I hope to have shown in this chapter, the periodic system developed through a process of gradual evolution rather than revolution, especially after Cannizzaro had published an accurate set of atomic weights. The discovery was made, essentially independently, by six diverse scientists who differed greatly in their fields of expertise and in their approaches. De Chancourtois, a French geologist, was unlucky to produce a rather complicated three-dimensional representation that suffered further at the hands of his publisher. But the fact remains that he made the first discovery of periodicity. In addition, many chemical mistakes led to the almost complete oblivion of his system. An English sugar chemist, Newlands, was the first to recognize the lawlike status of chemical periodicity but was somewhat ignored because, among other things, he compared periodicity to musical octaves. A more established English chemist, Odling, also designed successful periodic systems, but somewhat surprisingly denied the lawfulness of chemical periodicity. Hinrichs, a polymath working in the United States, developed a spiral periodic system using an extravagant form of Pythagoreanism in which he compared the dimensions of the solar system to the dimensions within the atom. Then came the fully mature periodic systems of Lothar Meyer and Mendeleev, two established chemistry professors in Germany and Russia, respectively, both of whom were engaged in writing chemistry textbooks. Lothar Meyer appears to have placed greater emphasis on physical properties of the atoms but hesitated to make predictions. Mendeleev, meanwhile, was the consummate chemist, familiar with the detailed chemical behavior of all the known elements and, as described in chapter 4, also ventured to make bold predictions concerning yet undiscovered elements.

MENDELEEV

Dimitri Ivanovich Mendeleev (figure 4.1) is the undisputed champion of the periodic system in at least two senses. First of all, he is by far the leading discoverer of the system. Although he was not the first to develop a periodic system, his version is the one that created the biggest impact on the scientific community at the time it was introduced and thereafter. His name is invariably and justifiably connected with the periodic system, to the same extent perhaps as Darwin's name is synonymous with the theory of evolution and Einstein's with the theory of relativity.

Although it may be possible to quibble about certain priority aspects of his contributions, there is no denying that Mendeleev was also the champion of the periodic system in the literal sense of propagating the system, defending its validity, and devoting time to its elaboration.[1] As discussed in chapter 3, there were others who produced significant work on the system, but many of them, such as Alexandre-Emile De Chancourtois, William Odling, and Gustav Hinrichs, moved on to other scientific endeavors. After publishing their initial ideas, these contributors devoted their attention to other fields and never seriously returned to the periodic system to examine its full consequences to the extent that Mendeleev did.

This is not to suggest that Mendeleev himself worked only on the periodic system. He is also known for many other scientific contributions, as well as for working in several applied fields, such as the Russian oil industry and as the director of the Russian institute for weights and measures. But the periodic system remained as Mendeleev's pride and joy throughout his adult life. Even toward the end of his life he published an intriguing essay in which he returned to the periodic system and, among other speculations, attempted to place the physicist's ether within the periodic system as a chemical element.

Much has been written on Mendeleev, and it would be impossible to do justice to his contributions in the space of a few pages.[2] Here I concentrate, as in other parts of this book, on the fundamental scientific and philosophical ideas that

FIGURE 4.1
Dimitri Ivanovich Mendeleev.
Photo and permission from
Emilio Segrè Collection.

underpinned the evolution of the system.[3] An important part of this investigation consists of trying to understand Mendeleev's conception of the nature of chemical elements. This issue forms the basis of what is perhaps the most philosophical aspect of the periodic system, and one that has been almost completely neglected by books and articles on Mendeleev and the periodic system generally.[4]

EARLY LIFE AND SCIENTIFIC WORK

Mendeleev was born in 1834 in the Siberian city of Tobolsk. He was the last child in a family of 14 children. His father died when he was very young, and his mother, who was devoted to encouraging his scientific studies, died when Dimitri was about 15 years old. Before her death, she went to great lengths and sacrifices to enroll her son at the Main Pedagogical Institute of St. Petersburg, where he took classes in chemistry, biology, and physics, as well as pedagogy. The last of these in particular was to have a profound influence on his scientific work, since it was in the course of writing a textbook for the teaching of inorganic chemistry that Mendeleev was to develop his periodic system.

Mendeleev's early scientific work involved a detailed examination of the chemical properties as well as the specific volumes of many substances. Between 1859–61 he spent some time working at Robert Bunsen's laboratory in Heidelberg, where he studied the behavior of gases and their deviations from the laws of perfect gases. By 1860, he had become sufficiently prominent a chemist to be invited to attend the Karlsruhe conference, where he met the likes of Alexandre Dumas, Charles-Adolphe Wurtz, and Stanislao Cannizzaro. The following year he published a textbook of organic chemistry, which enjoyed considerable success in his native Russia and for which he was awarded the prestigious Demidov Prize.

It was not until 1865 that Mendeleev defended his doctoral thesis, which was based on his study of the interaction between alcohol and water. At about this time, having already written a book aimed at systematizing organic chemistry, he began to consider the possibility of producing a book that would likewise attempt to systematize inorganic chemistry. These efforts eventually resulted in his discovery of the periodic system, which is now virtually synonymous with his name.

Although it is clear that Mendeleev's periodic system was conceived while he was writing his textbook *The Principles of Chemistry*, it is essential also to consider his shorter publications announcing the discovery of the periodic system, as well as earlier written evidence, in order to place this discovery in the wider context of his work. Many myths and legends have developed around the genesis of Mendeleev's periodic system, one of the most common being that he conceived of the idea in the course of a dream, or that it occurred to him while playing a game of patience with cards marked with the symbols of the elements. In fact, the idea took many years to mature and may have begun to do so around the time of the Karlsruhe conference, as long as 10 years before the publication of his famous table of 1869.

At the end of 1868, Mendeleev had completed the first volume of his textbook on inorganic chemistry, in which he made a systematic examination of different kinds of elements and compounds and dealt with the most common elements, such as hydrogen, oxygen, and nitrogen. He initially grouped the elements according to the valences they displayed when combining with hydrogen. This offered at least some means of organization, but at this stage there was no sign of any overarching organizing principle or any system of classification. Mendeleev ended volume 1 with a survey of the halogens and began volume 2 with a survey of the alkali metals. He was then faced with the question of which elements to treat next. As legend has it, he solved the problem in the course of a single day, in which he declined to fulfill an obligation to inspect a nearby cheese factory, instead working furiously on his new element scheme.

The question of an organizing principle had to be faced, and unlike the precursors of the periodic system, with which he was familiar, Mendeleev did not embrace either the concept of triads or the existence of a primary substance. Mendeleev knew the work of the Belgian chemist Jean-Servais Stas, for example, who had begun as an advocate of William Prout's hypothesis but, as noted in chapter 2, had become its strongest critic following a series of accurate atomic weight determinations he had himself undertaken. Mendeleev specifically refers to Staas in volume 1 of his book and expresses his distaste for Prout's hypothesis. Mendeleev's objection to a literal conception of triads is clear when he insists, also in volume 1, that rubidium, cesium, and thallium all belong to the alkali metals, along with the members of the original triad group, which are lithium, sodium, and potassium. Mendeleev is thus extending a group of elements previously thought to consist of just three elements to a group containing twice that number. In addition, he states that fluorine belongs to the halogens, thus extending the triad of chlorine, bromine, and iodine into a fourth member, a feat some others had resisted simply because it seemed to contradict the strict notion of a triad. Mendeleev thus explicitly freed himself from these pervasive general notions in order that his views might be judged on their own merit and so that the full originality of his work might be better appreciated.

On the other hand, although Mendeleev had grasped the importance of atomic weight early on, he had not fully embraced this means of characterizing the elements when he set out to write his textbook. The historian Donald Rawson, who has conducted a search of Mendeleev's views on atomic weights, finds that as early as 1855–1856, while an undergraduate at the Pedagogical Institute in St. Petersburg, Mendeleev was still using the atomic weights of Jacob Berzelius.[5] In his master's thesis, written shortly thereafter, Mendeleev had converted to the atomic weights of Charles Gerhardt, who had halved many of the values given by Berzelius. These values in turn also contained errors, including those for oxygen and carbon, which had been halved, thus resulting in the formula of water being considered H_2O_2 and that of benzene $C_{12}H_6$. Both of these are quite incorrect when compared with the modern formulas of H_2O and C_6H_6. Fortunately, Mendeleev readily abandoned Gerhardt's values when he attended the conference at Karlsruhe.[6]

Nevertheless, it took some further time before Mendeleev had fully converted to the atomic weights of Cannizzaro. In lecture notes written between the years 1864 and 1865, for example, Mendeleev listed 53 elements but still continued to use the more outdated equivalent weights for 13 of them. By 1868, when he began writing the second volume of his textbook, he was listing 22 elements, all of them given their new atomic weights according to Cannizzaro. Whether this is a coincidence or not, it implies that by the time Mendeleev had begun consciously to work on the classification of the elements, he had fully assimilated the use of the modern atomic weights, an approach that would prove to be so essential for his discovery.

It has been claimed that it was simply in seeking a quantitative justification for ordering the elements that Mendeleev arrived at the idea of using increasing atomic weights.[7] Although Mendeleev himself has written at some length on the genesis of his ideas, it is difficult to arrive at a clear and accurate picture of his motivations or even the course of the development of his thinking. For example, he steadfastly maintained in all subsequent writings that he did not see any of the systems developed by the five other discoverers of the periodic system, namely, De Chancourtois, Odling, Newlands, Hinrichs, and Lothar Meyer. This seems a little odd, given that he repeatedly acknowledged his debts to some earlier pioneers of the system, including Peter Kremers, Josiah Cooke, Max Pettenkofer, Alexandre Dumas, and Ernst Lenssen. Nor can it be supposed that this omission might have been due to isolation, since Russian chemistry, in particular, was rather advanced at this time, and Mendeleev had traveled in Europe and was well aware of the published literature in several languages.[8]

Also puzzling is the suggestion made by Mendeleev himself, as well as some later commentators, that it was the realization of the need to order the elements by atomic weight that was *the* bold and original step in the development of his system. Even if one grants that Mendeleev knew nothing of the work of the five other discoverers, surely the early precursors, whom Mendeleev so openly acknowledges, were already utilizing the concept of atomic weight in order to place the elements into some sort of order. One might rationalize this situation by recognizing that there is an important sense in which Mendeleev was indeed the first to recognize the *full* significance of the concept of atomic weight. This question is addressed after considering Mendeleev's actual discovery and the periodic tables he produced.

THE CRUCIAL DISCOVERY

We now consider the crucial steps that led Mendeleev to begin comparing elements horizontally (in the sense of the modern periodic table) in terms of atomic weights.[9] There is a letter in the Mendeleev archives, dated February 17, 1869, which is also the date of the famous first table he produced.[10] This letter, from one Alexei Ivanovich Khodnev, secretary of the Free Economic Society in St. Petersburg, to Mendeleev concerns arrangements regarding the visit to a cheese factory where Mendeleev was due to conduct an inspection. On the back of the letter, Mendeleev has made a comparison of the atomic weights of the following elements:

Na	K	Rb	Cs
Be	Mg	Zn	Cd

This is where Mendeleev is possibly trying to decide which elements to discuss after the alkali metals in his book. It could either be zinc and cadmium or the alkaline earth elements, or perhaps even both together as shown in the fragment periodic table above. In fact, this fragment may represent the first time that a horizontal comparison of the atomic weights of elements had been consciously carried out.[11]

Another early fragment periodic system that Mendeleev produced involves a comparison of three groups of elements:

F	Cl	Br	I				
Na	K	Rb	Cs			Cu	Ag
Mg	Ca	Sr	Ba	Zn	Cd		

On the same day, Mendeleev appears to have realized the need also to compare all the other groups of elements horizontally, thus allowing him to arrive at his first manuscript table, as shown in figure 4.2.

And so, in the space of a single day, February 17, 1869, Mendeleev not only began to make horizontal comparisons but also produced the first version of a full periodic table that included most of the known elements. There should be no doubt, therefore, that a sudden decisive step *did* indeed occur, even though the background ideas may have been developing over a period of about 10 years.

We turn now to Mendeleev's announcement of his discovery. Having arrived at a consistent periodic system, Mendeleev had 200 copies of his table printed and sent them to chemists in Russia and the rest of Europe. Nicolai Alexandrovich Menshutkin communicated the initial discovery to the Russian Chemical Society on March 6. Later in the same month, it appeared in print (in Russian) in the first volume of the journal of the newly formed Russian Chemical Society.[12] The full article contained several periodic tables, and a shorter abstract was published in German a few weeks later.[13]

This published periodic system of Mendeleev's (figure 4.3) contains divisions into main and subgroups. The first column of elements, for example, all show valences of 1 but are divided into the alkali metals, such as lithium, sodium, and potassium, and the noble metals, including copper, silver, and gold. Significantly, there are several vacant spaces in the table, and Mendeleev proceeds even in this first publication to make several predictions, specifically anticipating "many yet unknown elements e.g. elements analogous to aluminium and silicon with atomic weights 65–75."

Mendeleev's most famous predictions would be for the elements scandium, gallium, and germanium, all of which were anticipated in the periodic tables published in this 1869 paper. He made highly accurate entries for the expected atomic weights of two of these unknown elements in the form of "? = 68" and "? = 70" in the rows containing aluminum and silicon, respectively. (The atomic weights of these new elements turned out to be gallium = 69.2 and germanium = 72, respectively.) Moreover, his "attempt at a system" of 1869 contains an entry "? = 45", which turned out to correspond to scandium with an atomic weight of 44.6, although it has been the subject of some debate whether Mendeleev's early prediction of this element can be strictly identified with scandium.

Not only did Mendeleev predict the atomic weights of his famous three new elements as early as 1869, but he also made predictions of some of their other properties. In a talk to a Moscow Congress in that same year, he suggested that two elements missing from the system would show resemblances to aluminum and silicon and would have atomic volumes of 10 or 15 and specific gravities of about 6. In the following year, 1870, Mendeleev listed the expected atomic volumes of the elements that would become known as scandium, gallium, and germanium as 15, 11.5, and 13, respectively.[14]

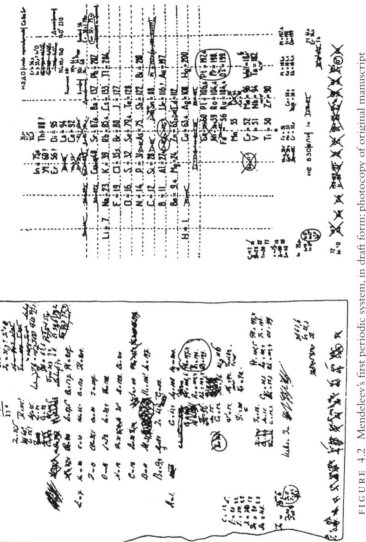

FIGURE 4.2 Mendeleev's first periodic system, in draft form: photocopy of original manuscript (left), and clarified version (right). D.I. Mendeleev, *Periodicheskii Zakon*, B.M. Kedrov (ed.), Izdatel'stvo Akademii nauk Soyuz sovietskikh sotsial'sticheskikh respublik, Moscow, 1958. Reproduced from J. van Spronsen, *The Periodic System of the Chemical Elements, the First One Hundred Years*, Elsevier, Amsterdam, 1969 (by permission from Elsevier).

Typische Elemente						
Li = 7	Na = 23	K = 39	Rb = 85	Cs = 133	Er = 178?	—
Be = 9,4	Mg = 24	Ca = 40	Sr = 87	Ba = 137	?La = 180?	—
B = 11	Al = 27,3	—	?Yt = 88?	?Di = 138?	Ta = 182	—
C = 12	Si = 28	Ti = 48?	Zr = 90	Co = 140?	W = 184	Tb = 281
N = 14	P = 31	V = 51	Nb = 94	—	—	—
O = 16	S = 32	Cr = 52	Mo = 96	—	Os = 195?	U = 240
F = 19	Cl = 35,5	Mn = 55	—	—	Ir = 197	—
		Fe = 56	Ru = 104	—	Pt = 198?	—
		Co = 59	Rh = 104	—	A. = 199?	—
		Ni = 59	Pd = 106	—	Hg = 200	—
		Cu = 63	Ag = 108	—	Tl = 204	—
		Zn = 65	Cd = 112	—	Pb = 207	—
		—	In = 113	—	Bi = 208	—
		As = 75	Sn = 118	—	—	—
		Se = 78	Sb = 122	—	—	—
		Br = 80	Te = 125?	—	—	—
		—	J = 127	—	—	—

H = 1

FIGURE 4.3 Mendeleev's published periodic system, of 1869. D.I. Mendeleev, Sootnoshenie svoistv s atomnym vesom elementov, Zhurnal Russkeo Fiziko-Khimicheskoe Obshchestv, 1, 60–77, 1869, p. 70.

Another evident feature of this system is the reversal of the elements tellurium and iodine, although, as mentioned in chapter 3, this step had already been taken by Odling and Lothar Meyer, regardless of whether or not Mendeleev might have been aware of this fact. In putting tellurium before iodine, Mendeleev was departing from his general approach of ordering the elements by atomic weight. As mentioned in chapter 3, tellurium has a higher atomic weight than iodine, and yet in terms of its valence, tellurium should occur before iodine in the ordering of the elements. But apart from this particular case, Mendeleev did not maintain the use of valence as a criterion for classification as had Lothar Meyer, for example, because many elements show variable valence and because of his philosophical preference for concentrating on elements as basic substances rather than elements with manifest chemical properties, as discussed below. Some aspects of this attitude is revealed in his writing:

> Not being susceptible to exact measurements, the above mentioned chemical properties can hardly serve to generalize chemical knowledge: They alone cannot serve as a basis for chemical considerations. However, the properties should not be altogether neglected as they explain a great number of chemical phenomena.[15]

Mendeleev consequently put more faith in his newly discovered criterion for ordering the elements according to atomic weight.

Some idea of the sophistication of Mendeleev's first article, as well as the German abstract, can be seen from the list of eight points with which he ends these publications:

1. The elements if arranged according to their atomic weights, exhibit an evident *periodicity* of properties.
2. Elements which are similar as regards their chemical properties have atomic weights which are either of nearly the same value (e.g. platinum, iridium, osmium), or which increase regularly (e.g. potassium, rubidium, caesium).
3. The arrangement of the elements, or of groups of elements, in the order of their atomic weights corresponds to their so-called *valences* as well as, to some extent, to their distinctive chemical properties—as is apparent among other series—in that of lithium, beryllium, barium, carbon, nitrogen, oxygen and iron.
4. The elements which are most widely diffused have *small* atomic weights.
5. The *magnitude* of the atomic weight determines the character of the element, just as the magnitude of the molecule determines the character of a compound body.
6. We must expect the discovery of many yet *unknown* elements, for example, elements analogous to aluminium and silicon, whose atomic weight should be between 65 and 71.

7. The atomic weight of an element may sometimes be amended by a knowledge of those of contiguous elements. Thus, the atomic weight of tellurium
   must lie between 123 and 126, and cannot be 128.

8. Certain characteristic properties of the elements can be foretold from their
   atomic weights. (All italics original.)

The manner and clarity with which Mendeleev makes these points are rather
striking in that he makes quite explicit what many of the codiscoverers only hinted
at. It also shows us clearly the depth of Mendeleev's chemical knowledge, a theme
that will be further explored in chapter 5, where the manner in which he placed
particular elements into his system is considered.

In the same year of 1869, Mendeleev also published a lesser known system in
which the separation into main and subgroups does not feature in any way whatsoever (table 4.1). For example, the elements lithium, sodium, potassium, copper,
rubidium, silver, cesium, and thallium are all just grouped together as the first horizontal row of the table. This publication, his second major one on the periodic
system, appeared as a report of a meeting of the Russian chemists and is dated
August 23, 1869.

In his third article, published in 1870, Mendeleev was already considering
the possibility that his system had been completed, at least in principle.[16] Among
the features displayed in this article are the relocation of uranium from the boron
group to the chromium group and a corresponding change in its atomic weight
from 116 to 240.[17] In addition, the atomic weight of indium is changed from 75
to 113, which allows Mendeleev to locate the element in the boron group rather
than merely leaving it ungrouped at the very top of his table as he did in 1869.[18]
Other changes included cerium being given a new atomic weight and its being
moved. Thallium is also given a new atomic weight. With the exception of that for
uranium, all of these changes are essentially correct from a modern perspective.[19]

TABLE 4.1
Mendeleev's Spiral Table of 1869

Li	Na	K	Cu	Rb	Ag	Cs	—	Tl
7	23	39	63.4	85.4	108	133	—	204
Be	Mg	Ca	Zn	Sr	Cd	Ba	—	Pb
B	Al	—	—	—	Ur	—	—	Bi?
C	Si	Ti	—	Zr	Sn	—	—	—
N	P	V	As	Nb	Sb	—	Ta	—
O	Si	—	Se	—	Te	—	W	—
F	Cl	—	Br	—	J	—	—	—
19	35.5	58	80	100	127	160	190	220

Redrawn from D.I. Mendeleev, *Zhurnal Russkeo Fiziko-Khimicheskoe Obshchestvo*, 1, 60–77, 1869. The
table is located in a footnote, which begins on p. 69 and ends on p.70.

MENDELÉEFF'S TABLE I.—1871.

Series.	GROUP I. R_2O.	GROUP II. RO.	GROUP III. R_2O_3.	GROUP IV. RH_4. RO_4.	GROUP V. RH_3. R_2O_5.	GROUP VI. RH_2. RO_3.	GROUP VII. RH. R_2O_7.	GROUP VIII. RO_4.
1	H=1							
2	Li=7	Be=9.4	B=11	C=12	N=14	O=16	F=19	
3	Na=23	Mg=24	Al=27.3	Si=28	P=31	S=32	Cl=35.5	
4	K=39	Ca=40	—=44	Ti=48	V=51	Cr=52	Mn=55	Fe=56, Co=59, Ni=59, Cu=63
5	(Cu=63)	Zn=65	—=68	—=72	As=75	Se=78	Br=80	
6	Rb=85	Sr=87	?Y=88	Zr=90	Nb=94	Mo=96	—=100	Ru=194, Rh=104, Pd=106, Ag=108
7	(Ag=108)	Cd=112	In=113	Sn=118	Sb=122	Te=125	I=127	
8	Cs=133	Ba=137	?Di=138	?Ce=140	....	....	....	....
9	....	....	....	....	....	....	....	....
10	....	....	?Er=178	?La=180	Ta=182	W=184	....	Os=195, Ir=197
11	(Au=199)	Hg=200	Tl=204	Pb=207	Bi=208	....	....	Pt=198, Au=199
12	....	....	....	Th=231		U=240	....	....

FIGURE 4.4 Mendeleev's table of 1871. Estestvennaya sistema elementov i primenie ee k ukaza-niyu svoistv neotkrytykh elementov, *Zhurnal Russkeo Fiziko-Khimicheskoe Obshchestvo*, 3, 25–56, 1871.

TABLE 4.2

Mendeleev's Long-Form Periodic Table of 1879

even elements						
I	II	III	IV	V	VI	VII
H						
Li	Be	B	C	N	O	F
Na						

even elements							VIII	odd elements						
I	II	III	IV	V	VI	VII	VIII	I	II	III	IV	V	VI	VII
—	—	—	—	—	—	—		—	Mg	Al	Si	P	S	Cl
K	Ca	—	Ti	V	Cr	Mn	Fe Co Ni	Cu	Zn	Ga	—	—	—	—
Rb	Sr	Yt	Zr	Nb	Mo	—	Ru Rh Pd	Ag	Cd	In	Sn	Sb	Te	J
Cs	Ba	La	Ce	—	—	—	—	—	—	—	—	—	—	—
—	—	Er	Di?	Ta	W	—	Os Ir Pt	Au	Hg	Tl	Pb	Bi	—	—
—	—	—	Th	—	U	—	—	—	—	—	—	—	—	—

Redrawn from D.I. Mendeleev, *Chemical News*, 40, 231–232, 231.

In 1871 Mendeleev published an extensive 96-page article in German containing tables in which he grouped elements vertically (figure 4.4) as well as horizontally. It was in this article that Mendeleev spelled out his detailed predictions that, when later confirmed, were to make him famous.

In all, Mendeleev published approximately 30 periodic tables and designed a further 30 tables which remained in manuscript form. These included horizontal tables, vertical tables, helical tables, and even long-form tables. The latter are popularly thought to have originated following the introduction of quantum mechanics into chemistry and yet an example of such a table, by Mendeleev from 1879, is shown in table 4.2.

THE NATURE OF THE ELEMENTS

This brief summary of the progression of Mendeleev's tables brings us to what I believe is the core philosophical idea at the heart of the periodic system. It is an idea so philosophically rich that it has hardly begun to be explored by modern scholars. It may perhaps be the key to many previously unanswered questions regarding the periodic system, such as why it was Mendeleev, above all others, who was prepared to venture forth to make bold predictions while others tended to be "intimidated" by the prevailing empirical data on the elements.

In the course of developing his system, Mendeleev acknowledged the question of how the elements manage to survive intact in any compound in which they might find themselves. One may consider the common example of sodium chloride, and the fact that the gray and poisonous metal sodium, and the green poisonous gas chlorine, apparently are nowhere to be found after their chemical combination to form the white crystalline compound sodium chloride.

In order to answer this question, Mendeleev appealed to a long-standing notion in chemical philosophy dating back to Aristotle. For Aristotle, the elements themselves were to be regarded as abstract even though they gave rise to the all the physical variety that is observed. The four elements (fire, earth, water, air) were considered as property bearers, responsible for the tangible features of substances although they were themselves unobservable.[20] The elements were immaterial qualities impressed on an otherwise undifferentiated primordial matter and were present in all substances. Thus, the proportion of the four elements present within a specific substance governed its properties.

This view was challenged, among others, by Antoine Lavoisier during the course of the chemical revolution in the eighteenth century, giving rise to a "new chemistry," which drew upon the Aristotelian tradition while making important modifications. The new chemistry introduced the concepts of *simple substance* and *material ingredient* of substances. A simple substance was one that could not be decomposed by any known means. The inclusion of the word "known" here is very important, since the scheme proposed that simple substances were to be regarded as such only provisionally, since they might lose this status following future refinements in analytical techniques. A major departure from Aristotle's scheme was that not all substances had to contain every one of these simple substances. There was no longer thought to be one undifferentiated primordial matter but instead a number of elementary constituents, or simple substances, now possessed of observable properties.

As a result of Lavoisier's work, it became a relatively simple experimental question to determine which substances were simple and which were not, and as mentioned in chapter 1, Lavoisier and his contemporaries created a list of the 37 simple substances known at the time. One consequence of Lavoisier's scheme, however, was that abstract elements did not necessarily correspond to particular known simple substances. Since it was possible that what was regarded as a simple substance at a particular stage in history might turn out to be decomposable, one would need to have perfect confidence in one's analytical techniques to be certain of the correspondence between a simple substance and an abstract element. To his credit, Lavoisier only provisionally identified abstract elements with those simple substances that had been isolated. Such caution began to fade toward the end of the nineteenth century, however, to the extent that simple substances began to be regarded as the only form of an element, and the abstract counterpart to each simple substance was largely forgotten.

And yet the abstract/metaphysical aspect of elements was not completely neglected, continuing to serve an explanatory function in nineteenth-century chemistry, though not necessarily as a microscopic explanation.[21] A chemist could be skeptical of atomistic explanations, as Mendeleev and many others were in the nineteenth century, and yet could readily accept a metaphysical explanation for chemical phenomena. In fact, one of the benefits of regarding the elements as having a metaphysical status is that it provides a way out of the apparent paradox, which Mendeleev was attempting to address, concerning the nature of elements when combined together in compounds. Again, with sodium chloride, one can ask in what sense the elements sodium and chlorine continue to exist in common salt. Clearly, the elements themselves, in the modern sense of the word, do not appear to survive or else they would be detectable, and one would have a mixture of sodium and chlorine that could show the properties of both these elements. The response available from the nineteenth-century element scheme is that simple substances do *not* survive in the compound, only abstract elements do.[22]

According to the nineteenth-century scheme, these abstract elements were believed to be permanent and responsible for the observable properties of simple bodies and compounds.[23] However, in a major departure from the Aristotelian view, the abstract elements were also regarded as being "material ingredients" of simple bodies and compounds. This concept of material ingredient thus served to link the metaphysical world of abstract elements and the observable, material realm of simple substances. For example, the stoichiometric relationships observed in chemical changes were explained in terms of amounts of abstract elements present in the reacting substances through the agency of the material ingredient.

There are thus three important concepts regarding elements carried over into the nineteenth century. The abstract element is a property bearer and owes its heritage to the Aristotelian element scheme.[24] In addition to being a property bearer, the abstract element is an indestructible material ingredient of substances, behaving according to Lavoisier's law of the conservation of matter. The third concept is that an abstract element is unobservable, whereas simple substances such as sodium, chlorine, and oxygen can be observed. It should be noted that in contemporary chemistry only the last notion seems to be retained, in that the term "element" is limited to what a nineteenth-century chemist would have called a simple substance.[25]

The culmination of the nineteenth-century element scheme was reached with the discovery of the periodic system and the work of Mendeleev, who begins his book by paying tribute to Lavoisier.[26] More than any other discoverer, Mendeleev was concerned with the philosophical status of the elements. It is an important and rather overlooked aspect of Mendeleev's approach to the periodic system that he distinguished carefully between what he terms "simple substance" and "element."[27] Unlike the periodic law itself, which seemed to have achieved full maturity only when Mendeleev had reached the end of the first volume,[28] the discussion of simple substance and abstract element occurs right at the beginning of the first volume and is revisited on several occasions in the course of the book:[29]

> It is useful in this sense to make a clear distinction between the conception of an element as a *separate* homogeneous substance, and as a *material* but invisible *part* of a compound. Mercury oxide does not contain two simple bodies, a gas and a metal, but two elements, mercury and oxygen, which, when free, are a gas and a metal. Neither mercury as a metal nor oxygen as a gas is contained in mercury oxide; it only contains the substance of the elements, just as steam only contains the substance of ice, but not ice itself, or as corn contains the substance of the seed but not the seed itself.[30]

For Mendeleev, the element was an entity, which was essentially unobservable but formed the inner essence of simple bodies. Whereas a particular "element" was to be regarded as unchanging, its corresponding simple body aspect could take many forms, such as charcoal, diamond, and graphite, in the case of carbon. In this respect, Mendeleev may be thought of as upholding the ancient philosophical tradition regarding the nature of elements as bearers of properties.[31] Mendeleev's genius now lay in recognizing that just as it was the "element" that survived intact in the course of compound formation, so atomic weight was the only quantity that survived in terms of measurable attributes. He therefore took the step of associating these two features together. An element (basic substance) was to be characterized by its atomic weight. In a sense, an abstract element had acquired a single measurable attribute that would remain unchanged in all its chemical combinations. Here is a profound justification for using atomic weight as the basis for the classification of the elements, quite unlike anything produced by other discoverers or precursors of the periodic system. How else is one to make sense of Mendeleev's otherwise rather naive-sounding claim that he had realized the need to order the elements according to atomic weight given that others had done so before him? The point is that he was providing a detailed account of why this was the correct approach to take.[32]

Mendeleev's periodic system was presented at the end of the first of the two volumes of his textbook on inorganic chemistry. His book was first published in Russian but then eventually translated into English, French, and German. The first English edition, a translation of the fifth Russian edition, appeared in 1891, that is, about 20 years after the first Russian edition.[33] Of course, most serious chemists in Europe first heard about Mendeleev's work through published articles rather than his book. Although Mendeleev never fully revised his textbook for its successive editions, the gradual evolution of his thoughts on atomic weight and the ordering of the elements can be traced through the voluminous footnotes that were added to it at various stages.

The Japanese historian Masanori Kaji has conducted a detailed survey of all eight successive editions of Mendeleev's book in the original Russian.[34] By studying the first Russian edition, which was never translated, Kaji argues that Mendeleev began his textbook by using the concept of valency as a means of ordering the elements. This is clearly revealed in the fact that Mendeleev considers the following elements in order: hydrogen, oxygen, nitrogen, and carbon, whose valences are 1, 2, 3, and 4, respectively.

Then Mendeleev turns to the halogens, beginning again with the valence of 1.[35] These are followed by a consideration of the alkali metals, also of valence 1, and then the divalent alkaline earths. As mentioned above, it was while making the transition between the alkali metals and the alkaline earths that Mendeleev appears to have made the crucial discovery that allowed him to produce the periodic system. Essentially, he realized that the key to classifying the elements was not valency but atomic weight. Now, of course, many previous chemists had been aware of this fact either implicitly or explicitly in proposing tables based on triads or differences between atomic weights of the elements. Nevertheless, Mendeleev added an important ingredient in realizing the possibility of comparing chemically dissimilar elements or, as one might say with hindsight, comparing elements placed horizontally in the present form of the periodic table.[36] As Mendeleev states:

> The purpose of my paper would be entirely attained if I succeed in turning the attention of investigators to the very relationships in the size of the atomic weights of nonsimilar elements, which have, as far as I know, been almost entirely neglected until now.[37]

Kaji claims that at least three noted authorities on the periodic system[38] have been mistaken in proposing that Mendeleev deliberately refrained from revising successive editions of his textbook in order to show his readers how his ideas evolved over time. Some of these authors have even proposed that we should ignore what Mendeleev himself says about the genesis of his periodic system and that we should trace the development of his ideas in the textbook itself, in all its permutations. But this suggestion is rather unconvincing since Mendeleev may have simply been too busy to undertake a thorough revision of the textbook, especially given his many and widely scattered interests. Perhaps we should also be less inclined to dismiss Mendeleev's own accounts of how he arrived at the periodic system. Clearly, this topic has not yet been sufficiently researched by Mendeleev scholars.

What still remains unexplained is why Mendeleev did not completely revise the first part of the book to comply with the way the elements were arranged in his newly discovered periodic system. In the third edition, the second of the two volumes was rearranged so that the discussion of the elements would follow the sequence in which they appear in the periodic system. There has been some debate as to whether this should be considered a major reorganization, but clearly the third edition bears some signs of the discovery of the periodic system. Indeed, it would also have been rather surprising if later editions of Mendeleev's book had borne absolutely no benefits from his discovery of the periodic system.

The fifth edition of *The Principles of Chemistry*, which appeared in 1889, is of particular importance to Western scholars since it was the first one to be translated into English, French, and German. It contains some changes from the previous editions, but still, these are not substantial enough to constitute a major revision. Three further editions, the sixth, seventh, and eight, were published, and a few changes were made to these editions. For example, in the seventh edition of 1903 the recently discovered element argon was incorporated into the periodic system but discussed in the course of the chapter on nitrogen and the air, presumably because argon was first isolated in small amounts from samples of nitrogen. Mendeleev also mentions the newly isolated element radium, while denying any possibility of transmutation of elements and while attempting to explain the phenomenon of radioactivity by appealing to the ether. Moreover, Mendeleev, who had struggled with the placement of the rare earth elements for an extended period of time, finally relinquished all attempts to do so to the Czech chemist Bohuslav Brauner, who contributed the chapter on the rare earths in the last edition of Mendeleev's book. In this, the eighth and final edition to be published during Mendeleev's lifetime, all footnotes are finally separated from the main text and placed in the second half of the book.

MAKING PREDICTIONS

Lothar Meyer and others preceded Mendeleev in predicting the existence of unknown elements, but it is beyond dispute is that Mendeleev made far more extensive predictions than any of the codiscoverers of the periodic system. Not only did he successfully predict new elements, but he also corrected the atomic weights of a number of known elements, as well as correctly reversing the positions of the elements tellurium and iodine. Why was it Mendeleev who was able to make such striking predictions and not Lothar Meyer or others? Is it simply that the others lacked the courage to do so, as many historians of science state?[39] I want to suggest that Mendeleev's advantage lay in his philosophical approach to chemistry, for it allowed him to arrive at insights his less philosophically minded contemporaries could not have entertained.

Mendeleev realized that abstract elements were to be regarded as more fundamental than simple substances. The explanation of why "elements" persist in their compounds was to be found in abstract elements and not simple substances, and as a consequence, if the periodic system were to be of fundamental importance, it would primarily have to classify the abstract elements. The predictions Mendeleev made were thus conceived of with the abstract elements in mind. If the available observational data on simple substances pointed in a certain direction, these features could be partly overlooked in the belief that the properties of the more fundamental abstract elements might be different from what had been observed up to that point in the form of a particular "simple substance." Of course, any prediction must eventually be realized by the isolation of a corresponding simple substance, precisely because "elements," in the more subtle sense of the term, are beyond observation. This requirement presented no problem to Mendeleev, however, for he believed that elements possess one significant and measurable attribute, namely, their atomic weight. In other words, his predictions of abstract elements could be identified empirically through their material ingredient in the form of their atomic weights.[40] As noted above, Mendeleev believed that atomic weight was the one property that does not change when an element combines to form compounds, whereas all the other properties of simple substances seem to be radically altered upon chemical combination.

Because he was attempting to classify abstract elements, not simple substances, Mendeleev was not misled by nonessential chemical properties. For example, the elements in the halogen group (fluorine, chlorine, bromine, and iodine) appear to be rather different from each other when one focuses on them as isolable simple substances, since they consist of two gases, a liquid, and a solid, respectively. The similarities among the members of the group are more noticeable when considering the compounds each one forms with sodium, for example, all of which are crystalline white powders. The point is that in these compounds, fluorine, chlorine, bromine, and iodine, are present not as simple substances but in a latent, or essential, form as basic substances.[41]

Thus, his view of the elements allowed Mendeleev to maintain the validity of the periodic law even in instances where observational evidence seemed to point against it.[42] Such boldness may have resulted from a deeply held beliefs that the periodic law applied to the abstract elements as basic substances and that this law was as fundamental and equal in status to Newton's laws of mechanics. Had he been more of a positivist, Mendeleev might easily have lost sight of the importance of the periodic law and might have harbored doubts about some of his predictions.

On one of the few occasions that Mendeleev allowed himself to express his philosophical views, he wrote of the relationship between "matter, force, and spirit." He claimed that contemporary philosophical problems stemmed from a tendency to search for one unifying principle, while he favored three basic components of nature: matter (substance), force (energy), and spirit (soul). Everything was composed of these three components, and no one category could be reduced to any of the others. According to Michael Gordin, Mendeleev's use of "spirit" amounts to the modern notion of essentialism, or that which is irreducibly peculiar to the object in question. Gordin also adds that Mendeleev's position is clearly metaphysical, thus removing him from the "companionship of positivists" and thus consistent with the position adopted by the present author.[43]

MENDELEEV AS REDUCTIONIST?

Whereas Mendeleev was clearly ahead of his competitors when it came to the prediction of elements, he does not seem to have fared so well with regard to his views on the reduction of chemistry. Textbooks often wax lyrical about the manner in which it is now believed the periodic system is dependent upon the electronic structure of atoms, whereas Mendeleev was concerned almost exclusively with chemical properties. Sometimes the fact that Mendeleev could construct the periodic system merely from considering chemical properties is marveled at in a rather patronizing fashion. Textbook accounts typically express surprise that he was able to deduce the periodic system from such apparently crude data. But as I have argued here, Mendeleev did not primarily classify the elements according to chemical properties.

More specifically, Mendeleev's denial of the reduction of chemistry has generally been held to be mistaken, especially in view of the subsequent discoveries of radioactivity and the structure of the atom. That historians of chemistry have reached such a conclusion is not at all surprising, especially given some of Mendeleev's own pronouncements on the subject. In his Faraday lecture, delivered at the Royal Institution in London, he said:

> [T]he periodic law . . . has been evolved independently of any conception as to the nature of the elements; it does not in the least originate in the idea of a unique matter; it has no historical connection with that relic of the torments of classical thought.[44]

Here Mendeleev is expressing his opposition to one kind of reductionism, namely, the reduction of all matter to one form of matter, as in Prout's hypothesis.

In other instances, Mendeleev appears to express views on an altogether different form of reductionism. This is the view that elements, or atoms of the elements, in modern terms, can be broken down:

> By many methods founded both on experiment and theory, has it been tried to prove the compound nature of the elements. All labour in this direction has as yet been in vain, and the assurance that elementary matter is not so homogeneous (single) as the mind would desire in its first transport of rapid generalization is strengthened from year to year.[45]

Not only did Mendeleev deny that all elements could be reduced to one form of matter, namely, hydrogen, as in Prout's view, but he also denied that the various elements would be found to be composed of more universal building blocks. But modern physics has revealed that the atoms of the elements do indeed have a "compound" nature, since they are composed of protons, neutrons, and electrons. Moreover, the nucleus of the atom, which in simple terms contains just protons and neutrons, has been found to give rise to a staggering 300 or so subnuclear particles. Needless to say, Mendeleev could not have known of these developments.

In emphasizing atomic weight as the key criterion for ordering the elements, Mendeleev also relegates chemical properties to a certain extent. Depending on how much importance one is prepared to place on this feature, Mendeleev might be viewed as a direct precursor to the modern reductionist tendency in chemistry. This is the tendency that reached greater heights in the 1920s and 1930s via the implementation of quantum mechanics, which continues to this day. Mendeleev's emphasis on atomic weight, above all else, might thus be regarded as a classic example of reductionism that places him in the vanguard of the twentieth-century approach to science rather than at the tail end of the classical chemical tradition, where some authors believe that he belongs.

There are many kinds of reductionism. Mendeleev may not have believed in the unity of all matter, but he was an influential proponent of the reduction of chemistry to physics in another sense, that is, in attaching great importance to physical data concerning the elements and especially their atomic weights. Indeed, it was the essence of his achievement that he elevated the ordering of the elements by atomic weight to the status of a law, protecting his emerging periodic system from the uncertainties of the chemical knowledge of his time. At the same time, Mendeleev's understanding of the individual chemical natures of the elements, and their compounds, was profound. This understanding gave him an intuitive sense of how the elements should be grouped. In fact, it can be argued that, although his views on the compound nature of elements and their atoms would turn out to be incorrect, Mendeleev would not have needed to change his position on the basis of current knowledge. As Fritz Paneth suggests,

> Yet I believe that something very essential in his [Mendeleev's] fundamental philosophical tenets would have remained untouched by the progress in physics and could be successfully defended even today; and it is just these "philosophical principles of our science" which he regarded as the main substance of his textbook.[46]

This resolution can be appreciated by realizing that Mendeleev adopted an intermediate position between realism and reduction to physics. Even though physics has revealed that atoms of the elements can be decomposed, it is still the case that chemists can continue to ignore this deeper structure for many chemical purposes.[47] This is the essence of Mendeleev's intermediate position, whereby it is more useful to regard the elements as having distinct identities and yet as also being decomposable into the same fundamental particles such as protons and electrons, in modern terms. It is the view that every science can decide for itself the level at which it should operate and that the deepest foundations are by no means always the best for every purpose.

As the French philosopher Gaston Bachelard, who began his career as a physical chemist, has written,

> La pensée du chimiste nous parait osciller entre le pluralisme d'une part et la reduction du pluralisme d'autre part. (The chemist's thinking seems to oscillate between pluralism on one hand, and the reduction of pluralism on the other hand.)[48]

Mendeleev, the creator of the periodic system of the elements, drew the philosophical distinction between basic substances (abstract elements) and simple substances. He cannot, therefore, be regarded as a naive realist. However, having arrived at the periodic classification by giving emphasis to abstract elements, he resisted the prevalent reductionist tendency of supposing the existence of a primary matter. He considered the elements as distinct individuals and adopted an intermediate position between realism and reduction.[49] This may be Mendeleev's true legacy. Perhaps it can provide the foundation of a genuine "philosophy of chemistry," which is as relevant today as ever, though it has been largely neglected.[50]

PREDICTION AND ACCOMMODATION
The Acceptance of Mendeleev's Periodic System

Although periodic systems were produced independently by six codiscoverers in the space of a decade, Dimitri Mendeleev's system is the one that has had the greatest impact by far. Not only was Mendeleev's system more complete than the others, but he also worked much harder and longer for its acceptance. He also went much further than the other codiscoverers in publicly demonstrating the validity of his system by using it to predict the existence of a number of hitherto unknown elements.

According to the popular story, it was Mendeleev's many successful predictions that were directly responsible for the widespread acceptance of the periodic system, while his competitors either failed to make predictions or did so in a rather feeble manner.[1] Several of his predictions were indeed widely celebrated, especially those of the elements germanium, gallium, and scandium, and it has been argued by many historians that it was such spectacular feats that assured the acceptance of Mendeleev's periodic system by the scientific community.

The notion that scientific theories are accepted primarily if they make successful predictions seems to be rather well ingrained into scientific culture, and the history of the periodic table has been one of the episodes through which this notion has been propagated. However, philosophers, and some scientists, have long debated the extent to which predictions influence the acceptance of scientific theories, and it is by no means a foregone conclusion that successful predictions are more telling than other factors.

In looking closely at the bulk of Mendeleev's predictions in this chapter, it becomes clear that, at best, only half of them proved to be correct. This raises a number of questions. First of all, why is it that history has been so kind to Mendeleev as a maker of predictions? As historian of chemistry William Brock has pointed out, "Not all of Mendeleev's predictions had such a happy outcome; like astrologers' failures, they are commonly forgotten."[2] To put the question another way, why is it that Mendeleev's successful predictions served to bolster the validity of his system while his unsuccessful ones failed to undermine it?

If we accept that it was his predictions that carried the most weight in the acceptance of Mendeleev's periodic system, then we are at a loss to answer this question. But perhaps, as some have argued, it is by no means established that prediction is the single most important factor in demonstrating the validity of a new scientific idea.[3] In fact, rather than proving the value of prediction, the development and acceptance of the periodic table may provide us with a powerful illustration of the importance of accommodation, that is, the ability of a new scientific theory to explain already known facts.

From the time he first published his mature periodic system, in 1869, Mendeleev began to predict the existence of specific unknown elements and also to correct the values of atomic weights of already known elements. Both of these forms of prediction were essential to the refinement of his system and are examined in the course of this chapter. Although the prediction of new elements and the correction of atomic weights of existing elements both represent forms of predictions, they are of a somewhat different character, an aspect that will be explored. The historian Stephen Brush has coined the apt phrase "contrapredictions" to describe the correction of already known elements.[4] He, too, believes that they represent a different category from the prediction of previously unknown elements.

The questions examined in this chapter are (1) whether the prediction of new elements by itself was such a decisive factor in the acceptance of Mendeleev's system, as the popular accounts would have it, and (2) whether successful predictions in general (new elements and contrapredictions) had significantly more impact than did successful accommodations (the fitting of elements into the periodic system).[5]

Mendeleev's extraordinary proficiency as a chemist, combined with his unwavering belief in atomic weight as the supreme ordering principle among the elements, guided him in developing his system. His genius lay in his ability to sift intuitively through the mass of correct and incorrect knowledge of the elements that had accumulated, to produce a system, an *idea*, that was both elegant and durable enough to withstand the chemical and physical discoveries that would follow its introduction.

MENDELEEV'S APPROACH

Mendeleev can be distinguished from his competitors by a devotion to, and love for, the individuality of the elements that went hand in hand with an intimate knowledge of their chemical characteristics. Whereas Julius Lothar Meyer, for example, seemed more concerned with physical properties in his own quest to arrive at a periodic classification, Mendeleev's approach can be described as a "natural history" of the elements.[6] The depth of knowledge of the elements that Mendeleev possessed was something most of today's chemists could not match.[7]

Contrary to many myths and legends, which would have us believe that Mendeleev arrived at the periodic system by juggling with playing cards and by tinkering with values of atomic weights, this was only a small part of the story.[8] The real work consisted in being very familiar with the chemical and physical properties of the building blocks from which the periodic system was to be fashioned. Mendeleev was expert in such matters and knew what kinds of salts all the elements were capable of forming and which reagents could be used to obtained precipitates from their salts. These and countless other details were synthesized and carefully weighed as evidence when he was deciding where to place any particular element.

It is important to understand Mendeleev's modus operandi regarding the placement of elements in the periodic system if we are to appreciate the motivation for many of his corrections of atomic weights and his predictions of unknown elements. Mendeleev considered a number of criteria in addition to atomic weight ordering, such as family resemblance among elements and the concept of the single occupancy of elements in any space in the periodic table. However, all these criteria could be, and often were, overridden as individual cases presented themselves to him.

Family resemblance among the elements was very important to Mendeleev. He looked for chemical similarities as revealed by reactions with other elements, the nature of salts, precipitation reactions, and the acid–base chemistry of the elements. In contrast to the approach of Lothar Meyer, Mendeleev believed chemical properties should take precedence over physical criteria, with the important exception of atomic weight, of course. Lothar Meyer had established his own periodic system by concentrating predominantly on physical resemblance in properties, such as atomic volumes, densities, and fusibilities.[9] For Mendeleev, it was so important that chemically similar elements be grouped together that he was willing to violate the concept of single occupancy, according to which each place in the periodic table may contain only a single element. This was the case in what he labeled group VIII[10] of his table, where sets of three elements, for example, iron, cobalt, and nickel, occupied what should have been several single spaces.[11]

The strictest criterion Mendeleev employed was that of the ordering of elements according to increasing atomic weight. As noted in chapter 4, he had stronger philosophical reasons than the other discoverers of the periodic system for insisting on the fundamental role of atomic weight, so much so that he was willing to try to bend nature to fit his grand philosophical scheme. He would occasionally seem to violate even this principle, however, in cases where the chemical character of an element seemed to demand it. An example is his placement of tellurium before iodine, as the atomic weight of tellurium has the higher value of the two elements. But while making this reversal, Mendeleev did not just disregard the issue of atomic weight, but rather insisted that the atomic weight of at least one of these elements had to have been determined incorrectly, and that future experiments would eventually reveal an atomic weight ordering in conformity with his placement of tellurium before iodine. Thus, Mendeleev's main guiding principle was that any apparent misplacement of an element in his original system, or those of others, was primarily the result of an incorrect atomic weight having been assigned to that element.

In some cases, Mendeleev would correct the atomic weights of a misplaced element, but there were also cases where he considered it sufficient to move an element in order to reflect more faithfully certain family resemblances, without changing the atomic weight in question. This is what he did with mercury, which he came to regard as an analogue of zinc and cadmium rather than of copper and silver, as he had done in his earliest tables.[12]

There are many elements over which Mendeleev deliberated for considerable periods of time and published several accounts. These include indium, erbium, and lanthanum, all of which involved subtle arguments having to do with atomic weight corrections, some of which are examined below.

CORRECTING ATOMIC WEIGHTS

Correcting the atomic weight of an element sometimes involved changing the multiple employed for obtaining its atomic weight from its equivalent weight. The repositioning of elements by this method of atomic weight change would prove to be particularly successful for Mendeleev. In adopting an alternative multiple of the equivalent weight, what Mendeleev was doing was adopting an alternative valence for certain elements, in view of the relationship

$$\text{atomic weight} = \text{valence} \times \text{equivalent weight}$$

In some cases, the valence of an element could be determined only indirectly through group resemblance. This approach was used by Mendeleev in the cases of beryllium, uranium, indium, and thorium, among others. Any suggested changes in group resemblance had to be carefully considered in order to determine whether the resulting change in valence, and hence atomic weight, was really warranted. For example, in the case of uranium, the element did not form any compounds with hydrogen, thus removing any possibility of arriving at its valence in the most direct manner.[13] With other elements uranium shows valences of 2, 3, 4, 5, 6, and even 8, variable valence being a common characteristic of transition metals. Mendeleev had to rely on uranium's other forms of chemical behavior to determine its group and to give it a primary valence of 4.

In other cases, Mendeleev called for a small adjustment in the equivalent weight, which in turn led to corresponding small changes in the value of the atomic weight on multiplication by the appropriate valence. Examples of this kind included titanium, tellurium, iodine, platinum, gold, cobalt, nickel, and potassium. The case of titanium is rather interesting because the element already had a secure place in the periodic table according to atomic weight ordering as well as in terms of family resemblances with other elements. Nevertheless, Mendeleev chose to alter its atomic weight from 50 to 48 in order to create greater regularity among differences between the values of consecutive elements ordered by atomic weight. Whether such regularity can be regarded as another criterion of Mendeleev's is open to question, given that he appears to have used it only in this single case. But the fact remains that Mendeleev was correct since the modern value for titanium is indeed closer to 48 than it is to 50. Mendeleev's uncanny sense for correcting atomic weights, which often seemed to defy logical reconstruction, served him well in this and other cases.[14]

BERYLLIUM

The placement of the metal beryllium provided one of the most severe tests for Mendeleev's system. Its case proved to be historically significant because it involved a controversy that lasted a considerable period of time, ending with the complete vindication of Mendeleev's position. The question was whether the element should be assigned a valence of 2 or 3, which would affect its atomic weight and thus would in turn govern the position it took in the periodic table.

Stanislao Cannizzaro's method for determining atomic weights was not easy to apply to the metallic elements, as it required volatile compounds. Instead, other methods continued to be used for metals. One important way of obtaining atomic weights was through the 1819 law of Pierre-Louis Dulong and Alexis-Thérèse Petit. As discussed in chapter 2, these authors had found an approximate relationship between the specific heat and atomic weight of a solid element to be[15]

$$\text{atomic weight} \times \text{specific heat} = \text{a constant} = 5.96$$

The measured specific heat of 0.4079 for beryllium indicated an atomic weight of 14.6, which would place the element in the same group as the trivalent aluminum.[16]

In addition to atomic weight, there were other reasons to place beryllium with aluminum. Clues to beryllium's valence could be obtained by combining it with oxygen to create an oxide. Metal oxides or hydroxides dissolve in water to form bases, while nonmetal oxides or hydroxides dissolve in water to form acids. Moreover, the chemical characteristics of the oxides generally provide an approximate indication of the valence of the metal concerned according to certain rules. These are summarized below for any metal in general, denoted as M:

Low-valence oxides	MO, MO_2	Strongly basic
Intermediate-valence oxides	M_2O_3	Weakly basic
High-valence oxides	M_2O_5, MO_3	Acidic

Beryllium oxide is weakly basic, with a metallic structure unlike that of magnesium, and beryllium chloride is volatile just like aluminum chloride. Taking these facts together, the association of beryllium with aluminum appears to be compelling.

In spite of all this evidence, Mendeleev supported the view that beryllium is divalent using arguments that were purely chemical, as well as arguments based on the periodic system. He pointed out that beryllium sulfate presents a greater similarity to magnesium sulfate than to aluminum sulfate and that, whereas the elements analogous to aluminum form alums, beryllium fails to do so. He also argued that if the atomic weight of beryllium were about 14, it would not find a place in the periodic system. Mendeleev noted that such an atomic weight would place beryllium near nitrogen, toward the right side of the table, where it should show distinctly acidic properties and have higher oxides of the type Be_2O_5 and BeO_3, which is not the case. Instead, Mendeleev argued that the atomic weight of beryllium might be approximately 9, which would place it between lithium (7) and boron (11) in the periodic table, and thus put it in group II.

In 1885, the issue was finally settled conclusively in favor of Mendeleev by measurements of the specific heat of beryllium at elevated temperatures. The specific heat of any element increases with temperature, and as a result, the constant value that appears in Dulong and Petit's law is achieved only if the measurements are carried out at high temperatures. This became appreciated soon after the discovery of Dulong and Petit's law and allowed more accurate measurements of atomic weight to be made. Further experiments with beryllium pointed to an atomic weight of 9.0, in reasonable agreement with Dulong and Petit's law, and supported the divalence of the element as Mendeleev had argued.[17]

URANIUM

One of Mendeleev's boldest atomic weight changes was in the case of uranium, where the atomic weight was changed by a whole multiple. The element uranium was first isolated in 1841 by Eugène Peligot in France. In his famous first table of 1869, Mendeleev placed the element, with its assumed weight of 116, between cadmium at 112 and tin at 118, thus making it a chemical analogue of boron and aluminum in group III of the table.

Mendeleev avoided using Cannizzaro's atomic weight of 120 for uranium, for such a value would not have allowed him to place it in the periodic table. If uranium had an atomic weight of 120, it would need to be placed between tin (118) and antimony (122). These two elements show valences of 4 and 3, respectively, and so the inclusion of uranium between them would have violated the gradual decrease in valence on moving across the elements in group IV through group VII.[18] In addition, the placement of both tin and antimony appeared quite secure and in little doubt. Tin was in the same group as silicon and lead, both of which show valences of 4, and antimony was in the same group as phosphorus, arsenic, and bismuth, all of which show valences of 3.

In an early manuscript table, Mendeleev designated uranium as "U 120" and listed it outside the table at the foot of the page. Later he crossed this out and replaced it with "U 116 ?" placed in the main body of the table between cadmium and tin. This place should have been filled by the element indium, but Mendeleev also initially misplaced this element because he wrongly assumed that its atomic weight was 75.6.

In the spring of 1869, Mendeleev personally undertook the experimental study of the atomic volume of uranium with the object of resolving the uranium problem. He decided that the element did not in fact fit between cadmium and tin, and considered that Cannizzaro's value of 120 might indeed be correct. Now he was back where he had started, with no place for uranium in the table, so he suggested again that perhaps there had been an error in the determination of its atomic weight. This time he proposed that the value should be doubled because the high density of uranium (18.4) was typical of heavy-atomic-weight elements such as platinum (197), osmium (199), and iridium (198). He then set his assistant Bohuslav Brauner the task of measuring the specific heat of uranium, but since the results were somewhat inconclusive, Mendeleev announced the atomic weight modification without the support of experimental evidence.

In conceptual terms, the doubling of Cannizzaro's value was not as great a leap as it might appear. Since atomic weight is a product of valence and equivalent weight, all that was required in doubling uranium's atomic weight was for its valence to be regarded as double of what it was previously thought to be. Based on its formation of tetravalent compounds, Mendeleev realized that uranium had a predominant valence of 4, as do such elements as chromium. He therefore began grouping uranium with chromium.

In late 1870, Mendeleev actually placed "U = 240" for the first time in the periodic table. Experimental support for the corrected atomic weight of uranium came later, in 1874, from Henry Roscoe in England. It took its place as a higher chemical analogue of chromium, molybdenum, and tungsten, where it remained throughout the rest of Mendeleev's life and indeed until the middle of the twentieth century. Eventually American chemist Glenn Seaborg's discovery of the actinide series prompted a major readjustment of the periodic table, which included the repositioning of uranium.[19]

TELLURIUM AND IODINE

The case of tellurium and iodine is one of only four pair reversals in the periodic system and the best known among them. Many historical accounts make a point of recounting how astute Mendeleev was to reverse the positions of these elements, thus putting chemical properties over and above the ordering according to atomic weight. In doing so, these accounts err in several respects. First of all, Mendeleev was by no means the first chemist to make this particular reversal. As mentioned in chapter 3, William Odling, John Newlands, and Lothar Meyer all published tables in which the positions of tellurium and iodine had been reversed, well before the appearance of Mendeleev's articles. Second, Mendeleev was not, in fact, placing a greater emphasis on chemical properties than on atomic weight ordering in this case. Mendeleev held to his criterion of ordering according to increasing atomic weight and repeatedly stated that this principle would tolerate no exceptions. Mendeleev's thinking regarding tellurium and iodine was rather that the atomic weights for one or both of these elements had been incorrectly determined and that future work would reveal that even on the basis of atomic weight ordering tellurium should be placed before iodine. On this point, as in many instances that tend to go unreported, Mendeleev was wrong.

But let us look into the historical sequence of events regarding tellurium and iodine, since it is only by examining the circumstances closely that the reader can obtain a clear notion of the nature of the work in which Mendeleev and other pioneers of the periodic system were engaged. At the time when Mendeleev proposed his first periodic systems, the atomic weights for tellurium and iodine were thought to be 128 and 127, respectively. Mendeleev's belief that atomic weight was the fundamental ordering principle meant that he had no choice but to question the accuracy of these two values. This was because it was clear that, in terms of chemical similarities, tellurium should be grouped with the elements in group VI and iodine with those in group VII or, in other words, that this pair of elements should be "reversed." Mendeleev continued to question the reliability of these atomic weights until the end of his life. This was one problem he was not able to solve.

Initially, he doubted the atomic weight of tellurium while believing that that of iodine was essentially correct. Mendeleev began to list tellurium as having an atomic weight of 125 in some of his subsequent periodic tables. At one time, he asserted that the commonly reported value of 128 was the result of measurements having been made on a mixture of tellurium and a new element he called eka-tellurium.[20] Prompted by these pronouncements, Bohuslav Brauner began a series of experiments in the early 1880s aimed at the redetermination of the atomic weight of tellurium. By 1883, he was able to report that the value for tellurium should be 125. Mendeleev was duly sent a telegram of congratulations by other participants present at the meeting at which Brauner had made this announcement. In response,

Mendeleev went as far as to list Brauner as one of the four "consolidators" of the periodic law in 1886. In 1889, Brauner obtained new results that seemed to further strengthen the earlier finding that the atomic weight of tellurium is 125.

But in 1895, everything changed as Brauner himself began reporting a new value for tellurium that was greater than that of iodine, thus returning matters to their initial starting point. Mendeleev's response was now to begin to question the accuracy of the accepted atomic weight value for iodine instead of tellurium. This time he requested a redetermination for the atomic weight of iodine and hoped that its value would turn out to be higher. In some of his later periodic tables, Mendeleev listed tellurium and iodine as both having atomic weights of 127. Clearly, the real story is far more complicated than is usually reported, and in the final analysis, it does not appear to further Mendeleev's reputation very much, since the atomic weight of tellurium simply *is* higher than that of iodine. The problem would not be resolved until 1913 and 1914 by Henry Moseley, who showed that the elements should be ordered according to atomic number rather than atomic weight. While tellurium has the higher atomic weight than iodine, it has a lower atomic number, and this is why it should be placed before iodine in agreement with its chemical behavior.

MENDELEEV'S PREDICTIONS

As is well known, Mendeleev successfully predicted the existence of several unknown elements. He arrived at his conclusions mainly through interpolation among atomic weights as well as among other chemical and physical properties. In a few cases he also used extrapolations, but only while warning of the less secure basis of this form of activity, since there is no guarantee that the trend shown among the measured data points will extend into regions where no measurements have been made.

In the case of interpolations, Mendeleev was attempting to fill in prescribed gaps in the table among elements that had already been placed and in many cases were well characterized. In his earliest periodic tables, of 1869, these unknown elements were represented by dashes or a predicted atomic weight value accompanied by a question mark, for it was clear to Mendeleev that there must be elements that would fill these gaps. As described below in some detail, Mendeleev was able to predict many of the characteristics of these unknown elements very successfully. By contrast, extrapolating the existence of unknown elements was a much more tenuous process, and there was no assurance that this was warranted. Mendeleev would later make use of such extrapolations and, not surprisingly, was rather unsuccessful in these cases.

Mendeleev first focused on two gaps in the periodic table, one below aluminum and one below silicon, and proposed to fill them with new elements. Such

gaps were more or less demanded by the vertical grouping of the known elements that surrounded them. The known elements could not be moved round
at will because they had to fit together according to chemical similarities. Gaps
in the horizontal sequence of increasing atomic weights might also suggest the
presence of a missing element, though not as reliably since the increase in atomic
weights is not perfectly uniform even among a complete sequence of known
elements.[21]

The first hint of these famous predictions was published along with his original table of 1869, when Mendeleev declared, "We must expect the discovery of
yet unknown elements, e.g. elements analogous to Al and Si, with atomic weights
65–75".[22] In a talk to the Moscow Congress of the Russian Scientists and Physicians a couple of months later, Mendeleev stated that "those two elements which
are still missing from the system and which show a resemblance to Al and Si, and
have atomic weights of about 70, will have atomic volumes of 10 or 15, i.e. will
have specific gravities of about 6,"[23] but he thought the lighter of these two might
be indium, a known element.

In the autumn of 1870, Mendeleev had also begun to look for an element analogous to boron, and he listed the atomic volumes of these three elements to be[24]

eka-boron	eka-aluminum	eka-silicon
15	11.5	13

Subsequent manuscripts listed the atomic weights for these three elements as 44,
68, and 74, respectively, and a little later as 44, 68, and 72.

In early 1871, Mendeleev published a list of detailed predictions on each element for the first time. It was also in this paper that he now referred to them provisionally as eka-boron (scandium), eka-aluminum (gallium), and eka-silicon (germanium). These were his most celebrated cases, and he was able to predict their
chemical and physical properties to an astonishing degree. It would take 15 years
from the time of these detailed predictions for all three of these new elements to
be isolated and characterized, but in the end Mendeleev would be almost completely vindicated.

Mendeleev could interpolate many of the properties of his predicted elements by considering the properties of the elements on each side of the missing
element and hypothesizing that the properties of the middle element would be
intermediate between its two neighbors. Sometimes he took the average of all
four flanking elements, one on each side and those above and below the predicted element. This interpolation in two directions was the method he used to
calculate the atomic weights of the elements occupying gaps in his table, at least
in principle.

In the various editions of his textbook, and in the publications dealing specifically with his predictions, Mendeleev repeatedly illustrates his method using
the known element selenium as an example. The atomic weight of selenium was

TABLE 5.1
The Predicted and Observed Properties of Eka-aluminum (Gallium)

Eka-aluminum	Gallium
General character: Properties should represent the mean of those of Zn and eka-silicon on the one hand, and those of Al and In on the other. More acidic than eka-boron.	Many properties do indeed represent a transition from those of Zn to those of Ge on the one hand, and from those of Al to those of In on the other. More acidic than scandium.
Atomic weight: ca. 68 (H = 1). Free element: A metal, which should be fairly easily obtained by reduction using C or Na. Its properties should in all respects represent a transition from those of metallic Al to those of metallic In. E.g., it will be more volatile than metallic Al, less so than metallic In. Specific gravity: ca. 6.0 (atomic volume: ca. 11.5) Further predictions in 1875: Metal easily obtained by reduction. Will melt at quite a low temperature. Almost involatile. Not oxidized on contact with air. When heated to red-heat, should decompose water. The pure and fused metal will be only slowly subject to the action of acids and bases.	Measured atomic weight: 69.2 (H = 1). Melting point 29.78 (lower than both In, 157°C, and Al, 660°C; I return to this point below). Boiling point is high, probably above 2,000°C; it probably falls between that of Al and In, but recorded figures are so discordant that this cannot be claimed with certainty. Specific gravity: 5.9 (atomic volume: 11.8) The metal is not oxidized in air at ordinary temperatures. Action on steam unknown. Gallium metal dissolves slowly in acids and alkalis.
Oxides and hydroxides: Formula of oxide, Ea_2O_3. The hydrous oxide will dissolve in KOH solution. Manuscript table of summer 1871 gives specific volume of oxide as "33?" Further Predictions in 1875: Specific Gravity of oxide, ca. 5.5. Basic properties more distinct than Al_2O_3, less than for ZnO; we should therefore expect it to be precipitated by $BaCO_3$ Soluble in strong acids. Should form an amorphous hydrate that is insoluble	The stable oxide is Ga_2O_3, gallic oxide. This is soluble in HCl, H_2SO_4, and aqueous alkali hydroxide and ammonia, but if it has been previously strongly heated, it dissolves in these media only extremely slowly. Barium carbonate precipitates the hydroxide from aqueous solutions of gallium salts. The hydroxide dissolves is aqueous acids and alkalis.

continued

TABLE 5.1

(*Continued*)

Eka-aluminum	*Gallium*
in water but that will dissolve in acids and alkalis.	
Halides: Should give a volatile anhydrous chloride, which is more soluble than aluminum salts. Eka-aluminum will certainly form alums. Its sulfide, Ea_2S_3, will be insoluble in water and will probably be precipitated by ammonium sulfide. It should give volatile organometallic compounds.	Anhydrous gallic chloride fumes in moist air; it is hydrolyzed by water with a hissing sound, though less violently than is aluminum chloride. Boiling point, 200°C.
1875: Eka-aluminum will form neutral and basic salts $Ea_2(OH,X)_6$, but not acidic salts.	$ZnCl_2$ boils at 730°C.
The alum $EaK(SO_4)_2 \cdot 12H_2O$ will be more soluble than the corresponding aluminum salt and have less tendency to crystallize. The sulfide Ea_2S_3 or oxysulfide Ea_2SO_3 should be precipitated by H_2S and will be insoluble in ammonium sulfide.	Gallic salts are even more strongly hydrolyzed in solution than are those of aluminum. Gallium forms alums. Ga_2S_3 is not precipitated by H_2S of $(NH_4)_2S$ in the absence of other metals in the solution.[2] It is, however, also quantitatively carried down with a number of other metallic sulfides (e.g., ZnS) when they are precipitated from alkaline or acetic acid solutions by H_2S. It is similarly precipitated with ZnS and other metallic sulfides by $(NH_4)_2S$.
	Gallium forms a basic sulfate in addition to the neutral sulfate. It does not form acid salts. Gallium gives volatile organometallic compounds.
Points relating to discovery:	
Eka-aluminum is likely to be discovered spectroscopically (on the grounds of its expected volatility), like In and Tl.	Gallium was indeed discovered spectroscopically.

Table reproduced from J.R. Smith, *Persistence and Periodicity*, unpublished Ph.D. thesis, University of London, 1975, 357–359.

known at the time and so could be used to test the reliability of his method. Given the position of selenium and the atomic weights of its four flanking elements:

	S (32)	
As (75)	Se ?	Br (80)
	Te (127.5)	

the flanking atomic weights can be averaged to yield approximately the correct value for the atomic weight of selenium:

$$(32 + 75 + 80 + 127.5)/4 = 79$$

However, Mendeleev did not always operate according to this clear procedure, even in the case of some of his most famous predictions. For example, if his method is applied to predicting the atomic weights, atomic volumes, densities, and other properties of gallium, germanium, and scandium, it produces values that differ significantly from those Mendeleev actually published. Employing Mendeleev's stated method of taking an average of the atomic weights of four flanking elements around gallium, using the atomic weights available at the time, gives a prediction of 70.9. In fact Mendeleev modified this value to "about 69" by means of a more complicated averaging method that he explained only briefly in a single German publication.[25] The accepted value of the atomic weight of gallium at the time of its discovery was 69.35.

Table 5.1 contains Mendeleev's predicted properties of eka-aluminum, subsequently named gallium, as well as the observed properties of the element.[26]

THE DISCOVERY OF GALLIUM

Eka-aluminum, or gallium as it was subsequently called, was discovered by French chemist Emile Lecoq De Boisbaudran in 1875. De Boisbaudran had been studying the spectra of the elements for a period of 15 years prior to making the discovery. He was aware of the fact that elements in the same family, or group, show the same general spectral features. De Boisbaudran did *not* discover gallium as a result of testing Mendeleev's prediction, however. Instead, he operated quite independently by empirical means, in ignorance of Mendeleev's prediction, and proceeded to characterize the new element spectroscopically. After working with 52 kg of zinc blende for a period of about 18 months, De Boisbaudran was able to observe a few spectral lines that had never been observed before, although he was unable to isolate the element. But following a further three months of work, using an additional several hundred kilograms of the same ore, De Boisbaudran isolated about a gram of the new element and reported his findings in the *Comptes Rendus de L'Académie des Sciences.*[27]

On reading a Russian translation of this paper, Mendeleev sent a note to the journal claiming that this was the element he had predicted and provisionally named eka-aluminum. De Boisbaudran at first reacted suspiciously to this claim, apparently believing that Mendeleev was asserting priority over the discovery of the element. He initially maintained that his own element had significantly different properties from those of the element predicted by Mendeleev, although he later changed his mind on this score.[28] De Boisbaudran did, however, continue to insist that his discovery of gallium involved empirical techniques quite separate from anything related to Mendeleev's work and that prior knowledge of it would, if anything, have hindered his discovery of the new element. He named the new element gallium after the Latin for France.[29] As it turned out, the remaining two of Mendeleev's three famous predictions would also result in new elements named from other European countries, namely eka-boron, which became scandium, and eka-silicon, which was named germanium.

In a note to the French journal,[30] Mendeleev repeated some of his earlier predictions and made several new ones. Interestingly, one of these newer predictions was rather dubious, and it is surprising that Mendeleev should have claimed it as a prediction. He claimed to predict that eka-aluminum oxide would be precipitated from aqueous solutions of eka-aluminum salts by barium carbonate. In fact, Mendeleev already knew that this was the case for the simple reason that it had already been reported by De Boisbaudran and, worse yet, Mendeleev himself had acknowledged this observation of the precipitation by $BaCO_3$ in a published note!

On a quite separate issue, Mendeleev had predicted in 1871 that eka-aluminum, or gallium, would "in all respects" have properties intermediate between those of the elements above and below it, namely, aluminum and indium. However, the melting point of gallium (30°C) is nowhere close to being intermediate between those of aluminum (660°C) and indium (155°C). In 1879, Mendeleev gave what appears to be an ad hoc rationalization of the anomalously low melting point for gallium. He first emphasized that gallium does indeed have an anomalously low melting point and that it can even melt in the hand. He then claimed that this was not unexpected since it could be rationalized by looking at trends within groups of elements on either side of the group containing gallium.

At this point, Mendeleev gives this fragment table:

Mg	Al	Si	P	S	Cl
Zn	Ga	...	As	Se	Br
Cd	In	Sn	Sb	Te	J

and claims that for the group containing magnesium, zinc, and cadmium, the element with the lowest atomic weight, magnesium, has the highest melting point. On the other hand, Mendeleev states that in the case of the group at the right-hand side of this fragment table, it is the element with the highest atomic weight, namely, iodine (J), that has the highest melting point.

Mendeleev then makes an almost comical claim that elements falling in a group between these two groups should show intermediate behavior in that it should be the middle element of the group that shows the highest melting point. This is supposed to explain why gallium, which lies in the middle column flanked by aluminum and indium, would be "expected" to show the highest melting point of the three. In his words:

> In a transitory group such as Al, Ga, In, *we must* expect an intermediate phe-
> nomenon; the heaviest (In) and the lightest (Al), should be less fusible than the
> middle one, which is as it is in reality. [31]

Not only had such ad hoc arguments never before been given by Mendeleev as a means of predicting trends in properties, but it also runs contrary to the spirit of his method of simple interpolation, which he used so successfully in many other instances. The ad hoc nature of the argument is compounded by the fact that it is by no means clear that the lesser fusibility of indium and aluminum truly represents "an intermediate phenomenon" with respect to the other groups mentioned. Nor is it clear why this somewhat contrived trend should begin at this particular place in the periodic table. In spite of his use of the word "must," there is nothing in the least bit compelling about Mendeleev's argument. Although nobody would consider denying Mendeleev his triumphs because of such indiscretions, there would probably be little harm caused if historical accounts were to mention some of Mendeleev's failings instead of merely concentrating on his spectacular successes.

SCANDIUM

Scandium was discovered in 1879 by the Swedish chemist Lars Frederick Nilson, in a mineral ore called euxenite. It was identified as Mendeleev's predicted eka-boron by a French chemist Pierre Clève. The discoverer, Nilson, promptly named the new element scandium after Scandinavia, where the ore had first been discovered.

Although it was not discovered spectroscopically, contrary to Mendeleev's prediction, the properties of the new element were very close to what Mendeleev had listed for eka-boron (table 5.2).

GERMANIUM

Another of Mendeleev's most successful predictions, eka-silicon, was discovered by Clemens Winkler in 1886 and named germanium. Winkler and other comfirmers of the periodic law, as Mendeleev called them, are shown in figure 5.1. The manner in which germanium was connected with Mendeleev's prediction of eka-silicon

TABLE 5.2
The Predicted and Observed Properties of Eka-boron (Scandium)

Properties predicted for eka-boron (Eb) by Mendeleev	Properties found for Nilson's scandium (Sc)
Atomic weight 44.	Atomic weight 44.
It will form an oxide Eb_2O_3 of specific gravity 3.5; more basic than alumina, less basic than yttria or magnesia; not soluble in alkalis; it is doubtful if it will decompose ammonium chloride.	Scandium oxide Sc_2O_3, has a specific gravity of 3.86; is more basic than alumina, less basic than yttria or magnesia. It is not soluble in alkalis; and does not decompose ammonium chloride.
The salts will be colorless and give gelatinous precipitates with potassium hydroxide and sodium carbonate. The salts will not crystallize well.	Scandium salts are colorless and give gelatinous precipitates with potassium hydroxide and sodium carbonate. The sulfate does not crystallize well.
The carbonate will be insoluble in water and probably be precipitated as a basic salt.	Scandium carbonate is insoluble in water.
The double alkali sulfates will probably not be alums.	The double alkali sulfates are not alums.
The anhydrous chloride, $EbCl_3$, should be less volatile than aluminum chloride, and its aqueous solution should hydrolyze more readily than that of magnesium chloride.	Scandium chloride, $ScCl_3$, begins to sublime at 850°C. Aluminum chloride begins to sublime at more than 100°C. In aqueous solution the salt is hydrolyzed.
Eka-boron will probably be discovered spectroscopically.	Scandium was not discovered spectroscopically.

Table adapted from P. Clève, Sur le Scandium, *Comptes Rendus des Seances de l'Académie des Sciences*, 89, 419–422, 1879, tables from pp. 421–422.

is rather interesting because it shows the complications that were involved in such cases compared with the sanitized historical accounts that one often encounters.

Germanium was not immediately identified with Mendeleev's eka-silicon. Winkler initially believed that he had discovered another of Mendeleev's predictions, eka-stibium, an element that was supposed to be placed between stibium (antimony) and bismuth. Mendeleev responded to this claim by publishing a paper[32] in which he gave a revised account of the properties expected of eka-stibium in order to argue that Winkler had not in fact found this element. Mendeleev believed that the new element was to be identified with yet another of his

FIGURE 5.1 The comfirmers of the periodic law. Clockwise from left: Nilson, De Boisbaudran, Winkler, and Brauner. Photo and permission provided by Gordon Woods.

predictions, eka-cadmium, which he believed would lie between cadmium and mercury. Victor von Richter, from Breslau, Germany, then wrote to Winkler to suggest that the new element might in fact be Mendeleev's eka-silicon. At about the same time, Lothar Meyer agreed with von Richter and further pointed out that this element coincided with his own predictions for a new element. So it was that Winkler went back to work on isolating larger quantities of the element and, on further characterization, was able to announce that it was indeed Mendeleev's

TABLE 5.3
The Predicted and Observed Properties of Eka-silicon (Germanium)

Property	Eka-silicon 1871 prediction	Germanium discovered 1886
Relative atomic mass	72	72.32
Specific gravity	5.5	5.47
Specific heat	0.073	0.076
Atomic volume	13 cm^3	13.22 cm^3
Color	Dark gray	Grayish white
Specific gravity of dioxide	4.7	4.703
Boiling point of tetrachloride	100°C	86°C
Specific gravity of tetrachloride	1.9	1.887
Boiling point of tetraethyl derivative	160°C	160°C

predicted eka-silicon. Table 5.3 summarizes the main predictions as well as the findings on this element.

Although Mendeleev had foreseen a number of properties, as table 5.3 shows, he had been wrong in thinking that the element would be difficult to liquify and difficult to volatilize, whereas Lothar Meyer's predictions on these points had been correct. Clearly, Mendeleev was spectacularly successful in these predictions, but perhaps not quite to the extent that is implied by the more selective tables of comparison that regularly appear in chemistry textbooks and even histories of chemistry.

MENDELEEV'S LESS SUCCESSFUL PREDICTIONS

In his later years, Mendeleev devoted considerable attention to elements occurring before hydrogen in the periodic table. He gave a number of reasons for taking such a possibility seriously: First of all, the discovery of a whole new series of elements, the noble gases, in the closing years of the nineteenth century led him to think that this series could be extended upward to earlier analogues of the first two noble gases, helium and neon. Second, the apparent success of the ether theory in optical physics suggested to him that ether should be identified as a new element, which he chose to call newtonium.[33] Third, ether would have to lack the ability for chemical combination since it was believed to permeate all substances. In addition, the notion of a completely unreactive element had become highly plausible after the discovery of the unreactive noble gases.

Mendeleev predicted the existence of two elements lighter than hydrogen, calling them elements x and y, based on numerical relations between atomic weight ratios in a periodic table, which he devised in 1904 (table 5.4).

In order to predict the atomic weight of the ether (newtonium), or element x, Mendeleev considered the atomic weight ratios of the known noble gas elements:

$$Xe:Kr = 1.56, \quad Kr:Ar = 2.15, \quad Ar:He = 9.5$$

From these figures, he extrapolated the ratio He:Newt = 23.6, thus giving an atomic mass of 0.17 for newtonium.[34]

To estimate the atomic weight for the element that he designated as y, Mendeleev considered the ratios of atomic weights for the first two members of adjacent groups in the periodic table. He noted that the value for this ratio decreased smoothly from left to right:

Ne:He	Na:Li	Mg:Be	Al:B	Si:C	P:N	S:O	Cl:F
4.98	3.28	2.67	2.45	2.37	2.21	2.00	1.86

TABLE 5.4

Fragment of Mendeleev's Periodic
Table of 1904 Showing the Positions of
Predicted Elements x and y

x	
y	H = 1.008
He = 4.0	Li = 7.03
Ne = 19.9	Na = 23.05
Ar = 38	K = 39.1
	Cu = 63.3
Kr = 81.8	Rb = 85.4

D.I. Mendeleev, *An Attempt Towards a Chemical Conception of the Ether*, Longmans, Green &
Co., London, 1904, table on p. 26.

Extrapolating from the atomic weight of newtonium and the additional ratio of Li:
H = 6.97, Mendeleev estimated that the ratio of He:y should be at least 10, from
which he deduced a value of at least 0.4 for element y. Thus, it would seem that
Mendeleev, who had earlier avoided any involvement with numerical relationships
concerning triads, had now also succumbed to a very similar form of numerology.[35] Indeed, he asserted this claim in the strongest possible terms:

> At the present time, when there remains not the slightest doubt that group I,
> which contains hydrogen, is preceded by a zero group containing elements of
> lesser atomic weights than the elements of group I, it seems to me *impossible* to
> deny the existence of elements lighter than hydrogen.[36]

But Mendeleev's elements x and y would never be found.

The discovery of the noble gases at the turn of the twentieth century also suggested to Mendeleev the possible presence of six new elements between hydrogen
and lithium, as he indicated in his periodic table of 1904. In one of these cases,
Mendeleev was more specific; namely, he predicted a possible analogue of the
halogen fluorine. He claimed that the new element would serve to restore symmetry to the table by making the number of halogens five, to coincide with the
five known alkali metals. Once again, we are forced to conclude that Mendeleev
was mistaken about these predictions, since none of the six elements was subsequently discovered.

Mendeleev made a number of other unsuccessful predictions. In two unpublished tables dated 1869, he made two entries indicating elements he thought
would be discovered: ? = 8 and ? = 22. As with some of his other predictions of
atomic weights of new elements, Mendeleev gave no indication of how he arrived
as these predicted values, which were later removed and never appeared again in

TABLE 5.5
List of Mendeleev's Major Predictions, Successful and Otherwise

Element as given by Mendeleev	Predicted atomic weight	Measured atomic weight	Eventual name
Coronium	0.4	Not found	Not found
Ether	0.17	Not found	Not found
Eka-boron	44	44.6	Scandium
Eka-cerium	54	Not found	Not found
Eka-aluminum	68	69.2	Gallium
Eka-silicon	72	72.0	Germanium
Eka-manganese	100	99	Technetium (1925)
Eka-molybdenum	140	Not found	Not found
Eka-niobium	146	Not found	Not found
Eka-cadmium	155	Not found	Not found
Eka-iodine	170	Not found	Not found
Eka-caesium	175	Not found	Not found
Tri-manganese	190	186	Rhenium (1925)
Dvi-tellurium	212	210	Polonium (1898)
Dvi-caesium	220	223	Francium (1937)
Eka-tantalum	235	231	Protactinium (1917)

Compiled by the author.

published form. Mendeleev also predicted the occurrence of elements with atomic weights of 2, 20, and 36, which, again, were never found.

In addition, he predicted lighter analogues of calcium and explicitly ruled out beryllium and magnesium as occupying these places.[37] This proved to be another mistake in that both of beryllium and magnesium are indeed the missing analogues of calcium, which Mendeleev had misplaced elsewhere in his original tables.

One cannot help speculating as to the cause of these unfortunate cases. It appears that Mendeleev was relying exclusively on atomic weight calculations and disregarding the many subtle chemical clues that had guided him so well in his successful cases. As Jan van Spronsen has aptly commented, Mendeleev's approach in these unsuccessful cases "stands as a warning to the investigators when applying the deductive scientific method exclusively."[38]

Unlike in physics, chemical reasoning does not generally proceed unambiguously from general principles. Chemistry is a more inductive science in which large amounts of observational data must be carefully weighed before reaching any conclusion, as Mendeleev had previously done when correcting atomic weights and predicting new properties by interpolation among known elements. The cases under consideration here seem to represent the speculations of an elderly and established scientist with nothing to lose. Here Mendeleev is not being guided by the chemical intuition that had served him so well in the past but is venturing into the less familiar field of attempting to produce new elements by deduction.

It is puzzling that Mendeleev's unsuccessful predictions do not seem to have counted against the acceptance of the periodic system.[39] There seem to have been as many as 10 failed predictions of elements by Mendeleev. In fact, if one considers all of Mendeleev's predictions, it appears that he was successful in only half of them. This fact has not been given much consideration, as it is much more common for scholars to be impressed by his dramatic successes.

Table 5.5 lists all of Mendeleev's firm predictions. It contains only the elements to which he gave provisional names.[40] Thus, it does not include elements such as astatine and actinium, which he predicted successfully but did not name. Neither does it include predictions that were represented just by dashes in Mendeleev's periodic systems. Among some other failures, not included in the table, is an inert gas element between barium and tantalum, which would have been called eka-xenon, although Mendeleev did not refer to it as such.

A success rate of half is clearly not outstanding by any stretch of the imagination. The fact that Mendeleev made as many failed predictions as successful ones seems to belie the notion that what counted most in the acceptance of the periodic system were Mendeleev's successful predictions.

THE ACCEPTANCE OF MENDELEEV'S PERIODIC SYSTEM

As mentioned in chapter 4, Mendeleev's mature periodic system first appeared in print in 1869 in the Russian chemical literature, and a German abstract of the article appeared in the same year. This was followed by a number of German translations of his articles in 1871. The first English announcement of an article by Mendeleev appeared in 1871 in the journal *Chemical News*.[41] French translations began appearing in 1875. Although his textbook *The Principles of Chemistry* did not appear in German until 1890, in English until 1891, or French until 1895, most European chemists would have heard of the new system much sooner through Mendeleev's various journal articles.

Many historians have argued that despite its prompt publication in the major European languages, Mendeleev's system did not attract much attention until the discovery of gallium by De Boisbaudran in 1875. Some point to this delay to suggest that it was Mendeleev's successful predictions that paved the way for the acceptance of his periodic system.[42] While there is no doubt that his predictions of gallium, germanium, and scandium, especially, received much attention, the question is whether these predictions and others greatly outweighed the system's many successful accommodations in bringing about its acceptance. In fact, a careful examination of the events following the system's first appearance in 1869 reveals that they may not have done so.

Mendeleev was awarded the prestigious Davy Medal in 1882, after gallium and scandium had been discovered. The philosophers Patrick Maher and Peter Lipton have recently pointed to this award as proof that it was not until Mendeleev's predicted elements had begun to be discovered that his system received the recognition it deserved. They take this to indicate that prediction weighed much more heavily than accommodation in the acceptance of the periodic system. In fact, Lipton goes so far as to say, "Sixty accommodations [the placement of the known elements] paled next to two predictions."[43]

Maher and Lipton both imply that there was a time lag between Mendeleev's accommodation of the known elements, in his constructing the periodic system, and his prediction of the three unknown elements. The existence of such a time lag is important to their argument. If it had not occurred, and the accommodations and predictions had been made in the same paper, it would be very difficult to ascertain whether the acceptance of Mendeleev's scheme rested primarily on its ability to accommodate or to predict.

In fact, Lipton, in paraphrasing Maher, claims quite specifically that when Mendeleev accommodated the 60 known elements (it should be 62), "the scientific community was only modestly impressed,"[44] thus clearly indicating a supposed time lag between the initial accommodation and later predictions. Maher implies such a time lag between accommodation and prediction by dating the predictions to 1871. Both of these authors are committing a historical fallacy relevant to the central issue, however, for although he did not give them names until 1871, Mendeleev left gaps for eka-boron, eka-aluminum, and eka-silicon, with their predicted atomic weights, when he first announced his periodic system in his famous paper of 1869. And in 1869 and 1870, he predicted their atomic volumes and specific gravities. Thus, the time lag that Maher and Lipton imply did not in fact take place.

The only justification Maher and Lipton might have for concentrating on the 1871 article is that it contained Mendeleev's first set of detailed predictions. Another factor may be that it was not until this paper that Mendeleev gave his predicted elements provisional names. But it is hard to imagine that Maher and Lipton would claim the 1871 predictions as more definitive than those of 1869 from the mere fact that Mendeleev was only then prepared to give the elements names, and provisional ones, at that. The whole question of prediction is fraught with problems. For example, should we consider predictions made in unpublished manuscripts or a talk given to a learned society? Similarly, one might well ask just how detailed the prediction itself should be to count as a true prediction.

In any case, Maher and Lipton were not the first to suggest there was such a time lag between Mendeleev's original announcement of the periodic system and his predictions of unknown elements. Other historians, too, have conveyed the impression that Mendeleev's predictions were decisive in the acceptance of the periodic table, but they regularly fail to cite reactions by chemists at the time that might support this view.[45] This is, of course, the crucial issue, namely, whether the scientific community values predictions above explanations of already known facts, and not what later historians might report. It would seem that these historians are merely reconstructing the course of events while incorporating the popular myth regarding predictions, and that Maher and Lipton have recently revived this view in the philosophical literature. Of course, the fact that Mendeleev's accommodations and predictions were published simultaneously does not rule out the Maher–Lipton position that scientists attach more importance to predictions. But to maintain their claim, these authors would need to cite historical evidence to the effect that scientists did indeed prefer the predictive aspects of the periodic system.

DAVY MEDAL CITATION

Since two eminent philosophers of science have cited the award of the Royal Society's Davy Medal to Mendeleev as evidence for the superiority of predictions in the acceptance of the periodic system, it is necessary to consider the citation of this award in full.

The Davy Medal has been awarded to Dimitri Ivanovich Mendeleeff and Lothar Meyer.

The attention of the chemists had for many years been directed to the relations between the atomic weights of the elements and their respective physical and chemical properties; and a considerable number of remarkable facts had been established by previous workers in this field of inquiry.

The labors of Mendeleeff and Lothar Meyer have generalized and extended our knowledge of those relations, and have laid the foundation of a general system of classification of the elements. They arrange the elements in the empirical order of their atomic weights, beginning with the lightest and proceeding step by step to the heaviest known elementary atom. After hydrogen the first fifteen terms of this series are the following, viz.:

Lithium	7	Sodium	23
Beryllium	9.4	Magnesium	24
Boron	11	Aluminum	27.4
Carbon	12	Silicon	28
Nitrogen	14	Phosphorus	31
Oxygen	16	Sulfur	31
Fluorine	19	Chlorine	35.5
Potassium	39		

No one who is acquainted with the most fundamental properties of these elements can fail to recognize the marvelous regularity with which the differences of property, distinguishing each of the first seven terms in the series from the next term, are reproduced in the next seven terms.

Such periodic re-appearance of analogous properties in the series of elements has been graphically illustrated in a very striking manner with respect to their physical properties, such as the melting points and atomic volumes. In the curve which represents the relations of atomic volumes and atomic weights analogous elements occupy very similar positions, and the same thing holds good in a striking manner with respect to the curve representing the relations of melting-points and atomic weights.

Like every great step in our knowledge of the order of nature, this periodic series not only enables us to see clearly much that we could not see before; it also raises new difficulties, and points to many problems which need investigation. It is certainly a most important extension to the science of chemistry.[46]

The first thing to emerge from an examination of this citation is that the medal is being jointly awarded to Mendeleev and Lothar Meyer. This feature has been conveniently omitted by Maher and Lipton, both of whom favor prediction over accommodation.[47] The very fact that the award is to both of these pioneers of the periodic system already argues strongly against the predictivist thesis since, according to the popular account, Mendeleev is given priority precisely because he made predictions, which were subsequently confirmed, whereas Lothar Meyer failed to make any significant predictions.

Second, the entire citation concerns the accommodation of chemical and physical phenomena of the elements and not the prediction of new elements, as Maher and Lipton's statements would seem to require. The only part of the citation that could remotely be linked with the prediction of gallium and scandium by Mendeleev is the phrase in the final paragraph that alludes to "seeing clearly much that we could not see before." However, this comment is too vague to allow such an interpretation, and even if it is a veiled reference to the prediction of the two new elements, it is clearly not stating that the medal is being awarded primarily as a result of them. So perhaps Maher and Lipton are mistaken in citing the award of the Davy Medal as an indication of the Royal Society's high regard for predictions, since the entire Davy award citation makes no mention whatsoever of the prediction of new elements.

CONTEMPORARY REACTIONS TO THE PERIODIC TABLE

Let us now examine the wider question of how scientists at large in this period of history reacted to the introduction of Mendeleev's periodic system, and whether they regarded predictions connected with the periodic table more favorably than the accommodation of already known elements.

The second successful prediction by Mendeleev concerned the element scandium, and this case offers a good opportunity for obtaining reactions of other scientists, since the identification of the newly discovered element was carried out by a third party, that is, neither the discover of the element nor Mendeleev. This third party was the French chemist Clève, who wrote,

> The great interest of scandium is that its existence had been predicted. Mendeleef in his memoir on the law of periodicity, had foreseen the existence of a metal which he named ekaboron, and whose characters agree fairly well with those of scandium.[48]

Clève is clearly attaching some importance to this prediction, although there is no indication that he regards the overall case for Mendeleev's periodic system to be strengthened by this finding.

In 1879, shortly after the above translation of Clève's article was published in the British journal *Chemical News*, the same journal undertook the serialization of Mendeleev's 1871 paper on the periodic law. The following interesting editorial appears along with a specially written introductory article by Mendeleev:

> Considerable attention having been drawn to M. Mendeleef's memoir 'On the Periodic Law of the Chemical Elements', in consequence of the newly discovered elements gallium and scandium being apparently identical with two predicted elements ekaluminum and ekaboron, it has been thought desirable to reproduce the entire article in CHEMICAL NEWS. . . .[49]

There followed a weekly serialization of Mendeleev's memoir in 17 parts. This may be among the strongest evidence that suggests that Mendeleev's predictions were indeed taken seriously at the time. Nevertheless, the above editorial gives no indication whether the successful predictions did anything to enhance the status of the periodic system.

In 1881, a year after the serialization of Mendeleev's memoir appeared, the famous priority dispute between Mendeleev and Lothar Meyer broke out in the pages of the same journal, *Chemical News*. After giving precise details as to the publication of his early papers in a note to the journal, Mendeleev adds the following more general remark concerning what he believes to be the essence of the priority question:

> That person is rightly regarded as the creator of a particular scientific idea who perceives not merely its philosophical, but its real aspect, and who understands so to illustrate the matter so that everyone can become convinced of its truth. Then alone the idea, like matter, becomes indestructible.[50]

Interestingly, Mendeleev does not specifically mention any of his predictions in arguing for his priority over Lothar Meyer.[51] His note is followed by one from Lothar Meyer, in which he, in turn, defends his own claim to priority with regard to the discovery of the periodic system. This note is followed by a third item by the well-known organic chemist Charles-Adolphe Wurtz, who is not impressed with the periodic system at all, let alone with Mendeleev's predictions. Wurtz grants that Mendeleev's proposition is a "powerful generalization and must in future be taken into account whenever we regard the facts of chemistry from a lofty and comprehensive point of view."[52] Nevertheless he points out that the system contains many imperfections, such as the way it reflects the [then] available knowledge of the rare earths. He discusses the problem with tellurium and iodine, whose atomic weight ordering is inconsistent with their chemical properties. Wurtz alludes to similar problems with cobalt and nickel, whose properties should coincide in view of their almost identical atomic weights.[53] Wurtz also points out the large chemical differences between such elements as vanadium and bromine, whose atomic weights are very closely related, which might therefore be expected to be chemically similar. He adds that the alleged gradations in properties do not in fact progress smoothly or regularly as Mendeleev would have us believe.

Wurtz then turns specifically to consider Mendeleev's predictions.

> In Mendelejeff's table we are chiefly struck with the gaps between two elements, the atomic weights of which show a greater difference that two or three units, thus marking an interruption in the progression of the atomic weights. Between zinc (64.9) and arsenic (74.9) there are two, one of which has been lately filled up by the discovery of gallium. But the considerations by which Lecoq de Boisbaudran was led in the search for gallium have nothing in common with the conception of Mendelejeff. Though gallium has filled up a gap between zinc and arsenic, and though other intervals may be filled up in future, it does not follow that the atomic weights of such new elements will be those assigned to them by this principle of classification. The atomic weight of gallium is sensibly different from that predicted by Mendelejeff. It is also possible that the future may have in reserve for us the discovery of a new element whose atomic weight will closely coincide with that of a known element, as do the atomic weights of nickel and cobalt. Such a discovery would not fill any foreseen gap. If cobalt were unknown it would not be discovered in consequence of Mendelejeff's classification.[54]

The inclusion of this rather severe criticism might be viewed as an attempt by the editor of *Chemical News* to temper his initial enthusiasm for Mendeleev's system, which had led to the 17-part serialization. Why he would otherwise choose to follow the priority dispute with this note is difficult to understand.

The fact that the successful predictions made by Mendeleev by no means gained universal acceptance for his periodic system can also be seen from further criticisms voiced by the likes of the chemist Marcellin Bertholet. In 1885, even after two of Mendeleev's predictions had been highly successful, Bertholet launched a highly critical attack on the periodic systems that had been introduced by Mendeleev and others. Not only was he unimpressed with Mendeleev's predictions, but Bertholet even refused to be seduced by the ability of the periodic system to accommodate what was already known about the elements.[55] Bertholet began by pointing out that the relations between atomic weights, atomic volumes, and physical and chemical properties had been known before the elements were placed into a periodic system. He claimed that, since these relations resulted from atomic weight relations, it was a coincidence that they reemerged when considered in the context of the periodic table. To Bertholet, this was not, therefore, proof of the existence of periodic series. He then turned to predictions and admitted that the periodic system should prove interesting in this respect. Bertholet also accepted that certain elements appeared to be missing but stressed that this was evident just from the gaps in the sequence of atomic weights. He claimed that, in their haste to fill such gaps, the authors of the periodic systems had made some mistakes, such as in the insertion of molybdenum between selenium and tellurium.

Similarly, Bertholet mocks the common grouping of hydrogen and lithium at one end of a group and copper, silver, and gold at the other end of the same group, as carried out by Mendeleev, as being "fanciful." He further accuses the authors of the periodic systems of making it too elastic in admitting elements that differ by no more than two units throughout the table. He suggests that *any* future discoveries could be accommodated if this was the case. Bertholet claims that there is no systematic means of predicting new elements from the periodic system or any means of synthetically forming the elements, thus referring to the hypothetical transmutation of the elements. Finally, Bertholet warns about the dangers of falling back into what he calls a mystical enthusiasm similar to that of the alchemists.

Up to this point, Bertholet's critique seemed to have been very reasonable, but in this final remark he gave Mendeleev an easy way of responding.[56] Mendeleev took issue in particular with Bertholet's reference to a "mystic enthusiasm" and responded by accusing Bertholet of confusing the idea of the law of periodicity with the ideas of William Prout, as well as with those of the alchemists and of Democritus on primary matter. It was in response to these criticisms by Bertholet that Mendeleev also made a much quoted remark, already cited in chapter 4, in which he emphasized that the periodic system owed nothing to the idea of a "unique matter" and had no connection with the "relic of the torments of classical thought. . . . "

A look at the historical record thus reveals that the acceptance of Mendeleev's system was not a simple matter, and certainly was not assured by either his accommodations or his successful predictions. Many of Mendeleev's contemporaries were impressed with the accommodations his system achieved; others, like Bertholet, seemed to not be impressed by either the predictions nor the accommodations. Thus, the question remains regarding the manner in which Mendeleev's periodic system did indeed take hold fairly quickly in the decades following its introduction and how it came to occupy the central position in chemistry it still holds today.

THE POWER OF AN IDEA

Although his chemical knowledge was extensive, Mendeleev was primarily considered to be a systematizer. He produced an *idea*, the periodic system, within which chemical phenomena could be systematized. He used this idea to sort through the mass of chemical data available in his time, and though he was not always correct, he demonstrated an uncanny ability to separate valid facts from irrelevant ones. It was because he could see patterns among the properties of the elements that he was able to predict not only the existence of new elements but also their chemical and physical characteristics. The historian of chemistry Brock has quoted Bonifatii Kedrov, the Russian historian of chemistry, as saying, "[T]he scientific world was astounded to note that Mendeleev, the theorist, had seen the properties of a new element more clearly than the chemist who had discovered it."[57] Meanwhile, the science historian Brush has posed an interesting question in asking whether theorists should be considered less trustworthy than observers. The reason one tends to give more credit to predictions than to accommodations is presumably because we suspect that a theorist might have designed his theory to fit the facts. But is it not equally possible, Brush asks, for observers to be influenced by a theory in their report of experimental facts? If so, then perhaps we should give greater consideration to observations obtained before a theory is announced than to observations produced in response to a theory.

Although Mendeleev was not above occasionally resorting to ad hoc arguments, as witnessed in his discussion of the melting point of gallium, there was nothing ad hoc about his atomic weight corrections, as unlikely as many of them may have appeared to be at the time. It is important to emphasize that there turned out to be independent empirical evidence for the new values assigned to these atomic weights. It was not that chemists simply came to accept the new values because they made elements fit Mendeleev's table better. The corrected value of the atomic weight of beryllium, for example, was confirmed independently of any consideration of its place in any table by Lars-Frederick Nilson and O. Pettersen's discovery of one of its compounds, beryllium chloride. This discovery meant that an evaluation of beryllium's atomic weight could be made using already accepted background knowledge.

In ordering the elements, Mendeleev was accommodating all that was known about them up to his time, their atomic weights, their physical properties, and their chemical character, in addition to being able to make dramatic predictions. Mendeleev did not have to correctly predict every element that would be discovered, and experimentalists were not restricted to looking only for elements that the system implied. The successful incorporation of the rare earths and the noble gases would ultimately do much to prove the validity of the periodic system. What is more, the system would be further strengthened by more general developments to come, such as the discovery of isotopes, atomic number, and later, quantum theory.

THE INERT GASES

The case of the inert, or noble, gases represents an interesting counterexample to the predictivist thesis in the sense that almost nobody, including Mendeleev, had predicted or even suspected the existence of an entire family of new elements.[58] Once they had begun to be discovered, it was immediately understood that the existence of the inert gases might pose a major threat to the periodic system. Indeed, a failure to incorporate them might have led to an abandonment of the periodic system, regardless of the earlier predictive successes achieved. As it turned out, the correct placement of the noble gases did not cause any harm to the periodic system, but did much to enhance it.

The first of the noble gases to be discovered, argon, was particularly difficult to place in the periodic system. Not only had it not been predicted from the periodic system, but there was the further difficulty that physical measurements suggested the gas was monoatomic, and this was regarded with some suspicion since the only other monoatomic gas then known was vaporized mercury.

The atomicity of argon was crucial to determining its atomic weight, which in turn was essential for its accommodation into the periodic system. As Mendeleev had repeatedly stressed, atomic weight was considered to be the one essential criterion on which the periodic law was founded. Further complications regarding the atomic weight of argon arose due to doubts over the purity of the gas. There was considerable debate as to whether it consisted of a mixture of gases, and whether what was being measured was actually an average atomic weight determined by the relative proportion of several components. While the interdependent issues of atomicity and the possibility of a mixture were still being discussed, a deeper and unsuspected complicating factor was operating to confuse the issue. The elements argon and the subsequent element potassium represent one of the very few examples of "pair reversals" in the periodic table.[59]

Argon was discovered in 1894 by Lord Rayleigh and William Ramsay (figure 5.2), who were studying nitrogen. Convinced by spectroscopic evidence that they had a new element on their hands, they set out to determine its properties. Because the gas was completely inert, they were forced to rely on physical measurements to determine its atomic weight. The determination of argon's specific heat was carried out through the measurement of the specific heat capacities of the gas at constant pressure and at constant volume, C_p and C_v, respectively.[60]

From these measurements, Rayleigh and Ramsay could determine the ratio of translational energy to kinetic energy, which would in turn reveal the atomicity of the gas. In general, the total kinetic energy of a molecule is made up of three contributions: translational, rotational, and vibrational energy. In the case of a monoatomic system, there is only translational motion, and so kinetic energy is equal to translational energy. Rudolf Clausius had shown in 1857 that

FIGURE 5.2
William Ramsay. Photo and permission
from Edgar Fahs Smith Collection.

if K is the translational energy of the molecules of a gas and H is the total kinetic energy, then

$$K/H = 3(C_p - C_v)/2C_v$$

If C_p/C_v is found to be 1.66, substitution into the equation shows that $K = H$, or in other words, all the kinetic energy of the molecules occurs in translational form. This means that the molecules are exhibiting no rotational or vibrational energy, which in turn implies the presence of an isolated atom, or monoatomicity.[61] The experimental result obtained by Rayleigh and Ramsay was that C_p/C_v was very nearly 1.66, from which they therefore inferred that argon is monoatomic. The account just given benefits from the knowledge of hindsight, since it is now well established that argon is indeed monoatomic. It took considerably more effort to arrive at this conclusion at the time when the mysterious gas was first discovered.

The public announcement of the argon problem took place on January 31, 1895, at a specially convened meeting of the Royal Society of London, and it was met with considerable debate by the leading chemists and physicists of the day. The meeting began with an exposition of the findings on the new substance given by Ramsay, including the specific heat ratio of nearly 1.66. Ramsay and Rayleigh interpreted this result to mean that the new constituent was either an element or a mixture of elements and was probably monoatomic. They admitted that the results were also consistent with diatomic or polyatomic molecules whose atoms acquire no relative motion, not even that of rotation. They added that the latter possibility seemed improbable, however, in that it would have required such a complex group of atoms to be spherical.

Rayleigh and Ramsay failed to take up a decisive position on the question of the purity of the new substance. They acknowledged that the spectral evidence provided by William Crookes, in a paper read the same evening, suggested a mixture. But they also pointed to the measurements on critical data, which indicated a sharp boiling point and melting point, as well as an observed constant pressure during boiling, all of which pointed to a single pure substance.

Their overall conclusion was that "the balance of evidence seemed to point to simplicity," meaning to a single element, but that this fact together with the probable monoatomicity suggested an atomic weight of 39.9. From this they were forced to conclude that such an element would find no place in the periodic table. On the other hand, Ramsay and Rayleigh proceeded to speculate that a 93.3% to 6.7% mixture of two unknown elements with atomic weights 37 and 82, one lying between chlorine and potassium and the other between bromine and rubidium, would also account for the observed density.

Henry Armstrong, the then president of the Chemical Society, politely congratulated Ramsay and Rayleigh on their researches but went on to make some criticisms. He suggested that the account of the probable nature of the new element was "of a wildly speculative character" and drew attention to the doubts expressed by Ramsay and Rayleigh over the interpretation of the specific heat data. He then proceeded to draw an analogy between nitrogen and argon. He pointed out that while nitrogen, as it occurs in the atmosphere in molecular form, is highly inert, so its constituent atoms are highly reactive. Similarly, he argued that argon, which is evidently even more inert than nitrogen, might consist of even more reactive constituent atoms. Their extreme reactivity would produce very strong interatomic bonding, thus producing a diatomic molecule so locked together that it would display translational motion without any form of relative motion between the constituent atoms.

Further support for Armstrong's view came from the Irish physicist George Fitzgerald, better known for his contributions to the theory of relativity, whose opinion had been communicated earlier in a letter to Rayleigh.[62] Like Armstrong, Fitzgerald was willing to contemplate a diatomic molecule in which the two atoms are so firmly bound together as to produce very little internal motion and added that this view would be in keeping with the chemical inertness of the new gas.

Lord Rayleigh, however, had difficulty imagining such a diatomic molecule, and expressed his reservations by saying,

> That argument is no doubt perfectly sound, but the difficulty remains how you can imagine two molecules joined together, which one figures roughly in the

mind, and I suppose not wholly inaccurately, as somewhat like two spheres put together and touching one another—how it would be possible such an excentrically-shaped atom as that to move about without acquiring a considerable energy of rotation.[63]

William Arthur Rücker, the president of the Physical Society, pointed out that such a diatomic molecule would actually have to be spherical to produce the observed ratio of specific heats of 1.66. He acknowledged that this would represent something of a problem, but he also admitted less concern about the problem of fitting the element into the periodic table, as he did not think Mendeleev's system to be so well established that overturning it would shake the foundations of chemistry.

Finally, the chair of the meeting, Lord Kelvin, added his own comment on the issue of the specific heat ratio of 1.66, also expressing reservations about the possibility of a diatomic molecule. "I do not admit that a spherical atom could fulfill that condition," he said. "A spherical atom would not be absolutely smooth." Kelvin also disputed the notion of a rigidly connected diatomic molecule since he felt that at least some relative vibrational motion would have been detected from such a mechanical system.[64]

About two months later, a report of Mendeleev's views on the accommodation of argon appeared in *Nature*.[65] Here, Mendeleev stated that the supposition that argon is a mixture "lies beyond all probabilities." He also considered it probable that the gas was an element due to its inert nature. He then moved on to a systematic consideration of atomic weights for the element, suggesting the following set of possible molecules:

$$A_1, A_2, A_3, \ldots A_n$$

Taking each of the possible atomicities in turn, he first discussed monoatomicity. Mendeleev was reluctant to accept the evidence for monoatomicity obtained from specific heat measurements on the grounds that there might be a chemical contribution to the kinetic energy of the molecule. He pointed to the difference between the values for C_p/C_v of 1.3 in the case of chlorine and 1.4 in the case of nitrogen, thus emphasizing the variation among diatomics and implying that even a value of 1.6, as observed in argon, might still belong to a diatomic molecule.

As might be expected, Mendeleev was most concerned with the problem of fitting the new element into his periodic system, but he dismissed monoatomicity on the grounds that there was no room in the periodic table for such an element. He reasoned that monoatomicity implied an atomic weight for argon falling between chlorine and potassium and that this would imply "an eighth group in the third series," something that he found inadmissible. This is a surprising error on the part of the master chemist, since there is in fact no fundamental reason why an eighth group should not be introduced. In fact, this is precisely how the problem was solved in due course.

Mendeleev raised similar objections against the notion of a diatomic argon molecule. Continuing with his original list of possible atomicities, he then settled on triatomicity, concluding that argon was nothing but a triatomic form of nitrogen and not a new element after all:

> If we suppose further that the molecule of argon contains three atoms, its atomic weight would be about 14, and in such case we might consider argon as condensed nitrogen N_3. There is much to be said in favour of this last hypothesis. . . .[66]

Among the reasons for favoring this molecule, Mendeleev argued that it would account for the "concurrent existence of nitrogen and argon in nature" and similarly that the inertness of argon might be related to the fact that it is derived from nitrogen. Molecules with higher atomicities, namely, 4 and 5, were ruled out because they would require atomic weights of 10 and 8, respectively, and could not thus be accommodated into the periodic system. As for hexatomicity, Mendeleev considered this plausible and indeed thought it to be the second most likely atomicity after triatomic N_3.

Meanwhile, the debate had spread to the larger scientific community. John Hall Gladstone, one of the pioneers of the relationship between refractive index and molecular structure, gave five reasons for why an element with an atomic weight of 20 would fail to fit into the periodic system. He then gave a further five reasons why he considered an element of atomic weight 40 would be well accommodated into the periodic table. One cannot help concluding that Gladstone was wrong on a total of 10 counts!

Within two years, terrestrial helium had been discovered, and the problem became one of accommodating two elements, namely, argon and helium. A further three years were to pass before the discovery of krypton, neon, and xenon. A whole new family of elements had been discovered without having been predicted, and the accommodation of these new elements into the periodic table was proving to be far from trivial. Indeed, it presented a severe threat to the survival of Mendeleev's system.

Mendeleev visited London in late 1895 and discussed the argon problem with Ramsay. He reported back to the Russian Physico-Chemical Society on his return to Moscow, "The subject has progressed little. There is little for its solution and the matter seems particularly obscure."[67] Two years later, he wrote that since no compounds could be formed from argon and helium, their atomic weights should be regarded as doubtful. For Mendeleev, the study of compounds played an essential role in the incorporation of elements into the periodic system. He was reluctant to accept argon and helium as new elements, as he would not entertain the possibility that an element could be completely inert.[68]

Finally, in the spring of 1900, at a meeting in Berlin, Ramsay suggested to Mendeleev that argon and its analogues should be placed in a new group between the halogens and the alkali metals. They would thus appear at the right-hand edge of the table and would serve to extend the length of each period by one element.[69] In spite of all his previous views on the inert gases, Mendeleev received this suggestion favorably and wrote of his response in 1902: "This was extremely important for him [Ramsay] as an affirmation of the position of the newly discovered elements, and for me as a glorious confirmation of the general applicability of the periodic law."[70] Mendeleev also adds that this step represented the "magnificent survival" of the periodic system in what had been a "critical test." Indeed, the periodic system had come through this test with "flying colors." Could it be that the eventual successful incorporation of the noble gases into the periodic table counted as much in favor of the periodic system's acceptance as the Mendeleev's celebrated predictions? My own belief is that it did, as did the eventual successful incorporation of the rare earth elements.

CONCLUSION

The claim is sometimes made that successful prediction gives more credit to a theory than does the accommodation of known facts. But it is difficult to find clear-cut evidence for this claim in the technical writings of scientists.[71] A successful prediction may yield much favorable publicity for a theory and thereby force other scientists to give it serious consideration. But subsequent evaluations of the theory in the scientific literature usually do not give greater weight to the prediction of novel facts than to the persuasive deductions of known facts.

This may be what happened with Mendeleev's periodic system. It was announced in 1869 to mixed reviews but seems to have received more favorable attention after the discovery of gallium in 1875. Rather than confirming that prediction of new elements was the overwhelming factor in the eventual acceptance of the system, the discovery of gallium, scandium, and germanium may have served simply to bring the system to the attention of the scientific community. From there, it appears that its many strengths began to be appreciated. In addition to the prediction of new elements, Mendeleev successfully predicted the correct atomic weights of many already existing elements, and these successes would have contributed to the acceptance of his periodic system.

But the placement of difficult elements such as beryllium, the accommodation of the newly discovered noble gases, and the ongoing struggle to position the rare earths all contributed to an atmosphere of productive debate that surrounded the periodic system. These factors may well have contributed just as much as the predictions to the eventual acceptance of the system, contrary to the popular myth that assigns the greatest credit to Mendeleev almost exclusively on the basis of his successful predictions. By 1890, Mendeleev's system was a permanent fixture on the landscape of chemistry. Almost all the lacunae of the magnificent edifice had been explored, revealing its profound elegance and propelling the research agenda for chemistry, and even physics, into the next century.

THE NUCLEUS AND THE PERIODIC TABLE
Radioactivity, Atomic Number, and Isotopy

Theories of the atom were reintroduced into science by John Dalton and were taken up and debated by chemists in the nineteenth century. As noted in preceding chapters, atomic weights and equivalent weights were determined and began to influence attempts to classify the elements. Many physicists were at first reluctant to accept the notion of atoms, with the tragic exception of Ludwig Boltzmann, who came under such harsh criticism for his support of atomism that he eventually took his own life.[1] But around the turn of the twentieth century, the tide began to turn, and physicists not only adopted the atom but transformed the whole of science by performing numerous experiments aimed at probing its structure. Their work had a profound influence on chemistry and, more specifically for our interests here, the explanation and presentation of the periodic table.

Beginning with J.J. Thomson's discovery of the electron in 1897, developments came quickly.[2] In 1911, Ernest Rutherford proposed the nuclear structure of the atom, and by 1920 he had named the proton and the neutron. All of this work was made possible by the discovery of X-rays in 1895, which allowed physicists to probe the atom, and by the discovery of radioactivity in 1896. The phenomenon of radioactivity destroyed the ancient concept of the immutability of the atom once and for all and demonstrated that one element could be transformed into another, thus in a sense achieving the goal that the alchemists had sought in vain.

The discovery of radioactivity led to the eventual realization that the atom, which took its name from the idea that it was indivisible, could in fact be subdivided into more basic particles: the proton, neutron, and electron. Rutherford was the first to try to "split the atom," something he achieved by using one of the newly discovered products of radioactive decay, the alpha particle.

In addition to its well-known medical applications, the earlier discovery of X-rays was to provide a powerful tool that could be used to study the inner structure of matter. By using these rays, Henry Moseley later discovered that a better ordering principle for the periodic system is atomic number rather than atomic weight. He did this by subjecting samples of many different elements to bombardment with X-radiation.

The dual discoveries of radioactivity and X-rays made possible the further discovery and identification of several new elements, such as radium and polonium, which needed to be accommodated, and thus provided further tests of the robustness of the periodic system and its ability to adapt to changes. Indeed, while it is the electron that is mainly responsible for the chemical properties of the elements, discoveries connected with the nucleus of the atom nevertheless have had a profound influence on the evolution of the periodic system. The exploration of the nucleus, along with further work on the nature of X-rays and radioactivity, led to the discovery of atomic number and isotopy, two developments that would together resolve many of the lingering uncertainties surrounding Dimitri Mendeleev's periodic system.

The discovery of isotopy initially presented certain dangers for the periodic system. The large number of new isotopes that were discovered suggested that there were many more "atoms," in the sense of smallest possible particles, of any particular element than had previously been recognized. Some chemists even suggested that the periodic table would have to be abandoned in favor of a classification system that included a separate place for every single isotope. Luckily, this idea was resisted since, as it turned out, isotopes of the same element showed identical chemical properties.[3]

The discovery of atomic number provided one of the most clear-cut modifications the periodic system had undergone since its foundation had been laid by the likes of Johann Döbereiner some 100 years previously. When the concept of atomic number was combined with the new understanding of isotopy, it became possible to appreciate why William Prout's hypothesis (that all elements are composites of hydrogen) had been so tantalizing to the early pioneers of the periodic system. Indeed, Prout's hypothesis could now be said to be valid in the somewhat modified form that all atoms in the periodic table were multiples of a single unit of atomic number or, as it was subsequently named, the proton. It also became possible to explain why triads had been so enticing and so instrumental in the early evolution of the periodic system.

The four main discoveries of X-rays, radioactivity, atomic number, and isotopy are examined in this chapter by following a roughly historical order, although it must be appreciated that there was much overlap among these four themes.

X-RAYS AND BECQUEREL RAYS

By the beginning of 1895 Röntgen had written forty-eight papers now practically forgotten. With his forty-ninth he struck gold.

Emilio Segrè, *From X-rays to Quarks*

This is how the Italian-born physicist Segrè has described the career of Wilhelm Conrad Röntgen before his momentous discovery of X-rays.

On November 8, 1895, Röntgen, a German physics professor, was working in his darkened laboratory in Wurzburg. His experiments focused on passing an electrical current into highly evacuated glass tubes known as Crookes tubes. To Röntgen's surprise, he noted that when one of his tubes was charged, an object across the room began to glow. This proved to be a barium platinocyanide-coated screen too far away to be reacting to the cathode rays coming from the tube. Over the next few days, Röntgen experimented in various ways with what he began to suspect might be a new form of emanation. Quite by accident, while holding materials between the tube and screen to test the new rays, he saw the bones of his hand clearly displayed on the screen in an outline of flesh. This was the first time anyone had ever seen a medical X-ray image. Röntgen plunged into seven weeks of intense and secretive experimentation in order to determine the nature of the mysterious rays. He worked in isolation, telling a friend that he had discovered something interesting but that he did not know whether his observations were correct.

On December 28, 1895 Röntgen gave his preliminary report to the president of the Würzburg Physical-Medical Society, accompanied by experimental radiographs, including an image of his wife's hand, which survives to this day. A few days later, he sent printed reports to physicist friends across Europe, and by January the world had been gripped by "X-ray mania." Röntgen was acclaimed as the discoverer of a medical miracle, and although he accepted the first Nobel Prize in physics in 1901, he decided not to seek patents or proprietary claims on X-rays.

One of the many scientists to whom Röntgen sent his X-ray images was Henri Poincaré. Poincaré in turn showed one of these radiographs to his colleagues at the Academy of Sciences in Paris, on January 20, 1896. Henri Becquerel, a professor at the Musée d'Histoire Naturelle and member of the academy, took note of a remark by Poincaré on the possible link between X-rays and luminescence. On returning to his laboratory, he designed an experiment to test the hypothesis that X-ray emission and luminescence are related. In order to see if a phosphorescent body emitted X-rays, he chose a hydrated salt of uranium that he had prepared some years before. On February 20, Becquerel placed a transparent crystal of the salt on a photographic plate wrapped between two thick sheets of black paper, and the experimental setup was exposed to sunlight for several hours. After development, the silhouette of the crystal appeared in black on the photograph, and Becquerel concluded that the phosphorescent substance emitted a penetrating radiation able to pass through black paper.

Unable to repeat such experiments in the following days because of a lack of sunshine, Becquerel put away his salt crystal, placing it by chance on an undeveloped photographic plate in a drawer. Later he developed the plate in order to determine the amount by which the phosphorescence had decreased. To his great surprise, he found that the phosphorescence had not decreased at all but was more intense than it had been on the first day. Noticing a shadow on the plate made by a piece of metal he had placed between it and the salt, Becquerel realized that the salt's activity had continued in the darkness. Clearly, sunlight had been unnecessary for the emission of the penetrating rays. Could it be that just one year after the discovery of X-rays, another new form of emanation was beginning to reveal itself?

Becquerel also found that the activity of his uranium salt did not diminish with time, even after several months. He also tried to use a nonphosphorescent uranium salt and found that the new effect persisted. He soon concluded that the emanation was due to the element uranium itself. Even after about a year had passed, from when he first began his experiments, the intensity of the new rays had shown no signs of decreasing. But Becquerel was soon to move onto other scientific interests, and it was left to others to explore the rays in greater detail.[4]

RADIOACTIVITY

The Polish born chemist-physicist Marie Curie (figure 6.1), neé Sklodowska, was the first to take up the study of Becquerel rays.[5] For her doctoral project, Marie Curie began to explore whether elements other than uranium might also produce Becquerel rays. She found that thorium, which occurs two places before uranium in the periodic table, also shows this form of behavior and coined the term "radio-activity" to describe this new property of matter.

While experimenting with pitchblende, an ore of uranium, she found that this material exhibited a more intense level of radioactivity than did pure uranium, which itself showed greater radioactivity than did uranium salts. She made the obvious deduction that a new element might be present in pitchblende and, work-ing in 1898 with Pierre Curie, quickly succeeded in extracting a substance with 400 times the activity of uranium, which she named polonium after her native country. This element would find its place at the foot of group VI of the periodic table, below selenium and tellurium and eight places before uranium.

Continuing to work with the pitchblende they had used to extract polonium, the Curies found that it showed traces of yet another substance, which after much separation and purification, displayed an even more pronounced level of radioac-tivity, amounting to about 900 times that of uranium. This element, also discovered in 1898, was named radium.

Meanwhile, other physicists were also exploring Becquerel's rays. After study-ing in Cambridge under Thomson, the New Zealand–born physicist Rutherford moved to McGill University in Montreal. There he undertook a line of inquiry connected with Becquerel's rays by investigating the nature of the radiation itself.

FIGURE 6.1
Marie Curie. Photo and permission from Emilio Segrè Collection.

In 1899, he showed that radioactivity, as displayed by uranium, for example, produced a species of rays that could easily be absorbed by a thin metallic surface. He called them alpha rays. He also discovered a more penetrating species of rays, which he termed beta rays. Rutherford's identification of alpha rays, which, as he eventually realized, consisted of atoms of helium stripped of its two electrons, was to provide a powerful tool for probing the structure of the atom.

Between the years 1900 and 1903, Rutherford began to study the chemistry of radioactive substances. Working with a young colleague, the chemist Frederick Soddy, who will feature prominently later in this chapter, he made an epochal discovery, remarkable especially for its impact on understanding of the nature of the chemical elements. Rutherford and Soddy were compelled to announce that, in the course of radioactive reactions, certain elements were transformed into completely new elements. While fully aware of the possible criticism that such a notion might bring, they went as far as to describe this new phenomenon as chemical transmutation, thus evoking the age-old dream of the alchemists.

Some authors believe that the interpretation of the properties of the elements passed from chemistry to physics as a result of the discovery of radioactivity. They speak of "the redefinition of Mendeleev's chemical element, which would lead to its appropriation by physics."[6] I believe this view to be overly reductionistic, as presumably did Fritz Paneth, who formulated his "intermediate position" in order to uphold the integrity of the chemical view of the elements and of the periodic system.[7]

THE DISCOVERY OF THE NUCLEUS

In 1911, following some experiments by his students Hans Geiger and Ernest Marsden, Rutherford revived the notion of a planetary atom in which electrons were believed to circulate around a central nucleus. As discussed in chapter 7, Jean Perrin and, in a somewhat different version, Hantaro Nagaoka were the first to propose such atomic models.[8] But the nuclear atom had since been eclipsed by the work of Thomson, which had suggested that the electrons were embedded in the main body of the atom.

Upon firing a stream of positively charged alpha particles at a thin foil made of gold, Geiger and Marsden observed that some of the particles were scattered at very large angles and some even appeared to rebound straight back toward the incoming direction. Such a set of findings was inexplicable in terms of Thomson's model, in which the positive charge of the atom was diffused throughout the atom. This model predicted that almost all of the alpha particles would pass through the foil. Rutherford was forced to conclude that the atoms in the foil must contain concentrations of positive charge intense enough to deflect some of the alpha particles. He thus discovered that the positive charge is localized in a very small volume at the center of the atom, while the negative charge is diffused throughout the entire volume.

On the basis of his analysis of the alpha-scattering experiments of Geiger and Marsden, Rutherford further concluded that the charge on an atom is approximately half of its atomic weight. Rutherford and his colleagues observed that the degree of scattering is proportional to the square of the atomic weight of any particular atom. This effect was checked in a number of different target elements ranging from aluminum to lead. This, in turn, led to the conclusion that the scattering was proportional to the square of the nuclear charge, given that it is charge, rather than weight, that causes scattering of charged alpha particles. From further analysis of the scattering data, they arrived at the following approximate relationship between charge (Z) and atomic weight (A):

$$Z \approx A/2$$

Another British physicist, Charles Barkla, had arrived at precisely the same conclusion, also in 1911, by analyzing the scattering of X-rays from various substances. Barkla found that heavier elements produced a greater scattering in amounts proportional to their atomic weights and concluded that "the number of scattering electrons per atom is about half the atomic weight in the case of the light atoms."[9] Since in the case of neutral atoms the number of positive charges is equal to the number of electrons, the conclusions of Rutherford and Barkla are identical.

ATOMIC NUMBER

Rutherford's and Barkla's work on atomic charge contributed to the discovery of atomic number, but it was not the main evidence in that brought it to the foreground. The discovery of atomic number provides the opportunity for a little digression on how the history of science is frequently rewritten and sanitized by subsequent commentators. The real discoverer was the amateur scientist Anton van den Broek (figure 6.2), whose contributions tend to be neglected. It is often thought that van den Broek merely summarized the work of physicists Rutherford and Barkla, but the true story is altogether different. It was van den Broek's close study of Mendeleev's periodic table, and his prolonged attempts to improve upon it, that led to his discovery of an ordinal number associated with each element. It also led to the identification of this number with the nuclear charge as well as the number of electrons in any atom.

Van den Broek trained in law and econometrics but published a number of influential articles in the leading scientific journals of his day. His first paper appeared in 1907 under the title "The α particle and the Periodic System of the Elements."[10] It took as its point of departure a paper of the previous year in which Rutherford had suggested various explanations for the nature of the α particle. One of these suggestions, and the one favored by van den Broek, was that this particle consisted of half of a helium atom with a charge of +1.[11] Van den Broek gave the name of alphon to this particle and proposed that it might take the place

FIGURE 6.2
Anton van den Broek. Photo and
permission from Jan van Spronsen.

of the hydrogen atom in Prout's theory that all elements are composites of one basic particle.

According to van den Broek's scheme, each particular number of aggregated alphons would thus correspond to a particular chemical element. Since the weight of the helium atom was known to be four units, the alphon would have a weight of 2, and all even numbers of composite alphons would correspond to the weights of the known elements. An atom of uranium, for example, with an atomic weight of 240, would be composed of 120 alphons, and so on. Since atomic weights are not exact multiples of each other, van den Broek realized that this suggestion would not be precise, but in the Pythagorean spirit of previous Proutian speculations, he was not unduly concerned by this aspect.

Of course, such a system required many more elements than were known to exist at the time if there were to be a total of 120, ending with uranium, which was the heaviest known element. Van den Broek made up for part of this deficit by incorporating many of the new radio-elements that had recently been discovered. These species would turn out to be isotopes rather than genuine new elements, but to enter into such matters now would be to get ahead of the story. At the time, the work on radioactivity that was emerging gave van den Broek confidence that the gaps in his table would be filled. In addition, the question of how to accommodate the rare earth elements into the periodic system had still not been settled, and it seemed plausible that several new rare earth elements awaited discovery.

In his paper of 1907, van den Broek included a periodic table constructed according to his scheme, showing a total of 41 gaps representing undiscovered elements that would have to be filled between hydrogen and uranium (figure 6.3). Each even number in the table corresponds to a chemical element, and the difference in atomic weight between any two adjacent elements is two units.

In 1911, van den Broek published a second paper, in which he claimed that Mendeleev had not sufficiently satisfied the requirements of chemical periodicity of the elements.[12] He also noted that Mendeleev had intended to devise a three-dimensional periodic system, which would have remedied the situation, and it was this task that van den Broek was proposing to take up now. Despite this assertion, the table that van den Broek published in this paper was in fact two dimensional, although he implied that it could be constructed in three dimensions (figure 6.4). The third dimension would consist of short series of three elements, each of which is shown diagonally on the two-dimensional table. In any case, none of the conclusions that van den Broek drew from this new table depended on its supposed three-dimensional nature.

In the 1911 article, the notion of the alphon had disappeared, but van den Broek retained the idea of successive elements differing by two units of weight, whereas in Mendeleev's table successive elements typically showed alternating mean differences of approximately two and four units.

TABLE 1

	VII	0	I	II	III	IV	V	VI
1	2* (α)	4 He	6 Li	8 Be	10 B	12 C	14 N	16 O
2	18 F	20 Ne	22 Na	24 Mg	26 Al	28 Si	30 P	32 S
3	34 Cl	36 Ar	38 K	40 Ca	42 Sc	44 Ti	46 V	48 Cr
4	50 Mn	52	54	56 Fe	58 Co	60 Ni	62	64
5	66	68	70 Cu	72 Zn	74 Ga	76 Ge	78 As	80 Se
6	82 Br	84 Kr	86 Rb	88 Sr	90 Y	92 Zr	94 Nb	96 Mo
7	98	100	102	104 Ru	106 Rh	108 Pd	110	112
8	114	116	118 Ag	120 Cd	122 Jn	124 Sn	126 Sb	128 Te
9	130 J	132 Xe	134 Cs	136 Ba	138 La	140 Ce	142 Nd	144 Pr
10	146	148	150 Sa	152	154 Gd	156	158 Tb	160
11	162	164	166 Er	168 Tu	170 Yb	172	174 Ta	176 W
12	178	180	182	184 Os	186 Ir	188 Pt	190	192
13	194	196	198 Au	200 Hg	202 Tl	204 Pb	206 Bi	208
14	210	212	214	216	218	220	222	224
15	226	228	230	232 Ra	234	236 Th	238	240 U

* Theoretical atomic weight.

FIGURE 6.3 Van den Broek table of 1907. From The α Particle and the Periodic System of the Elements, *Annalen der Physik*, 23, 199–203, 1907, p. 201. This version is from T. Hirosige, The Van den Broek Hypothesis, *Japanese Studies in the History of Science*, 10, 143–162, 1971, p. 148 (by permission from the publisher).

That same year, van den Broek also published a very brief, 20-line letter to *Nature* magazine.[13] This letter may represent the first anticipation of the concept of atomic number, given that John Newlands's much earlier suggestion of an ordinal number for each of the elements (see chapter 3) was rather more tenuous. Van den Broek began by drawing attention to the fact that two lines of experimental research, namely, Rutherford's and Barkla's, supported the view that the charge on an atom is approximately half its atomic weight, or to repeat an equation that appeared just above, $Z \approx A/2$. This evidence had provided support for his speculation of 1907 that atomic weight increases by approximately two units between each two consecutive elements.[14] He then referred to his new periodic table and his prediction that 120 elements exist altogether, ending with the words,

> If this cubic periodic system should prove to be correct, then the number of possible elements is equal to the number of possible permanent charges of each sign per atom, or to each possible permanent charge (of both signs) per atom belongs a possible element.

Van den Broek was suggesting that since the nuclear charge on an atom was half its atomic weight, and the atomic weights of successive elements increased in stepwise fashion by two, then the nuclear charge defined the position of an element in the periodic table. In other words, each successive element in the periodic table would have a nuclear charge greater by one than the previous element. In proposing this,

TABLE 2

		0			I			II			III			IV			V			VI			VII		
		1	2	3	1	2	3	1	2	3	1	2	3	1	2	3	1	2	3	1	2	3	1	2	3
A	1	He			Li			Be			B			C			N			O			F		
	2	Ne			Na				· Mg		Al			Si			P			S				Cl	
	3			Ar		K				Ca		Sc			Ti			V			Cr				Mn
B	1	Fe			Co			Ni			Cu			—			—			—				Zn	
	2	—			—			—			Ga			Ge				As			Se				Br	
	3		Kr				Rb			Sr			Y				Zr			Nb			Mo			Ru
C	1	Rh			Pb			—			—			Ag			—			—				Cd	
	2	—			—			—				In			Sn			Sb			Te				J	
	3		Xe			Cs			Ba			La			Ce				Nd			Pr			(Sm)
D	1	(Eu)			(Gd$_1$)			(Gd$_2$)			(Gd$_3$)			(Tb$_1$)			(Tb$_2$)			(Dy$_1$)			(Dy$_2$)		
	2	(Dy$_3$)			(Ho)			(Er)			(Tu$_1$)			(Tu$_2$)			(Tu$_3$)			(Yb)			(Lu)		
	3	—			—			—			—			—				Ta			W				Os
E	1	Ir			Pt			Au			Hg			Tl			Bi				Pb		—		
	2	—			—			—			—			—			—			—			—		
	3	—			—				Ra		—			Th			—				U		—		

FIGURE 6.4 Van den Broek table of 1911. From Das Mendelejeffsche "kubische" periodische System der Elemente und die Einordnuung der Radio-elemente in dieses System, *Physikalishe Zeitschrift*, 12, 490–497, 1911, p. 491. This version is from T. Hirosige, The Van den Broek Hypothesis, *Japanese Studies in the History of Science*, 10, 143–162, 1971, p. 149 (by permission from the publisher).

van den Broek was going beyond Rutherford and Barkla, neither of whom had been primarily concerned with elements in the periodic table. Whereas Rutherford and Barkla realized that $Z \approx A/2$, van den Broek also realized that $Z \approx A/2 =$ atomic number. As the well-known physicist and author Abraham Pais has commented, "Thus based on an incorrect periodic table and on an incorrect relation ($Z \approx A/2$), did the primacy of Z as an ordering number of the periodic table enter physics for the first time."[15]

But van den Broek's claim to fame does not lie just with this crude premonition of atomic number. By 1913, he abandoned his cubiform table and replaced it with an elaborate two-dimensional version (figure 6.5) and the clearly stated rule that "the serial number of every element in the sequence ordered by increasing atomic weight equals half the atomic weight and therefore the intra-atomic charge."[16] Although this step takes matters a little further by mentioning serial numbers for each of the elements, it is still somewhat incorrect in being tied rather firmly to atomic weight, albeit atomic weight divided in half. Nevertheless, van den Broek's article was cited by no less a person than Niels Bohr in his trilogy paper of 1913, in which he introduced quantum theory to the atom.[17]

Van den Broek's most significant contribution came in another short communication to *Nature* magazine, published in 1913, in which he explicitly connected the serial number with the charge on each atom and disconnected it from atomic weight: "The hypothesis [about the serial number of the elements being equal to

TABLE 3

0	I	II	III	IV	V	VI	VII	VIII		
2* He	3 Li	4 Be	5 B	6 C	7 N	8 O	9 F			
10 Ne	11 Na	12 Mg	13 Al	14 Si	15 P	16 S	17 Cl			
18 — 19 Ar	20 K	21 Ca	22 23 Sc —	24 Ti	25 V	26 Cr	27 Mn	28 Fe	29 Co	30 Ni
	31 Cu	32 Zn	33 Ga	34 Ge	35 As	36 Se	37 Br	38 —	39 —	40 —
41 — 42 Kr	43 Rb	44 Sr	45 46 Y —	47 Zr	48 Nb	49 Mo	50 —	51 Ru	52 Rh	53 Pd
	54 Ag	55 Cd	56 In	57 Sn	58 Sb	59 Te	60 J	61 —	62 —	63 —
64 — 65 Xe	66 Cs	67 Ba	68 69 La —	70 Ce	71 Nd	72 Pr	73 —	74 Sa	75 Eu	76 Gd
	77 Tb	78 (Tb_2)	79 Dy	80 Ho	81 Er	82 Ad	83 AcC	84 TuI	85 TuII	86 AcA
87 — 88 AcEm	89 AcX	90 TuIII	91 92 RAc Cp	93 Ct	94 Ta	95 Wo	96 —	97 Os	98 Ir	99 Pt
	100 Au	101 Hg	102 Tl	103 Pb	104 Bi	105 RaF	106 ThC	107 RaC	108 ThA	109 RaA
110 ThEm 111 RaEm	112 ThX	113 Ra	114 115 RTh Io	116 Th	117 UII	118 U	119 —	120 —	121 —	122 —

* Atomic number.

FIGURE 6.5 Van den Broek's table of 1913. From Die Radioelemente das Periodische System und die Konstitution der Atome, *Physikalische Zeitschrift*, 14, 32–41, 1913, table on p. 37. This version is from T. Hirosige, The Van den Broek Hypothesis, *Japanese Studies in the History of Science*, 10, 143–162, 1971, p. 152 (by permission from the publisher).

Z] holds good for Mendeleev's table but the nuclear charge is not equal to half the atomic weight."[18] Van den Broek was able to take this important liberating step on the basis of more scattering experiments by Geiger and Marsden, which he analyzed in detail and discussed in his short note. This contribution was praised by Soddy in the next issue of *Nature* and one week later also by Rutherford, who nevertheless privately resented the intrusion of amateurs. It was at this point that Rutherford actually coined the expression "atomic number":

> The original suggestion made by van den Broek that the charge on the nucleus is equal to the atomic number [i.e., the serial number in the periodic table] and not to half the atomic weight seems to me very promising.[19]

HENRY MOSELEY

So it was the amateur scientist van den Broek who confounded all the professional experts and first perceived the importance of atomic number as the ordering criterion for the elements. But as is often the case with scientific discoveries, it is the person who completes the task who is given the most credit, as seen in the case of Mendeleev and the discovery of periodicity.

Such is the case of Moseley (figure 6.6), who died in the First World War at the tender age of 26, before anyone outside the then very small circle of atomic physicists had heard of him. His subsequent fame lies in two brief articles that firmly established that atomic number, rather than atomic weight, was indeed a superior ordering principle for the elements. In addition, he was able to lay the groundwork by which others could settle conclusively that there were a total of 92 naturally occurring elements and could specify precisely where the remaining gaps were situated in the periodic table.[20]

Moseley went to work as a research student with Rutherford, who was at this time in Manchester. There, he was given a project connected with radioactivity.[21] In 1911, Moseley published an article with a fellow graduate student, the Polish born chemist Kasimir Fajans, concerning the measurement of half-lives of some radioactive products obtained from the element actinium.[22]

A year later Max von Laue in Zürich was investigating the nature of X-rays. Believing that they were extremely short electromagnetic rays and therefore should exhibit interference effects, von Laue was attempting to diffract X-rays by bouncing them off planes of atoms in crystalline substances such as sodium chloride.[23] By this time, it was known that X-rays came in two varieties. The first was a type originally observed by Röntgen, which were produced when electrons were stopped by some means, such as the glass walls of an evacuated Crookes tube. Second, Charles Barkla had discovered another kind of X-ray phenomenon brought

FIGURE 6.6 Henry Moseley. Photo from author's collection used by permission from Emilio Segrè Collection.

about when electrons struck targets made of metals. Each different metal would produce X-ray lines showing a characteristic frequency. It was these X-rays, the ones coming off a metal target, and in particular the so-called K_α rays, that Moseley would exploit in his own research.[24]

Moseley made it clear in his articles that he was essentially setting out to test van den Broek's speculation regarding the characterization of each element according to its atomic number. In addition, it is known that he had several meetings with the young Bohr while he, too, was a visitor in Rutherford's Manchester laboratory around 1912–1913. Bohr and Moseley discussed the question of the ordering of nickel and cobalt, an example of pair reversal. Bohr is known to have favored placing cobalt before nickel, to which Moseley is said to have responded "we shall see."[25] Moseley devised an ingenious apparatus in which many different metal plates could be rotated so that each one would become the target for a beam of electrons, and the emitted K_α X-rays would be measured.[26] He first experimented on 14 elements, nine of which, titanium to zinc, formed a continuous series in the periodic table.[27] Moseley discovered that a plot of the frequency of the lines of the K series of spectral lines of each element was directly proportional to the square of an integer representing the position of each successive element in the periodic table. He found that the frequency, n, of the K_α X-rays obtained from each sample target varied according to an expression of the form

$$n \propto Q^2,$$

where Q is a number that increases by a constant amount on moving through the elements.

Moseley had discovered a fundamental quantity that increased by regular intervals as he moved through his sequence of elements in the order in which they appeared in the periodic table. He quickly recognized this quantity as the positive charge on the nucleus, or van den Broek's atomic number. As he stated:

> We have here a proof that there is in the atom a fundamental quantity, which increases by regular steps as we pass from one element to the next. This quantity can only be the charge on the central positive nucleus, of the existence of which we already have definite proof.[28]

Moseley acknowledged the previous work of Barkla, van den Broek, and Bohr, all of whom had anticipated his own findings. He also showed that $Q = N - 1$, where N represents the number of unit charges in the nucleus and therefore the atomic number.

Moseley's second famous paper appeared in print in April 1914. He now reported measurements on a further 30 elements. By examining the K series from aluminum to silver, and the L series of spectral lines from zinc to gold, he found a general expression of the form,

$$n \propto A\,(N-b)^2,$$

where $b = 1$ for the K series and

$$A \propto (1/1^2 - 1/2^2)$$

Similarly, for the L series $b = 7.4$ and

$$A \propto (1/2^2 - 1/3^2)$$

As soon as Moseley had established the importance of atomic number experimentally, he began to apply this work in settling various questions regarding new elements that had been claimed by various chemists. A total of approximately 70 proposed new elements competed to fill the 16 gaps in Mendeleev's periodic table. Moseley succeeded in showing that many of these were spurious and was able to resolve some priority disputes regarding the discovery of certain elements.

For example, a Japanese chemist, Masataka Ogawa, claimed to have isolated an element that he called nipponium and that he believed to be Mendeleev's eka-manganese.[29] Moseley was able to show that this claim was unfounded since the sample provided by Ogawa did not show the required atomic number when subjected to Moseley's spectral analysis. Similarly, coronium, nebullium, casseopeium, and asterium, which appeared on many early periodic tables, often between hydrogen and lithium, could all be dismissed as spurious elements.[30]

Moseley's work was also used to settle the question of the placement of the rare earth elements, a task that had eluded Mendeleev and other early pioneers of the periodic table. Mendeleev had stated that the placement of the rare earths was one of the most difficult problems of all those confronting the periodic law. The rare earths were notoriously difficult to separate chemically. Since they appeared to differ only slightly in atomic weight and properties, no one had been able to find a satisfactory way of fitting them into the periodic table. According to William Crookes,

> The rare earths perplex us in our researches, baffle us in our speculations, and haunt us in our very dreams. They stretch like an unknown sea before us, mocking, mystifying, and murmuring strange revelations and possibilities.[31]

Georges Urbain, a French chemist known for his work on the isolation of rare earth elements, traveled to Oxford in order to meet Moseley after hearing of his groundbreaking work. As the story goes, Urbain handed Moseley a sample containing a mixture of rare earths and challenged him to identify which elements were present. After a matter of about one hour, Moseley is said to have surprised Urbain by correctly identifying the presence of erbium, thulium, ytterbium, and lutetium in the Frenchman's sample. The same feat had taken Urbain several months to achieve by chemical means. Urbain then asked Moseley to tell him the relative amounts of the various elements in the sample and was again astonished to receive an answer that coincided almost exactly with his own laborious chemical analysis.

Moseley's work clearly showed that successive elements in the periodic table have an atomic number greater by one unit. From this fact, Moseley and others could identify which gaps remained to be filled in the periodic system and found that there were a total of seven such cases still waiting to be discovered. Unlike previous lists of gaps, this list was now completely definitive and included the precise atomic numbers of the still elusive elements, which were 43, 61, 72, 75, 85, 87, and 91.[32]

The clarification that Moseley brought to the periodic table represents one of the finest examples of the reductive power of physics in the field of chemistry. Most lingering problems regarding pair reversals, such as those concerning tellurium and iodine, which had plagued Mendeleev throughout his career, were thereby resolved. Furthermore, Moseley's work made it easier to deal with the profusion of apparent new "elements" that emerged as a result of research on radioactive phenomena. Two substances could be regarded as being the same element if, and only if, they showed the same value of atomic number, which could be clearly measured by Moseley's method.

FILLING THE REMAINING GAPS

The seven remaining gaps in the periodic table were gradually filled, although not without further controversy in spite of the conclusive nature of Moseley's atomic number method. The first of these was element 91, discovered by Otto Hahn[33] and Lise Meitner in 1917.[34] The element behaved in the manner described by Mendeleev, who had given it the provisional name eka-tantalum. It now became known as protactinium but was not isolated until the year 1934 by Aristide Grosse in Germany.

Element 72, or hafnium, has a rich and controversial story associated with its discovery.[35] Several researchers, including Urbain, independently claimed to have discovered the element but were later found to have been mistaken. In 1923 Dirk Coster and György von Hevesy, working in the Niels Bohr Institute in Copenhagen, finally succeeded in isolating the element, naming it for *Hafnia*, Latin for Copenhagen. According to most accounts, Bohr had first made a theoretical prediction that the element would be a transition metal rather than a rare earth. In fact, some chemists already shared this opinion. As discussed in chapter 8, Charles Bury had also predicted that the element would be a transition metal and had even arrived at its electronic configuration before Bohr.

Element 75, called rhenium, was first discovered in Berlin in 1925 by the husband and wife team of Walter and Ida Noddack, who isolated it and used Moseley's X-ray method to confirm the presence of the new element. The Noddacks were also looking for element 43, Mendeleev's eka-manganese, and claimed to have found it in the same ores, calling the new element masurium after Walter Noddack's native region of Prussia. The Noddacks published X-ray data for masurium, but it was discredited by others for a variety of reasons.[36] The distinction of discovering element 43 went to Carlo Perrier and Segrè, who obtained it 12 years later, in 1937, at the University of California–Berkeley. They named the new element technetium to reflect the fact that it had been artificially synthesized as a byproduct in a nuclear reaction.

Recently, it has come to light that technetium may indeed have been isolated by the Noddacks and Walter Berg. This reassessment was carried out by the Dutch physicist Pieter Van Assche in the late 1980s on the basis of a careful reanalysis of the German team's X-ray data. The evidence marshaled by Van Assche is rather convincing and implies that the first isolation of element 43 involved a naturally occurring element. Had the discovery been recognized at the time, there would have been no need to name the element technetium since it seems that it does occur naturally.

In 1939, an element subsequently named francium, number 87, was discovered in Paris by Marguerite Perey, and in 1940 Segrè discovered astatine, element 85. The final piece of the jigsaw puzzle, element 61, promethium, was finally obtained as a byproduct in a nuclear reaction. The discoverers on this occasion were Jacob Marinsky, Lawrence Glendenin, and Charles Coryell.[37]

WHAT MOSELEY DID NOT ACHIEVE

As with so many scientific heroes, perhaps more so in this case because of his early death, the claims for what Moseley is supposed to have achieved far outpace the truth. The hagiography of Moseley is constantly propagated by science textbooks, sometimes in sincere attempt to simplify the account, but it also occurs in more detailed historical treatments. Contrary to most accounts, Moseley did not personally settle the question of how many naturally occurring elements exist. Nor did he even definitively resolve the question of how many elements exist between aluminum (13) and gold (79), which marked the boundaries of his own studies. By assuming that aluminum is the 13th element, Moseley argued that there could be only 79 elements up to and including gold.

Limiting the elements to 79 left only three remaining gaps in the periodic table, located at atomic numbers 43, 61, and 75. But Moseley could not be confident of this prediction since he did not have pure samples of some of the rare earths. X-ray spectra were not available for terbium (65), dysprosium (66), thulium (69), ytterbium (70), and lutetium (71), and this led him to assert incorrectly the existence of three forms of thulium, named thulium, thulium I, and thulium II.

This assignment in turn meant that such elements as ytterbium and lutetium were advanced by one place, so no vacant space was left at element 72. Moseley was unable to place Urbain's keltium, which eventually turned out to be the same as lutetium, discovered by Urbain a few years previously. When matters were resolved, after Moseley's early death, only one form of thulium remained and ytterbium and lutetium were found to have atomic numbers of 70 and 71, respectively. This meant that there was a vacant gap at 72 for a new element between lutetium and tantalum, the element that would eventually be named hafnium.

Since his experiments did not go beyond gold, or atomic number 79, Moseley certainly did not show that uranium is element 92, as is often claimed. This honor went to the spectroscopist Manne Siegbahn, working in Sweden, in 1916. Finally, even the central achievement invariably associated with Moseley, the realization that atomic number is equal to the number of positive charges in the nucleus, was not conclusively settled until some time later.

In 1920, James Chadwick, who 10 years later would discover the neutron, reanalyzed Moseley's work. He discovered that the choice of value for the constant $b = 7.4$ was not as inevitable as Moseley had claimed. It was still possible in principle for an atomic number not to equal the number of positive charges in the nucleus, and this in turn would have implied that there might be more than 13 elements from hydrogen to aluminum inclusive. Chadwick therefore decided to make some independent measurements of the charges on various nuclei using a refined version of Geiger and Marsden's experiment with alpha rays. Only after this work had been carried out and atomic charges been successfully measured by a second method did Chadwick announce the confirmation of Moseley's simple idea. Atomic number does indeed equal the number of positive charges in the nucleus of any atom.

PHILOSOPHICAL DEBATES REOPENED

Van den Broek's suggestion, and Moseley's experimental work on atomic number, had the effect of rehabilitating Prout's hypothesis, which had proposed that all elements were composites of hydrogen. The atomic numbers of all the elements were indeed exact multiples of the atomic number for hydrogen, which is 1. More generally, the work of van den Broek and Moseley revitalized some philosophical notions of the unity of all matter that had been so harshly criticized by Mendeleev, among others. By now, Thomson had shown that the electron was common to all elements and Rutherford had established that electric particles were present in the nuclei of all elements. Moseley had added the fact that all nuclei seemed to consist of an integral number of positive charges. There clearly seemed to be some form of underlying unity behind the apparent diversity of the elements. This view was strengthened further when Rutherford discovered that elements could be transmuted into each other through the use of radioactive techniques, thus once again recalling the ancient alchemical notion of the fundamental unity of all matter.

Nevertheless, the cause of the initial rejection of Prout's hypothesis, in its original form, had not been resolved. As described in chapter 2, this was the fact that certain elements such as chlorine (35.46) and lead (207.20) had nonintegral atomic weights. This particular puzzle had to await the discovery of isotopy, which is generally attributed to the chemist Soddy.

ISOTOPY

The idea that any element can consist of different kinds of atoms can be traced to a remark made by Crookes in 1886:

> I conceive that when we say the atomic weight of calcium is 40, we really explain the fact, while the *majority* of calcium atoms have an actual atomic weight of 40, there are not a few which are represented by 39 or 41, a less number by 38 or 42, and so on.[38]

The first clear elaboration of isotopy, however, came much later and belongs to the English chemist Soddy, who began as one of Rutherford's collaborators, like so many others who made important contributions to atomic science. After research in chemistry at Oxford, Soddy joined Rutherford at McGill University in Montreal in 1900 and participated in much of the work that established the concept of radioactive half-lives and the reality of radioactive transmutation.

During this time, other scientists were also exploring radioactivity and in the process were discovering new elements. The first of these were polonium, radium, actinium, and radon, followed by another 30 or so suspected new elements, most of which would turn out to be isotopes of existing elements.[39] Some, like van den Broek, tried desperately to incorporate all these new species into the periodic table. Others, including the Swedes Daniel Strömholm and Theodor Svedberg, realized that there were great similarities among many of these new species. As Jan van Spronsen notes in his book on the periodic table,[40] Strömholm and Svedberg can also be regarded as having anticipated the existence of isotopes. For example, they grouped together radium emanation, actinium emanation, and thorium emanation into one single place in their periodic system (table 6.1). Similarly, radium, actinium X, and thorium X were all made to occupy a single place.[41] Without explicitly stating the concept and drawing out its full consequences, Strömholm and Svedberg had realized that several species can occupy a single space in the periodic table, a concept Soddy was soon to name isotopy from the Greek *iso* (same) and *topos* (place).

TABLE 6.1
Strömholm and Svedberg's Fragment Table

	0 Row	1 Row	2 Row	3–4 Row
5 period	Xe	Cs	Ba	La –Yb
	Ra – Em	—	Ra	Ionium – (UX-Rad.U)
6 period	Akt – Em	—	Akt X	Rad.Akt.– Akt.
	Th – Em	—	Th X	Rad.Th – Mes.Th-Th

D. Stromholf, T. Svedberg, Untersuchungen über die Chemie der radioaktiven Grundstoffe. II. Die Aktiniumreihe *Zeits für Anorganischen Chemie*, 63, 197-206, 1909, table on p. 204. The abbreviation Rad stands for radio, a prefix used to denote suspected new elements, such as radio-thorium, most of which turned out to be isotopes. Em denotes the term emanation meaning substances emanating from a particular element. Akt was the German abbreviation for the modern symbol Ac for actinium.

Another strand of development came from several attempts to separate some of these new radio-elements chemically, which ended in failure. First of all, in 1907 Herbert McCoy and William Ross concluded that, in the case of thorium and radiothorium, "Our experiments strongly indicate that radiothorium is entirely inseparable from thorium by chemical processes,"[42] a comment Soddy considered the first definitive statement of the chemical inseparability of what were soon to be called isotopes. Soddy himself wrote in the same year that there seemed to be no known method of separating thorium X from mesothorium. They were in fact two isotopes of thorium. Similar cases began to multiply. Bertram Boltwood discovered the radio-element ionium, which could not be chemically separated from thorium. In another famous case, Hevesy and Paneth were asked by Rutherford to try to separate radio-lead from ordinary lead and likewise failed to do so, in spite of using 20 different chemical methods. Their work was not entirely in vain, however, since it led to the development of the use of radioactive tracers, which have become an indispensable tool in modern chemistry and biochemistry.

In 1911, Soddy wrote the following comments regarding the various series of very similar radioactive elements that had recently been discovered:

> The conclusion is scarcely to be resisted that we have in these examples no mere chemical analogues but chemical identities. . . . Chemical homogeneity is no longer a guarantee that any supposed element is not a mixture of several different atomic weights, or that any atomic weight is not merely a mean number. The constancy of atomic weight, whatever the source of the material, is not a complete proof of homogeneity.[43]

The first time Soddy actually used the term "isotope"[44] was in an article in 1913, where he wrote,

> The same algebraic sum of positive and negative charges in the nucleus, when the arithmetic sum is different, gives what I call "isotopes" or isotopic elements,

> because they occupy the same place in the periodic table. They are chemically
> identical, and save only as regards the relatively few physical properties which
> depend upon atomic mass directly, physically identical also.[45]

The first sentence in this quotation might appear to be a mistake, and in a sense it
is since there are no negative charges in the nucleus. What Soddy was referring to
was beta particles, which are identical to electrons but are created in the nucleus. In
modern terms beta decay involves the transformation of a neutron into a proton,
accompanied by the emission of a beta particle. Interestingly, the notion that beta
particles originate in the nucleus was first proposed by van den Broek and later
supported by several others, including Soddy.

Many aspects of the problem of inseparable elements were clarified when
Soddy and Fajans independently suggested what became known as the group displacement laws. They stated that the emission of an alpha particle from the atom of
an element produces an element located two places to the left, while the emission
of a beta particle resulted in a movement of one position to the right in the periodic table. It followed that the elements between lead and uranium in the periodic
table could exist as more than one kind of atom, differing in mass but displaying
the same chemical behavior. For example, if an atom of uranium-235 ($Z = 92$)
undergoes alpha decay, it forms an atom of thorium-231 ($Z = 90$). Meanwhile an
atom of actinium-230 ($Z = 89$) can undergo beta decay to form an atom of tho-
rium-230 ($Z = 90$). The products of both radioactive decays are atoms of the same
element but have different atomic weights.[46] Fajans coined the term "pleiad" to
mean a group of chemically identical atoms with different atomic masses, but his
term was not generally adopted.[47]

Only now did it become clear why such elements as tellurium and iodine, and
other pair reversals, had caused so much trouble for the pioneers of the periodic
system. Tellurium has a lower atomic number than iodine and so should genuinely
be placed before iodine, as Mendeleev and others had guessed. In addition, it was
now clear that the higher atomic weight of tellurium was due to a higher average
mass of the various isotopes that made up a terrestrial sample of this element.

The final piece of evidence that completed this episode regarding isotopes and
the periodic table was provided by the Harvard chemist Theodore W. Richards in
1914. Although the idea that isotopes of the same element, arising from different
mineral sources, might have different atomic weights had been discussed for a few
years, it had not been directly examined. Richards, an acknowledged expert on
atomic weight determination, was ideally placed to undertake this research. Since
the element lead had been found to be the end point in several radioactive decay
series, it was reasonable to expect this element to show some variation in atomic
weight. Any such variation would depend on the mineral source used, since any
lead found in earth might have resulted from one of several different elements by a
process of natural transmutation brought about by radioactive decay.

Fajans' and Soddy's independent work on the displacement series had shown that the stable end products of all three radioactive decay series, as well as common lead, were chemically indistinguishable or, in the new terminology, were all isotopes of lead. What Richards set out to discover was whether different naturally occurring mixtures of these isotopes might show different atomic weights, as one might expect. In the report by Richards and a young German student Max Lembert, they called their results "amazing." They had found atomic weights of lead differing from that of common lead by as much as 0.75 of an atomic weight unit, an amount that was several times larger than the error associated with their experimental method. Repeated purification of lead samples from various radioactive origins produced no changes in their atomic weights. Encouraged by this research, others tried to find other ores of lead in the hope of showing an even greater variation in atomic weights. These efforts eventually produced a lowest value of 207.05 and a highest value of 207.90.

POSTSCRIPT ON TRIADS

The discoveries discussed in this chapter effectively revived Prout's hypothesis as well as the related notion of the unity of matter, two philosophical ideas discussed in chapter 2. The other main theoretical notion discussed in chapter 2, that of triads, also received a form of rehabilitation. The discovery of triads had given the very first hint that groups of three elements were related to each other. These relationships were not just in chemical similarities but were also numerical, in that the atomic weight of one of the three elements in a triad was shown to be approximately the arithmetic mean of the weights of the other two.

This idea had been at the root of Döbereiner's work, which is often taken as marking the birth of the modern periodic system. However, there were limits to the applicability of the triad concept, and it is probably fair to say that much time was wasted by other researchers in trying to uncover triads where they simply did not exist. Some pioneers, including Mendeleev, made it a point to turn their backs on the two original concepts of Prout's hypothesis and the existence of numerical triads. This attitude certainly seems to have paid dividends for Mendeleev in that he made progress where others had failed to do so.

The problem with triads and, indeed, Prout's hypothesis is easy to discern in retrospect. It is simply that atomic weight, which both concepts draw upon, is not the most fundamental quantity that can be used to systematize the elements. Atomic weight such as that of lead, as just discussed, depend on the particular geological origin of the sample examined. In addition, the measured weight is an average of several isotopes of the particular element. Atomic number, on the other hand, is fundamental and correctly characterizes, as far as presently known, the distinction between one element and the next. Prout's hypothesis is brought back to life if one considers that all the elements are composites of the atomic number, or charge, of the element hydrogen, which has a value of 1.

In the case of triads, the adoption of atomic number has an intriguing consequence that has seldom been discussed: About 50% of all vertical triads based on atomic number, rather than atomic weight, are absolutely exact! This remarkable result is quite easy to appreciate by referring to the long form of the modern periodic table (figure 6.7).

By considering elements from rows 1, 2, and 3, such as helium, neon, and argon, a perfect atomic number triad is obtained:

$$
\begin{array}{ll}
\text{He} & 2 \\
\text{Ne} & 10 = (2 + 18)/2 \\
\text{Ar} & 18
\end{array}
$$

Or, from rows 3, 4, and 5:

$$
\begin{array}{ll}
\text{P} & 15 \\
\text{As} & 33 = (15 + 51)/2 \\
\text{Sb} & 51
\end{array}
$$

Or, from rows 5, 6, and 7:

$$
\begin{array}{ll}
\text{Y} & 39 \\
\text{Lu} & 71 = (39 + 103)/2 \\
\text{Lr} & 103
\end{array}
$$

Alternatively, any triads taken from combinations of elements in rows 2, 3, 4 or from rows 4, 5, 6, and so on, do not give perfect triads.

The reason why this works so perfectly, albeit in only 50% of possible triads, is because the length of each period repeats just once in the long-form periodic table, with the exception of the very first short period. The full sequence is 2, 8, 8, 18, 18, 32, 32, and so on. So, if one selects any element, then there is a 50% chance that the element above and below the selected element, in the same column of the periodic table, will have atomic numbers lying at an equal distance away from the original element. If this is the case, then it follows trivially that the second element in the sequence will lie exactly midway between the first and third elements. In numerical terms, its atomic number will be the exact mean of the first and third elements, or in other words, the atomic number triad will hold perfectly. All one needs to do is to pick a middle element from the first of a repeating pair of periods. Thus, half of all the elements are good candidates. This phenomenon falls out mathematically from the fact that all periods repeat (except for the first one) and that the elements are characterized by whole number integers. It would appear that the original discoverers had accidentally stumbled upon the fact that some periods of elements repeat twice. What held them back was that these repeat

H																															He
Li	Be																									B	C	N	O	F	Ne
Na	Mg																									Al	Si	P	S	Cl	Ar
K	Ca															Sc	Ti	V	Cr	Mn	Fe	Co	Ni	Cu	Zn	Ga	Ge	As	Se	Br	Kr
Rb	Sr															Y	Zr	Nb	Mo	Tc	Ru	Rh	Pd	Ag	Cd	In	Sn	Sb	Te	I	Xe
Cs	Ba	La	Ce	Pr	Nd	Pm	Sm	Eu	Gd	Tb	Dy	Ho	Er	Tm	Yb	Lu	Hf	Ta	W	Re	Os	Ir	Pt	Au	Hg	Tl	Pb	Bi	Po	At	Rn
Fr	Ra	Ac	Th	Pa	U	Np	Pu	Am	Cm	Bk	Cf	Es	Fm	Md	No	Lr	Rf	Db	Sg	Bh	Hs	Mt	Ds	Rg							

FIGURE 6.7 Long-form table showing that about 50% of all atomic number triads are exact. The heavily outlined elements represent examples of perfect atomic number triads.

distances vary in length and, of course, the fact that they were operating with the vagaries of atomic weight data.

It is somewhat amusing to think that the ancient notions of Prout's hypothesis and triads of elements, which were initially so productive and later so strongly criticized, have been shown to be essentially correct, and that the reason for their being essentially correct is now fully understood. In fact, the philosopher of science Imre Lakatos used the example of Prout's hypothesis to illustrate a theory making a "comeback" after being apparently refuted.[48]

CONCLUSION

This chapter examines the various lines of research on the nucleus of the atom that contributed to the evolution of the periodic system. This represents the first time that work in physics began to have a profound impact on the way the periodic system was understood. Perhaps the most important of these contributions has been the concept of atomic number, first argued for by van den Broek and first experimentally demonstrated by Moseley. The importance of this work is that, for the first time, chemists now had an unambiguous method for determining exactly how many elements were present and where in the periodic system any gaps might still remain to be occupied by new elements.

THE ELECTRON AND
CHEMICAL PERIODICITY

J.J. Thomson's discovery of the electron is one of the most celebrated events in the history of physics.[1] What is not so well known is that Thomson had a deep interest in chemistry, which, among other things, motivated him to put forward the first explanation for the periodic table of elements in terms of electrons.[2] Today, it is still generally believed that the electron holds the key to explaining the existence of the periodic table and the form it takes. This explanation has undergone a number of subtle changes. The extent to which the modern explanation is purely deductive or whether it is semiempirical is examined in this chapter.

While Dimitri Mendeleev had remained strongly opposed to any attempts to reduce, or explain, the periodic table in terms of atomic structure, Julius Lothar Meyer was not so averse to the reduction of the periodic system. The latter strongly believed in the existence of primary matter and also supported William Prout's hypothesis. Lothar Meyer did not hesitate to draw curves through the numerical properties of atoms, whereas Mendeleev believed this to be a mistake, since it conflicted with his own belief in the individuality of the elements.

This is how matters stood before the discovery of the electron, three years prior to the turn of the twentieth century. The atom's existence was still very much a matter of dispute, and its substructure had not yet been discovered. There appeared to be no way of explaining the periodic system theoretically.[3]

THE DISCOVERY OF THE ELECTRON AND
EARLY MODELS OF THE ATOM

Johnston Stoney first proposed the existence and name for the electron in 1891, although he did not believe that it existed as a free particle. Several researchers discovered the physical electron, including Emil Wiechert in Köningsberg, who was the first to publish his findings. Because these early researchers did not seriously follow up on their results, it was left to the British physicist Thomson to capitalize upon and establish the initial observations. These false starts show an interesting parallel with the discovery of the periodic system, where the essential idea of periodicity occurred to a number of scientists, including Emile De Chancourtois, John Newlands, and William Odling, none of whom was able to make much headway in establishing his insights.

While Wilhelm Röntgen discovered X-rays by experimenting with cathode rays, Thomson was one of several physicists who set out to explain the very nature of these cathode rays. The experiments he and others were carrying out typically involved the passage of an electric discharge of about 1,000 V through a gas held in a glass tube of about 300 cm in length and 3 cm wide at a pressure of about 0.01 mm of mercury. In 1869, the year of Mendeleev's famous periodic table, Johann Hittorf in Germany had observed that glowing rays were emitted from the cathode, or negative pole, of such an experimental apparatus. Some early workers, such as William Crookes, supported the notion that these "cathode rays" were particles projected from the negative pole and were themselves negatively charged. A number of others working in Germany, such as Heinrich Herz, came to believe that the cathode rays were a form of radiation. In 1897, Wiechert interpreted his own experiments by concluding, "We are not dealing with atoms known from chemistry, because the mass of the moving particles turned out to be 2000–4000 times smaller than the one of the hydrogen atom, the lightest of the known chemical atoms."[4] In the same year, Walter Kaufmann measured the charge-to-mass ratio of cathode rays and found it to be the same in every gas, a fact that puzzled him but did not lead him to draw the conclusion that the particle might be a universal constituent of all substances.

It was in this context that Thomson conducted his own research, which, according to the traditional account, led to the discovery of the electron and the realization that it was indeed a constituent of all matter. By now, it was known that cathode rays were negatively charged, and they appeared to be particulate. But there was yet no confirmation of their particle nature. This would be forthcoming only if it could be shown that cathode rays could be deflected by an electric field, something that had eluded all previous attempts. Thomson succeeded where others had failed by using an extremely high electric charge as well as by ensuring that the glass tube was under vacuum conditions. Under these conditions, the cathode rays finally showed a deflection due to an electric field in 1897.

Moreover, Thomson was able to measure the charge-to-mass ratio of cathode rays and found a value of 770 for cathode rays emanating from hydrogen atoms. This finding suggested three possibilities: the particles making up the cathode rays bore a very large charge, or they had a very small mass, or possibly a combination of the two effects. It was later found that cathode rays, or electrons, as they became known, had the same charge as hydrogen ions, although of opposite sign, but had a much smaller mass.[5] Last but not least, Thomson went beyond Wiechert in that he repeated his experiments with cathode rays produced from various different elements, concluding that the same particle was produced in each case, and that this particle was therefore a fundamental constituent of all matter. Thomson seems to have disliked Stoney's name for the particle, although it had been popularly adopted. He insisted on calling it the "corpuscle," only later capitulating to the popular usage of "electron."

MODELS OF THE ATOM

The newly discovered electron began to feature in several postulated models of the atom. The French physicist Jean Perrin, like Thomson in England, had conducted experiments on cathode rays. In fact, Perrin had been the first to obtain direct proof that the electron was negatively charged. This was carried out in experiments in which a metal cylinder was placed inside a vacuum tube in order to collect the charge. By drawing on this finding, in addition to the experimental evidence gathered by Thomson, Perrin suggested the first planetary conception of the atom in 1901. He proposed that each atom consisted of one or more highly charged positive bodies, much like a positive sun around which small negative planets, or electrons, were in orbit. Perrin also believed that the total negative charge in the atom would be exactly equal to the total positive charge, in apparent anticipation of current views on the structure of the atom. He stated his hypothesis thus:

> Each atom will consist of one or more highly charged positive bodies, a kind of positive sun whose charge is much higher than that of a corpuscle (electron), and also of a kind of small negative planets, all these bodies gravitating under the action of electrical forces, and with total negative charge exactly equal to the total positive charge, so that the atom is electrically neutral.[6]

A further proposal by Perrin in the same paper seems to foreshadow later work on the connection between the structure of the atom and spectral frequencies: "The gravitational periods of the different masses in the atom might correspond to the different wavelengths of light revealed in the rays of the emission spectrum."[7] In modern terms, the wavelengths of light revealed in the atomic spectra are not related to gravitational periods but to transitions between energy levels, which are characteristic of the various orbitals that the electrons can occupy.[8] In 1903, Hantaro Nagaoka in Japan independently proposed a Saturnian atom in which electrons move in one or more rings around a central body.[9] A translation of one of his lectures was published in 1904 in the *Philosophical* Magazine[10] and was subsequently quoted by leading physicists such as Ernest Rutherford and Henri Poincaré.

In that same year, Thomson began to think specifically about how the electrons might be arranged in the atom. He concluded that the solar system-like atoms of Perrin and Nagaoka would be unstable because the orbiting electrons would continuously radiate energy, eventually falling into the center of the atom. He suggested an alternative model in which the electrons were embedded in the nucleus, circulating within its positive charge. This became known as the "plum pudding" model of the atom. In a paper of 1904, Thomson also published the first set of electronic arrangements, or what today would be called electronic configurations.[11] In taking this step, Thomson went beyond Perrin and Nagaoka in conceiving of the electrons not just as moving around the atom but doing so in a structured manner.

Thomson based his configurations of electrons on the work of an American physicist Alfred Mayer, who had experimented with magnets that were attached to corks and floated in a circular basin of water above which was placed a current-bearing metal coil (figure 7.1). Meyer had found that when up to five magnets were floated, they would form a single ring, but that on the addition of a sixth magnet a new ring would be formed.[12] As more magnets were added, the phenomenon repeated: When a certain number of magnets was reached, the addition of a new magnet caused the formation of yet another ring, thus producing an arrangement of concentric rings. Thomson believed that the same kind of principle might operate in the case of electrons circulating in the atom and began to develop these views in an attempt to explain the periodic table in terms of the electron.

In many respects, Thomson can therefore be regarded as the originator of electronic configurations and of attempts to explain the periodic table in terms of them. Table 7.1 is an extract from one of Thomson's later articles showing how

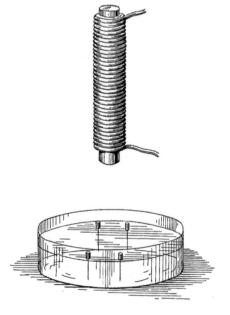

FIGURE 7.1
Mayer's floating magnets. From Alfred M. Mayer, On the Morphological Laws of the Configurations Formed by Magnets Floating Vertically and Subjected to the Attraction of a Superposed Magnet; With Notes on Some of the Phenomena in Molecular Structure Which These Experiments May Serve to Explain, *American Journal of Science*, 15, 276–277, 1878. This image from the original paper is reproduced from J.J. Thomson, *The Corpuscular Theory of Matter*, Archibald Constable, London, 1907, p. 111.

TABLE 7.1
J.J. Thomson's Electron Rings

Number of electrons	Rings	Number of electrons	Rings
5	5	16	5 + 11
6	1 + 5	17	1 + 5 + 11
7	1 + 6	18	1 + 6 + 11
8	1 + 7	19	1 + 7 + 11
9	1 + 8	20	1 + 7 + 12
10	2 + 8	21	1 + 8 + 12

Based on J.J. Thomson, *The Corpuscular Theory of Matter*, Archibald Constable, London, 1907, pp. 109–110.

his electron rings were arranged. As with Mayer's cork rings, the presence of five electrons in an atom results in the formation of one electron ring in Thomson's account. A second ring begins to form once the number of electrons reaches six, although after this happens, new electrons continue to be added to the first ring, just as in the case of the floating needles and corks. On reaching 10 electrons, a new electron suddenly appears in the second ring, and on reaching 17 electrons, a third ring begins to form. In each case, the additional, or differentiating, electron is generally being added to an inner ring rather than to an outer one.

From a modern point of view, these electronic arrangements have little merit in chemical terms since they suggest a nonexistent analogy between, for example, element 5, boron, and element 16, sulfur. It would be expected that, since they are assigned five electrons in their outermost shells according to this scheme, boron and sulfur would display similar chemical properties, which is not in fact the case. But it would be a mistake to criticize Thomson on this point, since in 1904 he and his contemporaries were not aware of the number of electrons in any particular atom. Not until Henry Moseley's work with atomic number was published 10 years later would it become clear that the serial number of an element in the periodic table, its atomic number, corresponds to the number of positive charges in the atom. In proposing his new scheme of electron rings, J.J. Thomson was merely suggesting the plausibility of explaining periodicity through similarities in electronic structures among different elements, something that remains valid to this day.

Although Thomson's atomic model would soon be discarded by Rutherford when he introduced his nuclear model of the atom,[13] it did succeed in establishing two important concepts. One was that the electron held the key to chemical periodicity, and the other was the notion that the atoms of successive elements in the periodic table differ by the addition of a single electron. Both of these ideas were to become important aspects of Niels Bohr's atomic theory of periodicity, which would soon be published.

THE QUANTUM THEORY OF THE ATOM

When Rutherford revived Perrin's and Nagaoka's planetary model of the atom following Hans Geiger and Ernest Marsden's alpha particle scattering experiments, he left the problem of the model's stability unresolved. According to James Clerk Maxwell's electromagnetic theory, any circulating charged body should lose energy through radiation, so the orbiting electrons would be expected to spiral into the nucleus. The nuclear model implied that any atom, and consequently all matter, would thus be unstable, contrary to the obvious facts of experience. Furthermore, Rutherford's model could not explain the discrete nature of the optical spectra of atoms that had been accurately recorded since the development of the spectroscope in 1859 and that had been used to identify many new elements.[14]

The pattern observed following the dispersion of light emitted from excited atoms is quite different from that of white light. Instead of a continuous spectrum ranging from red to violet frequencies, one observes a series of discrete lines of various colors. Some particular color frequencies are simply missing compared with the spectrum from a source of white light. The discrete nature of the spectrum in the case of atoms could not be explained by any of the atomic models that have been reviewed so far. In Rutherford's model, for example, the energies of the electrons are not restricted to particular values; consequently, all possible transition energies would be expected to occur, and the optical spectrum of any element would be continuous rather than discrete.

Both of these problems, the stability of atoms and the discrete nature of atomic spectra, were resolved by the Danish physicist Bohr (figure 7.2), who also provided one the first successful explanation of the periodic system in terms of arrangements of electrons in the atom.[15] Bohr obtained a Ph.D. in theoretical physics of the study of metals before undertaking a one-year postdoctoral fellowship with Rutherford at Manchester. Although other physicists had begun to establish the quantum theory in physics, Bohr was the first to apply these ideas in the context of atomic physics.

Bohr first came to prominence in 1913 when he published his quantum theory of the hydrogen atom. The notion of quanta, or packets of energy, had been introduced by Max Planck in 1900 to explain the details of observations made on the black-body radiation.[16] Bohr adopted Planck's notion of quantization[17] and applied it to the physics of atoms. His calculations led him to conclude that, in the planetary model of the atom, additional rings of electrons are formed outside already full rings, correcting Thomson's model of electron rings, in which electrons are added to inner rings. Most important, Bohr proposed that electrons would be stable if they remained in certain quantized orbits and would lose energy only on undergoing transitions from one orbit to another, more stable orbit.[18] Electrons in a discrete set of stable orbits around the nucleus of an atom were said to be in stationary states that would not radiate energy:[19]

FIGURE 7.2
Niels Bohr. Photo and permission
from Emilio Segrè Collection.

An atomic system can only exist permanently in certain series of states corresponding to a discontinuous series of values for its energy, and that consequently any change of the energy of the system, including emission and
absorption of electromagnetic radiation, must take place by a complete transition between two such states. These states are denoted as "stationary states" of
the system.[20]

Bohr was following Planck's lead in departing from classical electromagnetic
theory. In studying black-body radiation, which occurs at very short frequencies,
Planck had found it necessary to introduce a constant, h, also called the elementary
quantum of action, to explain its discontinuous nature. Such radiation could be
emitted or absorbed only in packets, or quanta, described by the formula $h\nu$, where
ν is the frequency of the radiation and h is Planck's constant. Bohr was suggesting
that the atom could likewise not be described adequately by the laws of classical
mechanics but that it required a quantum description.

Applying Planck's idea of how electrons move from one stationary state to
another, Bohr proposed that for the atom to pass from one energy state to another
it must emit or absorb one quantum, $h\nu$, of energy:

The radiation absorbed or emitted during a transition between two stationary
states is "unifrequentic" and possesses a frequency ν given by the relationship,

$$E' - E = h\nu$$

where h is Planck's constant, and E' and E are the values of the energy in the
two states under consideration.[21]

But this theory was limited in its application in that it gave an exact account only of the spectrum of hydrogen, the simplest case.[22] Atoms with more than one electron are much more complicated, since the various electrons exert influences on each other. Nevertheless, Bohr had sufficient confidence in his quantum theory of the atom to try to apply it to multielectron atoms in an approximate manner.

Although he first applied his quantum theory of the atom to the spectrum of the hydrogen atom, historians of physics John Heilbron and Thomas Kuhn have shown rather conclusively that the initial motivation for Bohr's theory was more comprehensive. Bohr was rather attempting to gain an understanding of the periodic table through electronic configurations[23] and to examine the stability of the electron rings with which Thomson had tried to explain the periodic table. In this same article of 1913, Bohr produced his first version of an electronic periodic table.[24] He assigned electronic configurations to the atoms of various elements in terms of the principal quantum number of each electron, which could be used to characterize its stationary or nonradiating states (table 7.2).

Bohr's general method, called the *aufbauprinzip* (which translates as "building up"), consisted in building up atoms of successive elements in the periodic table by the addition of an electron to the previous atom. On moving from one element to the next in the periodic table, Bohr supposed that an additional electron was added to the outermost shell, although there were exceptions to this rule, as discussed below. At specific stages in this process, a shell would become full, at which point a new shell would begin to fill. Contrary to the impression that he created in his published articles, however, Bohr was unable to deduce the maximum capacity of each electron shell, and he allowed himself to be guided almost entirely by chemical and spectroscopic data rather than theoretical calculations.

The fact that Bohr used essentially chemical considerations in producing these configurations can be seen clearly in his choice of configuration for certain elements. The population of electrons in the outermost ring is determined by chemical valence. These electrons are the most loosely bound to the nucleus and thus the most likely to bond with another atom. In the case of nitrogen, for example, Bohr was forced to rearrange an inner shell in order to make the configuration correspond to the element's known trivalence. This can be seen in table 7.2. Whereas from hydrogen to carbon the atoms have two inner electrons and a varying number of outer electrons, once nitrogen is reached the inner electron shell abruptly doubles in its number of electrons. This move appears a little odd until it is realized that it is made precisely to obtain the three outer electrons needed to correspond with the fact nitrogen forms three chemical bonds.

In altering his configurations to make them agree with experimental evidence, Bohr gave no theoretical arguments for why such a rearrangement should occur.[25] Such abrupt rearrangements can be seen in a number of places even just among the 24 configurations shown in the table 7.2, such as for the atoms of phosphorus and arsenic. The atoms of both of these elements show valences of 3, while oxy-

TABLE 7.2

Bohr's Original Scheme for Electronic Configurations of Atoms. Numbers of electrons is consecutive energy levels, beginning closest to the nucleus.

1	H	1				
2	He	2				
3	Li-	2	1			
4	Be	2	2			
5	B	2	3			
6	C	2	4			
7	N	4	3			
8	O	4	2	2		
9	F	4	4	1		
10	Ne	8	2			
11	Na	8	2	1		
12	Mg	8	2	2		
13	Al	8	2	3		
14	Si	8	2	4		
15	P	8	4	3		
16	S	8	4	2	2	
17	Cl	8	4	4	1	
18	Ar	8	8	2		
19	K	8	8	2	1	
20	Ca	8	8	2	2	
21	Sc	8	8	2	3	
22	Ti	8	8	2	4	
23	V	8	8	4	3	
24	Cr	8	8	2	2	2

N. Bohr, On the Constitution of Atoms and Molecules *Philosophical Magazine*, 26, 476–502, 1913, 497

gen and sulfur display valences of 2 and fluorine and chlorine display valences of 1, in accordance with the chosen configurations. Instead of rigorously deriving his atomic model from quantum theory, Bohr relied on intuition as well as spectroscopic and purely chemical considerations.[26]

Nevertheless, Bohr achieved two things with his theory. One was that he introduced the important idea that the differentiating electron should, in most cases, occupy an outer shell and not an inner one, as Thomson had thought was the case. Second, in spite of some arbitrary aspects, Bohr's scheme provided at least some correlation between electronic configurations and chemical periodicity. For example, the configuration of lithium is 2, 1, while that of sodium, which lies in the same group chemically, is 8, 2, 1. Similarly, beryllium and magnesium, which are found together in group II of the periodic table, share the property of having two outer-shell electrons. This is the origin of the modern notion that atoms fall into the same group of the periodic table if they possess the same numbers of outer-shell electrons, something that had already been hinted at by Thomson.[27]

Following this work, Bohr abandoned the question of periodicity for about a decade, and it was left to various chemists to try to improve upon the electronic version of the periodic table.[28] As discussed in chapter 8, there are some grounds for thinking that Bohr's later tables were directly influenced by the more detailed electronic configurations given by the chemist Charles Bury and that insufficient credit has been given to this pioneer of electronic configurations.

BOHR'S SECOND THEORY OF THE PERIODIC SYSTEM

In 1921, Bohr returned to the problem of atomic structure and the periodic table. In 1922 and 1923, he announced a new, improved version of the electronic periodic table.[29] Again he employed the *aufbauprinzip* to build up successive atoms in the periodic table, but this time he used two quantum numbers: n, the principal quantum number, and k, the second or azimuthal quantum number, which later became labeled as ℓ (table 7.3). The second quantum number had recently been discovered by Arnold Sommerfeld, a theoretical physicist in Munich.

Whereas Bohr had assumed the orbit of the hydrogen electron to be circular, Sommerfeld realized that it was elliptical. Since the angular momentum of an electron moving in an elliptical orbit would change continually, the orbit itself would precess, independently of the motion of the electron in its ellipse. Thus, the electron would have two degrees of freedom: the orbiting motion of the electron, and its precession. To describe the latter motion, Sommerfeld introduced a second quantum number, ℓ, the azimuthal quantum number, which depended on the principal quantum number and could adopt values of $n-1, n-2, \ldots, 0$.

When Bohr became aware of this discovery, he applied it to many-electron atoms and produced the set of more detailed electronic configurations shown in table 7.3. These numbers emerged from the quantization that was imposed mathematically on the system and served to identify the stationary states of the system, as they had in his earlier theory. According to this scheme, an atom of nitrogen, for example, with seven electrons, would have an electronic configuration of 2, 4, 1. It is interesting to see that in the case of nitrogen and a few other elements, Bohr's more detailed theory of 1922 seems to have taken a retrograde step, since contrary to the configuration he had given in 1913, the newer version did not accord well with the experimental fact that nitrogen forms three chemical bonds, a point taken up further below.

TABLE 7.3

Bohr's 1923 Electronic Configurations Based on Two Quantum Numbers. Numbers of electrons is consecutive energy levels, beginning closest to the nucleus.

H	1				
He	2				
Li	2	1			
Be	2	2			
B	2	3			
C	2	4			
N	2	4	1		
O	2	4	2		
F	2	4	3		
Ne	2	4	4		
Na	2	4	4	1	
Mg	2	4	4	2	
Al	2	4	4	2	1
Si	2	4	4	4	
P	2	4	4	4	1
S	2	4	4	4	2
Cl	2	4	4	4	3
Ar	2	4	4	4	4

N. Bohr, Linienspektren und Atombau, *Annalen der Physik*, 71, 228-288, 1923, p. 260

Following the early success of his theory of the hydrogen atom, Bohr was invited to give a series of seven lectures in 1922 at the University of Göttingen. Some of the physicists present in the audience for these lectures, which became known as the Bohrfest, included Werner Heisenberg, Wolfgang Pauli, Sommerfeld, and Max Born as well as Göttingen's leading mathematical physicist, David Hilbert. Throughout his career, Bohr was regarded more for his physical insight and his ability to synthesize ideas in atomic physics rather than for any special mathematical prowess, which he left to others like Heisenberg, Erwin Schrödinger, Pauli, and Paul Dirac. This lack of a formal mathematical approach was evident in Bohr's lectures at Göttingen, which produced questions from the audience regarding the mathematical justification for what Bohr was doing. It would appear that in many cases there were no such justifications.

As several of the Göttingen physicists who were exposed to these ideas by Bohr's own lectures later commented, the work rested on a mixture of ad hoc arguments and chemical facts without any derivations from the principles of quantum theory, to which Bohr frequently alluded. According to the German physicist Heisenberg,

It could very distinctly be felt that Bohr had not reached his results through calculations and proofs but through empathy and inspiration and it was now difficult for him to defend them in front of the advanced school of mathematics in Göttingen.[30]

Friederich Hund wrote:

After he had explained a simple spectrum he came to his crucial review of the structure of atoms with regard to their positions in the periodic system. In some respects this turned out to be obscure and not always easy to understand.[31]

In a book containing Bohr's famous 1923 paper on the *aufbauprinzip*, Pauli made a revealing marginal remark. In discussing the adding of the 11th electron to the closed shell of 10 electrons, Bohr says, "We must expect the eleventh electron to go into the third orbit." Pauli, obviously annoyed by this statement, writes hastily in the margin with an exclamation mark, "How do you know this? You only get it from the very spectra you are trying to explain!"[32] The notion that the periodic table was deduced from quantum theory by Bohr is thus something of an exaggeration.

Bohr claimed that his *aufbauprinzip*, by which he applied his theory of the atom to multielectron atoms, was based on an important principle of quantum theory called the adiabatic principle:[33]

Suppose that for some class of motions we for the first time, introduce the quanta. In some cases the hypothesis fixes completely which special motions are to be considered as allowed. This occurs if the new class of motions are derived by means of an adiabatic transformation from some class of motions already known.[34]

Introduced by Paul Ehrenfest in 1917, the adiabatic principle allows one to find the quantum conditions when an adiabatic or gradual change is imposed on a system.[35] However, it depends on the possibility of deriving the new motion from the known one by means of an adiabatic transformation. For example, the quantum states of a particular system are known, the new quantum states that result from a gradual change, such as the application of an electric or a magnetic field, can be calculated. The quantities that preserve their values after such a transformation are known as adiabatic invariants. Ehrenfest showed that for any arbitrary periodic motion, the following quantity is an adiabatic invariant:

$$2T/n,$$

where T is the time average of the kinetic energy and n is the frequency of motion.

There are stringent restrictions on the applicability of the adiabatic principle. Ehrenfest himself showed that it was applicable to simply periodic systems.[36] These are systems having two or more frequencies that are rational fractions of each other. In such systems, the motion will necessarily repeat itself after a fixed interval of time. Later, J.M. Burgers, a student of Ehrenfest, showed that it was also applicable to multiply periodic systems.[37] In these more general systems, the various frequencies are not rational fractions of each other, such that the motion does not necessarily repeat itself.[38] The hydrogen atom provides an example of a multiply periodic system, with its two degrees of freedom.

An even more general class of systems is termed aperiodic, and as far as is known, the adiabatic principle does not apply in such cases. Unfortunately for the field of atomic physics, all atoms larger than that of hydrogen constitute aperiodic systems. In the helium atom, for example, the motion of each of the two orbiting electrons changes according to the varying interaction with the other electron as their distance apart changes (in terms of the early Bohr theory). We may no longer speak of a constant period for either of the electrons.

Bohr was well aware of this limitation of the adiabatic principle but continued to use it even for many-electron atoms in the hope that it might still remain valid for such aperiodic systems. He repeatedly acknowledged this point in his writings:

> For the purposes of fixing the stationary states we have up to this point only considered simply or multiply periodic systems. However the general solution of the equations frequently yield motions of a more complicated character. In such a case the considerations previously discussed are not consistent with the existence and stability of stationary states whose energy is fixed with the same exactness as in multiply periodic systems. But now in order to give an account of the properties of the elements, we are forced to assume that the atoms, in the absence of external forces at any rate always possess sharp stationary states, although the general solution of the equations of motion for the atoms with several electrons exhibits no simple periodic properties of the type mentioned.[39]

Later in his 1923 article, he states:

> We shall try to show that not withstanding the uncertainty which the preceding conditions contain, it yet seems possible even for atoms with several electrons to characterize their motion in a rational manner by the introduction of quantum numbers. The demand for the presence of sharp, stable, stationary states can be referred to in the language of quantum theory as a general principle of the existence and permanence of quantum numbers.[40]

Bohr's attitude as expressed in these writings does not seem to be very rigorous, but more akin to the obscurantism that characterized some of his scientific work.[41] In the two above-quoted passages he appears to ignore the problems he himself elaborates and merely expresses the hope of retaining the quantum numbers even though one is no longer dealing with multiply periodic systems.

The main feature of the building-up procedure, as mentioned above, was Bohr's assumption that the stationary states would also exist in the next atom in the periodic table, obtained by the addition of a further electron. Bohr also assumed that the number of stationary states would remain unchanged from the atom of one element to the next, apart from any additional states pertaining to the newly introduced electron. He thus envisaged the existence of sharp stationary states, and their retention on adding both an electron and a proton to an atom.

Bohr's hypothesis of the permanence of quantum numbers came under attack from the analysis of the spectral lines under the influence of a magnetic field.[42] As is generally the case, the application of a magnetic field on the atoms results in a splitting, to produce more lines than occur in the absence of such a field. An atomic core consisting of the nucleus and inner-shell electrons showed a total of N spectroscopic terms in a magnetic field. If an additional electron having an azimuthal quantum number k were to be added, the new composite system would be expected to show $N(2k - 1)$ states, since the additional electron was associated with $2k - 1$ states. However, experiments revealed more terms. In general, the terms split into one type consisting of $(N + 1)(2k - 1)$ components and a second type consisting of $(N - 1)(2k - 1)$ components, giving a total number of $2N(2k - 1)$ components. This represents a violation of the number of quantum states, since a twofold increase seems to occur in the number of atomic states on the introduction of an additional electron. Bohr's response was to maintain adherence to the permanence of quantum numbers even in the face of this evidence. He merely alluded to a mysterious device, which he called a nonmechanical "constraint," to save the quantum numbers.

Bohr's account of the periodic table also came under attack from chemical evidence. The element nitrogen, for example, was attributed an electronic configuration of 2, 4, 1, as noted above. This grouping of electrons suggested that 1 or 5 electrons were more loosely bound than the others and implied either penta- or univalence, neither of which is the case in practice, as nitrogen is predominantly trivalent.

Despite the problems with his quantum theory, however, Bohr went on to make numerous other contributions to atomic physics and quantum mechanics in the course of his long life. Indeed, Bohr is probably the best-known physicist of the twentieth century, eclipsed only by Albert Einstein. After Bohr's theories of 1913 and 1922–1923, he remained at the heart of developments in quantum theory, although specific steps were often taken by others, including Heisenberg, Schrödinger, and Pauli. But throughout this period, and for many years later, Bohr played the role of godfather to quantum theory by founding an international institute in Copenhagen, which hosted many of the world's leading physicists as they continued to shape the new quantum mechanics. In addition, he served as the focal point for the discussions on the nature of quantum mechanics and had a profound influence on many of the physicists of his generation through his willingness to engage in debate.

EDMUND STONER

Shortly after introducing his second theory of the periodic system, Bohr began to believe that the assumptions on which it was based might be unfounded, but it was not until the work of Pauli a little later that the situation would begin to be clarified. In the meantime, another physicist, Edmund Stoner, was to provide the next missing piece of the puzzle of quantum numbers and the periodic table.

In 1924, British-born Stoner (figure 7.3), then a graduate student at Cambridge University, took the next step in using electronic configurations to explain the periodic table. His approach was based on using not merely two quantum numbers, but also a third one introduced by Sommerfeld shortly before.[43] The third, or inner, quantum number, j, refers to the precession of the orbital motion in the presence of a magnetic field. Its value is tied to the second quantum number such that j can take all values ranging from $-k$ to $+k$, increasing in integral steps.[44]

The occurrence of this third quantum number suggested additional stationary states in the atom, but Bohr did not extend his electronic configuration scheme accordingly. As mentioned above, Bohr was becoming increasingly interested in the deeper question of the existence of stationary states for individual electrons in

FIGURE 7.3
Edmund Stoner. Photo and
permission from University of Leeds.

TABLE 7.4

Stoner's Scheme for Assignment
of Electronic Configurations

n	k	j	Number of electrons
1	1	1	2
2	1	1	2
2	2	1	2
2	2	2	4
3	1	1	2
3	2	1	2
3	2	2	4

Based on E. Stoner, The Distribution of Electrons Among
Atomic Levels, *Philosophical Magazine*, 48, 719–736, 1924,
p. 720.

many-electron atoms. That is, he was concerned about the fact that, strictly speaking, the electrons in many-electron atom are not in stationary states. It is rather the atom as a whole that possesses stationary states. This "holistic" property denies the validity of the independent electron approximation wherein each electron is in a stationary state and can be characterized by its own set of quantum numbers.[45]

The young Stoner, undaunted by these theoretical problems, reexamined the experimental evidence on optical as well as X-ray spectra of atoms. Based on his studies, he suggested that the number of electrons in each completed level should equal twice the inner quantum number of that particular shell. This produced the scheme shown in table 7.4 for ascribing electrons to shells.

When Stoner applied this relationship to the three quantum numbers, he deduced the set of electron configurations shown in table 7.5. According to Stoner's scheme, the electronic configuration for the element nitrogen is 2, 2, 2, 1, where the last three numbers represent the outer-shell electrons. This configuration could account successfully for the valence state of 3 shown by the nitrogen, whereas Bohr's scheme could not. However, this new scheme could not resolve the above-mentioned problem of the violation of number of quantum states as was seen in the splitting of spectral lines in a magnetic field.

As the problems with what became known as Bohr's "old quantum theory" began to deepen, some physicists, such as Heisenberg and Pauli, started to question the reality of electron orbits. For example, in Pauli's correspondence with Bohr there is the following passage on this issue: "The most important question seems to be this: to what extent may definite orbits in the sense of electrons in stationary states be spoken of at all."[46]

TABLE 7.5

Stoner's Configurations of 1924 Based on Three Quantum Numbers

	Numbers of electrons in successive energy levels beginning closest to the nucleus.						
H	1						
He	2						
Li	2	1					
Be	2	2					
B	2	2	1				
C	2	2	2				
N	2	2	2	1			
O	2	2	2	2			
F	2	2	2	3			
Ne	2	2	2	4			
Na	2	2	2	4	1		
Mg	2	2	2	4	2		
Al	2	2	2	4	2	1	
Si	2	2	2	4	2	2	
P	2	2	2	4	2	2	1
S	2	2	2	4	2	2	2
Cl	2	2	2	4	2	2	3
Ar	2	2	2	4	2	2	4

Based on E. Stoner, The Distribution of Electrons Among Atomic Levels, *Philosophical Magazine*, 48, 719–736, 1924, p. 734

THE PAULI EXCLUSION PRINCIPLE

In 1923 Bohr wrote to Pauli (figure 7.4), asking him to try to bring order to the increasingly complicated situation in atomic physics[47] and to attempt to save the quantum numbers. Pauli responded with two papers that seemed to clarify matters, and in the process he developed his exclusion principle, which has become one of the central pillars of modern physics. Once again, the motivation for this work was partly an attempt to explain the periodic table of the elements.

Pauli's first main contribution was to challenge the view held at the time that the core of an atom possesses an angular momentum.[48] Alfred Landé[49] had proposed that the core of the atom, consisting of the nucleus plus the inner electrons, would explain the origin of the complex structure of atomic spectra. Pauli rejected this hypothesis and suggested that the spectral lines and their shifts in the presence of magnetic fields were due entirely to the presence of outer electrons. He went on to propose the assignment of a fourth quantum number, m_s, to each electron

FIGURE 7.4
Wolfgang Pauli. Photo from Author's
collection used by permission from
Emilio Segré Collection.

(table 7.6). This fourth number was due, according to Pauli, to a classically non-describable duplicity in the quantum theoretical properties of the optically active electron,[50] a property now called spin angular momentum.

Armed with four quantum numbers, Pauli found that he could obtain Stoner's classification of electronic configurations from the following simple assumption, which constitutes the famous exclusion principle in its original form: "It should be forbidden for more than one electron with the same value of the main quantum number n to have the same value for the other three quantum numbers k, j and m."[51] The principle is often stated as follows: no two electrons in an atom can have the same set of four quantum numbers. Meanwhile, Pauli justified the assignment of four quantum numbers to each electron by the following apparently clever argument. He supposed that if a strong magnetic field were applied, the electrons would cease to interact and could therefore be said to be in individual stationary states.

Of course, the periodic table arrangement must also apply in the absence of a magnetic field. In order to maintain the validity of the four-quantum-number assignment for each electron even in the absence of a field, Pauli appealed to what he called a "thermodynamic argument." He proposed an adiabatic transformation in which the strength of the magnetic field was gradually reduced such that, even in the absence of the field, the characterization of stationary states for individual electrons remained valid. This argument seemed to ensure the existence of sharp stationary states for individual electrons.

TABLE 7.6
Assignment of Electron Shells Based on Pauli's Scheme

n	ℓ	m_ℓ	m_s	Number of electrons
1	0	0	+1/2	
			−1/2	2
2	0	0	+1/2	
			−1/2	2
2	1	−1	+1/2	
			−1/2	2
2	1	0	+1/2	
			−1/2	2
2	1	1	+1/2	
			−1/2	2
3	0	0	+1/2	
			−1/2	2

Modern labels for the quantum numbers have been used instead of *k* and *j*. This does not alter any of the arguments presented here.

Based on W. Pauli, On the Connexion between the Completion of Electron Groups in an Atom with the Complex Structure of Spectra *Zeitschrift für Physik*, 31, 765–783, 1925. The table shown uses modern labels for the four quantum numbers, not those used by Pauli.

Pauli then considered how this proposal fared with regard to the experimental evidence showing a violation in the number of quantum states. As mentioned above, the problem was that a system expected to show $N(2k − 1)$ states on the addition of a single electron to the atomic core is in fact transformed into two sets of states numbering $(N + 1)(2\ k − 1)$ and $(N − 1)(2k − 1)$ states, or a total of $2N(2k − 1)$.

Pauli was able to resolve this problem very simply. According to his view, the additional electron possesses $2(2k − 1)$ states, in contrast to the former view of only $(2k − 1)$. The twofold increase in the number of observed states arises from the proposed duplicity of states of the new electron. The number of states of the atomic core therefore remains as N. Pauli's arguments appeared very persuasive and were received enthusiastically by the atomic physics community.

Not surprisingly, Bohr was pleased with Pauli's contribution, although both of them seemed to view it as a temporary measure. What they and everybody else failed to notice was that Pauli had committed a fallacy concerning the applicability of the adiabatic principle. A many-electron atom constitutes an aperiodic system to which the adiabatic principle does not apply, as previously emphasized by Bohr. Pauli merely changed the argument from the addition of an extra electron as in the *aufbauprinzip* to the case of gradually reducing the strength of a magnetic field.

TABLE 7.7
Quantum Numbers and Orbitals

n	Possible values of ℓ	Subshell designation	values of m_ℓ	Number of Possible orbitals in subshell
1	0	1s	0	1
2	0	2s	0	1
	1	2p	1, 0, −1	3
3	0	3s	0	1
	1	3p	1, 0, −1	3
	2	3d	2, 1, 0, −1, −2	5
4	0	4s	0	1
	1	4p	1, 0, −1	3
	2	4d	2, 1, 0, −1, −2	5
	3	4f	3, 2, 1, 0, −1, −2, −3	7

This does not alter the issue, however, since the system remains aperiodic, and Pauli was using the adiabatic principle where it did not strictly apply. But as often happens in science, taking a step that is not rigorous can often pay dividends, at least temporarily, as it did in this case.

Perhaps the reason why theoretical considerations were suspended was that Pauli's new scheme resolved some major problems. First, the notion of the existence and permanence of the quantum numbers could be retained, as Bohr had hoped. Second, the long-standing problem of the "closing of electron shells" in atoms was resolved. The question had been how to explain the series of whole numbers 2, 8, 18, 32, and so on, which characterizes the lengths of the periods in the periodic system of chemical elements. These numbers also correspond to the maximum number of electrons in each shell. Now the closing of the various shells could seen to be a consequence of Pauli's exclusion principle, which prohibits any two electrons from having the same four quantum numbers together with the assumption that the fourth number itself can adopt only two possible values. Meanwhile, all the previous rules for assigning the values for the second and third quantum numbers for a given value of the first quantum number were retained.

When the first quantum number, or n, takes the value of 1, the second quantum number can only be 0, and likewise the third quantum number (table 7.7). According to Pauli's principle, the first shell can therefore contain a maximum number of two electrons that differ just in the value of the fourth quantum number.

For the $n = 2$ shell, the situation is more complicated, since there are two possible values for the second quantum number: 0 and 1. As noted above, when the second quantum number is 0, the third quantum number also adopts a 0 value and, since the fourth quantum number can adopt two possible values, two electrons are accounted for. When the second quantum number in the second shell takes a value of 1, the third quantum number may take on three possible values: $-1, 0$, and $+1$. Each of these possibilities can show two values for the fourth quantum number, thus accounting for a further six electrons. Considering the second shell as a whole, a total of eight electrons is therefore predicted, in accordance with the well-known short period length of eight elements.

Similar considerations for the third and fourth shells predict 18 and 32 electrons, respectively, once again in accordance with the arrangement of the elements in the periodic table.

This scheme is still widely regarded as the explanation for the periodic table, and some version of it is found in virtually every textbook in chemistry or physics. But it is only a partial explanation. It relies for its success on using experimentally observed data in order to determine at what point, in the sequence of the elements, any particular electron shell begins to be filled. The explanation provided by Pauli and most textbooks is only an explanation of the maximum number of electrons successive electron shells can accommodate. It does not explain the particular places in the periodic table at which periodicity occurs. This is to say that Pauli's explanation alone does not explain the lengths of *periods*, which is the really crucial property of the periodic table.

The more important aspect of the periodic system, namely, the lengths of the periods, and their explanation, is taken up again in chapter 9. Just to anticipate matters a little, it will emerge that even present-day physics has not provided a deductive explanation of the closing of the periods, although some promising candidate explanations are becoming available. This is a situation that is seldom acknowledged in textbooks or even in the research literature. Such sources give the impression that quantum physics provides a fully deductive explanation of the closing of the periods, or the particular atomic numbers at which each period is terminated.

CHAPTER

8

ELECTRONIC EXPLANATIONS
OF THE PERIODIC SYSTEM
DEVELOPED BY CHEMISTS

Given the advances in explanations of the periodic system provided by physicists in the first quarter of the twentieth century, described in chapter 7, it is interesting to consider what advances, if any, were achieved by chemists during the same period. Unlike physicists, chemists were working largely inductively with experimental data on the elements and not via any theoretical arguments.

However, in many instances, the electronic configurations proposed by chemists were superior to those postulated such physicists as Niels Bohr and Edmund Stoner. This is not entirely surprising given the chemist's familiarity with the properties of the elements. Inductive arguments based on macroscopic behavior of elements were often more fruitful than the deductive arguments based on physical principles. Moreover, as described in chapter 7, even physicists' routes to electronic configurations were not always as deductive as they were claimed to be by their authors.

The starting point for the chemical contributions to the assignment of electronic configurations can be regarded as J.J. Thomson's discovery of the electron in 1897, since without the existence of this particle there could be no electronic configurations. In 1902, the American chemist Gilbert Newton Lewis (figure 8.1) began speculating about the electronic structure of atoms, although he did not publish his views due to the prevailing empiricist climate in U.S. chemistry, which was rather hostile toward theoretical approaches.[1] Considerably later, Lewis recalled his early thoughts on the constitution of atoms:

> In the year 1902 (while I was attempting to explain to an elementary class in chemistry some of the ideas involved in the periodic law) becoming interested in the new theory of the electron, and combining this idea with those implied in the periodic classification, I formed an idea of the inner structure of the atom which, although it contained certain crudities, I have ever since regarded as representing essentially the arrangement of electrons in the atom.

FIGURE 8.1 Gilbert Newton Lewis. Photo and permission from
Edgar Fahs Smith Collection.

Some dated fragments of this work still survive, including a diagram in which
Lewis depicts the electronic structures of the elements from helium up to fluorine
(figure 8.2). Lewis envisaged the electrons as being arranged at the corners of a cube
and that when the number eight was exceeded, another cube would be formed to
build up a series of concentric cubes around the nucleus of any element.

It should be realized that from purely chemical evidence Lewis had succeeded
in deducing the correct number of electrons for all but one (helium) of the first
dozen or so elements in the periodic system, whereas, as discussed in chapter 7, this
had been a major stumbling block for the physicist Thomson.[2] The latter's electronic account of the periodic table showed only that it might be possible to relate
elements in the same group in terms of analogous configurations. Lewis succeeded
in explaining the formation of polar or, as more commonly termed, ionic compounds such as sodium chloride by means of his cubic atom concept. According
to his model, the sodium atom, which possessed one electron on the corner of a
cube, could lose this electron to form a positive sodium ion with no outer electrons on any of the corners. Meanwhile, the cube around a chlorine atom would
begin with seven of its eight corners occupied with electrons and would gain the
spare electron from sodium. This would give the chlorine a full complement of
eight electrons, thereby forming a negative chlorine ion, which would be attracted
to the positive sodium ion, as in figure 8.2.

However, while this model could explain the formation of polar compounds,
it could not address how nonpolar organic compounds, such as methane, might be
formed. This serious limitation may have been another reason why Lewis did not
publish his early ideas on the cubic atom. It was only in 1916 that Lewis did so.

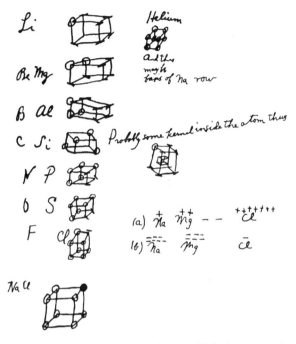

FIGURE 8.2 Lewis's sketch in his unpublished memorandum of March 28, 1902. Lewis Archive, University of California–Berkeley.

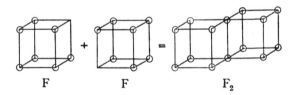

FIGURE 8.3 The formation of an ionic bond between cubes representing the sodium and chloride ions. Reproduced from A. Stranges, *Electrons and Valence*, Texas A&M University Press, College Station, TX, 1982, p. 212, with permission from the publisher.

These ideas on the cubic atom and pairs of electrons were to lead Lewis to formulate one of the most influential ideas in the whole of modern chemistry: the notion of a covalent bond as a shared pair of electrons. Until this time, chemical bonds had been regarded exclusively as involving the transfer of electrons and the formation of ionic bonds.[3] It is interesting to realize that the concept of a covalent bond thus began, like so many important developments in modern science, with research connected to the periodic system of the elements. Moreover, Lewis's 1916 article, titled "The Atom and the Molecule," has turned out to be one of the most influential works in modern chemistry.[4] What Lewis also suggested was that the two kinds of bonding were essentially forms of the same behavior, namely, the sharing of electrons between atoms. In the case of polar compounds, this sharing could be regarded as being very uneven, whereas in nonpolar compounds the electrons could be more or less equally shared by adjacent atoms of different elements.

In this chapter I concentrate on what Lewis had to say on the periodic system and the structure of atoms. Lewis began by paying tribute to the work of German chemist Richard Abegg, who in 1902 had proposed a rule of valence and contravalence:

> *The total between the maximum positive and negative valences of an element is frequently eight and in no cases more than eight.*

Abegg's maximum positive valence corresponds to the group number of the element in question in the periodic table. The normal valence is whichever of these two valences is less than 4 while the contravalence, or "other valence," is displayed less commonly and gives rise to less stable compounds. The elements in group IV of the periodic table show no natural preference with respect to valency and were called amphoteric, a term that was first coined by Abegg and is still in chemical use today, although with a somewhat different meaning.[5]

Abegg explained why the sum of the valence and contravalence was eight by assuming that this number represented the number of electron attachment sites on any atom. However, he did not venture to speculate why each atom could attach precisely this number of electrons. This step was provided by Lewis, who devised his cubical atom on the basis of Abegg's rule. Each of the eight corners of a cube represented a site at which an electron could be attached. Indeed, whereas Abegg merely accepted the stability of eight electrons to an atom, for Lewis it was a simple consequence of the cubic structure of his hypothetical atom. As historian Anthony Stranges has suggested, Lewis's cubic atom appears to be a geometric representation of Abegg's arithmetical rule.[6]

Lewis gave a number of postulates:

1. In every atom is an essential kernel which remains unaltered in all ordinary chemical changes and which possesses an excess of positive charges corresponding in number to the ordinal number of the group in the periodic table to which the element belongs.
2. The atom is composed of a kernel and an outer atom or shell, which, in the case of the neutral atom, contains negative electrons equal in number to the excess of positive charges of the kernel, but the number of electrons in the shell may vary during chemical change between 0 and 8.
3. The atom tends to hold an even number of electrons in the shell, and especially to hold eight electrons which are normally arranged at the eight corners of a cube.

4. Two atomic shells are mutually impenetrable.
5. Electrons may ordinarily pass with readiness from one position in the outer shell to another. Nevertheless they are held in position by more or less rigid constraints, and these positions and the magnitude of the constraints are determined by the nature of the atom and of such other atoms as are combined with it.
6. Electric forces between particles which are very close together do not obey the simple law of inverse squares which holds at greater distances.[7]

As a comment on postulate 3, Lewis points out that among the "tens of thousands of known compounds," only a few of them do not have an even number of electrons in their valence shells. In every compound in which the element uses either its highest or lowest valence, the total number of valence electrons is a multiple of eight. Examples given by Lewis include ammonia (NH_3), water (H_2O), and potassium hydroxide (KOH), all of which show a total of eight valence electrons, and magnesium chloride ($MgCl_2$), where the total is 16, and sodium nitrate ($NaNO_3$), where it is 24. The few exceptions to the notion of even numbers of valence electrons include NO (11), NO_2 (17), and ClO_2 (19), but Lewis adds that these molecules are highly reactive, forming more stable molecules where the number of valence electrons is once again even, such as in the case of the dimerization of NO_2 to form N_2O_4.

Postulate 4, which allows cubic atoms to interpenetrate, forms the basis of the electron sharing mechanism. In this way, one electron or more could belong to two atoms simultaneously without being gained or lost by either of the atoms. This simple idea, stemming directly from cubic arrangements of electrons, seems to be the origin of the now ubiquitous concept of electron sharing in chemistry. Lewis illustrated the interpenetration of cubes with the formation of a diatomic molecule held together by a single bond (figure 8.4). The concept could be extended to explain double bonds in such molecules as diatomic oxygen, which Lewis represented as two cubes sharing a common face (figure 8.5).

Postulate 6 implies an abandonment of Coulombic repulsion in the case of two closely lying electrons and is essential if one is to contemplate the existence of pairs of electrons as an integral part of the model. Interestingly, Lewis qualifies this postulate further by saying that electrons can act as small magnets that, when correctly oriented, can account for the stability of the shared pair of electrons.[8] This statement, and later elaborations, have been interpreted by many as an anticipation of the concept of electron spin, which, as discussed in chapter 7, was not formally introduced until 1925.[9]

What is curious about this article of 1916 is that Lewis begins with a detailed account of his cubic atom but in the very same article shows that it is necessary to go beyond this model. One of the shortcomings of the simple model is its inability to explain the formation of triple bonds such as in a molecule of ethyne (acety-

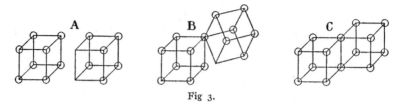

Fig 3.

FIGURE 8.4 Single bond formation with cubic atoms. Reproduced from A. Stranges, *Electrons and Valence*, Texas A&M University Press, College Station, TX, 1982, p. 212, with permission from the publisher.

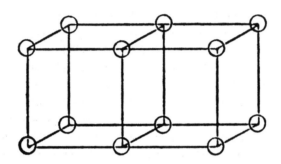

FIGURE 8.5 Double bond formation with cubic atoms. Reproduced from A. Stranges, *Electrons and Valence*, Texas A&M University Press, College Station, TX, 1982, p. 213, with permission from the publisher.

FIGURE 8.6
Lewis's tetrahedral atom. The electron pairs are now on the corners of a tetrahedron that has been superimposed on the earlier cubic structure.

lene), with formula H–C≡C–H, or the diatomic nitrogen molecule N≡N. There seems to be no way that two cubes of electrons can share three pairs of electrons, and Lewis concludes that he needs to assume a somewhat different way of arranging the electrons in order to overcome this problem. The solution he offers is to place the electrons on the four corners of a tetrahedron that has been superimposed on the earlier cubic structure (figure 8.6). This new tetrahedral atom can accommodate the formation of a triple bond by assuming that two adjacent atoms share a common tetrahedral face.

TABLE 8.1
Lewis's Outer Electronic Structures for 29 Elements

1	2	3	4	5	6	7
H						
Li	Be	B	C	N	O	F
Na	Mg	Al	Si	P	S	Cl
K	Ca	Sc		As	Se	Br
Rb	Sr			Sb	Te	I
Cs	Ba			Bi		

The number at the head of each column represents the positive charge on the atomic kernel and also the number of outer-shell electrons for each atom. Compiled by the author from G.N. Lewis, The Atom and the Molecule, *Journal of the American Chemical Society*, 38, 762–785, 1916. Lewis does not give a table.

Lewis's article also introduces the use of a pair of dots to denote the shared pair of electrons, a form of notation that survives to this day:[10]

$$H : Cl \qquad H : H \qquad Na : H$$

Whereas in 1902 Lewis had thought that helium possessed a complete cube of eight electrons, he corrected its electronic structure in his 1916 paper to include just two electrons, as had been revealed by the work on Henry Moseley.[11] Table 8.1 is based on the positive charges on the atomic kernels of a number of elements as given by Lewis in the same article.

IRVING LANGMUIR

The next step in the evolution of chemically motivated electronic configurations was taken by another American, Irving Langmuir, who spent his life as an industrial research chemist and was responsible for making many of Lewis's ideas widely known.[12] In 1919, Langmuir published an article that begins with an insightful comment, especially in view of the kinds of questions raised in the present book, regarding the relationship between chemistry and physics in the evolution of the periodic system:

> The problem of the structure of atoms has been attacked mainly by physicists who have given little consideration to the chemical properties, which must ultimately be explained by a theory of atomic structure. The vast store of knowledge of chemical properties and relationships, such as is summarized in the periodic table, should serve as a better foundation for a theory of

atomic structure than the relatively meager experimental data along purely physical lines. [13]

Langmuir remarks that Lewis confines his attention to only 35 of the 88 known elements and that his theory does not apply at all satisfactorily to the remaining elements, thus highlighting the limitations of Lewis's scheme especially for the transition elements. Langmuir then briefly reviews the work of Walther Kossel in which electrons are regarded as being in concentric rings and comments on the fact that while Kossel's theory considers elements up to cerium, that is, a total of 57 elements, just like Lewis's theory, it is unable to deal with the transition metals. [14]

Langmuir bases his own theory on the number of electrons in the atoms of the noble gases:

$$He = 2, Ne = 10, Ar = 18, Kr = 36, Xe = 54, niton^{15} = 86.$$

By means of "constant checking against the periodic table and the specific properties of the elements," Langmuir proceeds to elaborate seven postulates, which enable him to assign electronic configurations for all the naturally occurring elements up to uranium or $Z = 92$. He then provides a classification of the elements according to their arrangement of electrons in what is essentially a short-form periodic table in which the numbers of outer-shell electrons are displayed for each element along with their atomic number (figure 8.7).

Unlike Lewis, and Kossel before him, Langmuir does not hesitate to assign configurations to the transition element atoms. For example, the first transition series is depicted as follows:

Sc	Ti	V	Cr	Mn	Fe	Co	Ni	Cu	Zn
3	4	5	6	7	8	9	10	11	12

where the numbers denote the number of outermost electrons. [16]

The 10th postulate given by Langmuir is significant for two reasons. It includes the statement that there can be no electrons in the outer shell of an atom until all the inner shells contain the maximum number of electrons that each one can accommodate. In addition, he states that a second outer-shell electron can occupy a cell[17] only when all other cells contain at least one electron. The first of these statements requiring that electron shells be filled in a strictly sequential order was subsequently abandoned. Indeed, the precise order of shell and subshell filling has become an important question to assess the degree to which modern physics can explain the periodic system. [18]

But while, in the light of current knowledge, Langmuir's first statement appears to be incorrect, the second statement seems to anticipate an important aspect of modern quantum mechanical configurations, namely, the existence of Hund's rule. This rule states that when electrons are distributed among a number of orbitals

TABLE I.

Classification of the Elements According to the Arrangement of Their Electrons.

Layer.	N	$E=0$	1	2	3	4	5	6	7	8	9	10
I			H	He								
IIa	2	He	Li	Be	B	C	N	O	F	Ne		
IIb	10	Ne	Na	Mg	Al	Si	P	S	Cl	A		
IIIa	18	A	K	Ca	Sc	Ti	V	Cr	Mn	Fe	Co	Ni
			11	12	13	14	15	16	17	18		
IIIa	28	Niβ	Cu	Zn	Ga	Ge	As	Se	Br	Kr		
IIIb	36	Kr	Rb	Sr	Y	Zr	Cb	Mo	43	Ru	Rh	Pd
			11	12	13	14	15	16	17	18		
IIIb	46	Pdβ	Ag	Cd	In	Sn	Sb	Te	I	Xe		
IVa	54	Xe	Cs	Ba	La	Ce	Pr	Nd	61	Sa	Eu	Gd
			11	12	13	14	15	16	17	18		
IVa			Tb	Ho	Dy	Er	Tm	Tm$_2$	Yb	Lu		
		14	15	16	17	18	19	20	21	22	23	24
IVa	68	Erβ	Tmβ	Tm$_2$β	Ybβ	Luβ	Ta	W	75	Os	Ir	Pt
			25	26	27	28	29	30	31	32		
IVa	78	Ptβ	Au	Hg	Tl	Pb	Bi	RaF	85	Nt		
IVb	86	Nt	87	Ra	Ac	Th	Ux$_2$	U				

FIGURE 8.7 Langmuir's periodic table of electronic configurations. I. Langmuir, Arrangement of Electrons in Atoms and Molecules, *Journal of the American Chemical Society*, 41, 868–934, 1919, p. 874.

with equal energies, they are placed in separate orbitals until the pairing of electrons into single orbitals becomes unavoidable.[19]

Langmuir's configurations appeared in pictorial form a couple of years later in a chemistry textbook written by Washburn as shown in figure 8.8.[20]

CONTRIBUTIONS OF CHARLES BURY

As mentioned above, Langmuir assumed that electron shells fill in a strictly sequential order. The first chemist to challenge this idea was Charles Bury, working at the University College of Wales in Aberystwyth.[21] In his article,[22] Bury states that the configurations he is proposing give a better explanation of the chemical properties

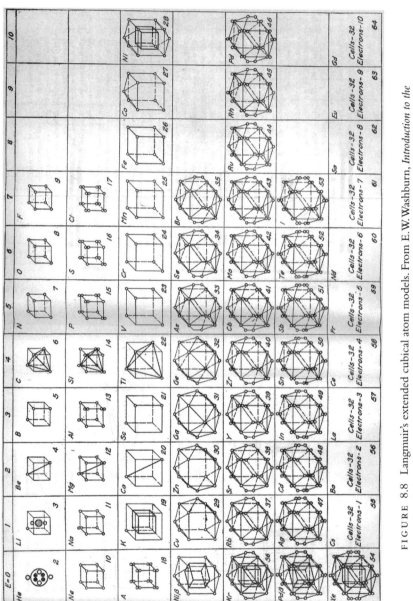

FIGURE 8.8 Langmuir's extended cubical atom models. From E. W. Washburn, *Introduction to the Principles of Physical Chemistry*, 2nd ed., McGraw-Hill, New York, 1921, p. 470.

TABLE 8.2
Bury's Configurations for Some First Transition Series Atoms

Ti	(2, 8, 8, 4)	(2, 8, 9, 3)	(2, 8, 10, 2)		
V	(2, 8, 8, 5)	(2, 8, 9, 4)	(2, 8, 10, 3)	(2, 8, 11, 2)	
Cr	(2, 8, 8, 6)	(2, 8, 11, 3)	(2, 8, 12, 2)		
Mn	(2, 8, 8, 7)	(2, 8, 9, 6)	(2, 8, 11, 4)	(2, 8, 12, 3)	(2, 8, 13, 2)
Fe	(2, 8, 10, 6)	(2, 8, 12, 4)	(2, 8, 13, 3)	(2, 8, 14, 2)	
Co	(2, 8, 13, 4)	(2, 8, 14, 3)	(2, 8, 15, 2)		
Ni	(2, 8, 14, 4)	(2, 8, 15, 3)	(2, 8, 16, 2)		
Cu	(2, 8, 17, 2)	(2, 8, 18, 1)			

C.R. Bury, Langmuir's Theory of the Arrangement of Electrons in Atoms and Molecules · *Journal of the American Chemical Society*, 43, 1602-1609, 1921, table from p. 1603.

of the elements than do those of Langmuir. He also notes that his own configurations dispense with the need to postulate cells as Langmuir does.

Contrary to Langmuir, Bury suggests that the number of electrons in the outer shell cannot be more than eight. Furthermore, he claims that an inner stable group of eight electrons can change into one containing 18 electrons, or that one of 18 can change into one of 32, in the course of the development of a transition series of elements (table 8.2). These transition elements are supposed by Bury to have more than one electronic structure depending on their state of chemical combination. Although Bury's postulates do not lead to any disagreement with Langmuir's regarding the first few configurations, differences begin to appear in the first long period ranging from potassium to krypton.

Bury clearly states that electron shells do not fill sequentially when he writes, "Since eight is the maximum number of electrons in an outer layer K, Ca and Sc must form a fourth layer although their third is not complete. Their structures will be 2, 8, 8, 1, 2, 8, 8, 2, 2, 8, 8, 3." Furthermore, he infers that the elements from titanium to copper " form a transition series in which the incomplete group of eight in the third layer is changed to a saturated group of eighteen."[23] However, as mentioned above, Bury considers the elements from titanium to copper to possess more than one configuration, depending on what compounds they find themselves in, a suggestion that has not survived the test of time.

THE CASE OF HAFNIUM (ELEMENT 72)

The prediction and eventual confirmation that element 72 is not a rare earth element is widely regarded as a triumph for Bohr's theory of the periodic system. It is often argued that, while chemists believed that this element was a rare earth, Bohr drew on quantum theory to suggest otherwise. This is therefore presented as an early case of reduction of chemistry by quantum theory as distinct from the later quantum mechanics.

However, both parts of this commonly held view are partly mistaken. First, only a minority of chemists believed that hafnium should be a rare earth, and second, Bohr's prediction was not very conclusive and was based on a highly empirical theory of electron shells rather than on any deep theoretical principles. According to Bohr's theory, the rare earths are characterized by the building up of the N group or the fourth electron shell from the nucleus. On this view, the first rare earth is cerium with a fourth shell of the following configuration:[24]

$$\text{cerium (58): } (4_1)^6 \, (4_2)^6 \, (4_3)^6 \, (4_4)^1$$

And the last rare earth is lutetium with the following configuration:

$$\text{lutetium (71): } (4_1)^8 \, (4_2)^8 \, (4_3)^8 \, (4_4)^8$$

The completion of the fourth shell represents the end of the rare earth series, and the next, as yet undiscovered, element is expected to be a transition metal and a homologue of zirconium, showing a valence of 4. The general approach used by Bohr in his assignment of electron shells is to ensure an overall agreement with the known periodic table. The form of the periodic table in fact guided Bohr to electronic configurations, as he sometimes admitted.

It emerges that the notion that element 72 was not a rare earth was the commonly held view among a number of chemists. Element 72, or at least the vacant space that it was supposed to fill, was often placed beyond the rare earth block in published periodic tables prior to Bohr's theory, for example, by Bury.[25] In fact, the prediction that hafnium is not a rare earth element can be obtained quite simply by counting and is by no means dependent on assuming the existence of electron shells. This can be illustrated as follows: It had been known for some time that the number of elements in each period follows a definite sequence given by 2, 8, 8, 18, 18, 32 (probably followed by 32), and so on. By adding the first six of these numbers, one arrives at the conclusion that the sixth period terminates with a noble gas of atomic number 86. It is a simple matter to work backward from this number to discover that element 72 should be a transition metal and a homologue of zirconium, which shows a valence of 4. This procedure depends on the plausible assumption that the third transition series should consist of 10 elements, as do the first and second transition series:

72, 73, 74, 75, 76, 77, 78, 79, 80, 81	82, 83, 84, 85, 86
..........10 transition metals	IV V VI VII VIII
1st 2nd 3rd 4th 5th 6th 7th 8th 9th 10th	main group elements

The only chemists who did indeed believe that element 72 was a rare earth were the few who specialized in obtaining these types of elements by painstaking separation of some mineral ores that were first found in Sweden. In 1879, Charles

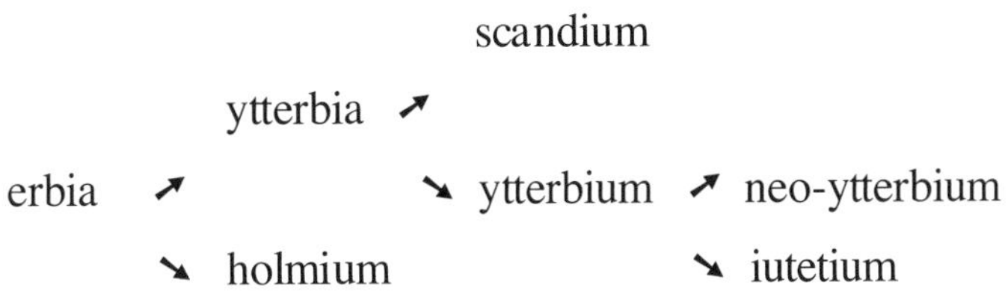

scandium

ytterbia ↗

erbia ↗ ↘ ytterbium ↗ neo-ytterbium

↘ holmium ↘ iutetium

FIGURE 8.9 Sequence of discovery of some rare earth elements.

Marignac showed that the rare earth erbia could be separated into two rare earths, ytterbia and another element later called holmium (figure 8.9).[26] A year later, ytterbia was separated into two distinct elements named scandium and ytterbium. The next step was taken by Georges Urbain and Auer von Welsbach, who independently found that ytterbium itself could be separated into neo-ytterbium and lutetium. It was only natural that these workers should suspect the possibility of discovering further new elements by repeated separations of the same minerals.

Urbain and von Welsbach both believed that ytterbium contained small amounts of a third rare earth that would possibly turn out to be element 72. Indeed, Urbain announced what he believed to be a positive spectroscopic identification of element 72 in 1911,[27] although this claim could not be confirmed by Moseley using his X-ray method. Urbain then abandoned his claim for 11 years, after which time he announced that together with Alexandre Dauvillier that he had used a more accurate X-ray experiment and had detected two weak lines whose frequencies corresponded approximately to those expected for element 72 on the basis of Moseley's law. But this claim, too, turned out to be unfounded.

BACK TO BOHR

As mentioned in chapter 7, when Bohr presented his method for ascribing electron shells, most physicists were puzzled by the manner in which he obtained his results. Letters to Bohr following the publication of his theory of the periodic system in *Nature* magazine[28] contain passages such Ernest Rutherford's: " Everybody is eager to know whether you can fix the rings of electrons by the correspondence principle or whether you have recourse to the chemical facts to do so."[29] And from Paul Ehrenfest: "I have read your article in Nature with eager interest. . . . Of course I am now even more interested to know how you saw it all in terms of correspondence."[30] Several years later, Hendrik Kramers wrote:

> It is interesting to recollect how many physicists abroad thought, at the time of
> the appearance of Bohr's theory of the periodic system, that it was extensively

supported by unpublished calculations which dealt in detail with the structure of individual atoms, whereas the truth was, in fact, that Bohr had created and elaborated with a divine glance a synthesis between results of a spectroscopical nature and of a chemical nature.[31]

This remark seems to be particularly true regarding element 72, for which Bohr never produced any mathematical arguments or any other form of argument resting specifically on quantum theory. Bohr's predictions regarding the electronic arrangements of the rare earths and that of hafnium were as follows.

First of all there are vague arguments based on "harmonic interaction," correspondence, and symmetry, such as the following:

> Even though it has not been possible to follow the development of the group [rare earths] step by step, we can even here give some theoretical evidence in favour of the occurrence of a symmetrical configuration of exactly this number of electrons. I shall simply mention that it is not possible without coincidence of the planes of the orbits to arrive at an interaction between four sub-groups of six electrons each in a configuration of simple trigonal symmetry, which is equally simple as that shown by these sub-groups. These difficulties make it probable that a harmonic interaction can be attained precisely by four groups containing eight electrons the orbital configurations of which exhibit axial symmetry.[32]

In a somewhat less obscure fashion, Bohr uses a counting argument not unlike one mentioned above in order to arrive at the conclusion that element 72 should not be a rare earth:

> As in the case of the transformation and completion of the 3-quanta orbits in the fourth period and the partial completion of the 4-quanta orbits in the fifth period, we may immediately deduce from the length of the sixth period the number of electrons, namely 32, which are finally contained in the 4-quanta group of orbits. Analogous to what applied to the group of 3-quanta orbits it is probable that, when the group is completed, it will contain eight electrons in each of the four subgroups.[33]

According to Bohr, the element that represents the completion of the four-quanta groups and therefore marks the end of the rare earths is lutetium, with the following grouping of four-quanta electrons:

$$\text{lutetium (71) } (4_1)^8 \ (4_2)^8 \ (4_3)^8 \ (4_4)^8$$

Bohr writes:

> This element therefore ought to be the last in the sequence of consecutive elements with similar properties in the first half of the sixth period, and at the place 72 an element must be expected which in its chemical and physical properties is homologous with zirconium and thorium.[34]

He then adds: "This which is already indicated on Julius Thomsen's old table, has also been pointed out by Bury."[35]

In his Göttingen lectures on the periodic table, Bohr alluded to a calculation concerning the rare earth configurations. But this form of calculation was never produced by Bohr, nor has anything of the sort ever been found in the Bohr archives. Bohr also expressed a certain amount of doubt over his prediction that hafnium would not be a rare earth. Having explained the filling of the four-quanta groups as described above, he says: "However the reasons for indicating this arrangement are still weaker than in the case of the 3-quantic group and the preliminary closure of the 4-quantic group in silver."[36]

When Urbain and Dauvillier claimed to have discovered element 72 and that it was a rare earth, Bohr's initial response was to doubt his own prediction that it lay beyond the rare earths. This wavering was expressed in letters to colleagues as well as in the appendix of a book on atomic constitution.[37] He wrote to James Franck:

> The only thing I know for sure about my lectures in Göttingen is that several of the results communicated are already wrong. A first point is the constitution of element 72, which, as shown by Urbain and Dauvillier, contrary to expectations has turned out to be a rare earth element after all.[38]

And to Dirk Coster: "The question is apparently rather clear but one must of course always be prepared for complications. These may arise from the circumstance that we have to do with a simultaneous development of two inner electron rings."[39]

Bohr soon returned to his original claim about element 72. In doing so, he provided further examples of his essentially chemical arguments for his views on the missing element. In referring to the claim by Urbain and Dauvillier, Bohr points out that if element 72 is a rare earth, it should have a valence of 3 in common with other members of this group. Moreover, Bohr mentions that element 73, tantalum, is known to have a valence of 5: "This would mean an exception to the otherwise general rule, that the valency never increases by more than one unit when passing from one element to the next in the periodic table."[40]

Meanwhile, in response to the claims of Urbain and Dauvillier, which they believed to be unfounded, György Hevesy and Coster, working at Bohr's institute in Copenhagen, began a search for element 72 in the ores of zirconium. Even their first attempt proved to be unexpectedly successful. After some further concentration of the new element, they obtained six clear X-ray lines whose frequencies were in very good agreement with Moseley's law applied to element 72.[41] The new element was named hafnium after *Hafnia*, Latin for Copenhagen, where the element had been discovered. The commonly cited story that Bohr instructed his colleagues to look for the new element in the ores of zirconium, where they discovered hafnium, is simply untrue. The suggestion to search for the new element within zirconium ores was made by a chemist, Fritz Paneth, based on purely chemical arguments.

No doubt the controversy over Urbain's claim and Bohr's theory stimulated this eventual true discovery of element 72, but in view of all the factors described above, it can still be doubted whether this development represents a successful chemical prediction on the basis of quantum theory. Had Bohr's theory been correct in making a prediction that had been contrary to the beliefs of the chemists of the day, this would have made Bohr's triumph all the greater, but this was not the case.

It might be more accurate to say that the view held by most chemists that hafnium would not be a rare earth was rationalized by Bohr's quantum theory of periodicity, which was partly empirical in origin. In addition, as noted above, Bury had already obtained the same correct prediction that hafnium was a transition metal, and not a rare earth, on the basis of purely chemical arguments.

JOHN DAVID MAIN SMITH

John David Main Smith is another chemist who succeeded in producing more detailed electronic configurations and a more accurate explanation of the periodic system than did Bohr, but without receiving much credit at the time or in subsequent accounts of the development of the periodic system.

In March 1924, Main Smith, at Birmingham University in England, published an article in the somewhat obscure journal *Chemistry and Industry*, in which he corrected an important feature in Bohr's electronic configurations, discussed in chapter 7.[42] Bohr's scheme of 1923 assumed that all subgroups of electrons were equally populated when full. For example, Bohr assumed that the second shell consisted of a total of eight electrons distributed equally into two subgroups, each containing four electrons. Main Smith challenged this notion on the basis of chemical, as well as X-ray diffraction, evidence.

Let us pause to consider the chemical arguments, especially in view of the general theme of the present book. In his 1924 book *Chemistry and Atomic Structure*, Main Smith begins a chapter titled "Atomic Structure and the Chemical Properties of Elements" with the following remark, which may seem obvious to chemists but perhaps not to physicists:

> Bohr's theory of atomic structure is strictly a theory relating to single atoms, neutral or ionised, far removed from the influence of other atoms. The fact that it is an interpretation of the periodic classification of the elements, largely based on the properties of atoms in combination, indicates that it must be valid for atoms in combination, at least so far as the broad outlines of the theory are concerned.[43]

He then proceeds to mount a sustained critique of the electronic configurations as assigned by Bohr on the basis that they do not take into account the chem-

TABLE 8.3

Bohr's Subgroups Consisting of Equal Subdivisions of Electrons

Quantum number	1	2	3	4
Number of electrons	2	8	18	32
Subgroups	1_2	$2_1, 2_2$	$3_1, 3_2, 3_3$	$4_2, 4_2, 4_2, 4_2$

N. Bohr, *Theory of Spectra and Atomic Constitution*, Cambridge University Press, Cambridge, UK, 1924, table based on p. 113.

ical behavior of many of the elements. Main Smith points out that the elements of group III of the periodic table commonly display two distinct valences, a fact that cannot be explained from Bohr's configurations. He points out that the elements in group IV, and to some extent those of groups V, VI, and VII, commonly show two different valences. Main Smith remarked:

> This may be interpreted as evidence that all elements containing more than two valency electrons have two electrons, which, further interpreted in terms of orbits, indicates that that two electrons in the outer structure of atoms are in quantum orbits the energies of which are different from that of other outer electrons.[44]

From the point of view of our current knowledge, Main Smith had discovered that every main shell begins with an orbital containing just two electrons. However, another conclusion that is apparently equally drawn from chemical facts has turned out to be incorrect in the light of contemporary knowledge of electronic configurations. Main Smith argued that all the elements of groups V, VI, and VII have a marked tendency to form compounds with a coordination number of four and that this suggests that valence electrons in excess of four are "equally feebly attached to atoms." He interpreted this behavior to mean that any valence electrons in excess of four are all in similar quantum orbits. In fact, it is the six electrons in excess of two that form another shell in the atoms of successive elements, or in modern terminology, the following six electrons enter into three equivalent p orbitals.

Be that as it may, Main Smith concludes that the subgroups, in the second main shell, for example, should contain 2, 2, and 4 electrons, respectively, starting with the subgroup closest to the nucleus (table 8.4). He also states categorically that "This evidence shows conclusively that Bohr's subgroup scheme, of two subgroups of four electrons for the two-quanta group, cannot be maintained."[45]

According to Bohr's set of configurations, the third shell consists of three subgroups, each of which contain six electrons. If this were indeed the case, Main Smith reasoned, we should expect to observe a single transition between levels

TABLE 8.4

Subgroups of Electrons in the Second Main Shell
for Elements in the Second Period According to
Main Smith and Bohr

	Bohr	Main Smith
Lithium	1	1
Beryllium	2	2
Boron	3	2, 1
Carbon	4	2, 2
Nitrogen	4, 1	2, 2, 1
Oxygen	4, 2	2, 2, 2
Fluorine	4, 3	2, 2, 3
Neon	4, 4	2, 2, 4

Compiled by the author.

3_1 and 3_2, for example, in the X-ray spectrum of any element that possesses outer electrons in these levels. In fact, the X-ray evidence in the sodium atom spectrum, which is associated with the transition between these levels, is not a single line but the well-known sodium doublet. Main Smith concluded that there were additional levels between Bohr's levels 3_1 and 3_2. The subgroups of electrons for the elements in the second period according to Bohr and Main Smith, respectively, are compared in table 8.4. Main Smith proceeds to extend his list of configurations (figure 8.10), and with far greater detail than Bohr had published, for many of the elements up to gold ($Z = 79$).

Not surprisingly, these configurations have not withstood the test of time, since subgroups of four or eight electrons are not admitted in the modern scheme, which contains 2, 6, 10, or 14 in successive subshells of electrons.[46] However, these little known articles by Main Smith are of great historical interest because they show that some chemists not only understood Bohr's theory of periodicity but were prepared to grapple with its details, including the physical evidence such as that obtained from X-ray spectroscopy. Moreover, several chemists were able to deduce more comprehensive explanations of the periodic system in terms of electronic configurations, as in the case of Main Smith.

As mentioned in chapter 7, the next major step after Bohr's two–quantum–number account was given by Stoner. It emerges that Stoner's scheme is almost identical to the one by Main Smith just described. The two scientists arrived at their respective versions independently, although Main Smith was the first to publish. Both of them drew upon a detailed analysis of X-ray spectra to arrive at their electron subgroups.

Total quantum number $n=$	1	2	3							
Azimuthal quantum number $k=$	1	112	112	23	23	23	23	23	23	23
Valency of ion	—	—	—	1	2	3	4	5	6	7
Sc (21)	2	224	224	—	—	00				
Ti (22)	2	224	224	—	02	01	00			
V (23)	2	224	224	—	03	02	01	00		
Cr (24)	2	224	224	—	04	03	—	01	00	
Mn (25)	2	224	224	—	14	13	12	—	10	00
Fe (26)	2	224	224	—	24	23	—	—	20	
Co (27)	2	224	224	—	25	24	23			
Ni (28)	2	224	224	—	26	—	24			
Cu (29)	2	224	224	46	45	44				

FIGURE 8.10 Configurations of some transition elements, extracted from more complete tables of configurations. J.D. Main Smith, *Chemistry and Atomic Structure*, Ernest Benn Ltd., London, 1924, p. 196.

Main Smith correctly considered that his contributions had not been properly acknowledged and wrote the following published letter to the editor of *Philosophical Magazine*:

[A]ttention to the fact that the distribution of electrons in atoms characterized by the subgroupings 2; 2,2,4; 2,2,4,4,6; 2,2,4,4,6,6,8 did not originate with Mr. E. Stoner, as within recent months various papers published in your journal have suggested.[47]

It is gratifying to note that Stoner fully conceded Main Smith's priority when, a couple of years later in a textbook of atomic physics, he wrote,

I have since found that my distribution had already been proposed, primarily on the basis of chemical arguments by Main Smith. It is very satisfactory that two different lines of attack should have led to the same conclusions.[48]

This is a fitting conclusion to this chapter, which has demonstrated how chemistry was by no means eclipsed by the discoveries in atomic physics, as is sometimes implied. The cases of Main Smith and Bury, in particular, show that chemists not only were able to compete with the atomic physicists on their own terms but also arrived at more detailed configurations before the physicists could do so.

CONCLUSION

In a matter of a few years, several chemists, including Lewis, Langmuir, and Bury, had obtained detailed electronic configurations for all the known elements, including the more complicated transition elements. Bury had realized that the atoms of the transition elements do not fill their electron shells in sequential order and had predicted that element 72 would be a transition metal that would show chemical similarities with zirconium. All this work was achieved without any arguments based on theoretical physics or, more specifically, without using quantum theory. The chemists' configurations were obtained inductively on the basis of the chemical properties of the elements. This aspect of the history of the periodic system is seldom emphasized, with most accounts promoting the view that electronic configurations resulted entirely from the work of theoretical physicists such as Bohr. In truth, Bohr had also reached electronic configurations inductively, frequently drawing on chemical evidence, as the chemists themselves had done. Bohr's configurations were frequently less detailed in that he specified only those of the closed-shell atoms of the noble gases and did not cite those of the intervening series of elements in each period of the table.

The popular story according to which Bohr predicted the chemical nature of element 72, subsequently named hafnium, has been criticized and can no longer be sustained by anyone who examines the historical evidence surrounding this case.[49] Indeed, Bohr seems to have arrived at many configurations by appeal to chemical as well as other experimental data that he then dressed up in quantum mechanical language through his characteristically obscure style of writing.

It is also quite possible that Bohr could have relied quite heavily on the work of Bury, who had predicted that element 72 would be a transition element. Bury's priority in this matter is even conceded by Bohr himself in his articles, although he does not seem to attach much significance to Bury's prediction, presumably because it is not couched in terms of the quantum theory. In addition, whereas Bohr initially gave configurations only for the noble gases, he appears to have begun listing those of the intervening elements only after the appearance of Bury's detailed versions published in 1921.

The case for Bury's priority over Bohr in these matters has been valiantly argued by one of Bury's former students, Mansel Davies, who published a number of articles to this effect.[50] This claim has been taken up and amplified by Keith Laidler, the distinguished chemical kineticist and historian.[51] While it may be that Bury's work has been highly neglected, perhaps a more conservative conclusion may be more appropriate. The reason for the neglect of Bury's work may not due to any duplicity by Bohr, or his supporters, but rather because Bury gave chemical arguments for his own assignment of configurations, whereas the prevailing reductionist climate implied that quantum mechanics inevitably provides a more fundamental explanation for the periodic system.[52] A fuller discussion of this issue is given in chapter 9, which includes an analysis of the reduction of the periodic system using quantum mechanics.

And finally, the chemist Main Smith, drawing on the same X-ray data as the atomic physicist Stoner, as well as chemical evidence, was able to arrive at the conclusion that the number of electrons in each subgroup was twice the value of the inner quantum number. He thus obtained the same electron subgrouping several months before Stoner, whose discovery of the same concept is far better acknowledged by historians of science.

CHAPTER 9

QUANTUM MECHANICS AND THE
PERIODIC TABLE

In chapter 7, the influence of the old quantum theory on the periodic system was considered. Although the development of this theory provided a way of reexpressing the periodic table in terms of the number of outer-shell electrons, it did not yield anything essentially new to the understanding of chemistry. Indeed, in several cases, chemists such as Irving Langmuir, J.D. Main Smith, and Charles Bury were able to go further than physicists in assigning electronic configurations, as described in chapter 8, because they were more familiar with the chemical properties of individual elements. Moreover, despite the rhetoric in favor of quantum mechanics, that was propagated by Niels Bohr and others, the discovery that hafnium was a transition metal and not a rare earth was not made deductively from the quantum theory.[1] It was essentially a chemical fact that was accommodated in terms of the quantum mechanical understanding of the periodic table.

The old quantum theory was quantitatively impotent in the context of the periodic table since it was not possible to even set up the necessary equations to begin to obtain solutions for the atoms with more than one electron. An explanation could be given for the periodic table in terms of numbers of electrons in the outer shells of atoms, but generally only after the facts. But when it came to trying to predict quantitative aspects of atoms, such as the ground-state energy of the helium atom, the old quantum theory was quite hopeless. As one physicist stated, "We should not be surprised . . . even the astronomers have not yet satisfactorily solved the three-body problem in spite of efforts over the centuries."[2] A succession of the best minds in physics, including Hendrik Kramers, Werner Heisenberg, and Arnold Sommerfeld, made strenuous attempts to calculate the spectrum of helium but to no avail.

It was only following the introduction of the Pauli exclusion principle and the development of the new quantum mechanics that Heisenberg succeeded where everyone else had failed. In fact, Heisenberg performed the calculation using both his own matrix mechanics and Erwin Schrödinger's wave mechanics as discussed below. In terms of wave mechanics, Heisenberg interpreted his result as showing the need for the overlap between the wavefunctions of the two electrons in helium. This overlap, which he called an "exchange term," was due entirely to the indistinguishability of the two electrons. This meant that the terms in the equation had to be written in two ways, the second of which involved the exchange of labels to account for the fact that both electrons are identical. Such exchange terms[3] are highly nonclassical and follow from Wolfgang Pauli's discovery of the exclusion principle.

This discovery was to be the beginning of the use of exchange terms in the quantum mechanics of atoms and molecules. It became the key factor that shortly afterward allowed Walter Heitler and Fritz London to obtain the first successful quantum mechanical calculation of the covalent bond in the simplest case of a diatomic hydrogen molecule. Exchange terms would also pave the way for the notion of quantum mechanical resonance and the development of the quantum mechanical theories of bonding by Linus Pauling and many others.[4]

Perhaps the key advance that quantum mechanics provided, compared with the old quantum theory, was that the quantization itself seemed to arise in a more natural manner. In the old quantum theory, Bohr had been forced to postulate that the angular momentum of electrons was quantized, while the advent of quantum mechanics showed that this condition was provided by the theory itself and did not have to be introduced by fiat. For example, in Schrödinger's version of quantum mechanics, the differential equation is written, and certain boundary conditions are applied, with the result that quantization emerges automatically.

A conceptual grasp for how the application of boundary conditions to waves leads naturally to quantization can be obtained from the following analogy. Suppose that a string is tied at both ends and made to vibrate. It turns out that the string can adopt one of many possible standing wave patterns where certain points along the string remain stationary. As shown in figure 9.1, the string can vibrate either as a whole or with a number of so-called nodes, each of which represents a stationary point along the string.

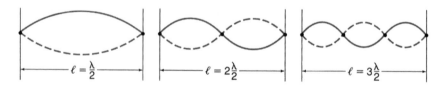

FIGURE 9.1 The imposition of boundary conditions produces quantization in a vibrating wave. From R. Chang, *Physical Chemistry for the Chemical and Biological Sciences*, University Science Books, Sausalito, CA, 2000, p. 576. By permission from the publisher.

In other words, the mere presence of waves that are bound at both ends immediately implies quantization of the form described above. The string can vibrate only in a number of well-defined ways that have 0, 1, 2, 3, and so on, nodes. No other intermediate vibrational nodes can exist, and this is the essence of quantization associated with any kind of standing wave phenomenon.

FROM BOHR'S OLD QUANTUM THEORY TO QUANTUM MECHANICS

What Bohr had been doing in his explanations of the periodic table was not deducing electronic configurations from first principles, as he led his readers to believe; rather, he was essentially working backward from chemical and spectroscopic facts and showing that these facts were consistent with a quantum theoretical description.

But the old quantum theory was only the beginning of quantum mechanics, which is the most powerful physical theory that has ever been devised. The transition between the old quantum theory and the new quantum mechanics is examined in this chapter, as is the impact that the updated theory had on attempts to understand the periodic table. As I argue here, the effect has been considerable, but surprisingly still incomplete, from the fundamental point of view of trying to provide a deeper explanation of the periodic system. Nevertheless, many forms of more accurate calculations can now be carried out in quantum chemistry than were even dreamt of at the time of the old quantum theory.

Although it is not my intention to give a history of the transition between Bohr's old quantum theory and the later quantum mechanics, it is necessary to sketch some of the steps that were taken. In fact, one of the connecting steps between the old and the new versions of quantum theory is mentioned in chapter 7, because it provided the culminating step in the explanation of the periodic system as it is still generally understood. This was the introduction by Pauli of the fourth quantum number and his subsequent discovery of the Pauli exclusion principle, which dictates that no two electrons in an atom can possess the same four quantum numbers. What follows from this assumption is an elegant explanation of the possible lengths of any period but only provided that one is willing to admit some experimental information into the explanation.

As noted in chapter 7, an explanation can be given for the maximum possible number of electrons in any shell around the nucleus. The formula $2n^2$, which had been recognized for some time as summarizing the number of elements in any particular period, is thus given an apparent theoretical underpinning. But there is one aspect, the order of shell filling, that has not yet been deduced from first principles. This issue cannot be avoided if one is to really ask whether quantum mechanics explains the periodic system in a fundamental manner.

THE ADVENT OF QUANTUM MECHANICS

The old quantum theory reached a crisis point around 1924–1925, at which time it was realized that a more radical theory would be needed in order to settle a number of outstanding problems in physics. One of these problems of particular importance to the story of the periodic table was the attempt to calculate the properties of helium, the second atom in the periodic table following hydrogen. Whereas the old quantum theory provided a means of obtaining an exact solution to the calculation of the energy of the hydrogen atom, the move to considering helium appeared to cause insurmountable problems. It was not that the solution of this problem was just difficult in the old quantum theory. It was not even possible to formulate the necessary equations adequately.

Only following the advent of quantum mechanics, as distinct from the old quantum theory, was there a possibility of calculating the energy of helium, and even then only approximately. Developments initially occurred along two distinct lines. First of all, Heisenberg, a very young German, developed an approach that eventually became known as matrix mechanics. Heisenberg's original motivation appears to have been the complete abandonment of unobservable features of the world, such as atomic orbits.[5] This followed the realization that atomic orbits were quite different from the orbits of planets and other macroscopic objects. They were eventually renamed "orbitals" instead of orbits, a name intended to mean a form of motion without a definite trajectory. Unfortunately, the change in terminology is too subtle, with the result that many chemists, in particular, still seem to maintain some form of pathlike visualization.[6]

Heisenberg intended to build a theory centered on observable quantities such as spectral frequencies. The theory that he developed was highly counterintuitive and required physicists to invest much time and effort in learning a new branch of mathematics dealing with the manipulation of matrices. In addition, the attempt to reject unobservable quantities that Heisenberg had hoped for was not realized.

At about the same time, Schrödinger developed what came to be known as wave mechanics. Already in 1924, the French physicist Prince Louis De Broglie had suggested an analogy to Albert Einstein's earlier discovery that light waves have a particulate nature as well as their expected wave nature. De Broglie made the association run in the opposite sense. Why not suppose that particles such as electrons could likewise display wavelike properties? The test for this idea would be to demonstrate experimentally that electrons produce diffraction and interference effects just like classical waves, such as waves on the surface of water.[7]

Two physicists, Clinton Davisson and Lester Germer, successfully carried out just such an experiment in 1927, thus giving experimental support to De Broglie's proposal.[8] With this discovery, theorists such as Schrödinger received the impetus to further pursue the mathematical analogies between classical waves and electron waves.[9] Whereas Heisenberg's approach was mathematically abstract, that of

Schrödinger was more familiar to physicists because it dealt principally with wave motion. Unlike Heisenberg, Schrödinger had not originally tried to break with realistic notions of the microscopic world and, in fact, had hoped that his method would retain strong connections with classical physics and physical visualization.

As it turned out, neither Heisenberg's nor Schrödinger's hopes materialized fully. The quantum mechanics that emerged after a few years of intense debate was not based solely on observable properties, and nor was it possible to retain a realistic view of matter waves as Schrödinger had originally hoped. Moreover, the two forms of quantum mechanics were shown to be equivalent.[10]

The new theory became centered on the wavefunction for an atom or molecule. This wavefunction could be expressed with a number of terms called "atomic orbitals." As mentioned above, the name was derived from atomic orbits of the old quantum theory but without any intended connection with a definite trajectory for the electron. Such orbitals inhabit a multidimensional Hilbert space in quantum mechanics, thus further denying their visualizability in familiar three-dimensional space. Moreover, wavefunctions and their component building blocks consisting of such orbitals are themselves complex mathematical functions in the sense that they contain factors involving the square root of -1.

What is observable in the case of wavefunctions, as it emerged a little later, is the square of the wavefunction,[11] which is called the electron density.[12] In addition, even the square of the wavefunction cannot be obtained for a single electron at a specific point. The interpretation of quantum mechanics calls for a statistical view in which one can know only the probability of an electron residing in a certain region of space.

HARTREE–FOCK METHOD

When it comes to calculating the properties of atoms, the new quantum mechanics provides a way in which the problem can be attacked by means of approximation methods.[13] The basis of the most widely used approximation for solving quantum mechanical equations for atoms is called the Hartree-Fock method after Douglas Hartree (figure 9.2) and Vladimir Fock, an English and a Russian physicist, respectively.[14]

The main assumption made in the Hartree-Fock model is that any given electron moves in a field resulting from the attraction of the nucleus added to the field that results from the sum of all the remaining electrons. This approach avoids dealing directly with individual electron–electron repulsion terms, and instead, one recovers a situation not altogether unlike that of the hydrogen atom in which one electron is moving is a spherically symmetrical field. In the many-electron case, the field consists of those of the nucleus and of all the other electrons lumped together. The only difference is that instead of one equation for one electron, there

FIGURE 9.2
Douglas Hartree. Photo
and permission from
Emilio Segrè Collection.

are now as many equations as there are electrons in the atom. In addition, the solution for each electron must be consistent with those for all the other electrons, thus requiring a self-consistent iteration procedure that is typically carried out on a computer.

But let us return to the question of the explanation of periodicity, which opened this chapter. The Pauli exclusion principle and the use of four quantum numbers only provide a deductive explanation of the total number of electrons that any electron shell can hold. The correspondence of these values with the number of elements that occur in any particular period is something of a coincidence. The lengths of successive periods have not yet been strictly deduced from the theory.[15] However, most chemistry and physics textbook authors do not emphasize or even mention this point. As a result, they imply that quantum mechanics does indeed provide a perfectly satisfactory deductive explanation of the periodic system. This, in turn, fuels the general impression that chemistry is fully explained by quantum physics and has a negative effect on chemical education. Instead of starting from chemical facts, and the properties of the elements, the modern tendency is to expose students to the rules for electronic configurations in the belief that the chemistry will somehow follow.[16] Nevertheless, the number of electrons contained in any shell, as opposed to the lengths of periods, does emerge directly from the rules for combining the four quantum numbers. This part of the explanation for periodicity is completely satisfactory, as shown in the next section.

WRITING ELECTRONIC
CONFIGURATIONS FOR ATOMS

The assignment of electronic configurations to the atoms in the periodic table proceeds according to three principles:

1. The *aufbau* principle (*aufbauprinzip* in chapter 7): Orbitals are occupied in order of increasing values of $n + \ell$. For example, the 4s orbital for which $n + \ell = 4$ is filled before the 3d orbital for which $n + \ell = 5$. This rule is often accompanied by a diagram like the one shown in figure 9.3, which represents the Madelung or $n + \ell$ rule.

2. The Hund principle: When electrons fill orbitals of equal energies, they occupy as many different orbitals as possible.

3. The Pauli exclusion principle: Only two electrons can occupy a single orbital, and if they do so they must orient their spin angular momenta in opposite directions.[17]

Several points need to be made about these principles. The first principle does not, in fact, refer to the ordering of energies of atomic orbitals. What it really refers to is the order of filling of the various orbitals. These are related but separate issues. But there is more involved in the occupation of orbitals than their individual energies, as discussed further below. The $n + \ell$ rule has not yet been derived from the principles of quantum mechanics. This failure has been described as one of the outstanding problems in quantum mechanics by the leading quantum chemist Per-Olav Löwdin.[18]

It emerges that all three of these principles are essentially empirical, and none of them has been strictly derived from the principles of quantum mechanics.[19] Pauli's principle, for example, takes the form of an additional postulate to the main postulates of quantum mechanics. Despite strenuous efforts on the part of many physicists, including Pauli himself, it has never been possible to derive the principle from the postulates of quantum mechanics and/or relativity theory.[20] So, rather

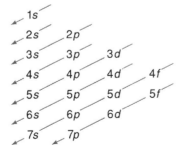

FIGURE 9.3
Madelung (or $n + \ell$ rule) for the order of filling of orbitals. From R. Chang, *Physical Chemistry for the Chemical and Biological Sciences*, University Science Books, Sausalito, CA, 2000, p. 601. By permission from the publisher.

than providing an explanation for electronic configurations, the three commonly used rules are really statements that summarize what is known to happen from experimental data on atomic spectra.

And now let us turn to the explanation for the number of electrons in each shell and its connection to the number of elements in each subsequent period of the periodic table. These facts are usually explained in terms of the relationship between the four quantum numbers, which can be assigned to any electron in a many-electron atom. The relationship between the first three quantum numbers is rigorously deduced from the Schrödinger equation for the hydrogen atom. The first quantum number, n, can adopt any integral value starting with 1.[21] The second quantum number, which is given the label ℓ, can have any of the following values related to the values of n:

$$\ell = n - 1, \dots 0$$

In the case when $n = 3$, for example, ℓ can take the values 2, 1, or 0. The third quantum number, labeled m_ℓ can adopt values related to those of the second quantum numbers:

$$m_\ell = -\ell, -(\ell + 1), \dots 0 \dots (\ell - 1), \ell$$

For example, if $\ell = 2$ the possible values of m_ℓ are $-2, -1, 0, +1$, and $+2$. Finally, the fourth quantum number, labeled m_s can take only two possible values, either $+1/2$ or $-1/2$ units of spin angular momentum. There is therefore a hierarchy of related values for the four quantum numbers, which are used to describe any particular electron in an atom.

As a result of this scheme, it is clear why the third shell, for example, can contain a total of 18 electrons. If the first quantum number, given by the shell number, is 3, there will be a total of $2 \times (3)^2$ or 18 electrons in the third shell.[22] The second quantum number, ℓ, can take values of 2, 1, or 0. Each of these values of ℓ will generate a number of possible values of m_ℓ and each of these values will be multiplied by a factor of 2 since the fourth quantum number can adopt values of $1/2$ or $-1/2$.

But the fact that the third shell can contain 18 electrons does not strictly explain why it is that some of the periods in the periodic system contain 18 places. It would be a rigorous explanation of this fact only if electron shells were filled in a strictly sequential manner. Although electron shells begin by filling in a sequential manner, this ceases to be the case starting with element 19, potassium. Since the configuration of element 18, argon, is

$$1s^2, 2s^2, 2p^6, 3s^2, 3p^6,$$

one might expect the configuration for element 19, potassium, would be

$$1s^2, 2s^2, 2p^6, 3s^2, 3p^6, 3d^1.$$

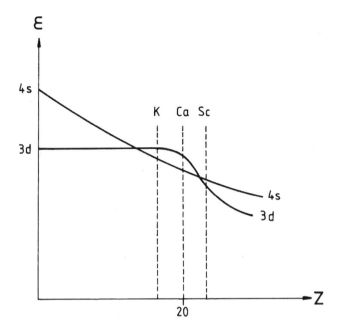

FIGURE 9.4 The relative ordering of the 3d and 4s energy levels. L.G. Vanquickenborne, K. Pierloot, D. Devoghel, *Journal of Chemical Education*, 71, 469–471, 1994, p. 469. By permission from the publisher.

This would be expected because up to this point the pattern has been one of adding the differentiating electron to the next available orbital at increasing distances from the nucleus. However, experimental evidence shows that the configuration of potassium should be denoted as

$$1s^2, 2s^2, 2p^6, 3s^2, 3p^6, 4s^1.$$

As many textbooks explain, this can result from the fact that the 4s orbital has a lower energy than the 3d orbital for the atoms of potassium and calcium (see figure 9.4).

In the case of element 20, calcium, the new electron also enters the 4s orbital. But in the case of element 21, scandium, the orbital energies have reversed so that the 3d orbital has a lower energy. Textbooks typically claim that since the 4s orbital is already full, the next electron necessarily begins to occupy the 3d orbital. This pattern is supposed to continue across the first transition series of elements, apart from the elements chromium and copper, where further anomalies occur (table 9.1).

In fact, this explanation for the configuration of the scandium atom, and most other first transition elements, is inconsistent.[23] If the 3d orbital has a lower energy

TABLE 9.1

Electronic Configurations for
First Transition Metals

Sc	$4s^2 3d^1$
Ti	$4s^2 3d^2$
V	$4s^2 3d^3$
Cr	$4s^1 3d^5$
Mn	$4s^2 3d^5$
Fe	$4s^2 3d^6$
Co	$4s^2 3d^7$
Ni	$4s^2 3d^8$
Cu	$4s^1 3d^{10}$
Zn	$4s^2 3d^{10}$

than 4s starting at scandium, and if one were indeed filling the orbitals with electrons in order of increasing energy, one would expect that all three of the final electrons would enter 3d orbitals. The argument that most textbooks present is incorrect since it should be possible to predict the configuration of an element from a knowledge of the order of its own orbital energies. One should not have to consider the configuration of the previous element and assume that this configuration is somehow carried over intact on moving to the next element.

What seems to make this issue more mysterious is that, whereas all transition elements show a preferential occupation for an s orbital, it appears that the s electrons are also the easiest to ionize. This situation may be represented by two diagrams, one for relative occupation and one for relative ionization of orbitals in the first transition series, to give the sets of ordered levels shown in figure 9.5.

The apparent paradox is as follows: If the 4s orbital is preferentially occupied by electrons, this suggests that it has greater stability after all interactions have been properly taken into account.[24] However, the right diagram in figure 9.5, which depicts the relative ease of losing electrons, suggests that the 4s electrons are overall less stable since they, rather than the 3d electrons, are more easily removed. Many complicated analyses of this situation have been published in recent years in order to try to resolve the apparent paradox.[25]

There is perhaps a simple solution, or perhaps a dissolution, of the problem. In posing the paradox regarding the 4s and 3d orbitals, many authors appear to have overlooked one very important feature, which makes the comparison problematic. In considering the buildup of atoms across the periodic table, one is concerned with the successive addition of one proton and one electron to each previous atom.[26] However, in considering the ionization of any particular atom, one is concerned only with the successive removal of electrons and not the removal

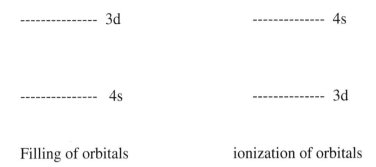

FIGURE 9.5 Ordering of orbital energies as implied by relative filling and relative ionization. The lower the level, the more stable the atom.

of protons. As a result, the comparison of the two diagrams in figure 9.5 does not constitute a comparison of like with like. The long-standing puzzle, which has exercised the minds of generations of students and their instructors, can be dissolved at a stroke. The question of why 4s fills first but also empties first is an illegitimate question in some respects.

AN EXPLANATION FOR SHELL CLOSING BUT NOT FOR PERIOD CLOSING

As suggested above, there is a problem with the claim that the periodic table is deductively explained by quantum mechanics. A feature that seems to generally go unnoticed is the need to assume the empirical order of shell filling rather than trying to derive it from the theory. The order in which orbitals are occupied with electrons is not derived from first principles. It is justified *post facto* and by some complex calculations.

Suppose, for example, that the Hartree-Fock method is used to compare the energies of the scandium atom with two alternative configurations: [Ar] $4s^2 3d^1$ and [Ar] $4s^1 3d^2$. This can be carried out using ordinary nonrelativistic quantum mechanics or, alternatively, by including relativistic effects. The results obtained are as shown in table 9.2.[27] In each case, the more negative the calculated value of the energy, the more stable the configuration.[28] Clearly, the inclusion of relativistic effects serves to reduce the energy from the nonrelativistic value, as one would expect. In the case of scandium, it appears that both nonrelativistic and relativistic ab initio calculations correctly compute that the $4s^2$ configuration has the lower energy, in accordance with experimental data. Similar calculations do not fare so well in the case of the chromium atom, however (table 9.3). In this case, it appears that both nonrelativistic and relativistic calculations fail to predict

TABLE 9.2

Nonrelativistic and Relativistic Calculated
Energies for Two Configurations of Scandium
(in Hartree units)

$4s^2\ 3d^1$ configuration	
Nonrelativistic	−759.73571776
Relativistic	−763.17110138
$4s^1\ 3d^2$ configuration	
Nonrelativistic	−759.66328045
Relativistic	−763.09426510

One Hartree is equal to 4.3597×10^{-18} J.

Compiled by the author. These results were obtained using
the Internet web pages designed by Charlotte Froese-Fischer,
one of the leading pioneers in the field of Hartree-Fock calculations. http://atoms.vuse.vanderbilt.edu/

TABLE 9.3

Nonrelativistic and Relativistic Calculated
Energies for Two Configurations of Chromium
(in Hartree units)

$4s^1\ 3d^5$ configuration	
Nonrelativistic	−1043.141755
Relativistic	−1049.24406264
$4s^2\ 3d^4$ configuration	
Nonrelativistic	−1043.17611655
Relativistic	−1049.28622286

One Hartree is equal to 4.3597×10^{-18} J.

Compiled by the author. These results were obtained using
the Internet web pages designed by Charlotte Froese-Fischer,
http://atoms.vuse.vanderbilt.edu/

which of these two configurations is the correct experimentally observed ground
state, namely, $4s^1\ 3d^5$.

Looking at the calculated energies for the copper atom in table 9.4 shows
that a nonrelativistic calculation sometimes gives the correct result for the lowest
energy configuration. However, it also emerges that by carrying out the calculation
to a greater degree of accuracy by including relativistic effects the prediction can
in some cases deteriorate in that one predicts the opposite order of stabilities than
observed experimentally. The lowest energy configuration for copper cannot yet be
successfully calculated from first principles, at least at this level of approximation.

TABLE 9.4

Nonrelativistic and Relativistic Calculated
Energies for Two Configurations of Copper
(in Hartree units)

$4s^1 3d^{10}$ configuration	
Nonrelativistic	−1638.9637416
Relativistic	−1652.66923668
$4s^2 3d^9$ configuration	
Nonrelativistic	−1638.95008061
Relativistic	−1652.67104670

One Hartree is equal to 4.3597×10^{-18} J.

Compiled by the author. These results were obtained using
the Internet web pages designed by Charlotte Froese-Fischer,
http://atoms.vuse.vanderbilt.edu/

The fact that copper has a $4s^1 3d^{10}$ configuration rather than $4s^2 3d$ is an experimental fact. The theory is, strictly speaking, accommodating what is already known experimentally. For example, the first of the two periods of 18 elements is not due to the successive filling of 3s, 3p, and 3d electrons but due to the filling of 4s, 3d, and 4p. It just so happens that both of these sets of orbitals are filled by a total of 18 electrons. This coincidence is what gives the generally given explanation its apparent credence. It does not seem to be appreciated that these are not the same 18 electrons that are "doing the occupying" as one traverses the periodic table.[29]

The Nickel Atom

The case of nickel turns out to be more interesting (table 9.5). According to nearly every chemistry and physics textbook, the configuration of this element is given as $4s^2 3d^8$. However, the research literature on atomic calculations invariably quotes the configuration of nickel as $4s^1 3d^9$. The difference occurs because in more accurate work one considers not just the lowest possible component of the ground-state term but the average of all the components arising from a particular configuration. Nickel is somewhat unusual in that, although the lowest energy term arises from the $4s^2 3d^8$ configuration, the average energy of all the components arising from this configuration is higher than the average energy of all the components arising from the $4s^1 3d^9$ configuration. As a consequence, the $4s^2 3d^8$ configuration is regarded as having the lowest energy, and it is this average energy that is compared with experimental energies. When this comparison is made, it emerges that the quantum mechanical calculations using a relativistic Hartree-Fock approach give an incorrect ground state.

TABLE 9.5

Nonrelativistic and Relativistic Calculated
Energies for Two Configurations of Nickel
(in Hartree units)

$4s^2 3d^8$ *configuration*	
Nonrelativistic	−1506.87090774
Relativistic	−1518.68636410
$4s^1 3d^9$ *configuration*	
Nonrelativistic	−1506.82402795
Relativistic	−1518.62638541

One Hartree is equal to 4.3597×10^{-18} J. Note that theory
predicts a $4s^2\ 3d^8$ configuration.

Compiled by the author. These results were obtained using
the Internet web pages designed by Charlotte Froese-Fischer,
http://atoms.vuse.vanderbilt.edu/

Of course, the calculations can be improved by adding extra terms until this
failure is eventually corrected, but these additional measures are taken only after
the fact. Moreover, the lengths to which theoreticians are forced to go to in order
to obtain the correct experimental ordering of terms does not give one too much
confidence in the strictly predictive power of quantum mechanical calculations in
this context.[30]

Back to Hund's Rule

Let us now consider the Hund principle and the manner in which it is used to try
to justify the configurations of elements in the first, second, and third transitions.
The elements in the first transition are generally believed to show two "anomalous" configurations, which include a $4s^1$ orbital occupation, rather than the more
common $4s^2$ configuration.[31] These atoms are those of chromium and copper,
which are taken to have respective configurations of $4s^1\ 3d^5$ and $4s^1\ 3d^{10}$. The justification for the adoption of the first of these configurations is frequently given by
appeal to Hund's rule of maximum spin multiplicity. It is argued that this configuration is more stable than any alternatives because it involves a half-filled d subshell.
However, if the configurations of the elements in the second transition series are
considered, it is clear that this form of explanation is rather ad hoc in the sense that
it cannot be generalized to other transition series.

For example, the configurations of the elements in the second transition series
are shown in table 9.6. Once again, this set of configurations is primarily arrived at
from experimental data, although these ground-state configurations are supported

TABLE 9.6
Configurations of Outermost
Two Orbitals of Elements in
Second Transition Series

Y	$5s^2 4d^1$
Zr	$5s^2 4d^2$
Nb	$5s^1 4d^4$
Mo	$5s^1 4d^5$
Tc	$5s^1 4d^6$
Ru	$5s^1 4d^7$
Rh	$5s^1 4d^8$
Pd	$5s^0 4d^{10}$
Ag	$5s^1 4d^{10}$
Cd	$5s^2 4d^{10}$

by theoretical calculations in most cases. But if the possession of half-filled orbitals is the explanation for why chromium adopts a $4s^1$ configuration in the first transition series, some other factors must be operating in many cases of the second transition. This is because many of these atoms likewise show an s^1 configuration even though they do not possess a half filled d subshell.[32] Hund's principle is essentially an empirical result. In spite of many attempts, nobody has yet succeeded in deriving the principle from quantum mechanics.[33] Of course, some plausible arguments can be given for its effectiveness, such as the claim that one is thereby minimizing the contribution from exchange terms involving repulsions between electrons. For example, a calculation can be carried out to show that, in the case of the helium atom, the triplet state (one involving two unpaired electrons) has lower energy than the singlet state where the two electrons are paired. But contrary to the standard account one encounters in textbooks, it has been shown that the reason for the greater stability of the helium triplet state is not reduced electron–electron repulsion but the greater electron–nucleus attraction that occurs in the triplet state.[34]

Choice of Basis Set

There is yet another general problem that mars any hope of claiming that electronic configurations can be predicted theoretically and that quantum mechanics thereby provides a purely deductive explanation of what was previously only obtained from experiments. In most of the configurations considered above, it has been possible to use quantum mechanics to calculate the particular configuration that possesses the lowest energy. However, in performing such calculations, the candidate configurations that are subjected to the calculation are themselves obtained from the *aufbau* principle and other rules of thumb such as Hund's principle, or by straightforward appeal to experimental data. Theoretical calculations cannot actually predict the electronic configuration for any element. There is a very simple reason for this state of affairs, which is often overlooked. The quantum mechanical calculations on ground-state energies involve the initial selection of a basis set, which in simple terms is the electronic configuration of the atom in question. Quantum mechanical calculations do not actually generate their own basis sets.[35] So, whereas the correct ground-state electronic configurations can in many cases be correctly calculated among a number of plausible options, the options themselves are not provided by the theory. This is another weakness of the present claims to the effect that quantum mechanics fully explains the periodic system, although this limitation is being addressed in some recent work.

THREE POSSIBLE APPROACHES TO THE REDUCTION OF THE PERIODIC TABLE

This section attempts to take stock of the various senses of the claim that the periodic system is reduced, or fully explained, by quantum mechanics.

Qualitative Reduction/Explanation of Periodic Table in Terms of Electrons in Shells

In broad terms, the approximate recurrence of elements after certain regular intervals is explained by the possession of a certain number of outer-shell electrons. This form of explanation appears to be quantitative because it deals in number of electrons but, in fact, turns out to be rather qualitative in nature. It cannot be used to predict quantitative data such as the ground-state energy of any particular atom. In order to do so, one needs to go beyond the ground-state configuration of the atom in question, and it is essential to assume that electrons also find themselves in higher energy orbitals that are not considered in the textbook configuration of the element.

In addition, it emerges that the possession of a particular number of outer-shell electrons is neither a necessary nor a sufficient condition for an element's being in any particular group. It is possible for two elements to possess exactly the same outer electronic configuration and yet not to be in the same group of the periodic system. For example, the inert gas helium has two outer-shell electrons and yet is not generally placed among the alkaline earth elements such as magnesium, calcium, and barium, all of which also display two outer-shell electrons.[36]

Conversely, there are cases of elements that do belong in the same group of the periodic table even though they do not have the same outer-shell configuration. In fact, this occurrence is rather common among the transition metal series. Consider this interesting example:[37]

Ni	[Ar] $4s^2\ 3d^8$
Pd	[Kr] $5s^0\ 4d^{10}$
Pt	[Xe] $6s^1\ 4f^{14}\ 5d^9$

In addition, the very notion of a particular number of electrons in a particular shell stands in violation of the Pauli exclusion principle, which states that electrons cannot be distinguished.[37] The indistinguishability of electrons implies that one can never state that a particular number of electrons are in any particular shell, although it is frequently useful to make this approximation. Indeed, the independent-electron approximation, as it is known, represents one of the central paradigms in modern chemistry and physics. To state the electronic configuration of an atom is to operate within this level of approximation. For example, one might state that the configurations of two randomly chosen elements are as follows:

Carbon	$1s^2, 2s^2, 2p^2$
Fluorine	$1s^2, 2s^2, 2p^5$

This kind of activity could only be considered as fully satisfactory and as indicating a theoretical deduction if such configurations themselves could be derived from quantum mechanics. However, as discussed above, electronic configurations such as those for carbon and fluorine are arrived at essentially by means of the *aufbau* principle, which is experimentally based. The configurations can be justified in terms of calculations in some cases, but they cannot be derived from first principles because the basis set, consisting of a particular set of atomic orbitals, is generally selected before any calculation can be carried out.

Ab Initio Calculations

The second approach to be considered is a far better candidate for the claim to explain the periodic table from quantum mechanics. Even if the crude notion of a particular number of outer-shell electrons for any particular atom fails to give a fundamental explanation, it should be possible to carry out detailed calculations that allow atoms to have more complicated configurations. Going to such a deeper level than the notion of a particular number of electrons in shells might thus provide a more successful explanation of the periodic system.

Ab initio calculations aim to calculate the properties of atoms and molecules starting from the fundamental equation of quantum mechanics, the Schrödinger equation for the system. The various methods utilized vary in the extent to which they are genuinely ab initio. In some cases, the methods incorporate semiempirical aspects. For example, certain integration terms that are too difficult to evaluate are replaced by quantities derived from experimental data. But the type of approach considered here is the purer variety of such calculations, where no semiempirical aspects are incorporated. My aim is to examine the extent to which such ab initio approaches provide a reduction of the periodic system.

Such an approach represents an improvement and is a better contender for the claim of a full explanation of the periodic system. In order to illustrate both the power and the pitfalls of the method, I focus on the ab initio calculation of ionization energies of atoms. In this approach, the notion of electrons in shells is used instrumentally with the knowledge that such an approximation represents only a first-order approach to calculations. If one wishes to still think in terms of electrons in orbitals, these calculations can be thought of as regarding the atom as existing in many different electronic configurations simultaneously. The ground-state configuration, so beloved of chemistry and physics textbooks, is just the leading term in an algebraic expansion for the wavefunction of the atom in question.[39]

At this level of approximation, the fact that certain elements fall into the same group of the periodic table is not explained by recourse to the number of outer-shell electrons. Instead, the explanation lies in calculating the magnitude of a property such as the first ionization energy and seeing whether the expected periodicity is recovered in the calculations. Figure 9.6 shows schematically the experimental first-ionization energies for elements 3–53 in the periodic table, along with the values calculated using ab initio quantum mechanical methods. As is readily apparent from the figure, the periodicity is captured remarkably well, even down to portions of the graph occurring between elements in groups II and III and between groups V and VI in each period of the table. Clearly, the calculation of atomic properties can be achieved by the theory to a high degree of accuracy. The quantum mechanical explanation of the periodic system within this approach represents a far more impressive achievement than merely claiming that elements fall into similar groups because they share the same number of outer electrons.

And yet in spite of these remarkable successes, such an ab initio approach may still be considered to be semiempirical in a rather specific sense. In order to obtain the calculated points shown in figure 9.6, the Schrödinger equation must be solved separately for each of the 50 atoms concerned. The approach there-fore represents a form of "empirical mathematics,"[40] where one solves 50 indi-vidual Schrödinger equations in order to reproduce the well-known pattern in the periodicities of ionization energies. It is as if one had performed 50 individual experiments, although the "experiments" in this case are all iterative mathematical

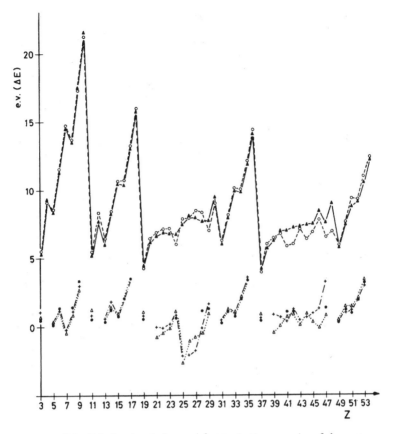

FIGURE 9.6 Calculated and observed first ionization energies of elements 3–53. Ionization energy is plotted against atomic number. Circles represent experimental values; triangles are calculated values From E. Clementi, *Computational Aspects of Large Chemical Systems, Lecture Notes in Chemistry*, vol. 19, Springer-Verlag, Berlin, 1980, p. 12. By permission from the publisher.

computations. This is still, therefore, not a general solution to the problem of the electronic structure of atoms.

Density Functional Approach

The third kind of approach to reducing the periodic table does not suffer from the drawback just mentioned in the case of ab initio calculations, at least not in principle. In 1926, the physicist Llewellyn Thomas proposed treating the electrons in an atom by analogy to a statistical gas of particles. No electron shells are envisaged in this model, although electrons still possess angular momentum values as they do in the electron-shell model. This method was independently rediscovered by Italian physicist Enrico Fermi two years later, and is now called the Thomas-Fermi method. For many years, it was regarded as a mathematical curiosity without applications since the results it yielded were inferior to those obtained by the method based on electron orbitals. The appeal of the Thomas-Fermi method comes from the fact that it treats the electrons around the nucleus as a perfectly homogeneous electron gas and that the mathematical solution for this system is "universal" in the sense that it can be solved once and for all. This represents an improvement over any method in which one seeks to solve a Schrödinger equation for every separate atom as in the wavefunction approach illustrated in figure 9.6.

Gradually, the Thomas-Fermi method or its modern descendants, which are known as density functional theories, have become equally powerful compared to methods based on orbitals and wavefunctions and in many cases can outstrip the wavefunction approaches in terms of computational accuracy. The solution is expressed in terms of the variable Z, which represents atomic number, the crucial feature that distinguishes one kind of atom from any other element. One does not need to repeat the calculation separately for each atom, but this advantage applies only in principle, as discussed below.

There is another important conceptual, or even philosophical, difference between the orbital/wavefunction methods and the required density functional methods. In the case of orbitals, the theoretical entities are completely unobservable, whereas electron density, which is featured in density functional theories, is a genuine observable.[41] Experiments to observe electron densities have been routinely conducted since the development of X-ray and other diffraction techniques.[42] Orbitals cannot be observed either directly or indirectly since they have no physical reality, a state of affairs dictated by quantum mechanics. The orbitals used in ab initio calculations are just mathematical constructs that exist in a multidimensional Hilbert space,[43] while electron density is altogether different, as indicated, since it is a well-defined observable and exists in real three-dimensional space.[44]

DENSITY FUNCTIONAL THEORY
IN PRACTICE

Most of what has been described so far concerning density theory applies in theory rather than in practice. The fact that the Thomas-Fermi method is capable of yielding a universal solution for all atoms in the periodic table is a potentially attractive feature but has not been realized in practice. Because of various technical difficulties, which are not described here, the attempts to implement the ideas originally due to Thomas and Fermi have not materialized.[45] This has meant a return to the need to solve a number of equations separately for each individual atom as one does in the Hartree-Fock method and other ab initio methods using atomic orbitals. In addition, most of the more tractable approaches in density functional theory also involve a return to the use of atomic orbitals in carrying out quantum mechanical calculations since there is no known means of obtaining the functional based directly on electron density.[46] Researchers therefore fall back on using basis sets of atomic orbitals that yield the electron density when squared.

To make matters worse, the use of a uniform gas model for electron density does not enable one to carry out accurate calculations. Instead, "ripples" must be introduced into the uniform electron gas distribution. The way in which this has been implemented has typically been in a semiempirical manner by working backward from the known results on a particular system, usually taken to be the helium atom. In this way, it has been possible to obtain an approximate set of functions that also give successful approximate calculations in many other atoms and molecules. By carrying out this combination of a semiempirical approach and retreating from the pure Thomas-Fermi ideal of a uniform gas, it has actually been possible to obtain computationally better results, in many cases, than with conventional ab initio methods using orbitals and wavefunctions.[47]

If anything, the early promise and hope offered by quantum mechanics and Paul Dirac's famous dictum that all of chemistry can be calculated from first principles has turned out to be only partly fulfilled.[48] Although calculations have become increasingly accurate, one realizes that they include considerable semiempirical elements at various levels. From the purist philosophical point of view, this implies that not everything is being explained from first principles.

As time has progressed, the best of both approaches have been blended together with the result that many computations are now performed using a mixture of wavefunction and density approaches within the same computations. This feature brings with it advantages as well as disadvantages. The unfortunate fact is that, as yet, there are no pure density functional methods that are tractable for performing calculations. The philosophical appeal of a universal solution for all the atoms of the periodic system, based on electron density rather than fictitious orbitals, has not yet borne fruit.[49]

CONCLUSION

The aim of this chapter has not been trying to decide whether or not the periodic system is explained by quantum mechanics *tout court*, since the situation is more subtle. It is more a question of the extent of reduction or extent of explanation that has been provided by quantum mechanics.

Whereas most chemists and educators seem to believe that the reduction is complete, perhaps there is some benefit in pursuing the question of how much is strictly explained from the theory. After all, it is hardly surprising that quantum mechanics cannot yet fully deduce the details of the periodic table, which gathers together a host of empirical data from a level far removed from the microscopic world of quantum mechanics.

It is indeed something of a miracle that quantum mechanics explains the periodic table to the extent that it does at present. But we should not let this fact seduce us into believing that it is a deductive explanation. One thing that is clear is that the attempt to explain the details of the periodic table continues to challenge the ingenuity of quantum physicists and quantum chemists and that the periodic table will continue to present a test case for the adequacy of new methods developed in quantum chemistry.[50]

Our story has now been brought up to date. From its humble beginnings as a set of isolated triads of elements, the periodic system has grown to embody well more than 100 elements and has survived various discoveries such as that of isotopes and the quantum mechanical revolution in the study of matter. Rather than being swept aside, it has continued to provide a challenge to the development of ever more accurate means of calculating the basic properties of the atoms of the chemical elements. The central role of the periodic system in modern chemistry has been consolidated rather than eroded.

The reduction of chemistry to quantum mechanics has neither failed completely, as some philosophers of science have claimed,[51] nor has it been a complete success, as some contemporary historians have claimed.[52] The reductive enterprise has been highly successful but not to the extent of deposing the chemical facts or the quintessential discovery of chemical periodicity made by De Chancourtois, Newlands, Odling, Hinrichs, Lothar Meyer, and most significantly, Mendeleev. Rather than undermining chemical periodicity, modern quantum physics has literally re-presented the periodic system and has provided it with a theoretical justification. More important, quantum physics has achieved this feat without assuming the imperialistic role that it is sometimes attributed.

CHAPTER 10

ASTROPHYSICS, NUCLEOSYNTHESIS, AND MORE CHEMISTRY

Having now examined attempts to explain the nature of the elements and the periodic system in a theoretical manner, it is necessary to backtrack a little in order to pick up a number of important issues not yet addressed. As in the preceding chapters, several contributions from fields outside of chemistry are encountered, and the treatment proceeds historically.

So far in this book, the elements have been treated as if they have always existed, fully formed. Nothing has yet been said about how the elements have evolved or about the relative abundance of the isotopes of the elements. These questions form the contents of the first part of this chapter. It also emerges that different isotopes show different stabilities, a feature that can be explained to a considerable extent by appeal to theories from nuclear physics.

The study of nucleosynthesis, and especially the development of this field, is intimately connected to the development of the field of cosmology as a branch of physical science.[1] In a number of instances, different cosmological theories have been judged according to the degree to which they could explain the observed universal abundances of the various elements.[2] Perhaps the most controversial cosmological debate has been over the rival theories of the big bang and the steady-state models of the universe. The proponents of these theories frequently appealed to relative abundance data, and indeed, the eventual capitulation of the steady-state theorists, or at least some of them, was crucially dependent upon the observed ratio of hydrogen to helium in the universe.[3]

Later parts of this chapter go on to examine further aspects of the chemistry of the elements and, in particular, the aspects that do not fit neatly into the simple idea that trends occur just within groups and within periods of the periodic system. The complexity that is seen on examining these more esoteric yet, in many cases, well-known effects raises new questions for the overall understanding of the periodic system. Finally, the fact that scientists working in many different fields such as geology, metallurgy, physics, and chemistry have developed their own versions of the periodic table is a feature that is discussed in a quest to reach a global or philosophical understanding of the periodic system.

EVOLUTION OF THE ELEMENTS

Chapters 2, 3, and 6 discussed Prout's hypothesis, according to which all the elements are essentially made out of hydrogen. Although the hypothesis was initially rejected on the basis of accurate atomic weight determinations, it underwent a revival in the twentieth century. As mentioned in chapter 6, the discoveries of Anton van den Broek, Henry Moseley, and others showed that there is a sense in which all elements are indeed composites of hydrogen. This is so if one focuses on the fact that hydrogen contains one proton while all other elements contain a particular number of protons bound together in their nuclei. In this chapter I concentrate on the second sense in which Prout's hypothesis may be said to have made a comeback. The elements are now believed to have literally evolved from hydrogen by various mechanisms. One of the first people to take this possibility seriously was the English scientist William Crookes, who was also the founder and editor of the influential journal *Chemical News*.

Crookes belongs among the pioneers of the periodic system, although his name is less frequently encountered in this context than are those of precursors such as Johann Döbereiner or discoverers such as John Newlands and Dimitri Mendeleev. Crookes began by studying chemistry under A.W. Hoffmann at the Royal College of Chemistry and then under Michael Faraday at the Royal Institution, while initially working in the field of spectroscopy. Among other accomplishments, Crookes seems to have anticipated the discovery of isotopes, as demonstrated in a quotation from him in chapter 6.

In 1861, Crookes announced the discovery of a new element, thallium, which he identified through a prominent green line in its spectrum.[4] But the most important contribution made by Crookes, for the purposes of the present chapter, was his advocating the inorganic evolution of the elements:

> In the very words selected to denote the subject that I have the honour of bringing before you, I have raised a question which may be regarded as heretical. At the time when our modern conception of chemistry first dawned upon the scientific mind, the average chemist as a matter of course accepted the elements as ultimate facts.
>
> I venture to say that our commonly received elements are not simple or primordial, that they have not arisen by chance and have not been created in a desultory and mechanical manner but have been evolved from, simple matters—or perhaps indeed from one sole kind of matter. . . .[5]

Crookes made a spectroscopic study of gases at low pressure that were subjected to high-voltage electric discharges. In 1879, he speculated that the plasma present in the gases treated in this manner, as well as in the stars, consisted of a fourth state of matter. Under such conditions, Crookes argued, the atoms of the elements existed as primary matter that he identified with William Prout's protyle. Seven years later, at a meeting of the British Association for the Advancement of Science, Crookes announced his theory that the chemical elements had evolved in the stars, as they cooled from such a plasma state, through the oscillation of giant electrical forces analogous to those he had studied in discharge tubes (figure 10.1). He claimed that the main oscillating electrical force had an amplitude that corresponded to a period in Mendeleev's periodic system, for example, from hydrogen from fluorine and beyond it. A subsidiary oscillation would act in such a manner as to separate the electropositive elements from the electronegative ones. According to Crookes, the elements were formed in increasing order of atomic weight as the cosmic plasma cooled down. Such giant electrical oscillations would recur to form all the elements in the periodic table, including some that occupied new vacant spaces for elements that were still not known. Each successive amplitude became smaller, with further cooling and with increasing atomic weight, with the result that heavier elements would have more similar properties among each other than would the lighter ones.

This mechanism was illustrated by a three-dimensional pretzel-shaped periodic system, created after the discovery of the noble gases, which is still displayed at the Science Museum in London. This double helical model shows hydrogen at the top and moves downward toward the final element, uranium (figure 10.2).[6] When the noble gases were discovered in the 1890s Crookes was quick to point out that these elements were all implied by his periodic system since they represented the centers of the giant electrical oscillations that he had published in 1886, such as the place between fluorine and sodium on his diagram (figure 10.1).

Mendeleev, however, was critical of evolutionary schemes such as these, declaring in his Faraday lecture of 1889, "The periods of the elements have a character very different from those which are so simply represented by the geometrers. . . . [T]hey correspond to points, to numbers, to sudden changes of the masses, and not to a continuous evolution."[7] A large number of chemists were involved in founding the field of nucleosynthesis in addition to Crookes.[8] Among them was Richard Tolman, a Caltech chemist who was also an expert in the theory of relativity and statistical mechanics. Another was Jean Perrin, a physical chemist who contributed crucially to the acceptance of atoms as real physical entities and whose early atomic model was mentioned in chapter 7. Svante Arrhenius was a Nobel Prize–winning chemist and one of the founders of physical chemistry at the turn of the twentieth century. He developed a cosmological theory in which he speculated, unsuccessfully as it turned out, that the universe would not have to suffer a heat death.[9]

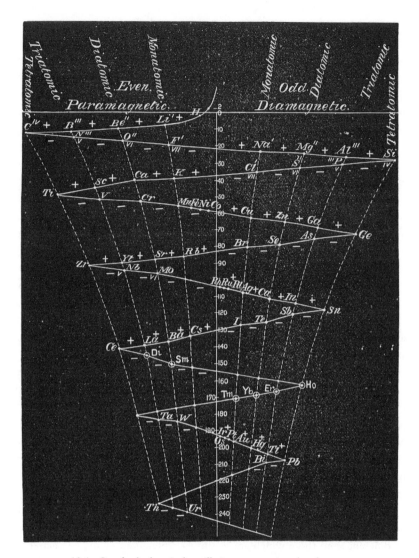

FIGURE 10.1 Crookes's electrical oscillations generating the elements.
W. Crookes, *Chemical News*, 45, 115–126, 1886, figure on p. 120.

Similarly, Walther Nernst, another of the founding figures of physical chemistry
and the discoverer of the third law of thermodynamics, speculated that radioactive
atoms could be created in the ether, which was in turn associated with the zero-
point energy that had recently been discovered through the new quantum mechanics. He hoped that a mechanism of continuous recycling would prevent the dreaded
heat death of the universe that is generally predicted from thermodynamics.

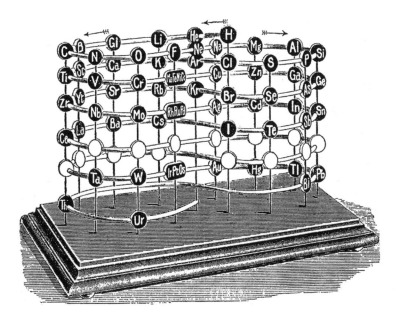

FIGURE 10.2 Crookes's periodic system. W. Crookes, *Proceedings of the Royal Society of London*, 63, 408–411, 1898, figure on p. 409.

The idea of the evolution of the elements was serious taken up again by the astronomer Arthur Eddington, who was intrigued by Prout's hypothesis. Eddington started by suggesting that four hydrogen atoms could combine together to form atoms of helium. In an article published in *Nature* in 1920, Eddington speculated that the artificial transmutation of elements, which Rutherford had recently discovered by bombarding nuclei with protons, might also take place in the interiors of stars: "What is possible in the Cavendish Laboratory [to make atomic nuclei react] may not be too difficult in the sun."[10]

Big bang cosmology, although it did not originally bear this name, originated with the Belgian priest Georges Lemaître, who was the first to discuss the universe as having been created at a particular moment in time. At first his theory was not taken seriously because it seemed to conflict with Albert Einstein's view of a static universe and because, as some suggested, it seemed to border on theology, especially given Lemaître's declared religious affiliations.[11]

Gradually, astronomical observations showed that the universe was indeed expanding, but whether this was occurring as a result of an initial moment of creation remained controversial.[12] The person who placed the big bang theory on a more secure foundation and, coincidentally for our story, a physicist who made the first major contribution to the theory of nucleosynthesis was the Ukrainian born

George Gamov. Gamov was also the first to bring a knowledge of nuclear physics into cosmology and, as it turned out, to some considerable advantage. Broadly
speaking, Gamov, along with colleagues Ralph Alpher and Hans Bethe, was able
to show that the hypothesized conditions, which prevailed just after the big bang,
were consistent with the synthesis of the light elements from hydrogen to beryllium.[13] They argued that the birth of the elements did not take place under equilibrium conditions but as a result of what subsequently became known as the big
bang creation of the universe. The mechanism that the authors appealed to was
one of neutron absorption by hydrogen atoms followed by beta decay, which could
in principle be repeated to form all the elements successively. This notion depends
on the fact that beta decay involves the conversion of a neutron into a proton and
a beta particle, which is essentially a fast-moving electron created in the nucleus.
The absorption of a neutron thus results in the formation of an element with one
more proton than the previous one:

$$\; _{0}^{1}n \rightarrow \; _{1}^{1}p + \; _{-1}^{0}\beta$$

In spite of some early successes, Gamov's theory quickly encountered a couple of
major stumbling blocks concerning the formation of nuclei of masses 5 and 8. The
abundance of both of these nuclei is almost completely negligible. This fact may
seem harmless enough until it is appreciated that it puts a bottleneck on the formation of nuclei larger than ^{4}He by the absorption of neutrons or even protons.
And if the possibility of forming successive elements by the addition of protons
to lighter nuclei in the sequence is interrupted, it becomes difficult to explain the
occurrence of any nuclei whatsoever with a mass heavier than 4. In addition, the
nonoccurrence of nuclei of mass 8 suggests that it is impossible for two helium
nuclei (mass 4) to combine together to form a composite nucleus. Of course, the
gaps at mass 5 and mass 8 leave open the possibility of completely different mechanisms for the formation of heavier nuclei, but none that were even contemplated
by Gamov's theory.

   This impasse was partly surmounted in 1952 by Edwin Saltpeter at Cornell
University. His suggestion was that a "triple alpha" mechanism could provide a
means of building nuclei beyond mass number 4, as summarized in the following
equation:

$$3 \; ^{4}He \rightarrow \; ^{12}C + 2\,\gamma + 7.3 \text{ MeV}$$

Saltpeter argued that this process could very well take place in the interior of stars.
In addition, he deduced that even if ^{8}Be could exist only for a small fraction of a
second, this would be enough time to enable the formation of ^{12}C by an additional
mechanism:

$$2 \; ^{4}He \rightarrow \; ^{8}Be$$
$$^{4}He + \; ^{8}Be \rightarrow \; ^{12}C^{\star} \rightarrow \; ^{12}C + 2\,\gamma$$

FIGURE 10.3　From left to right, Margaret and Geoffrey Burbidge, William Fowler, and Fred Hoyle, coauthors of the B^{2}FH paper (see text).

Saltpeter also suggested that the ^{12}C formed by both mechanisms (triple alpha and double alpha) could then go on to capture some further alpha particles to yield ^{16}O and ^{20}Ne in accordance with the observed higher abundances of these two particular isotopes. However, Saltpeter did not have much to say on the nature of what he labeled as ^{12}C*, and his theory created little impression in the astrophysical community.

The problem of how ^{12}C is formed was solved by the enigmatic British physicist Fred Hoyle (figure 10.3), who perhaps has made the greatest contributions to the question of nucleosynthesis of any person to date, as well as being one of the three architects of the steady-state cosmological theory.[14] Before describing how Hoyle solved the missing link in the triple alpha mechanism, it is necessary to return to an influential article that he published in 1946.

Although Eddington had suggested that element formation could take place in the interior of stars, the temperatures for such processes were far higher than the temperatures that were assumed to exist inside of most typical stars. For example, our sun has a core temperature of a few million degrees. While these conditions can support the burning of hydrogen to form helium, they cannot begin to support the fusion of helium atoms (helium burning), which requires temperatures in the billions of degrees.[15]

Another way to appreciate the situation is to consider the following argument: In order for two nuclei to fuse together, they must approach each other to a distance approximately equal to the sum of their radii. However, such an approach is counteracted by a strongly repulsive Coulomb force, which would seem to render this process impossible. Only following the advent of quantum mechanics was it realized that such a close encounter between nuclei could still occur by means of the phenomenon of quantum mechanical tunneling, which is now believed to take place in stars. In a paper published in 1946, Hoyle sketched the essential pathways through which stellar nucleosynthesis takes place.[16]

In the course of this work, Hoyle also uncovered many important features of how stars change in the course of their lifetimes. For example, a middle-age star fuses hydrogen into helium and, in the process, loses heat as radiant light energy. Two effects then compete to determine the eventual fate of the star. On one hand, the star contracts due to the effect of the gravitational force, while on the other hand, the high temperature generated at the core of the star opposes the contraction. As the star loses its hydrogen fuel, less hydrogen burning can occur, and consequently, the temperature starts to decrease. At this point, the gravitational force begins to dominate and causes the star to contract. However, the compression that occurs causes a new increase in temperature, which acts to halt the further collapse of the star. In addition, the newly established temperature, which is invariably higher than it was previously, allows for new fusion reactions to take place. This reestablished equilibrium is only temporary, however, since the new nuclear reactions eventually run out of fuel, leading to a further contraction phase and consequently another increase in temperature.

This cycle repeats itself many times over, and each time the temperature is higher such that increasingly heavier nuclei can be made to fuse together. The essential details of Hoyle's scheme are shown in table 10.1 for a star of approximately 25 solar masses, although his calculations extended to various types of stars. In this way, different elements could be formed at different stages in the course of a star's life, culminating with the formation of the most stable nuclei of them all, those of iron.

When all the nuclear fuel is consumed, the core collapses in a very short time, followed by an explosion of the star in the form of a supernova. The explosion and the conditions generated by it lead to the formation of many heavy elements and the expulsion of this material into space. All this takes place in the outer parts of the star, while the inner core undergoes an implosion or collapse. During the collapse phase, the nuclei of iron are broken down to form neutrons and the entire star forms a neutron star in cases where the mass is up to two to three solar masses. In heavier stars, not even the Pauli exclusion principle can halt the further collapse of the star to form a black hole.[17]

Hoyle thus obtained an almost complete solution to the problem of nucleosynthesis. What remained was to find an explanation of the second step shown in

TABLE 10.1

Conditions Needed for Different Stages in Nucleosynthesis
According to Hoyle's Calculations

Burning stage	Density g/cm³	Temperature (°C)	Time scale
Hydrogen	5	4×10^7	10^7 years
Helium	700	2×10^8	10^6 years
Carbon	2×10^5	6×10^8	600 years
Neon	5×10^5	1.2×10^9	1 year
Oxygen	1×10^7	1.5×10^9	6 months
Silicon	3×10^7	2.7×10^9	1 day
Core collapse	3×10^{11}	5.4×10^9	0.25 seconds
Core bounce	4×10^{14}	2.3×10^{10}	0.001 seconds
Explosive	Variable	About 10^9	10 seconds

Based on S. Singh, *Big Bang*, Harper Collins, New York, 2004, table on p. 388.

table 10.1. How could helium atoms fuse together to form carbon? As mentioned above, this was a problem that had already been confronted by Gamov and later by Saltpeter: the lack of any plausible mechanism to form atoms of ^{12}C. Without such a mechanism, all the subsequent steps in Hoyle's table would have remained in the realm of wishful thinking.

But Hoyle succeeded in solving even this problem in an unequivocal and dramatic fashion. He predicted that a ^{4}He nucleus would combine with a nucleus of ^{8}Be to form a high-energy state (or resonance) of the carbon nucleus, contrary to all the then-known evidence on the resonance states of carbon. Hoyle was able to predict the mass and hence the energy of this new excited state by means of a wonderfully simple argument: If the mass of a ^{4}He nucleus is added to that of ^{8}Be, one obtains the mass of the hypothetical new state of carbon that can subsequently decay to form the more common ground state of carbon. The result of this calculation yields an energy of 7.68 MeV above the carbon ground state. While on a sabbatical leave at Caltech, Hoyle eventually persuaded the experimental nuclear physicist William Fowler to try to detect the new resonance state. When the experiments were conducted, they indeed revealed a new state at energy at precisely 7.68 ± 0.03 MeV above the carbon ground state![18] Hoyle's triumph was complete and became further solidified when he published an even more widely cited article along with Fowler and the husband and wife team of Margaret and Geoffrey Burbidge, which became subsequently known as the B²FH paper (figure 10.3).[19]

Returning to the formation of elements heavier than iron, these authors found that they formed through two main processes. First, there is a slow process of neutron capture, known appropriately as the s-process, which takes place over thousands of years, typically in red giant stars. Nuclei of zinc, for example, absorb neutrons and, following beta decay, produce nuclei of higher atomic numbers:

$$^{68}Zn \xrightarrow{n} {}^{69}Zn \xrightarrow{\beta} {}^{69}Ga \xrightarrow{n} {}^{70}Ga \xrightarrow{\beta} {}^{70}As \ldots$$

Nuclei with masses of 230 and greater, however, are formed in the course of multiple neutron absorptions followed by multiple beta decays. This so-called r-process occurs very rapidly and in the course of supernova explosions. Elements that are ejected in supernova explosions are later incorporated into new stars, generation after generation. In fact, the presence of certain heavy elements in the sun, and the fact that solar conditions cannot support the formation of these elements, has led to the conclusion that the sun is at least a second-generation star.

ASTROPHYSICS AND COSMOLOGY: THE CURRENT VIEW

The universe is now generally believed to have come into being about 13.7 billion years ago in the course of a cataclysmic explosion, involving matter of density 1,070 g/cm^3 and whose temperature has been set at 10^{32} K (table 10.2). This hot big bang produced matter and energy, of which just 4% is ordinary matter and the rest is present as "dark energy" and "dark matter."[20] Of the 4% of ordinary matter, 75% consists of hydrogen and 24% of helium; just 1% consists of all the other elements put together. It is therefore a remarkable fact that all the elements other than hydrogen and helium make up just 0.04% of the universe. Seen from this perspective, the periodic system appears to be rather insignificant. But the fact remains that we live on the earth, which consists entirely of ordinary matter, as far as we know, and where the relative abundance of elements is quite different from the overall cosmic abundance. But before coming to the elements on the earth, it is interesting to consider solar abundances for a moment.

The sun is a good deal younger than the universe as a whole, being 4.55 billion years old. The percentage of hydrogen in the sun is a little less than for the entire universe at 70%, while helium is a little higher at 28%; all the remaining elements account for 2% of the sun. The planets, including the earth, vary widely in chemical composition. While the inner planets have lost most of their gaseous atmospheres, the outer, more massive ones continue to exert an attraction on their gaseous envelopes. Jupiter, Saturn, and Neptune are often called the "gas giants" due to their predominantly gaseous compositions. On earth, hydrogen ranks as only the 11th element in terms of abundance, or just 0.12% by mass, while helium is present only in trace amounts.

Stages in Big Bang Cosmology

	Time	Temperature (K)
Big bang	0	10^{32}
Protons and neutrons form	Few seconds	10^{10}
Nuclei form	3 minutes	10^9
Atoms form	3×10^5 years	3000

STABILITY OF NUCLEI AND COSMIC ABUNDANCE OF ELEMENTS

The stability of nuclei can be estimated through their binding energy, a quantity given by the difference between their masses and the masses of their constituent particles. This difference in mass gives a measure of the energy released any particular nucleus is formed, via Einstein's famous equation $E = mc^2$. If the binding energy is divided by the mass number of any particular nucleus, one obtains the binding energy per nucleon, which provides a better means of comparing the stability of nuclei. A plot of this quantity against mass number is shown in figure 10.4. Attempts to understand this curve theoretically have been made by appealing to theories from nuclear physics.

An approximate understanding can be gained through the liquid drop model of the nucleus, as developed by Bethe, Carl Weizächer, Niels Bohr, and others. In this model, the nucleus is assumed to be of uniform density like any drop of a uniform liquid. The objective is to explain the rapid rise in binding energy per nucleon, up to a maximum value of between 7 and 8 MeV, which occurs for iron, the most stable of all nuclei. Beyond this mass number of $A = 56$, a slow decrease occurs, indicating that nuclei become progressively less stable. Indeed, the formation of nuclei lighter than iron proceeds via exothermic processes in which energy is released. This is why it is favorable for stars to form progressively heavier elements starting from hydrogen and helium, since the energy evolved provides energy to sustain the star. Beyond iron, however, the formation of heavier nuclei occurs via endothermic processes that do not contribute to the power output of the stars.

A nucleus is stable only if the attractive nuclear force within it outweighs the repulsive force between the positive protons. The strong nuclear force, unlike the repulsive Coulomb force, operates equally between protons and neutrons and has a short range with an effect that does not exceed about 2×10^{-15} m. The observed

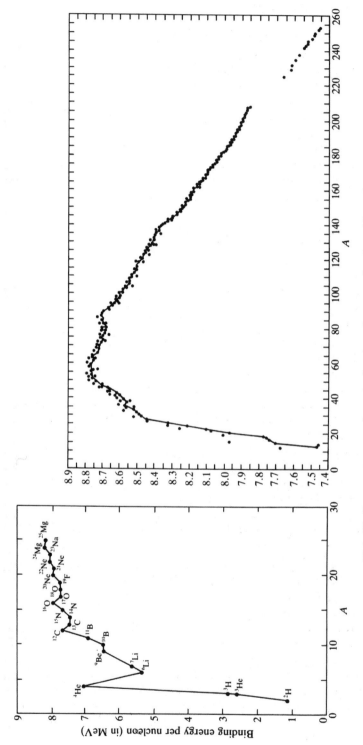

FIGURE 10.4 Binding energy per nucleon as a function of mass number for stable nuclei. Reproduced from G. Friedlander, J.W. Kennedy, E.S. Macias, J.M. Miller, *Nuclear and Radiochemistry*, John Wiley & Sons, New York, 1981, pp. 26, 27 (with permission).

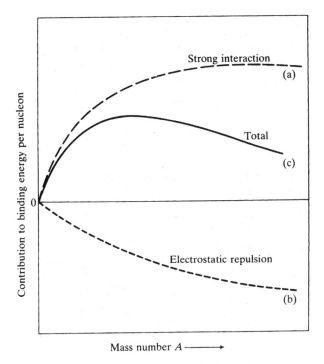

FIGURE 10.5 Separate and combined effects of strong nuclear force and repulsive Coulomb force in the nucleus. Reproduced from P.A. Cox, *The Elements*, Oxford University Press, Oxford, 1989, figure from p. 33 (with permission).

curve in figure 10.4 can be explained in qualitative terms as the net result of combining the strong force (a) and the repulsive Coulomb force (b) in any nucleus, as shown in figure 10.5.[21]

However, the liquid drop model is powerless to explain the more detailed features within the binding energy per nucleon curve, such as the various discontinuities that are superimposed on it, reflecting the enhanced stabilities of nuclei of ^{4}He, ^{12}C, ^{16}O, ^{20}Ne, and ^{24}Mg. To explain these more subtle features, we need to consider the quantum mechanical nuclear-shell model, which bears a number of similarities to the electron–shell model as described in chapters 7 and 9.

The irregularities shown in figure 10.4 can be more easily appreciated by plotting a curve of the difference in binding energy between successive nuclei. This is carried out in a separation energy plot, which gives the energy required to remove a nucleon from any nucleus (figure 10.6). Such plots can be separately drawn for protons or neutrons and show similar general characteristics. They provide plots analogous to those of first-ionization energy plotted against atomic number, as shown in figure 9.6.

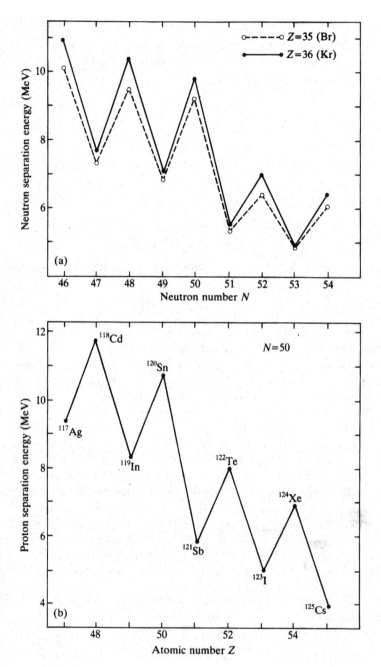

FIGURE 10.6 Separation energy plot giving energy required to remove a nucleon from any nucleus. Reproduced from P.A. Cox, *The Elements*, Oxford University Press, Oxford, 1989, p. 38 (with permission).

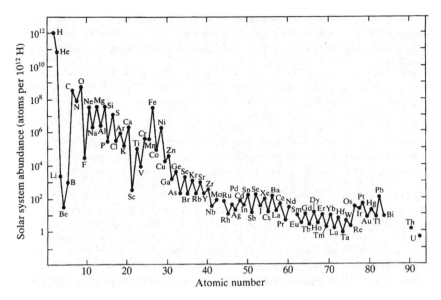

FIGURE 10.7 Relative abundance of elements in the solar system, including contributions from meteorites. Reproduced from P.A. Cox, *The Elements*, Oxford University Press, Oxford, 1989, figure from p. 17 (with permission).

The separation energy curve for a number of nuclei, all having 50 neutrons in this case, shows a distinctive sawtooth pattern with nuclei displaying alternatively more or less stable values depending on whether the number of protons is even or odd, respectively. In addition to the sawtooth pattern, there is an overall decrease in stability following the value of $Z = 50$ as atomic number increases. If this diagram is extended to all known nuclei, it reveals a series of maxima corresponding to especially stable nuclei at Z and $N = 2, 8, 20, 28, 50, 82,$ and $126,$ the so-called magic numbers.[22] To some extent, the magic numbers for protons also correspond to the maxima in the plot of solar abundance of elements (figure 10.7). These elements are $_2$He, $_8$O, $_{20}$Ca, $_{28}$Ni, $_{50}$Sn, $_{82}$Pb.[23]

The nuclear-shell model approaches this problem by approximating the forces present in the nucleus by means of a central-field potential.[24] As in the case of electrons in an atom, solving the Schrödinger equation for the nucleus yields a number of distinct energy-level solutions. The labels used for the nuclear levels are similar to those for electrons: s, p, d, and f. But there are also a number of differences in that, for example, the lowest p and d levels in the nuclear case are labeled 1p and 1d, respectively, although such combinations do not occur in the case of electrons. The nuclear energy levels can be thought of as being progressively occupied by nucleons just like the electronic levels are progressively occupied with electrons. However, the energy levels predicted by using only the central field approxima-

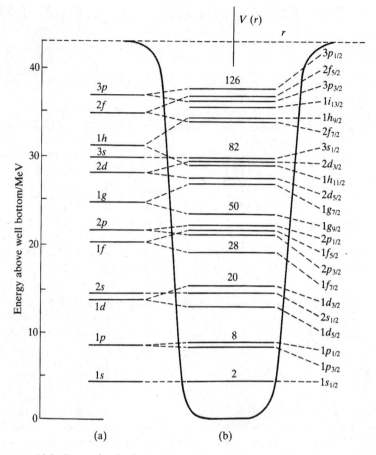

FIGURE 10.8 Energy levels obtained in nuclear-shell theory with inclusion of spin-orbit coupling as given by the Goeppert-Mayer–Jensen–Suess model. Numbers on the right represent the number of protons or neutrons in each level, with cumulative totals on extreme right. Reproduced from L. Pauling, *General Chemistry*, Dover, New York, 1970, p. 855 (with permission).

tion starting with 1s < 1p < 1d < 2s, and so on, do not explain the occurrence of the magic numbers. The latter feat was achieved by Maria Goeppert-Mayer, Hans Suess, and Hans Jensen in the 1950s by taking into account the effect of spin–orbit coupling present between all nucleons (figure 10.8).[25]

The introduction of spin–orbit coupling between nucleons results in the splitting of energy levels. In addition, there is considerable overlap in these newly formed levels to produce the sequence shown on the right side of figure 10.8. The filling of nuclear energy levels thus proceeds in the order $1s_{1/2} < 1p_{3/2} < 1p_{1/2} <$

$1d_{5/2}$ < and so on. Finally, the number of nucleons that can fill any particular level is $2j + 1$ for any given angular momentum j value.

In both the electronic and nuclear-shell theories, one is dealing with a many-body problem for which there is no analytical solution. As a result, the explanations provided in both cases are approximate and rely to some extent on empirical evidence, such as the precise ordering of levels. These relative orderings of levels have not been deduced from first principles, contrary to the impression created by some presentations. Indeed, problems are more severe in the nuclear case, in view of the greater complexity of the nucleus. Just as the $n + \ell$ rule is obtained empirically in the electronic case, as described in chapter 9, so the nuclear ordering by the *aufbau* principle is also obtained by appeal to empirical data.[26]

The explanation of the magic numbers by nuclear-shell theory is nevertheless a remarkable achievement in that the number of nucleons per level as well as the relationship between the various quantum numbers is deduced from first principles even if the ordering of levels is not.

MORE CHEMISTRY

The trends within rows and columns of the periodic table are quite well known and are not repeated here.[27] Instead, I concentrate on a number of other chemical trends, some of which challenge the form of reductionism that attempts to provide explanations based on electronic configurations alone.[28] In the case of one particular trend described here, the knight's move, the chemical behavior defies any theoretical understanding whatsoever, at least at the present time.

Diagonal Behavior

As is well known to students of inorganic chemistry, a small number of elements display what is termed diagonal behavior where, in apparent violation of group trends, two elements from adjacent groups show greater similarity than is observed between these elements and the members of their own respective groups (figure 10.9). Of these three classic examples of diagonal behavior, let us concentrate on the first one to the left in the periodic table, that between lithium and magnesium. The similarities between these two elements are as follows:

1. Whereas the alkali metals form peroxides and superoxides, lithium behaves like a typical alkaline earth in forming only a normal oxide with formula Li_2O.
2. Unlike the other alkali metals, lithium forms a nitride, Li_3N, as do the alkaline earths.

Li	Be		B	
	Mg		Al	Si

FIGURE 10.9 Elements that display diagonal behavior: lithium and magnesium, beryllium and aluminum, and boron and silicon.

3. Although the salts of most alkali metals are soluble, the carbonate, sulfate, and fluorides of lithium are insoluble, as in the case of the alkaline earth elements.

4. Lithium and magnesium both form organometallic compounds that act as useful reagents in organic chemistry. Lithium typically forms such compounds as $Li(CH_3)_3Br$, while magnesium forms such compounds as CH_3MgBr, a typical Grignard reagent that catalyzes nucleophilic addition reactions. Organolithium and organomagnesium compounds are very strong bases that react with water to form alkanes.

5. Lithium salts display considerable covalent character, unlike its alkali metal homologues but in common with many alkaline earth salts.

6. Whereas the carbonates of the alkali metals do not decompose on heating, that of lithium behaves like the carbonates of the alkaline earths in forming the oxide and carbon dioxide gas.

7. Lithium is a considerably harder metal than other alkali metals and similar in hardness to the alkaline earths.

Although some good explanations for this behavior are available, they serve to undermine the simplistic physicist's notion that chemical behavior is governed just by the electronic configuration of atoms.

The diagonal effect can be explained as the outcome of several opposing trends.[29] As one moves down any group, electronegativity, to consider just one property, decreases. But, as one moves across the table, the same quantity increases. If one moves diagonally the two trends cancel each other out, and there is little change in electronegativity. Similarly, ionization energy and atomic radii trends are such that a diagonal movement results in little change in these properties that, like electronegativity, govern a great deal of the chemistry of the elements. The broader implication is that the electronic configurations of the gas-phase atoms are of little relevance in trying to understand chemical properties. Or to state matters a little differently, the influence of any particular configuration seems to be outweighed by other properties such as those that have just been mentioned, for example, electronegativity and ionization energy.

A rather useful means of discussing the diagonal effect is to appeal to the charge density of the ions of the elements in question, a property that consists of the charge of an ion divided by its volume. The ions of elements that show a diagonal relationship typically have similar charge densities. However, in the case of the boron–silicon relationship, that option is not even available since these elements do not typically form ions.

Similarities between Group (*n*) and Group (*n* + 10)

The chemical similarities of this type have already been mentioned in passing in chapters 1 and 3. They are similarities that were well known to the pioneers of the periodic table in the nineteenth century and that were embodied in the short-form periodic table (figure 1.6). Unfortunately, many of these trends have been forgotten as a result of the widespread adoption of the medium-long–form table, which does not point to them in any obvious manner.

The first significant example of an *n*, *n* + 10 effect is observed in the elements magnesium and zinc, which belong respectively in groups 2 and 12 according to the modern International Union of Pure and Applied Chemistry (IUPAC) numbering scheme. Both elements form water-soluble sulfates and water-insoluble hydroxides as well as carbonates. Also, their chlorides are hygroscopic and predominantly covalent.

Moving one step to the right across the periodic table one comes to an even more pronounced example of this kind of behavior in the case of aluminum and scandium. In fact, the Canadian chemist and metallurgist Fathi Habashi has suggested that there are grounds for moving the position of aluminum from group 13- into the scandium group or group 3-.[30] Indeed, one can even make a good argument for this repositioning on electronic grounds. A comparison of the +3 ions of the elements aluminum, scandium, and gallium suggests that the first two of these elements might be grouped together since they share a noble gas configuration, whereas Ga^{3+} does not.[31]

Some similarities between aluminum and scandium can be seen in table 10.3. The standard electrode potential for aluminum and the other elements would seem to point clearly in favor of a repositioning to the scandium group. The implications from the melting point data are not quite as suggestive, but it does appear as though the high melting-point value of 660°C for aluminum is somewhat anomalous for an element in this group but not so out of place in the scandium group.[32] In addition, Al^{3+} and Sc^{3+} both hydrolyze to produce acidic solutions, both of which contain unusual polymeric hydroxo species. Yet another similarity concerns the reactions of Al^{3+} and Sc^{3+} with hydroxide ions, resulting in the formation of gelatinous precipitates that redissolve in excess hydroxide ions to form such anions as $[Al(OH)_4]^-$. Also, both aluminum and scandium form

TABLE 10.3

Comparison of Melting Points and Standard Electrode Potentials (E°)
of Elements in Groups 3 and 13

Group 3			Group 13		
Element	Melting point (°C)	E° (V)	Element	Melting point (°C)	E° (V)
			Al	660	−1.66
Sc	1,540	−1.88	Ga	30	−0.53
Y	1,500	−2.37	In	160	−0.34
La	920	−2.52	Tl	300	+0.72
Ac	1,050	−2.6	—	—	—

Based on G. Rayner Canham, The Richness of Periodic Patterns, in D. Rouvray, R.B. King, (eds.), *The Periodic Table: Into the 21st Century*, Research Studies Press, Bristol, UK, 2004, 161-187, table on p. 169.

isomorphous compounds of the general type Na_3MF_6, where M is aluminum or scandium.

Meanwhile, aluminum differs significantly from gallium, the element lying directly below it in group 13. Whereas aluminum forms a polymeric solid hydride with a formula $(AlH_3)_x$, gallium forms the gaseous and monomeric hydride Ga_2H_6. Nevertheless, it must also be recognized that the halides of aluminum resemble those of gallium more than they do those of scandium. Nothing is ever simple as far as the elements are concerned.

The case of silicon and titanium is interesting for a somewhat different reason (table 10.4). Here, the similarity is greatest between elements that are related as n and $n + 10$ but that belong to different periods (3 and 4). Moreover, these two elements provide one of the closest similarities between any two elements in different groups, even more so than the classic cases of diagonal relationships mentioned above.

The same kinds of similarities can be seen in cases where $n = 5, 6, 7$, and 8, although these are not all examined here.[33] The case of groups 8 and 18, for example, is interesting since it is surprising that an element like xenon among the noble gases (group 18-) should bear any resemblance to any other elements in the periodic table. And yet both osmium and xenon, from group 8- and 18-, respectively, form covalent compounds in which they show the +8 oxidation state such as in the cases of OsO_4 and XeO_4, both of which occur as yellow solids. Furthermore, both elements form other sets of analogous compounds, including OsO_2O_4 and XeO_2O_4 as well as OsO_3F_2 and XeO_3F_2.

In fact, the only case in which there are no similarities between elements in group n and $n + 10$ are those of group 1 and 11-. The alkali metals (group 1) such

TABLE 10.4

Comparison of Titanium (Group 4) with Tin (Group 14)

Oxides (TiO$_2$ and SnO$_2$)

    Isomorphous structures

    Both show rare property of turning yellow when heated (thermochromism)

Chlorides (TiCl$_4$ and SnCl$_4$)

    Similar melting points and boiling points:

        TiCl$_4$—melting point, $-24°C$; boiling point, $136°C$

        SnCl$_4$—melting point, $-33°C$; boiling point, $114°C$

    Both tetrachlorides act as Lewis acids in hydrolysis:

$$\text{TiCl}_{4(l)} + 2\ \text{H}_2\text{O}_{(l)} \rightarrow \text{TiO}_{2(s)} + 4\ \text{HCl}_{(g)}$$

$$\text{SnCl}_{4(l)} + 2\ \text{H}_2\text{O}_{(l)} \rightarrow \text{SnO}_{2(s)} + 4\ \text{HCl}_{(g)}$$

Nitrates of Sn and Ti or M(NO$_3$)$_4$ are isomorphous

Based on G. Rayner Canham, The Richness of Periodic Patterns, *The Periodic Table: Into the 21st Century*, Research Studies Press, Bristol, UK, 2004, 161-187, see p. 170.

as sodium and potassium show pronounced *dissimilarities* from such elements as copper, silver, and gold (group 11-). The alkali metals are soft, low-density metals that react vigorously with water, whereas the so-called noble metals of group 11- are hard, display high density, and, particularly in the legendary case of gold, show a great reluctance to react with water and many other reagents.

The fact that these two particular groups should display such an anomaly regarding n, $n + 10$ behavior serves to highlight further the complexity of the elements, which, as described here, sometimes defies the reductionist's desire for regimentation. In many of these cases, the reductionist can point to the obvious similarity in electronic configurations between an atom from group n and one from group $n + 10$, such as the example of magnesium and zinc discussed above. However, as Geoffrey Rayner Canham, a leading advocate of teaching inorganic chemistry in a qualitative manner, has written, the similarities shown far exceed any expectations on electronic grounds.

Early Actinoid Relationships

The relationship concerning members of the actinide series was mentioned is passing in chapter 1. Prior to the work of Glenn Seaborg, the similarities between the transition elements and the early actinides were used to determine the placement of the early actinides in the periodic table (figure 1.9). The modern tendency to separate out the actinides has its merits in terms of electronic configurations but serves, not for the first time, to obscure some undeniable chemical similarities among a number of pairs of elements.[34]

4	5	6
Ti	V	Cr
Zr	Nb	Mo
Hf	Ta	W

Th	Pa	U

FIGURE 10.10
Early actinides that show
analogies with transition
metals. Numbers denote the
IUPAC group labels.

A number of analogies have been noted between thorium and the members of group 4- headed by titanium, between protactinium and members of group 5- headed by vanadium, and between uranium and the members of group 6- headed by chromium (figure 10.10). For example, uranium, which is assigned to the actinide series and not regarded as a transition metal these days, forms a yellow ion, $U_2O_7^{2-}$, while chromium forms the well-known oxidizing ion with an orange color of formula $Cr_2O_7^{2-}$. The analogy with chromium is further displayed in the compounds UO_2Cl_2 and CrO_2Cl_2, respectively. But in other respects, uranium resembles tungsten, such as in the formation of the hexachlorides UCl_6 and WCl_6, neither of which is analogously formed by chromium or molybdenum.

Although relativistic effects may well play a role in these matters, they must be outweighed by other factors since these analogies cease quite abruptly beyond group 6, whereas the influence of relativistic effects is known to increase regularly as a function of atomic number. On the other hand, a comparison of the actinides with respective lanthanides lying directly above in the medium-long–form or long-form periodic table reveals little similarity, except for thorium and cerium, in spite of similarities in electronic configurations between members of these two series.

Secondary Periodicity

This behavior was first described in a 25-page paper by the Russian chemist Evgenii Biron. He noted that various chemical and physical properties show a zigzag or alternating behavior instead of the expected regular trend as one descends any group of elements. For example, the elements in group 15 display the zigzag pattern in their common oxidation states, as shown in table 10.5. Whereas the elements in rows 3 and 5, phosphorus and antimony, show a valence of 5, the other three elements predominantly show trivalence.

TABLE 10.5

Secondary Periodicity among Elements in Group 15

Element	Oxidation state	Row	Configuration
Nitrogen	+3	2	[He] $2s^2\,2p^3$
Phosphorus	+5	3	[Ne] $3s^2\,3p^3$
Arsenic	+3	4	[Ar] $4s^2\,3d^{10}\,4p^3$
Antimony	+5	5	[Kr] $5s^2\,4d^{10}\,5p^3$
Bismuth	+3	6	[Xe] $6s^2\,4f^{14}\,5d^{10}\,6p^3$

Compiled by the author from E.V. Biron, *Zhurnal Russkogo Fiziko-Khimicheskogo Obschestva, Chast'Khimichevskaya,* 47, 964–988, 1915.

The traditional explanation for this behavior has been to invoke the additional electron screening due to the $3d^{10}$ electrons in the case of atoms in row 4, such as arsenic, and the even greater screening in atoms in row 6, such as bismuth. The notion is that whereas phosphorus can readily lose five electrons, at least formally, to form a +5 ion, arsenic cannot do so because of d electron screening, which acts to "separate" the outermost p electrons from the outermost s electrons. A similar argument can be made for the change from antimony to bismuth. The removal of the five outermost electrons in bismuth is prevented by the even greater separation in energy between the 6s and 6p outer-shell electrons due to the intervening $4f^{14}$ electrons, which are absent in the atoms of antimony.[35]

The drop in the sum of the first five ionization energies between arsenic and antimony is at first sight surprising since both sets of outermost p and s electrons are screened to the same extent by 10 d orbital electrons in each case. But an overall decrease in ionization is the normal behavior observed on descending most groups in the periodic table.[36]

Ralph Sanderson, an author who has published extensively on the periodic system, has listed some further interesting examples of secondary periodicity:[37]

Group 13

B_2H_6 and Ga_2H_6 are volatile, whereas the intervening $(AlH_3)_x$ is not.
Stable borohydrides are formed by aluminum but not by boron itself or gallium.
$Al(CH_3)_3$ and $Al(C_2H_5)_3$ are both dimeric in the vapor phase, whereas the analogous boron and gallium compounds are monomeric.

Group 14

Germanium resembles carbon more than silicon does. For example, SiH_4 is far more readily oxidized than is either GeH_4 or CH_4.

TABLE 10.6

Variation in Atomic Radii (Å) of Free Atoms for
Transition Metal Groups

Group	Elements (atomic radii)
3	Sc (1.570), Y (1.693), Lu (1.553)
	La (1.915) is anomalous
4	Ti (1.477), Zr (1.593), Hf (1.476)
5	V (1.401), Nb (1.589), Ta (1.413)
6	Cr (1.453), Mo (1.520), W (1.360)
7	Mn (1.278), Tc (1.391), Re (1.310)
8	Fe (1.227), Ru (1.410), Os (1.266)
10	Cu (1.191), Ag (1.286), Au (1.187)
11	Zn (1.065), Cd (1.184), Hg (1.126)

Table re-drawn from Chistyakov, but using IUPAC group numbering. E.V. Chistyakov, Biron's Secondary Periodicity of the Side
d-Subgroups of Mendeleev's Short Table, *Zhurnal Obshchei Khimii*
(Engl. Ed.), 38, 213-214, 1968.

Group 15

Phosphorus and antimony form a pentachloride, whereas arsenic does not.

While N(+5) and As(+5) compounds act as good oxidizing agents, P(+5) does
not.

N(+3) and As(+3) compounds are far weaker reducing agents than are compounds of P(+3).

Interestingly, secondary periodicity is not confined to main-group elements
but occurs even more consistently among transition metals, so much so that this
behavior has been used to argue for an alternative placement of lutetium and lawrencium in the periodic table.[38] In 1968, V.M. Chistyakov presented data showing
that secondary periodicity occurs in most transition metal groups (table 10.6), suggesting that the scandium group should consist of scandium, yttrium, and lutetium,
rather than scandium, yttrium, and lanthanum, as is more frequently assumed in
published periodic tables.

Finally, several authors who seek to provide group-theoretical explanations for
the periodic system have claimed that their approaches also "predict" secondary
periodicity.[38]

Knight's Move Relationship

The knight's move relationship is perhaps the most mysterious one among all the
unusual relationships involving the periodic table (figure 10.11). It takes its name
from the knight's move in the game of chess, meaning a move of one step in any

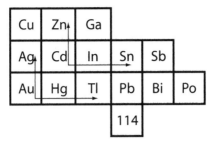

FIGURE 10.11 Elements that show knight's move relationships. For example, zinc and tin or silver and thallium.

direction followed by two steps in a direction at right angles to the first movement.[40] The South African chemist Michael Laing discovered such a relationship among the elements[41] and has described it in detail in a number of articles.[42]

The examples of the knight's move relationship so far discovered are located at the heart of the medium-long–form table among metallic elements. Consider the elements zinc and tin. Both are commonly used for plating steel such as in the case of food cans. Not only do layers of both metals successfully delay the onset of corrosion in the iron, but they are also nonpoisonous, unlike many other metals lying close to them in the periodic table.[43] Zinc and tin are not merely nonpoisonous but also appear to be biologically important. Zinc is an essential element for many living organisms because it occurs in a variety of important enzymes. Tin is not essential to humans although it may be so for some living organisms, a fact that has yet to be settled. The compounds of tin are generally regarded as being nontoxic with the exception of organotin compounds such as trimethyl tin.[44] Nevertheless, tin is found in many medicines and even in toothpaste in the form of stannous fluoride, which it is claimed can prevent tooth cavities.

Zinc and tin share another important property: their ability to form alloys with the element copper. Whereas they fail to form an alloy with each other and do not form any intermetallic compounds, zinc alloys with copper to form brass and tin alloys with copper to form bronze, both of which have been known since antiquity.[45]

Cadmium and lead, on the other hand, are both poisonous, which is not too surprising when it is realized that they, too, stand in a knight's move relationship to one another. Further similarities include some closely lying boiling and melting points among their chlorides, bromides, and iodides, as shown in table 10.7. There is also a striking similarity between $PbCrO_4$ and $CdCrO_4$, both of which are yellow substances that are insoluble in water. Table 10.7 also shows aspects of the knight's move relationship between silver and thallium as well as between gallium and antimony and provides further evidence for the zinc–tin relationship discussed

TABLE 10.7

Melting and Boiling Points That Support Knight's Move
Relationships Among Pairs of Elements

Silver and thallium

AgCl	Melting point	445
TlCl	Melting point	429
AgBr	Melting point	430
TlCl	Melting point	456

Cadmium and lead

CdI_2	Melting point	385
PbI_2	Melting point	412
$CdCl_2$	Boiling point	980
$PbCl_2$	Boiling point	954
$CdBr_2$	Boiling point	863
$PbBr_2$	Boiling point	916

Zinc and tin

$ZnCl_2$	Melting point	275
$SnCl_2$	Melting point	247
$ZnBr_2$	Boiling point	650
$SnBr_2$	Boiling point	619

Gallium and antimony

GaF_3	Melting point	77
SbF_3	Melting point	73
$GaCl_3$	Boiling point	200
$SbCl_3$	Boiling point	221
$GaBr_3$	Boiling point	279
$SnBr_3$	Boiling point	280

Based on M. Laing, The Knight's Move in the Periodic Table *Education in Chemistry*, 36, p. 160-161, 1999, table 1.

above. The elements silver and thallium form another knight's move pair. Among their similarities is the fact that their monochlorides AgCl and TlCl are both photosensitive and insoluble in water.

Laing has considered possible theoretical explanations of the knight's move relationship but concludes that none is forthcoming. He ends one of his articles by making predictions concerning element 114, which has been observed in trace amounts but has yet to be named.[46] The significance of this element is that lies in the middle of the so-called "island of stability" among the superheavy nuclei, and this leads one to suppose that it may eventually be possible to synthesize enough of the element to examine its macroscopic properties. Given the knight's move relation to mercury, Laing has predicted that element 114, or eka-lead, should possess a moderate density of around 16 g/cm^3, will have a very low melting point, and will possibly be a liquid at room temperature.[47]

First-Member Anomaly

It has long been recognized that the first members of groups, especially main-group elements, are anomalous with respect to other members of their groups. This applies equally to physical and chemical properties. For example, hydrogen is a gas, unlike the other members of group 1. Similarly, nitrogen and oxygen occur as gases at room temperature, whereas all the remaining members of their respective groups are found as solids.

In chemical terms, the first members of each group fail to achieve higher oxidation states; that is, they fail to expand their octet of electrons. For example, oxygen shows a maximum oxidation state of just +2 by contrast to the following members, starting with sulfur, which commonly display oxidation states of +4 and +6. Such behavior in the higher members has usually been explained in electronic terms by invoking available d orbitals that allow the atoms to expand their octets. While nitrogen forms only NCl_3, phosphorus is said to form PCl_5 as a result of promotion of two electrons into available d orbitals and the associated hybridization of the five unpaired electrons. More recently, such explanations have been called into doubt, however. Following some theoretical calculations, it has been argued that the d orbital contribution to the bonding in compounds in which octet expansion occurs is highly insignificant.[48]

In addition to the kind of first-member anomaly that has just been described, there is a more specific observation that has independently been made by William Jensen and Henry Bent, two inorganic chemists and chemical educators.[49] This effect is such that the extent of the first-member anomaly is greatest in the s block of the periodic table, followed by a moderate effect in the p block, and progressively less noticeable in the d and f blocks, respectively. Thus, hydrogen is vastly different from its analogues in group 1, namely, the alkali metals, such as sodium and potassium.[50] The first-member anomalies in the case of the p block elements include the well-known cases mentioned above and involving such elements as nitrogen, oxygen, and fluorine. In the d block, the first members of each group, such as scandium and titanium, show less pronounced anomalies compared to the other elements in their groups, and finally in the case of the f block, the lanthanides show even less difference from the actinides.[51]

But whereas Bent and Jensen have agreed to share the credit for the discovery of this more detailed aspect of first-member anomaly, they draw surprisingly different conclusions regarding the noble gases. For Jensen, helium remains a noble gas, whereas Bent takes the radical step of moving helium to the alkaline earth group and champions the use of the left-step table, as discussed below.

Other Relationships

Aluminum and iron in their +3 oxidation states show a number of curious similarities. They are especially curious from the point of view of their electronic configurations, which show no hint of any similarity:

Aluminum   [Ne] $3s^2\ 3p^1$ iron (III)   [Ar]$4s^2 3d^6$

And yet both of these elements form analogous hydrated ammonium sulfates: $(NH_4)Al(SO_4)_2 \cdot 12H_2O$ and $(NH_4)Fe(SO_4)_2 \cdot 12H_2O$. Their chlorides exist as dimers in the gas phase with formulas Al_2Cl_6 and Fe_2Cl_6. Their anhydrous chlorides act as Friedel-Crafts catalysts to introduce alkyl groups into aromatic compounds. The active species have been identified as $AlCl_4^-$ and $FeCl_4^-$, respectively. Finally, the cation of both elements hydrolyzes in water to produce acidic solutions.

Another unexpected behavior consists in the close similarity between the combination of boron and nitrogen in certain compounds and compounds consisting of carbon bonded to itself. First of all, boron nitride has a structure analogous to graphite. In addition, as in the case of graphite, the application of very high pressure to boron nitride produces an extremely hard substance that behaves like diamond. Even more intriguing is the analogy between the benzene ring with its characteristic aromatic chemistry and the chemistry of the boron–nitrogen analogue consisting of $B_3N_3H_6$, called borazine.[52] A post hoc explanation for these similarities is that the sum of the number of outer electrons in an atom of boron plus those from an atom of nitrogen is eight. Similarly, the total number of outer electrons in two carbon atoms is also eight. One wonders, however, whether such similarities, or others like it, could have been predicted.

Ions That Imitate Elements

There are some examples of polyatomic ions whose behavior mimics that of an ion of a group of elements in the periodic table. This is the case with the ammonium ion, NH_4^+, which in some respects behaves like an alkali metal ion.[53] On the one hand, this may be explained by the remarkable similarity between the charge densities of NH_4^+, which is 151 C/m^3, and K^+ at 152 C/m^3. Nevertheless, the chemistry of the ammonium ion more closely resembles that of Rb^+ and Cs^+. Among the similarities are reactions with the $[Co(NO_2)_6]^{3-}$ anion, which give precipitates in the case of NH_4^+, K^+, Rb^+, and Cs^+.

Superatom Clusters

The recent discovery of superatom clusters threatens to disturb the peaceful order of the periodic table in a radical manner. Some chemical elements present in the form of clusters or "superatoms" can take on the properties of entirely different elements that are completely unrelated in terms of their grouping. Indeed, there are cases of a single element that can be made to mimic several different elements according to the precise number of atoms present in its cluster. In the 1980s, Thomas Upton at Caltech discovered that a cluster of six aluminum atoms could catalyze the splitting of hydrogen molecules, thus mimicking the behavior of the element ruthenium. Moreover, a superatom consisting of 13 atoms of aluminum behaves as the analogue of noble gas atoms with its full outer shell of electrons. If an electron is removed to form Al_{13}^+, the properties of this superatom ion are similar to those of halogen atoms. More specifically, Al_{13}^+ behaves like Br^-. Furthermore, just as Br^- can react with I_2 to form BrI_2^-, the analogous reaction occurs between Al_{13}^+ and I_2 to form $Al_{13}I_2^-$. Even more curiously, a cluster of 14 aluminum atoms mimics the behavior of alkaline earth atoms such as calcium and magnesium.[54] It has been suggested that there might be a new kind of periodic table waiting to be discovered and that the customary two-dimensional table that has been known since the 1860s might require another dimension to take atomic clusters into account.[55]

VARIETY OF PERIODIC TABLES: IS THERE ONE MOST FUNDAMENTAL PERIODIC TABLE?

It would be a pity to conclude a book on the periodic table without broaching the subject of the variety of tables and systems currently on offer. In addition, this final section serves to revisit some philosophical strands that may have been left hanging in preceding chapters, such as the question of elements as basic substances.

The number of periodic systems that have been proposed probably exceeds 1,000, and are certainly more than the 700 that were carefully analyzed by Edward Mazurs in his classic book of 1974.[56] The rapid proliferation and growth of the Internet and new means of representing data and information on the elements has also fueled the development of what may be called the "periodic table industry." There are now dozens of websites devoted to all aspects of the periodic table, from its history to etymological derivations of the names of elements and tabulations of chemical and physical data.[57]

There seems to be considerable agreement among chemists, in particular, that there is no one "best" periodic table and that one's choice depends on what particular aspect of periodicity or of the elements one is most interested in depicting. In the present account, I subject this view to further analysis and suggest that there may, in fact, be a most fundamental table, regardless of the immediate utility that such a table might possess.

In 2003, the geologist Bruce Railsback published what he termed "An Earth Scientist's Periodic Table" of the elements and their ions. He claims that such a table arranges lithophiles, siderophiles, and chalcophiles[58] into distinct groups unlike the conventional chemist's periodic table. Railsback also seeks to group elements together into naturally occurring sets depending on whether they might be concentrated in the mantle, in seawater, or in soil.[59]

Meanwhile, as mentioned above, the metallurgist Habashi has proposed a periodic table in which the element aluminum is moved to the top of the scandium group.[60] Moving to chemists, Rayner Canham has published an "Inorganic Chemist's Table" (see figure 10.12), in which he highlights a number of unusual relationships, several of which have been reviewed in this chapter. One particularly unusual feature of his table is the inclusion of the ions CN^- and NH_4^+ because of their similarities to certain elements.

Of course, one would not want to deny the utility of these and dozens of other tables that could have been mentioned. But from a philosophical point of view, there is value in asking whether there might be one periodic table that represents the truth about the elements and how they are related to each other, rather than focusing only on the utility of tables to any particular discipline. Pursuing this line of thought raises the question of realism about the elements and realism about grouping them together into columns, at least as far as the conventional representation is concerned. Is the grouping of the elements a matter of objective fact, or is it merely a matter of convention? Should one be a realist or an antirealist about the periodic system and the grouping of elements? My suggestion is that one should adopt a realistic attitude. It is proposed that the arrangement of elements into chemically similar groups is a matter of fact and not a question of choice. If this is indeed the case, then in addition to the huge variety of tables that may be useful to geologists, metallurgists, chemists, and so on, there might nevertheless be one objective periodic system that most closely approximates the truth about the elements.

Back to Elements as Basic Substances

In some preceding chapters, the subject of elements as basic substances compared with elements as simple substances is discussed. As described in chapter 4, Mendeleev placed greater emphasis on the elements as basic substances than on elements as simple substances when he produced his periodic classification. The main criterion of basic substances was their atomic weight. When atomic number took

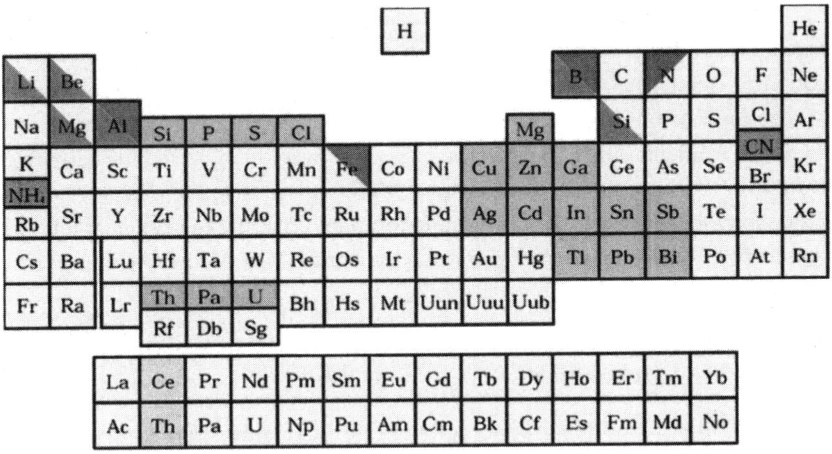

FIGURE 10.12 The inorganic chemist's periodic table, designed by G. Rayner Canham. Shadings indicate relationships: (n) and ($n + 10$) relationships; diagonal relationships; knight's move relationships; aluminum–iron link; actinoid relationship; lanthanide–actinide relationship; combination elements; pseudoelements.

over the role as the ordering criterion for the elements, Fritz Paneth in particular took it upon himself to redefine basic substances as being characterized by their atomic numbers.[61]

Moreover, in the 1920s, Paneth drew on the metaphysical essence of elements as basic substances in order to save the periodic system from a major crisis. Over a short period of time, many new isotopes of the elements had been discovered, such that the number of "atoms" or most fundamental units suddenly seemed to have multiplied. The question was whether the periodic system should continue to accommodate the traditionally regarded atoms of each element or whether it would be restructured to accommodate the more elementary isotopes that might now be taken to constitute the true "atoms." Paneth's response was that the periodic system should continue as it had before, in that it should accommodate the traditional chemical atoms and not the individual isotopes of the elements.[62] Paneth regarded isotopes as simple substances in that they are characterized by their atomic weights, while elements as basic substances are characterized in his scheme by atomic number alone.[63]

Moreover, Paneth, along with György Hevesy, provided experimental evidence in support of this choice for chemists.[64] They showed that the chemical properties of isotopes of the same element were, for all intents and purposes, identical.[65] As a result, chemists could regard the isotopes of any element as being the same simple substance even though such atoms might occur in different isotopic forms.

It is worth noting that, in the case of this isotope controversy, Paneth's recommendation for the retention of the chemist's periodic table depended on the notion of elements as basic substances and not as simple substances. If the chemists had focused on simple substances, they would have been forced to recognize the new "elements" in the form of isotopes that were being discovered in rapid succession. By choosing to ignore these "elements" in favor of the elements as basic substances, chemists could continue to uphold that the fundamental units of chemistry, or its natural kinds, remained as the entities that occupied a single place in the periodic system.

Elements and Groups of Elements as Natural Kinds?

Elements defined by their atomic numbers are frequently assumed to represent "natural kinds" in chemistry.[66] The general idea is that the elements represent the manner in which nature has been "carved at the joints." On this view, the distinction between an element and another one is not a matter of convention. The question arises as to whether groups of elements appearing in the periodic table might also represent natural kinds. Could it be that there is some objective feature that connects all the elements that share membership to a particular group in the periodic system?

It would seem that the criterion for membership to a group is by no means as clear-cut as that which distinguishes one element from another. In the case of groups of the periodic table, it is the electronic configuration of gas-phase atoms that seems to provide a criterion, although in neither a necessary nor a sufficient manner.[67] However, one may also argue that the placement of the elements into groups is not a matter of convention. If periodic relationships are indeed objective properties, as I argue here, it would seem to suggest that there is one ideal periodic classification, regardless of whether or not this may have been discovered. This in turn would have a bearing on some recent questions regarding the placement of some elements within the periodic system. And if electronic configurations do not perfectly capture the fact that groups are natural kinds, this may merely indicate the limitations of the concept of electronic configurations.[68]

The Placement of Hydrogen and Helium in the Periodic System

There has been considerable debate within chemistry in recent years as to the placement of the elements hydrogen and helium within the periodic system.[69] For example, hydrogen is similar to the alkali metals in its ability to form single positive ions. However, hydrogen can also form single negative ions, thus suggesting that the element might be placed among the halogens, which also display this type of ion formation. Helium is traditionally regarded as a noble gas in view of its extreme inertness and is thus placed among the other inert gases in group 18 of the periodic system. However, in terms of its electronic configuration, helium has just two outer electrons and might therefore be placed among the alkaline earth metals such as magnesium and calcium. Many periodic tables appearing in physics books do just that, as do many spectroscopic periodic systems.

Peter Atkins and Herbert Kaesz have proposed a modification to the periodic table concerning the placement of the element hydrogen.[70] Contrary to its usual placement at the top of the alkali metals, and its occasional placement among the halogens, Atkins and Kaesz choose to position hydrogen on its own, floating above the table. In addition, they place helium alongside hydrogen, thus also removing it from the main body of the periodic table.

Rather than considering the relative virtues of these placements in chemical terms, the argument for the removal of hydrogen and helium from the main body of the table can be examined from the perspective of the elements as basic substances. The widely held belief among chemists is that the periodic system is a classification of the elements as simple substances that can be isolated and whose properties can be examined experimentally. However, as emphasized in the present book, there is a long-standing metaphysical tradition of also regarding the elements as unobservable basic substances.[71]

I suggest that our current inability to place hydrogen and helium in the periodic table in an unambiguous manner should not lead us to exclude them from the periodic law altogether, as Atkins and Kaesz seem to imply. Hydrogen and helium are surely as subject to the periodic law as are all the other elements. Perhaps there is a "fact of the matter" as to the optimum placement of hydrogen and helium in the main body of the table. Perhaps this question is not a matter of utility or convention that can be legislated, as most authors have argued.

Surprising as it may seem, some chemists have even proposed chemical evidence for placing helium in this manner. Such arguments are based on the first-element rule, as discussed above, which in its simplest form states that the first element in any group of the periodic system tends to show several anomalies when compared with successive members of its group. For example, in the p block, all the first-member elements show a reluctance to expand their octets of outer-shell electrons while subsequent group members do so quite readily. In addition, there is a more sophisticated version of this first-member rule that also specifies the extent to which the first elements in the various blocks of the periodic table display anomalies.

But once again, rather than relying on specific properties of the elements as simple substances, I suggest that we should concentrate on elements as basic substances. Perhaps one should seek some form of underlying regularity in order to settle the question of the placement of any element. Such a possibility is discussed below, along with the question of the best possible form for the periodic system.

IS THERE A BEST FORM FOR THE
PERIODIC TABLE?

The periods in the currently most popular representation of the periodic system, the so-called medium-long form, are arranged so that each one begins with a new value of n, or the first quantum number (figure 1.4). This value denotes the main shell of the most energetic electron in each case, in terms of the *aufbau* principle, that is used to "build up" the configuration of any particular atom. In more macroscopic chemical terms, the medium-long–form table places the reactive metals such as the alkalis and alkaline earths on the left side of the periodic table and the reactive nonmetals on the right side.

The conventional medium-long form displays the periods as though the main-shell number is the dominant criterion for the buildup of successive periods. But, as is well known, this form of display leads to a somewhat confusing layout whereby in several cases a main shell begins to fill, followed by an interruption due to a transition metal series in which a penultimate shell is filled. Only after such interruptions, which are more pronounced in the case of periods that also include inner transition elements, does the filling of the main shell resume.

Many authors have suggested that a more satisfactory representation can be obtained by basing the start of periods on $n + \ell$ instead of n.[72] Such a table requires that the s block be shifted to the right of the p block elements, which leads more specifically to at least two particular periodic tables. The first one is the so-called left-step periodic table (figure 10.13). The second is a modified form of the pyramidal periodic system that likewise places the s block elements on the right-hand edge of a pyramid (figure 10.14).[73] Both of these tables display two short periods of two elements, thus satisfying the desire for regularity that many authors, including some group theorists, believe might lie at the heart of the periodic system.[74]

Both of these alternative representations of the periodic system display the elements in a continuous manner with no break between any sets of elements, contrary to what is encountered in the currently accepted medium-long form. But these tables also contain a feature that causes many chemists some concern, in that the element helium is firmly located among the alkaline earth elements.

However, as I argued in the preceding section, such worries are alleviated once one acknowledges that the periodic system is primarily intended to classify the elements as basic substances and not simple substances. Although one can partly agree with the view that different representations can help to convey different forms of information, I believe that one may still maintain that one particular representation reflects chemical periodicity, regarded as an objective fact, in the best possible manner.

It may seem odd to the reader that the suggested periodic systems are ones that appear to rest rather heavily on a reductionist view in favor of the importance of electronic configurations of atoms. In addition, these considerations, and more

																																n+ℓ
																														H	He	1
																														Li	Be	2
																								B	C	N	O	F	Ne	Na	Mg	3
																								Al	Si	P	S	Cl	Ar	K	Ca	4
														Sc	Ti	V	Cr	Mn	Fe	Co	Ni	Cu	Zn	Ga	Ge	As	Se	Br	Kr	Rb	Sr	5
														Y	Zr	Nb	Mo	Tc	Ru	Rh	Pd	Ag	Cd	In	Sn	Sb	Te	I	Xe	Cs	Ba	6
La	Ce	Pr	Nd	Pm	Sm	Eu	Gd	Tb	Dy	Ho	Er	Tm	Yb	Lu	Hf	Ta	W	Re	Os	Ir	Pt	Au	Hg	Tl	Pb	Bi	Po	At	Rn	Fr	Ra	7
Ac	Th	Pa	U	Np	Pu	Am	Cm	Bk	Cf	Es	Fm	Md	No	Lr	Rf	Db	Sg	Bh	Hs	Mt	Ds	Rg										8

FIGURE 10.13 The left–step or Janet table. Numbers on the right represent values of $n + \ell$.

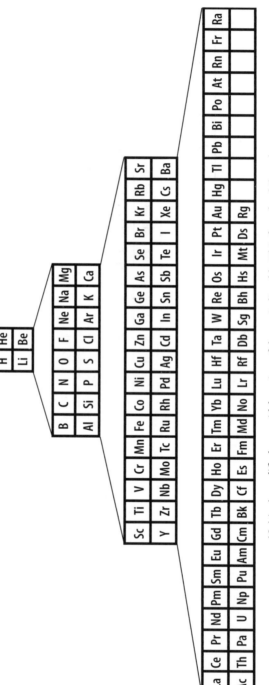

FIGURE 10.14 A modified pyramidal version of the periodic table. This form has bilateral symmetry.

specifically the $n + \ell$ rule concerning the order of filling of atomic orbitals, are being placed above current wisdom concerning the chemical nature of helium, which dictates that it should be regarded as a noble gas par excellence and not as an alkaline earth element.

My response to such worries is to point out that throughout this book I have sought to examine the limits of reductionism in chemistry and have not been critical of reductionism as a general approach. As mentioned at the outset, reductionism has provided an undeniably successful approach to the acquisition of scientific knowledge. The thrust of this book has been directed against exaggerated claims made on behalf of reductionism, for example, Bohr's claim that he had predicted the chemical nature of the element hafnium from first principles or the claim that all aspects of the periodic table have been strictly deduced from the later quantum mechanics. It is rather the limitations of reductionism that are of interest to philosophers of science and that should be taken more seriously by science educators.

A CONTINUUM OF PERIODIC TABLES?

The metaphysical notion of the elements as basic substances and as the bearers of properties has been historically important in the case of Mendeleev's establishment of the periodic system and Paneth's resolution of the fate of the periodic system in light of the discovery of isotopes.

I have suggested that the notion of elements as basic substances can cast some light on the question of the optimal representation of the periodic system. As in the case of the distinction between elements as basic substances and as simple substances, the aim should be to obtain a classification that primarily classifies elements as basic substances, while also recognizing aspects of the elements as simple substances. This optimal classification will not be obtained by behaving as naive inductivists and agonizing over the minutiae of the properties of hydrogen, helium, or other problematic elements.[75] It is suggested that an optimal classification can be obtained by identifying the deepest and most general principles that govern the atoms of the elements, such as the $n + \ell$ rule, and by basing the representation of the elements on such principles.[76]

But I conclude with a less controversial proposal. Let us imagine that the various representations of the periodic system lie on a continuum. At one end of this continuum is the "unruly" Rayner Canham table (figure 10.14) that attempts to do justice to many unusual relationships of the kind that have been highlighted in this chapter. At the other end of the continuum lies what I call the Platonic periodic table, or what is usually called the left-step or Janet periodic table[77] (figure 10.13). Somewhere near the middle of this continuum of representations, one can locate the currently popular medium-long representation. It is not altogether surprising that this form has been so popular since it appears to capture the correct

FIGURE 10.15
Dufour's 3-D periodic tree.
Photo and permission from
Fernando Dufour.

balance between utility and the display of order and regularity. While it sacrifices many of the unusual chemical and physical relationships that Rayner Canham's table features, it embodies the physics and chemistry of the elements as simple substances as well as basic substances. At the same time, the medium-long form stops short of adopting a fully reductionist approach that puts the highest premium on electronic configurations, which would commit one to the placement of helium among the alkaline earths.

The left-step table, I suggest, embodies the elements entirely as basic substances since it relegates the chemical and physical properties of elements such as helium and places greater importance upon more fundamental aspects. From a philosophical point of view, I believe that the left-step table may provide an optimal periodic system in showing the greatest degree of regularity while also adhering to the deepest available principles relating to the elements as basic substances.

It would be gratifying to think that principles of beauty and elegance[78] as embodied in the system shown in figure 10.15, for example, or indeed, a philosophical version of the periodic table may eventually become the standard form of the periodic system. But the argument between the relative virtues of utility and beauty in science is not an easy one to resolve, and I do not propose to do so here. It is with some trepidation that I advocate the general adoption of the left-step periodic system since I am well aware of the resistance that this proposal will meet, especially from the chemical community, which, rightly or wrongly, regards itself as the sole proprietor of the periodic system.[79]

NOTES

Acknowledgments

1. M.P. Melrose, E.R. Scerri, Why the 4s Orbital Is Occupied before the 3d, *Journal of Chemical Education*, 73, 498–503, 1996; E.R. Scerri, J. Worrall, Prediction and the Periodic Table, *Studies in the History and Philosophy of Science*, 32, 407–452, 2001.

2. At about the same time, the official journal of the society, *Foundations of Chemistry* was also started.

Introduction

1. There have only ever been two conferences specifically on the periodic table. The first was in 1969 as part of the celebrations commemorating the centenary of Mendeleev's famous table of 1869 [M. Verde (ed.), *Atti del Convegno Mendeleeviano*, Accademia delle Scienze di Torino, 1971]. The second was held as recently as 2003 in Banff, Canada [D. Rouvray, R.B. King, *The Periodic Table: Into the 21st Century*, Science Studies Press, Bristol, 2004].

2. J. Van Spronsen, *The Periodic System of the Chemical Elements, the First One Hundred Years*, Elsevier, Amsterdam, 1969. I find it embarrassing to make this criticism given the enormous debt I owe to Van Spronsen's wonderful book on the periodic system.

3. F.P. Venable, *The Development of the Periodic Law*, Chemical Publishing Co., Easton, PA, 1896.

4. E. Mazurs, *The Graphic Representation of the Periodic System During 100 Years*, University of Alabama Press, Tuscaloosa, 1974. In addition, Mazurs has given some arguments for the adoption of a symmetrical representation of the periodic system.

5. For a bibliography of secondary articles on the periodic system, which emphasizes philosophical works, see E.R. Scerri, J. Edwards, Bibliography of Literature on the Periodic System, *Foundations of Chemistry*, 3, 183–196, 2001.

6. P.W. Atkins, *The Periodic Kingdom*, Basic Books, New York, 1995. Also see E.R. Scerri, A Critique of Atkins' Periodic Kingdom and Some Writings on Electronic Structure, *Foundations of Chemistry*, 1, 287–296, 1999.

7. R.J. Puddephatt, P.K. Monaghan, *The Periodic Table of the Elements*, Oxford University Press, Oxford, 1985.

8. D.G. Cooper, *The Periodic Table*, Plenum Press, New York, 1968.

9. J.S.F. Pode, *The Periodic Table; Experiment and Theory*, Wiley, New York, 1973.

10. R.T. Sanderson, *The Periodic Table of the Chemical Elements*, School Technical Publishers, Ann Arbor, MI, 1971.

11. P. Strathern, *Mendeleyev's Dream*, Thomas Dune Books, New York, 2001; O. Sacks, *Uncle Tungsten*, Alfred Kopf, New York, 2001; R. Morris, *The Last Sorcerers: Atoms, Quarks and the Periodic Table*, Walker & Co., New York, 2003.

12. M. Gordin, *A Well-Ordered Thing*, Basic Books, New York, 2004.

13. B.J.T. Dobbs, M.C. Jacob, *Newton and the Culture of Newtonianism*, Humanity Books, Amherst, NY, 1998.

14. Principe, *The Aspiring Adept: Robert Boyle and His Alchemical Quest: Including Boyle's "Lost" Dialogue on the Transmutation of Metals*, Princeton University Press, Princeton, NJ, 1998, quoted from p. 220.

15. A number of detailed studies on philosophical aspects of scientific experiments now exist, including David Gooding, Trevor Pinch, Simon Schaffer, *The Uses of Experiment: Studies in the Natural Sciences*, Cambridge University Press, New York, 1989; and Allan Franklin, *The Neglect of Experiment*, New York, Cambridge University Press, 1986.

16. E.g., the book on scientific models by the philosopher Nancy Cartwright, *How the Laws of Physics Lie*, Oxford University Press, Oxford, 1983.

17. D. Shapere, Scientific Theories and their Domains, in F. Suppe (ed.), *The Structure of Scientific Theories*, Illinois University Press, Urbana, 518–599.

18. The literature in this area has grown tremendously in recent years. For discussions, see P.R. Gross, N. Levitt, *Higher Superstition*, Johns Hopkins University Press, Baltimore, MD, 1994; A. Sokal, Transgressing the Boundaries: Towards a Transformative Hermaneutics of Quantum Gravity, *Social Text*, 46–47, 217–252, 1996; J.A. Labinger, H. Collins, *The One Culture?* University of Chicago Press, Chicago, 2001.

19. G. Bodner, M. Klobuchar, D. Geelan, The Many Forms of Constructivism, *Journal of Chemical Education*, 78, 1107–1134, 2001. For a critical appraisal, see E.R. Scerri, Philosophical Confusion in Chemical Education Research, *Journal of Chemical Education*, 80, 468–474, 2003.

20. This is not to imply that Kuhn himself or anyone else that I am aware of has argued that the development of the periodic system *did* represent a scientific revolution.

21. Related claims about Kuhn's conservatism are made in Steve Fuller, *Thomas Kuhn: A Philosophical History for Our Times*, University of Chicago Press, Chicago, 2000; and Mara Beller, *Quantum Dialogue: The Making of a Revolution*, University of Chicago Press, Chicago, 1999.

22. Readers interested in scholarly research on Mendeleev in particular should consult the work of a number of excellent contemporary historians of science, including N. Brookes, Developing the Periodic Law: Mendeleev's Work during 1869–1871, *Foundations of Chemistry*, 4, 127–147, 2002; M. Gordin, *A Well-Ordered Thing*, Basic Books, New York, 2004; M. Kaji, Mendeleev's Discovery of the Periodic Table, *Foundations of Chemistry*, 5, 189–214, 2003.

23. Over the past 20–30 years, there has been a great deal of debate among historians of chemistry regarding Lavoisier's role in the chemical revolution and whether this was indeed a revolution or just the culmination of previous work begun by the likes of Georg Stahl. See articles by Gough, Siegfried, Perrin, and Holmes in A. Donovan (ed.), Chemical Revolution, Essays in Reinterpretation, *Osiris*, 2nd series, vol. 4, 1988.

24. The popular story found in most books is that Bohr was primarily concerned with explaining the spectrum of the hydrogen atom. But as Kuhn and Heilbron have convincingly argued, Bohr was not even aware of the problems with atomic spectra when he began applying quantum theory to the structure of the atom. T.S. Kuhn, J. Heilbron, The Genesis of the Bohr Atom, *Historical Studies in the Physical Sciences*, 3, 160–184, 1969.

25. J.D. Trout, in R. Boyd, P. Gaspar, J.D. Trout (eds.), Reduction and the Unity of Science, *The Philosophy of Science*, MIT Press, Cambridge, MA, 1992, 387–392.

26. A few modern commentators appear to disagree in this respect and regularly advertise their support for the "dis-unity of science": J. Dupré, *The Disorder of Things, Metaphysical Foundations of the Disunity of Science*, Harvard University Press, Cambridge, MA, 1993; N. Cartwright, *How the Laws of Physics Lie*, Clarendon Press, Oxford, 1983; N. Cartwright, *Nature's Capacities and Their Measurement*, Oxford University Press, Oxford, 1989; P. Galison, D. Stump, *The Disunity of Science*, Stanford University Press, Palo Alto, CA, 1996.

27. For a more detailed treatment of this modern approach to the reduction of chemistry, see E.R. Scerri, Popper's Naturalized Approach to the Reduction of Chemistry, *International Studies in Philosophy of Science*, 12, 33–44, 1998.

28. K.R. Popper, Scientific Reduction and the Essential Incompleteness of All Science, in F.L. Ayala, T. Dobzhansky (eds.), *Studies in the Philosophy of Biology*, Berkeley University Press, Berkeley, CA, 1974, pp. 259–284.

29. J. LaPorte, Chemical Kind Terms, Reference and the Discovery of Essence, *Noûs*, 30, 112–132, 1996; J. LaPorte, *Natural Kinds and Conceptual Change*, Cambridge University Press, New York, 2004.

30. J. van Brakel, *Philosophy of Chemistry*, Leuven University Press, Leuven, Belgium, 2000.

31. I believe that this criticism can be countered by appeal to the elements as basic substances in the sense used by Mendeleev. E.R. Scerri, Some Aspects of the Metaphysics of Chemistry and the Nature of the Elements, *Hyle*, 11, 127–145, 2005.

32. P. Bernal, Foundations of Chemistry: Special Issue on the Periodic System (editor-in-chief Eric R. Scerri), *Journal of Chemical Education*, 79, 1420, 2002.

33. E.g., see an editorial in *Foundations of Chemistry* for the special issue on the Periodic System, vol. 3, 97–104, 2001.

Chapter 1

1. There has been some debate concerning the precise criteria used by Lavoisier, given that several of the entries fail to meet his stipulation that simple substances are the final products of chemical analysis. R. Siegfried, B.J. Dobbs, Composition, A Neglected Aspect of the Chemical Revolution, *Annals of Science*, 29, 29–48, 1982.

2. With the exception of fire, which is a process rather than a substance.

3. Lavoisier's simple substances were defined in opposition to the classical view of the elements as abstract entities or principles. However, Lavoisier was not entirely consistent in that some of his simple substances appear to be more akin to the older principles. This issue concerning the dual nature of elements is resumed in later chapters.

4. Until recently, it was believed that the element technetium with atomic number 43 does not occur naturally. It has now been established that this element does occur terrestrially and that the initial reports of its discovery by the Noddacks in 1925 may have been correct. I. Noddack, W. Noddack, Darstellung und Einige Chemische Eigenschaften des

Rheniums, *Zeits für Physikalishe Chemie,* 125, 264–274, 1927. Promethium is therefore the only element within the first 92 that does not occur naturally.

5. M.E. Weeks, H. Leicester, *The Discovery of the Elements,* 7th ed., Journal of Chemical Education, Easton, PA, 1968.

6. H.M. Van Assche, The Ignored Discovery of Element Z = 43, *Nuclear Physics A,* A480, 205–214, 1988. See also note 4.

7. For a very informative article on the naming of compounds, from which I have drawn liberally for this section, see V. Ringnes, Origin of the Names of Chemical Elements, *Journal of Chemical Education,* 66, 731–738, 1989.

8. Primo Levi, *The Periodic Table,* 1st American ed., Schocken Books, New York, 1984.

9. Levi took his own life after surviving the holocaust and after writing several books, the best known of which remains *The Periodic Table.* He is believed to have acted out of survivor guilt.

10. O. Sacks, *Uncle Tungsten,* Alfred Knopf, New York, 2001.

11. The classic book on the discovery of the individual elements remains M.E. Weeks, H. Leicester, *The Discovery of the Elements,* 7th ed., Journal of Chemical Education, Easton, PA, 1968.

12. Other elements that take their names from mythology are vanadium, niobium, and tantalum.

13. The element names of rubidium, indium, and thallium are also derived from colors.

14. These scientists were all famous physicists, with the exception of Mendeleev and Seaborg, who were chemists.

15. The most recently approved names, at the time of writing, are darmstadtium and roentgenium for elements 110 and 111, respectively.

16. The unofficial reason is that many members of the IUPAC committee objected to the idea of naming an element after the person who had synthesized such deadly substances as plutonium, which was used in one of the atomic bombs dropped on Japan during the Second World War.

17. The modern author Ruth Sime has written several articles and a book about the way in which the Nobel Prize committee overlooked Lise Meitner's work. R.L. Sime, *Lise Meitner: A Life in Physics,* University of California Press, Berkeley, 1996.

18. The others are potassium (K), iodine (I), yttrium (Y), phosphorus (P), tungsten (W), boron (B), uranium (U), and vanadium (V).

19. These Latin names are *cuprum* (Cu), *natrium* (Na), *ferrum* (Fe), *plumbum* (Pb), *hydrargyrum* (Hg), *argentum* (Ag), *kalium* (K), and *aurum* (Au).

20. An excellent and detailed account of the discovery and properties of all the individual elements can be found in John Emsley, *Nature's Building Blocks: An A-Z Guide to the Elements,* Oxford University Press, Oxford, 2001.

21. The story of this development is mentioned by the discoverer of the compound, Paul Chu, in the article, Yttrium, *Chemical & Engineering News,* Special Issue on the Elements, September 8, 2003, p. 102.

22. The groups headed by copper and zinc are labeled as IB and IIB, respectively, in both the U.S. and European systems. Also note that in the US and European systems three groups are collectively labeled as VIIIB and VIIIA respectively.

23. According to some observers, the IUPAC recommendation is really the European system in disguise since it numbers the groups sequentially from left to right regardless of whether they might be main-group elements or transition elements. The IUPAC numbering proposal also raises the question of what should be done if the rare earth elements are

incorporated into the periodic table rather than being displayed as a footnote. This would strictly necessitate a numbering of all groups from 1 to 32.

24. In chapter 10, however, the IUPAC system is used.

25. With the exception of the first member of the group, lithium, which is rather unreactive, although it, too, forms an alkaline solution and even reacts vigorously with nitrogen when heated.

26. One example of a loss is the correspondence between the numbers in the group labels and the maximum oxidation state of the element in question.

27. Fullerenes, also known as buckyballs or buckminsterfullerenes are a recently discovered form of carbon. The simplest such molecule contains a total of 60 carbon atoms arranged in the shape of a soccer ball, that is, a set of interlocking hexagons and pentagons.

28. But there are variations, in that the valence of 2 is increasingly the more predominant one as the group is descended. In the case of lead, compounds displaying a valence of 2 such as $PbCl_2$ are actually more stable than their 4-valent analogues such as $PbCl_4$ in this case.

29. However, in many countries cesium (melting point 28.5°C) and gallium (melting point 29.8°C) are also liquid at room temperature.

30. One further element in this group, astatine, has been discovered but only a handful of atoms of it have ever been isolated. Its macroscopic properties, such as the color of the element, therefore remain unknown.

31. It is the abstract element that survives when elements form compounds, and so there are more similarities among the compounds of the elements than in the isolated elements or elements as simple substances.

32. The actual dates are helium, 1895; neon, 1898; argon, 1894; krypton, 1898; xenon, 1898; and radon, 1900.

33. Hundreds of chemical compounds of krypton and xenon are now known. Only helium and neon still resist all attempts at making them combine with other elements.

34. This way of counting period length counts the first element up to and including the element that represents the approximate repetition of the first one.

35. D.W. Theobald, Some Considerations on the Philosophy of Chemistry, *Chemical Society Reviews*, 5, 203–213, 1976.

36. The growing interest in the philosophy of chemistry, and specifically in the autonomy of chemistry, makes holding such views increasingly more plausible. E.R. Scerri, L. McIntyre, The Case for Philosophy of Chemistry, *Synthese*, 111, 213–232, 1997.

37. Whether or not a circular-shaped display would count as a table is debatable, although one sometimes encounters the term "circular periodic table."

38. Some time ago, in his highly popular book *The Tao of Physics* (Shambala, Berkeley, CA, 1975), F. Capra argued that modern physics shares many similarities with the Taoist philosophy of complementary opposites. It has been suggested that chemistry lends itself far more directly to such analogies. E.R. Scerri, The Tao of Chemistry, *Journal of Chemical Education*, 63, 100–101, 1986.

39. Much chemical history has been condensed into this sentence. The quantities used were first equivalent weight, then followed by a period of confusion in which atomic weights and equivalent weights (confusingly defined in several different ways) were used. After the Karlsruhe conference in 1860, atomic weights began to be used more exclusively, and finally atomic weight was replaced by atomic number as the main ordering criterion for the elements.

40. I am greatly oversimplifying the situation. As Alan Rocke and many others before him have argued, the use of equivalent weights in preference to atomic weights carried

out by William Wollaston and others was motivated by the notion of avoiding theoretical assumptions as well as the existence of atoms. However, these chemists still needed to assume formulas for the compounds they were considering, and as a result, what they were calling equivalent weights were operationally equivalent to atomic weights. A. Rocke, Atoms and Equivalents, The Early Development of Atomic Theory, *Historical Studies in the Physical Sciences*, 9, 225–263, 1978; A.J. Rocke, *Chemical Atomism in the Nineteenth Century*, Ohio State Press, Columbus, 1984.

41. Historian Alan Rocke argues that this change was already in the air at the time and would have taken place regardless of the Karlsruhe meeting. A. Rocke, *Chemical Atomism in the Nineteenth Century*, Ohio State Press, Columbus, 1984.

42. This is a generalization. E.g., it is not clear that Gustav Hinrichs's system followed this form of ordering. Also, some of the discoverers of the periodic system such as John Newlands began by using equivalent weights and later changed to using atomic weights.

43. The fact that atoms contain protons and neutrons also explains the concept of isotopes, which is of crucial importance in the story of the periodic table. Atoms of an element that differ in the number of neutrons they contain are said to represent different isotopes of that element. E.g., the element carbon has three most common isotopes: carbon-12, carbon-13, and carbon-14. Each of these contains six protons (which identifies them as carbon) but also six, seven, or eight neutrons, respectively. Each is said to have a different mass number given by the sum of protons and neutrons. The atomic weight of carbon is given by a weighted average of the masses of all the isotopes of the element, meaning an average, which takes account of how much of each isotope occurs in any given natural sample. Until atomic number was understood, the existence of isotopes made it difficult to fit some elements into a periodic scheme in what would appear to be the proper order based on their chemical properties.

44. E.g., see F. Habashi, A New Look at the Periodic Table, *Interdisciplinary Science Reviews*, 22, 53–60, 1997. This article presents a periodic table from the point of view of metallurgy. An interesting geologist's table can be found in L.B. Railsback, An earth scientist's periodic table of the elements and their ions, *Geology*, 31, 737–740, 2003.

45. After publishing articles in *American Scientist* and *Scientific American* on the development of the system, the author received about 50 models, diagrams, and letters outlining new designs from passionate advocates of some particular version. He still frequently receives messages and letters from well-meaning enthusiasts asking him to comment on a new theory or a new form of representation of the periodic table.

46. The final comment is rather controversial, with many chemists believing that there is no one best representation. These authors consider representation to be a secondary issue, which is dictated by convention. The present author takes issue with this view and supports a more realist interpretation whereby the grouping of troublesome elements such as hydrogen has an objective aspect and is not merely a manner of convenience. See E.R. Scerri, The Best Representation of the Periodic System: The Role of the $n + \ell$ Rule and the Concept of an Element as a Basic Substance, in D. Rouvray, R.B. King (eds.), *The Periodic Table: Into the 21st Century*, Science Studies Press, Bristol, 2004, 143–160.

47. P. Armbruster, F.P. Hessberger, Making New Elements, *Scientific American*, 72–77, September 1998. This article is followed by one on the history of the periodic system: E.R. Scerri, The Evolution of the Periodic System, *Scientific American*, 78–83, September 1998.

48. W.B. Jensen, Classification, Symmetry and the Periodic Table, *Computation and Mathematics with Applications*, 12B, 487–509, 1986; H. Merz, K. Ulmer, Position of lanthanum and lutetium in the Periodic Table, *Physics Letters*, 26A, 6–7, 1967; D.C. Hamilton, M.A. Jensen,

Mechanism for Superconductivity in lanthanum and uranium, *Physical Review Letters*, 11, 205–207, 1963; D.C. Hamilton, Position of lanthanum in the Periodic Table, *American Journal of Physics*, 33, 637–640, 1965.

49. The contemporary author on the periodic system, W.B. Jensen, uses the terms primary and secondary kinship, which should not be confused with the terms primary and secondary classification as used by the present author. A primary kinship as termed by Jensen results from secondary classification according to the terminology used in this book. W.B. Jensen, *Computation and Mathematics with Applications*, 12B, 487–509, 1986.

50. This is an important issue. The helium question is at the center of recent attempts to revolutionize the way in which the periodic system should be represented. See, e.g., Gary Katz, The Periodic Table: An Eight Period Table for the 21st Century, *The Chemical Educator*, 6, 324–332, 2001.

51. Evidence for higher elements is not yet conclusive, and some claims have, in fact, been withdrawn.

52. Strictly speaking, the electron is regarded as much as a delocalized wave as an orbiting particle, as explained a little later in the text. The use of the phrase "rapidly moving electrons" should therefore be regarded as a classical approximation in this context. Relativistic effects, more properly speaking, arise from the greater nuclear masses of the "heavier" elements.

53. There are several relatively accessible articles on relativistic effects in atoms. L.J. Norrby, Why Is Mercury Liquid? *Journal of Chemical Education*, 68, 110–113, 1991; M.S. Banna, Relativistic Effects at the Freshman Level, *Journal of Chemical Education*, 62, 197–198, 1985; D.R. McKelvey, Relativistic Effects on Chemical Properties, *Journal of Chemical Education*, 60, 112–116, 1983. For a more technical account, see P. Pyykkö, Relativistic Effects in Structural Chemistry, *Chemical Reviews*, 88, 563–594, 1988.

54. This idea was first proposed by the discoverer of the electron, J.J. Thomson.

55. The change in terminology from orbit to orbital is considered to be rather unfortunate by many, since the similarity in the two words does not begin to convey the radical change in the way that electron motion is regarded in the new quantum mechanics.

56. The detailed evolution of the concept of electronic configurations is given in chapter 7.

57. I am not disputing the approximate nature of the current explanation of the periodic system, as B. Friederich seems to believe, but am rather referring to the fact that the important $n + \ell$ rule has not yet been deduced from first principles. B. Friederich, *Foundations of Chemistry*, 6, 117–132, 2004; this is a response to the present author's article Just How Ab Initio is Ab Initio Quantum Chemistry?, *Foundations of Chemistry*, 6, 93–116, 2004.

58. R. Hefferlin, H. Kuhlman, The Periodic System for Free Diatomic Molecules III, *Journal of Quantitative Spectroscopy and Radiation Transfer*, 24, 379–383, 1980. Hefferlin, who is named in the text, is also the author of a book on the subject, R. Hefferlin, *Periodic Systems of Molecules and their Relation to the Systematic Analysis of Molecular Data*, Edwin Mellin Press, Lewiston, New York, 1989.

59. At the time of writing, only 111 of these elements have been confirmed, although preliminary reports claiming the synthesis of the heavier elements have been published. Elements 116, 117, and 118 have yet to be synthesized. Early reports of the synthesis of elements 116 and 118 were withdrawn about a year after their initial announcement. In addition, a member of the research team that had initially made the claim was dismissed on the grounds that he had fabricated data.

Chapter 2

1. The role of physics cannot be eliminated, however. John Dalton was concerned with the physics of the air in the early research that led to his atomic theory.

2. B. Bensaude-Vincent, A Founder Myth in the History of Sciences? The Lavoisier Case in L. Graham, W. Lepenies, P. Weingart (eds.), *Functions and Uses of Disciplinary Histories*, Reidel, Dordrecht, 1983, pp. 53–78.

3. This is not to suggest that Lavoisier was the first to use the chemical balance, as some popular accounts would have it. There is some evidence that the balance was already being used by Johann van Helmont, one of the leading alchemists of his day. W.R. Newman, L.M. Principe, *Alchemy Tried in the Fire: Starkey, Boyle and the Fate of Helmontzian Chemistry*, Chicago University Press, Chicago, 2002.

4. However, Lavoisier seems to have regarded oxygen as a principle rather than a simple substance, thus diminishing his break with the past.

5. These aspects have been stressed by J.B. Gough, Lavoisier and the Fulfillment of the Stahlian Revolution, *Osiris*, 2nd series, vol. 4, 15–33, 1988.

6. R. Siegfried, M.J. Dobbs, Composition, A Neglected Aspect of the Chemical Revolution, *Annals of Science*, 24, 275–293, 1968.

7. But as I describe in chapter 4, Dmitri Mendeleev was to later resuscitate the notion of abstract elements, which would serve a crucial role in his periodic classification.

8. E.g., several of Lavoisier's alleged simple substances appear to be more akin to abstract elements, or principles, even by Lavoisier's own descriptions. This is true of *lumière* and *calorique*, as described in chapter 1, and even *oxygène*, *azote* (nitrogen), and *hydrogène*.

9. For a short account of atomism and its origins in Greek philosophy, see William R. Everdell's review of Bernard Pullman's book *The Atom in the History of Human Thought*, In *Foundations of Chemistry*, 1, 305–309, 1999.

10. B. Pullman, *The Atom in the History of Human Thought*, Oxford University Press, New York, 1998.

11. I. Newton, *Opticks*, Query 31, London, 1704; also see A.R. Hall, *An Introduction to Newton's Opticks*, Clarendon Press, Oxford, 1993.

12. Some authors use the term "positivist" to describe Lavoisier's philosophical approach. I will avoid this in view of the more technical sense of the word positivism, which followed Auguste Comte's later usage of the term.

13. As discussed in chapter 4, the Russian chemist Mendeleev held a similar view of the existence of distinct elements, whereas his leading competitor, Julius Lothar Meyer, believed in the essential unity of matter.

14. Dalton's reasons for arriving at his atomic theory have a complicated and incomplete history. This is because many of his papers were lost in a fire during the Second World War.

15. F. Greenaway, *John Dalton and the Atom*, Heinemann, London, 1966; A.J. Rocke, *Chemical Atomism in the Nineteenth Century*, Ohio State University Press, Columbus, 1984.

16. Dalton's assumption of binary formulas for water and ammonia is taken up later in this chapter.

17. This conclusion is not quite as inevitable as I imply here, however. For a different opinion, see, e.g., P. Needham, Has Daltonian Atomism Provided Chemistry with Any Explanations? *Philosophy of Science*, 71 (2004), 1038–1047. A closely related article is P. Needham, Paul , When did Atoms Begin to do any Explanatory Work in Chemistry?, *International Studies in the Philosophy of Science*, 18 (2004), 199–219. Needham's view is based, in part, on that of Pierre Duhem. See Needham's translation of Duhem's article, Atomic Notation and

Atomistic Hypotheses Translated by Paul Needham *Foundations of Chemistry*, 2, 127–180, 2000.

18. As mentioned in chapter 1, the notion that equivalent weights are purely empirical is problematic, although perhaps more true of Richter's early tables involving just reactions between acids and metals.

19. This statement is not quite true since two oxygen atoms can also combine with two atoms of hydrogen to form hydrogen peroxide. Fortunately, this complication did not arise in Dalton's time, since hydrogen peroxide had not yet been discovered when he began his work. The value of 8 for the atomic weight of oxygen is more in keeping with the modern value. Dalton's original estimate of the atomic weight of oxygen was 7 rather than 8 because of experimental inaccuracies. Nevertheless, he soon corrected it from 7 to 8 when better experimental became available.

20. Dalton was well aware of the arbitrary nature of his rule of simplicity. In 1810, he discussed the possibility that water could be a compound of three atoms, two of hydrogen and one of oxygen, in which case the atomic weight of oxygen would be 14. In fact, water does consist of these three atoms, and the modern atomic weight of oxygen is closer to 16 than 14. At the same time, Dalton also discussed the possibility that water might consist of two atoms of oxygen and one of hydrogen, which would give oxygen an atomic weight of 3.5. As mentioned in note 19, Dalton originally believed that the atomic weight of oxygen was 7 rather than 8, hence explaining his estimate of 14 rather than 16 for the atomic weight of oxygen when using the correct formula for water.

21. J.F. Gay-Lussac, *Alembic Club Reprints*, No. 4, Edinburgh, reprinted 1923.

22. In modern terms we would write the reaction as

$$2H_2 + O_2 \rightarrow 2H_2O$$

Since the right-hand side includes a substance containing a single oxygen atom, it appears that the oxygen molecule on the left side must be divisible; hence, it is written as O_2. Up to this point the notion of 'diatomic molecules' that we have today had not been envisaged.

23. It appears that Dalton had implicitly used the EVEN hypothesis in estimating the relative weights of the atoms of different gaseous elements from their relative densities.

24. Anonymous, On the Relation between the Specific Gravities of Bodies in the Gaseous State and the Weights of Their Atoms, *Annals of Philosophy (Thomson)*, 11, 321–330, 1815.

25. Anonymous, Correction of a Mistake in the Essay on the Relation between the Specific Gravities etc., *Annals of Philosophy (Thomson)*, 12, 111, 1816.

26. The classic historical and philosophical study on Prout is William Brock, *From Protyle to Proton*, Adam Hilger, Boston, 1985.

27. *Collected Works of Sir Humphry Davy*, edited by his brother John Davy, Smith, Elder & Co., London, 1839–1840, vol. 5, p. 163.

28. Berzelius's own values underwent many revisions in successive published versions. E.g., the following selected elements show how his later values contained many nonintegral values:

	Mg	Al	Cl	Zn
1815 value	50	55	70	128
1828 value	25.4	27.4	35.5	64.6

29. Berzelius's weights are relative to the oxygen standard, whereby O = 16.

30. J. Berzelius, Tafel über die Atomengewichte der elementaren Körper und deren hauptsächlichsten binairen Vebindungen, *Annalen der Physikalishe Chemie*, 14, 566–590, 1828.

31. L. Gmelin, *Handbuch der theoretischen Chemie*, Frankfurt, 1827, as cited on p. 23 of F.P. Venable, *The Development of the Periodic Law*, Chemical Publishing Co., Easton, PA, 1896.

32. These whole number coincidences involving carbon, oxygen, and nitrogen were not evident from the early published values of equivalent and atomic weights. Carbon initially showed values of 12.2 and 12.3, and it was not until 1843 that a value of 12 was obtained. Oxygen was assigned values of 5.5, 7, and 7.6 in various early tables. The value of 16 emerged in 1815. Finally, nitrogen appeared as 14.2 until 1843, when Charles Gerhard's table gave the value of 14.

33. J.S. Stas, Researches on the Mutual Relations of Atomic Weights *Bulletin de l'Académie Royale de Belgique*, 10, 208–350, 1860.

34. J.S. Stas, Sur les Lois des Proportions Chimiques, *Memoires Academiques de L'Académie Royale de Belgique*, 35, 24–26, 1865.

35. R.J. Strutt, On the Tendency of the Atomic Weights to Approximate to Whole Numbers, *Philosophical Magazine*, 1, 311–314, 1901.

36. Many common elements show one predominant isotope, and as a result, their atomic weights are very close to integral multiples of the weight of the hydrogen atom. Hydrogen itself consists of about 99.99% one particular isotope. Carbon is 98.89% carbon-12, nitrogen is 99.64% nitrogen-14, oxygen is 99.76% oxygen-16, sulfur is 95.0% sulfur-32, and fluorine is 100% fluorine-19.

37. Modern writers on the periodic system have tended to downplay Prout's hypothesis, perhaps due to Whiggish tendencies and the fact that it has turned out to be incorrect. One exception is F.P. Venable's classic history of the early stages of the periodic system, in which he makes the following highly laudatory remark about Prout's idea: "Probably no hypothesis in chemistry has been so fruitful of excellent research as this much discussed hypothesis of Prout" (*The Development of the Periodic Law*, Chemical Publishing Co., Easton, PA, 1896, p. 3).

38. Why Döbereiner chose to begin his work with oxides is not known. These compounds had recently been isolated in England by Davy and might thus have aroused general interest. In addition, working with the oxides would not have required the isolation of the elements and would therefore present an easier experimental option.

39. It is worth emphasizing that, contrary to the accounts still found in many chemistry textbooks, Döbereiner's discovery of triads, whose middle member has approximately the mean weight of the two flanking members, did not in fact concern elements but instead their compounds.

40. These values were recalculated by Johannes van Spronsen using a correct atomic weight for oxygen of 16 instead of the value of 7.5 that Döbereiner used.

41. This also tends to be omitted from accounts of the evolution of the periodic system, presumably because transmutation is now known not to occur. In fact, nuclear reactions *can* result in transmutation of elements, in a different sense, as first discovered in the twentieth century by the physicist Ernest Rutherford.

42. A printer's error was probably responsible for this small error in the calculated mean, which should be 80.97.

43. Döbereiner was working with incorrect formulas for the oxides of these elements, MO instead of M_2O, with the result that his atomic weights appear to be about twice the currently accepted values.

44. This seems to be another printer's error and more serious this time since the mean should be 84.241.

45. There is a sense in which the chemical properties can be regarded as more basic since the purpose of the exercise is to obtain a chemical classification. Numerical data serve

to formalize the system and to sometimes resolve cases that may be difficult to decide on the basis of chemical properties. Nevertheless, this question of the relative importance to be attached to chemical and numerical properties is in itself an important issue that will recur in our story.

46. I am not claiming that this system is unknown to other authors, only that it has been highly neglected.

47. Rather than being literally a "hand" book, this work amounted to a massive 18 volumes.

48. Gmelin obtained a value of 20.5 for calcium by direct determination.

49. If one examines the later system of John Hall Gladstone, which is almost exclusively based on Gmelin's system under discussion, this reveals that Gmelin did in fact order most elements according to trends in their atomic weights. J.H. Gladstone, On the Relations Between the Atomic Weights of Analogous Elements, *Philosophical Magazine*, 5(4), 313–320, 1853.

50. Nevertheless, Gmelin's ordering within a family of chemically similar elements is explicitly based on the earlier concept of electronegativity.

51. Whether to present chemistry inductively or deductively ultimately depends on each author's philosophical taste. It is by no means clear that Mendeleev's apparent decision to proceed inductively is the only correct option.

52. Several subsequent volumes continue this detailed survey of the chemical properties of the elements.

53. This placement cannot be blamed just on Gmelin since it recurs in many later periodic systems.

54. Surprisingly, the excellent book on the early history of the periodic system written by F.P. Venable (*The Development of the Periodic Law*, Chemical Publishing Co., Easton, PA, 1896) fails to even mention Gmelin's system.

55. J. van Spronsen, *The Periodic System of the Chemical Elements, the First 100 Years*, Elsevier, Amsterdam, 1969, p. 70.

56. This is all the more surprising given that van Spronsen even mentions this fact himself on the same page, although in a different context. Van Spronsen also criticizes Gmelin for not seeming to arrange the elements in order of increasing atomic weight. But Gmelin may well have based his system on atomic weights, although perhaps a little erratically. It is difficult to see how he can be said to have "demoted" atomic weight in producing his system, as van Spronsen claims.

57. Helga Hartwig (chief ed.), *Gmelin Handbook of Inorganic Chemistry*, 8th ed., Springer-Verlag, Berlin, 1988.

58. For a detailed historical account of the evolution of the concept of electronegativity, see W.B. Jensen, Electronegativity from Avogadro to Pauling: Part 1: Origins of the Electronegativity Concept, *Journal of Chemical Education*, 73, 11–20, 1996; and Jensen, Electronegativity from Avogadro to Pauling: II. Late Nineteenth-and Early Twentieth-Century Developments, *Journal of Chemical Education*, 80, 279–287, 2003.

59. Chlorine forms acids, which Gmelin gives as ClO_3 and ClO_4 in addition to ClO_5 and ClO_7. The fact that all mineral acids contain hydrogen had not yet been realized.

60. In contemporary terms, one might also see the differences between fluorine and the other halogens as resulting from the phenomenon of first-member anomaly, whereby the uppermost element in the main-group elements shows anomalous behavior when compared with other group members.

61. The question of the placement of tellurium takes on some importance in view of its being one of the few elements that belong to a reversed pair, the other element in this case being iodine. Many pioneers of the periodic system reversed the positions of tellurium and

iodine in order to better reflect their respective chemical analogies. It would appear that this was more easily decided for iodine than for tellurium, given Gmelin's apparent uncertainty in the chemical analogies of tellurium.

62. The omission of nitrogen from the group that includes phosphorus, arsenic, and antimony may be because nitrogen alone occurs as a gas while the other three elements mentioned are all solids at room temperature. In addition, the properties of nitrogen are somewhat anomalous, in keeping with the phenomenon of first-member anomalies, once again from the perspective of contemporary knowledge. Similarly, Gmelin did not include oxygen, a gas at room temperature, with sulfur and selenium, two solids with which it is grouped in the modern periodic table.

63. The other elements that would eventually join this trio of elements are rubidium and cesium, discovered in 1860 and 1861, respectively.

64. Gmelin places beryllium in a neighboring group along with cerium and lanthanum. This would be regarded as a mistake from the perspective of the modern table, in which beryllium is a group II main-group element while the other two are rare earths. Radium had not yet been discovered.

65. As noted above, Gmelin himself had discovered that magnesium, calcium, and barium form a triad.

66. Formulas given by Gmelin for the oxides were LiO, NaO, CaO, and BaO. In modern terms, only the third and fourth of these are correct. The first two should read Li_2O and Na_2O, respectively.

67. For some interesting remarks on the nature of predictions and its relationship to scientific laws as seen by Mendeleev, see M. Gordin, *A Well-Ordered Thing*, Basic Books, New York, 2004.

68. This lecture was subsequently published in the scientific journals of various European countries and thus exerted considerable influence on chemists worldwide.

69. Michael Faraday, *A Course of Six Lectures on the Non-metallic Elements. Before the Royal Institution*, Royal Institution, London, 1852.

70. Of course, *some* of these elements do have things in common, namely O, S, and Se, all of which are grouped together in group 16 of the modern periodic table.

71. The subject of secondary relationships is taken up in chapter 10. This is a feature that is revealed in the older short-form tables as well as pyramidal displays of the periodic system but not, unfortunately, in the currently popular medium-long form.

72. E. Lenssen, Uber die gruppirung der elemente nach ihrem chemisch-physikalischen charackter, *Annalen der Chemie Justus Liebig*, 103, 121–131, 1857.

73. Many articles have appeared on this issue, including S.J. Brush, The Reception of Mendeleev's Periodic Law in America and Britain, *Isis*, 87, 595–628, 1996; R. Campbell, T. Vinci, Novel Confirmation, *British Journal for the Philosophy of Science*, 34, 315–341, 1983; P. Maher Prediction, Accommodation and the Logic of Discovery, in A. Fine, J. Leplin (eds.), *PSA 1988*, vol. 1, Philosophy of Science Association, East Lansing, MI, 1988, 273–285; J. Worrall Fresnel, Poisson and the White Spot: The Role of Successful Prediction in the Acceptance of Scientific Theories, in D. Gooding, T. Pinch, S. Schaffer (eds.): *The Uses of Experiment*, Cambridge University Press, Cambridge, 1989, pp. 135–157; E.R. Scerri, J. Worrall, Prediction and the Periodic Table, *Studies in History and Philosophy of Science*, 32, 407–452, 2001.

74. This could be regarded as a form of nonrational development in science, but not in the sense implied by Thomas Kuhn, for whom rival scientific theories cannot be strictly compared because they speak different languages such that translation is never quite possible.

75. There are 21 elements that show just one single isotope. They include sodium, cesium, beryllium, aluminum, phosphorus, arsenic, bismuth, fluorine, iodine, manganese, cobalt, and gold. Also see note 36. (Throughout this book I use the American spelling for the elements whose symbols are Al and Cs rather than the official IUPAC spelling of aluminium and caesium.)

76. In addition, hydrides did not lend themselves to very accurate analysis. E.g., for several years water was reported to contain 13.27% hydrogen until Pierre Dulong corrected this measurement to 11.1%.

77. The following four chemists used four different values: Thomas Thomson, 1; William Wollaston, 10; Berzelius, 100; Stas, 16. The final value most closely approximates the modern value and is consistent with an approximate value of 1 for hydrogen.

78. The law of Dulong and Petit failed for a number of elements such as carbon, silicon, and boron due to variations in their specific heats with increasing temperature. This issue is taken up again in chapter 5, where the determination of the atomic weight of beryllium is discussed in more detail.

79. S. Mauskopf, Crystals and Compounds: Molecular Structure in Nineteenth Century Century French Science, *Transactions of the American Philosophical Society*, 66, pt 3, 5–82, 1976.

80. Dalton had incorrectly assigned an atomic weight of 40 to selenium, while Berzelius and others did not venture any value whatsoever in their early tables of atomic weights. In 1827, Berzelius adopted Mitscherlich's value, which he cited as 79.1.

81. J.B.A. Dumas, Memoire sur Quelques Points de la Theorie Atomique, *Annales de Chimie et Physique*, 33(2), 334–414, 1826.

82. J.B.A. Dumas, *Leçons sur la philosophie chimique*, 1837 (reprint, Editions Culture et Civilization, Brussels, 1972), p. 249.

83. The problem with the anomalous elements was solved by Gaudin, who suggested that molecules of different elements might contain different numbers of atoms. E.g., sulfur would be hexa-atomic while mercury was monoatomic. M.A.A. Gaudin, *Annales de Chimie et Physique*, 52(2), 113, 1833. A discussion of a more recent method of determining atomicity using the kinetic theory is given in chapter 5 in connection with the noble gases.

Chapter 3

1. In saying this, I am essentially agreeing with Jan van Spronsen's analysis of the developments (chapter 5). I am also following van Spronsen rather faithfully in saying that there were six discoverers. However, I would *not* want to emphasize the occurrence of definite periods in the history of the discovery of the periodic system, as van Spronsen seems to favor. Nevertheless, I have accepted van Spronsen's terms "precursors" and "discoverers" partly as a means of presenting the material in a more coherent fashion. It must be realized that this is something of a conventionalist strategy and not meant to be taken too literally.

2. L.M. Ampère, Lettre de M. Ampère à M. le comte Berthollet, sur la détermination des proportions dans lesquelles les corps se combinent d'après le nombre et la disposition respective des molécules dont leurs particules intégrantes sont composées, *Annales de Chimie*, 90, 43–86, 1814.

3. C. Gerhardt, Recherches sur la Classification Chimique des Substances Organiques, *Comptes Rendus*, 15, 498–500, 1842.

4. The importance of the Karlsruhe conference in connection with the rationalization of atomic weights and the concept of the molecule is disputed by historian Alan Rocke,

who claims that even without this meeting the changes would have occurred quite quickly. A. Rocke, *Chemical Atomism in the Nineteenth Century*, Ohio State Press, Columbus, 1984.

5. The question of whether Cannizzaro was committed to chemical or physical atomism is the subject of Alan Chalmers, Cannizzaro's Course of Chemical Philosophy Revisited (forthcoming). Chalmers believes that Cannizzaro would only be committed to the former.

6. For a detailed exposition of Cannizzaro's method see, J. Bradley, *Before and after Cannizzaro*, Whittles Publishing Services, North Humberside, UK, 1992.

7. S. Cannizzaro, *Il Nuovo Cimento*, 7, 321–366, 1858 (English translation, Alembic Club Reprints, no. 18, Ediburgh, 1923), quoted from p. 11.

8. A. Naquet, *Principes de Chimie*, F. Savy, Paris, 1867.

9. The further inclusion of uranium in this group is incorrect in the light of modern knowledge.

10. The modern medium-long form of the periodic system separates out the transition metals even though they show the same valence of 4, as in the case of zirconium and titanium.

11. It would also be possible to make a case for simultaneous discovery by these six researchers.

12. Inquirer, Numerical Relations of Equivalent Numbers, *Chemical News*, 10, 156, 1864; Studiosus, Numerical Relations of Equivalent Numbers, *Chemical News*, 10, 11, 1864; Studiosus, Numerical Relations of Equivalent Numbers, *Chemical News*, 10, 95, 1864.

13. P.J. Hartog, A First Foreshadowing of the Periodic Law, *Nature*, 41, 186–188, 1889; P.E. Lecoq De Boisbaudran, A. Lapparent, A Reclamation of Priority on Behalf of M. De Chancourtois Referring to the Numerical Relations of the Atomic Weights, *Chemical News*, 63, 51–52, 1891.

14. A.E. Béguyer De Chancourtois, *Comptes Rendus de l'Academie des Sciences*, 54, 1862, 757, 840, 967.

15. Among the recipients of De Chancourtois's privately published system was Prince Napoleon.

16. A.E. Béguyer De Chancourtois, Mémoire sur un Classement Naturel des Corps Simples ou Radicaux Appelé Vis Tellurique, *Comptes Rendus de l'Académie des Sciences*, 54, 757–761, 840–843, and 967–971, 1862.

17. P.J. Hartog, A First Foreshadowing of the Periodic Law, *Nature*, 41, 186–188, 1889 186–188, quoted from p. 187.

18. J.A.R. Newlands, *Journal of the Chemical Society*, 15, 1862, 36.

19. C.J. Giunta, J.A.R. Newlands, Classification of the Elements: Periodicity, but No System (1), *Bulletin for the History of Chemistry*, 24, 24–31, 1999.

20. As described below, Mendeleev repeated the first of Newlands's incorrect predictions concerning an element with an atomic weight in the region of about 170, along with making a number of other failed predictions of his own.

21. The chemical association of tellurium with sulfur and selenium and that of iodine with the halogen elements were well known on qualitative chemical grounds.

22. Newlands beat Odling by just four months in terms of publication dates: July 1864, as compared with October 1864 for Odling. I disagree with Carmen Giunta's denial of this anticipation and especially with the reasons that he gives for taking this stance. C. Giunta, J.A.R. Newlands' Classification of the Elements: Periodicity But No System, *Bulletin for the History of Chemistry*, 24, 24–31, 1999. A response to this article is E.R. Scerri, A Philosophical Commentary on Giunta's Critique of Newlands' Classification of the Elements, *Bulletin for the History of Chemistry*, 26, 124–129, 2001.

23. Both of the qualitatively based systems of families of Gmelin and Naquet, described earlier in this chapter, grouped lithium and sodium together.

24. A periodicity of 8 was correct for the chemistry known at the time. Today the periodicity is actually 9, counting from the first element up to and including the first analogous element (e.g., from lithium to sodium), as discussed in chapter 1.

25. J.A.R. Newlands, On the Law of Octaves, *Chemical News*, 12, 83, August 18, 1865, emphasis original.

26. Wendell H. Taylor, J.A.R. Newlands: A Pioneer in Atomic Numbers, *Journal of Chemical Education*, 26, 152–157, 1949.

27. W. Odling, On the Proportional Numbers of the Elements, *Quarterly Journal of Science*, 1, 642–648, October 1864, quoted from p. 648.

28. J.A.R. Newlands, On the Law of Octaves, *Chemical News*, 13, 130–130, 1866.

29. On the other hand, Newlands can be faulted for omitting gaps for as yet undiscovered elements in the manner that Mendeleev later did.

30. Here I am considering the distance between the number of successive similar elements to be consistent with the Newlands quotation and not as in other parts of this book when considering one element up to and including its analogue.

31. J.A.R. Newlands, *On the Discovery of the Periodic System and on Relations among the Atomic Weights*, Spon, E&FN, London, 1884. Many copies of this book were published, and the copy owned by the Science Museum in London is still displayed on open shelves and signed by Newlands himself.

32. W. Odling, On the Proportional Numbers of the Elements, *Quarterly Journal of Science*, 1, 642–648, October 1864, quoted from p. 642.

33. Ibid, quoted from p. 643.

34. W. Odling, On the Proportional Numbers of the Elements, *Quarterly Journal of Science*, 1, 642–648, 1864, quoted from p. 644.

35. Van Spronsen correctly praises Odling (pp. 112–116) in my view, for being the first to recognize this feature, although I differ somewhat regarding the details, as I argue in the main text.

36. Van Spronsen's claim (p. 113) that Odling had anticipated the separation of transition metals in the modern table would therefore need to be qualified somewhat.

37. Carl A. Zapffe, Hinrichs, Precursor of Mendeleev, *Isis*, 60, 461–476, 1969.

38. Ibid, p. 464.

39. In conversation with Bill Jensen, one very rainy evening in a Cincinnati restaurant, the author's car having broken down while en route to a conference in South Carolina.

40. H. Cassebaum, G. Kauffman, The Periodic System of the Chemical Elements: The Search for Its Discoverer, *Isis*, 62, 314–327, 1971.

41. The connection is altogether different from that postulated by Hinrichs, however.

42. Clearly, the correspondence with the astronomical distances is only approximate.

43. Isaac Newton is credited with first performing a similar experiment with sunlight, which he dispersed into its component colors, also by means of a glass prism.

44. It is said that Bunsen never once referred to the work of his former students Mendeleev and Lothar Meyer, either in writings or in lectures. This was in spite of the fact that both of these former students acquired considerable fame for their respective systems of classifying the elements.

45. According to the atomic weights used by Hinrichs, calcium has a weight of 20 and barium a weight of 68.5.

46. One cannot exclude the possibility that he designed his system to fit the known facts.

47. The element didymium (Di) was included in many systems, including several of Mendeleev's tables. It eventually turned out to be a mixture of rare earth elements praseodymium and neodymium.

48. In many ways, the earlier table of Newlands, published in 1863, is more similar to that of Hinrichs in terms of groupings. The table of 1864 by Odling also shows very similar groupings to Hinrichs's spiral table.

49. Admittedly, Newlands's grouping of all these elements together makes sense in terms of secondary periodicity relationships as embodied in many short-form periodic tables. E.g., each of these elements shows a valence of 1.

50. G. Hinrichs, *The Elements of Chemistry and Mineralogy*, Davenport, Iowa; Day, Egbert & Fidlar, 1871; G. Hinrichs, *The Principles of Chemistry and Molecular Mechanics*, Davenport, Iowa, Day, Egbert & Fidlar, 1874.

51. G. Hinrichs, *The Pharmacist*, 2, 1869, 10.

52. K. Hentschel, *Why Not One More Imponderable? John William Draper's Tithonic Rays, Foundations of Chemistry*, 4, 5–59, 2002.

53. Part of the motivation for Cannizzaro's work on atomic weights lies with the earlier work of Avogadro, as mentioned above.

54. Lothar Meyer in his editorial on the papers of Cannizzaro in Oswald's *Klassiker der Exacten Wissenschaften*, vol. 30, Arbriss eines Lehrganges der theoretischen Chemie, vorgetragen von Prof. S. Cannizzaro, Leipzig, 1891.

55. Thus, the compounds in any homologous series can be defined by a formula, and the molecular weight of successive members of such a series varies by a characteristic constant value. E.g., the compounds CH_4, C_2H_6, C_3H_8, C_4H_{10}, etc., are members of the alkane homologous series, and they all conform to the general formula of C_nH_{2n+2}. The radicals of this series, $CH_3 = 15$, $C_2H_5 = 29$, $C_3H_7 = 43$, and $C_4H_9 = 57$, increase in weight in intervals of 14, as do the compounds themselves.

56. Given the subsequent discovery of atomic substructure, this view may be considered one of the instances in which Lothar Meyer's thinking was more advanced than that of Mendeleev.

57. The modern atom may be said to be composite in the sense that it consists of smaller parts such as protons, neutrons, and electrons.

58. The term "horizontal relationship" may be a little ambiguous given that some tables show chemical groups vertically and others horizontally. I am using the term here in the sense mentioned in connection with Ernst Lenssen (see chapter 2) to mean relationships between elements that are not chemically analogous, or elements with steadily increasing atomic weights. These relationships appear horizontally as periods in the modern table and, indeed, in many but not all tables of the Lothar Meyer–Mendeleev period.

59. In the modern table, one sees an initial increase from valence 1 to valence 4 followed by a decrease down to 1 again once the halogens are reached. Lothar Meyer's table differs from the modern one simply in that he chooses to begin with the modern group 14. In addition, the noble gases had not yet been discovered in 1864, and the modern group 13 had not yet been recognized as a separate group.

60. E.g., van Spronsen makes this criticism (page 126).

61. The main reason for separating these elements must also be the valency relationships among them as well as more specific chemical similarities.

62. The original sense of the term "transition metal" referred to elements such as iron, cobalt, and nickel, which represented a "transition" between successive periods in the short-form table. In the modern sense, the term denotes a transition between the s and p blocks of

the long-form tables, either medium-long or long forms. There are a total of 54 transition metals in the modern sense, up to and including element 112.

63. In his famous table of 1969, Mendeleev wrongly placed mercury with copper and silver, he misplaced lead with calcium, strontium, and barium, and he also misplaced thallium among the alkali metals. For a more detailed set of comparisons, see J. van Spronsen, *The Periodic System of the Chemical Elements, the First One Hundred Years*, Elsevier, Amsterdam, 1969, pp. 127–131. The misplacement of mercury with silver is perhaps not altogether surprising given that *hydrargyrum*, the Latin name for mercury, means "liquid silver."

Chapter 4

1. In an earlier article in *Scientific American*, I implied that Mendeleev had spent the remainder of his life in elaborating the periodic system. This view has now been corrected by Michael Gordin. See M. Gordin, *A Well Ordered Thing*, Basic Books, New York, 2004.

2. M. Gordin, *Historical Studies in the Physical Sciences*, 32, 183–196, 2001; M. Gordin, *A Well Ordered Thing*, Basic Books, New York, 2004; N.M. Brookes, Dimitrii Mendeleev's *Principles of Chemistry and the Periodic Law of the Elements,* in B. Bensaude Vincent, A. Lundgren (eds.), *Communicating Chemistry: Textbooks and Their Audiences 1789–1939*, Science History-Publications, Canton, MA, 2000, pp. 295–309.

3. Not being a reader of the Russian language, I have not been able to consult the primary literature, as have contemporary Mendeleev scholars.

4. The question of the nature of elements is mentioned in Jan van Spronsen's book *The Periodic System of the Chemical Elements, the First One Hundred Years*, Elsevier, Amsterdam, 1969. One of the few articles to examine the issue is a two-part paper by F.A. Paneth, The Epistemological Status of the Chemical Concept of Element, *British Journal for the Philosophy of Science*, 13, 1–14, 144–160, 1962. Another analysis has been given in an unpublished Ph.D. thesis written in French by Bernadette Vincent-Bensaude, *Les Pièges de l'Elémentaire*, Université de Paris, 1981 (I am grateful to the author for sending me a copy).

5. D.C. Rawson, The Process of Discovery: Mendeleev and the Periodic Law, *Annals of Science*, 31, 181–193, 1974.

6. Rawson refers to a letter from Mendeleev to his mentor, Aleksandr Voskresenskii, in which he describes how he is impressed by Cannizzaro's system, which is based on Amedeo Avagadro's hypothesis.

7. The Russian expert on the periodic table, Bonifatii Kedrov, has made this claim.

8. It may even be that Mendeleev consciously avoided mentioning his immediate precursors and competitors, although I have no evidence to support this notion.

9. For a more detailed account, see M. Gordin's recent book, *A Well Ordered Thing*, Basic Books, New York, 2004.

10. At this date, Russia was still using the Julian calendar of the Roman Empire. Most other European countries had switched to the Gregorian or reformed calendar, according to which the date would have been March 1st.

11. As mentioned in chapter 2, Kremers also did something of this kind, but it appears that the comparison was not made consciously.

12. D.I. Mendeleev, Sootnoshenie svoistv s atomnym vesom elementov, *Zhurnal Russkeo Fiziko-Khimicheskoe Obshchestvo*, 1, 60–77, 1869.

13. The German abstract of Mendeleev's famous first paper on the periodic system appeared in the *Journal für Praktische Chemie*, 1, 251, 1869, and a longer article summarizing Mendeleev's first article appeared in *Berichte der deutschen chemischen Gesellschaft*, 2, 553, 1869.

14. Mendeleev, Manuscript Table 19 (M13) dating from summer–early fall of 1870 (the numbering denotes the 13th manuscript table and the 19th table in the overall sequence of 65 tables of all forms).

15. D. Mendeleev, The Periodic Law of the Chemical Elements, *Chemical News*, 40, 243–244, November 21, 1879, quoted from p. 243. Mendeleev specifically mentions combination with oxygen and hydrogen. The quote appears in one of 18 sections of a serialization which appeared in Chemical News, 1879, 40, 231–232, 243–244, 255–256, 267–268, 279–280, 291–292, 303–304, ibid. 1880, 41, 2–3, 27–28, 39–40, 49–50, 61–62, 71–72, 83–84, 93–94, 106–108, 113–114, 125–126.

16. D.I. Mendeleev, Über die Stellung des Ceriums in System der Elemente, *Bulletin of the Academy of Imperial Science (St. Petersburg)*, 16, 45–51, 1870.

17. The basis on which Mendeleev carried out this change is analyzed in chapter 5.

18. Lothar Meyer had also moved indium in this way.

19. Mendeleev's grounds for making these changes varied considerably. E.g., his moving the element uranium from the boron group to the chromium group is discussed in detail in chapter 5.

20. See E.R. Scerri, Realism, Reduction and the Intermediate Position, in N. Bhushan, S. Rosenfeld (eds.), *Of Minds and Molecules*, Oxford University Press, 2000, pp. 51–72.

21. In using the word "metaphysical," I am following the work of Fritz Paneth on this question. Some contemporary philosophers of chemistry, e.g., Paul Needham and Robin Hendry, deny any metaphysical notion when discussing the question of basic substances.

22. The more prosaic explanation given in contemporary chemistry is that what survives of each of the elements is the number of protons, in other words, the nuclear charge of the atoms of sodium and chorine. This would also be the case in rather extreme examples, e.g., the Na^{11+} and Cl^{17+} ions. Although this response is correct, it also seems a little unsatisfactory for the identity of chemical elements to depend on the nucleus of their atoms given that all the chemical properties are supposed to be determined by the configurations and exchanges in the electrons around the nucleus.

23. This is perhaps why Mendeleev, a great defender of the nineteenth-century element scheme, was so reluctant to accept the notion of the transmutation of the elements discovered by Ernest Rutherford at the turn of the twentieth century.

24. The term refers to substance in the philosophical sense as discussed by Aristotle and then as recently as Spinoza and Kant, although each of there authors has rather different views on the question.

25. This is despite the fact that our present element scheme, which we owe to Paneth, was arrived at partly by his insistence on the distinction between abstract element and simple substance.

26. There is a certain irony here in that Mendeleev is really breaking away from Lavoisier in upholding the importance of the elements as basic substances, something that Lavoisier considered a sterile concept.

27. For a discussion of this point, see F.A. Paneth, The Epistemological Status of the Chemical Concept of Element, *British Journal for the Philosophy of Science*, 13, 1–14, 144–160, 1962. Also, the mere fact that Mendeleev and considerably later Paneth continue to maintain a dual nature for the elements attests to the fact that the chemical revolution did not eliminate the metaphysical view of elements.

28. My comments apply specifically to the first English translation of Mendeleev's book, or the fifth Russian edition.

29. The French translation lends itself more readily to making the distinction between element and simple substance, whereas in the English translation the word "element" is fre-

quently used to mean simple body. It is not surprising, therefore, that the only extensive philosophical analysis of this distinction has been made by the French philosopher-historian Bernadette Bensaude Vincent. The only other such analysis in modern times has been by Paneth, who used the German translation of Mendeleev's book, which likewise preserves the spirit of the distinction by speaking of simple body rather than using the word "element" indiscriminately.

30. D.I. Mendeleev, *The Principles of Chemistry*, 5th Russian ed., vol. 1, 1889 (1st English trans., by G. Kemensky, Collier, New York, 1891), p. 23.

31. The phrase "elements as principles" is also frequently used in the literature on the nature of elements.

32. It is worth noting that this question has taken on a rather mundane sense in today's chemistry, namely, that every element, by which one actually means simple body, is characterized by a particular atomic weight or atomic number.

33. The dates of the eight Russian editions published during Mendeleev's lifetime are as follows: 1st ed., 1868–1871; 2nd ed., 1872–1873; 3rd ed., 1877; 4th ed., 1881–1882; 5th ed., 1889; 6th ed., 1895; 7th ed., 1903; 8th ed., 1906. A further five posthumous editions have been published in Russian, with some additions to cover more recent discoveries. The translations of Mendeleev's book are as follows: 1st English trans., 1891 (of the 5th Russian ed.); 2nd English trans., 1897 (of the 6th Russian ed.); 3rd English trans., 1905 (of the 7th Russian ed.). In addition, the fifth Russian edition was translated into German in 1890 and the sixth Russian edition into French in 1895.

34. M. Kaji, Mendeleev's Conception of the Chemical Elements and the Principles of Chemistry, *Bulletin for the History of Chemistry*, 27, 4–16, 2002; M. Kaji, Mendeleev's Discovery of the Periodic Law: The Origin and the Reception, *Foundations of Chemistry*, 5, 189–214, 2003. Of course, many Russian scholars have conducted such surveys before Kaji, but I am not aware of any that have been translated into English.

35. It is interesting to note that this is the first time in his book that Mendeleev actually treats a group of elements together in the same chapter, namely, chapter 11.

36. Even on this point, there were precursors, as mentioned in chapter 2. The first person to consider what might be termed horizontal relationships was Kremers.

37. Mendeleev, 1869, quoted in translation in H.M. Leicester and H.S. Klickstein, Dmitrii Ivanovich Mendeleev, *A Sourcebook in Chemistry, 1400–1900*, Harvard University Press, Cambridge, MA, 1952, pp. 439–444, quoted from p. 442.

38. Kaji cites Henry Leicester, William Brock, and Bernadette Bensaude.

39. E.R. Scerri, J.W. Worrall, Prediction and the Periodic Table, *Studies in History and Philosophy of Science*, 32, 407–452, 2001.

40. I do not mean to imply that the weights of individual atoms can be directly observed. I intend this remark in the chemist's sense that stoichiometric reactions can be rationalized by appeal to atomic weights of participating elements.

41. As a matter of fact, the grouping together of the halogens had already been anticipated on chemical grounds in spite of their obvious visual differences. I merely cite this as an example of the unreliability, in general, of classification based on simple substances. J.H. Kultgen, one of very few philosophers to try to analyze Mendeleev's underlying assumptions, has supported the general philosophical approach I am emphasizing, in some respects. J.H. Kultgen, Philosophical Conceptions in Mendeleev's Principles of Chemistry, *Philosophy of Science*, 25, 177–183, 1958.

42. Many other chemists had already realized that chlorine, bromine, and iodine belong together in one group.

43. M. Gordin, *A Well Ordered Thing*, Basic Books, New York, 2004, p. 228.

44. D.I. Mendeleev, The Periodic Law of the Chemical Elements, *Journal of the Chemical Society*, 55, 634–658, 1889, Quoted from p. 644.

45. D.I. Mendeleev, *The Principles of Chemistry*, 3rd English translation, of 7th Russian edition, 1905, vol. 1, Longmans, London, p. 20.

46. F.A. Paneth, Chemical Elements and Primordial Matter, in H. Dingle, G.R. Martin (eds.), *Chemistry and Beyond*, Wiley, New York, 1965, pp. 53–72, quoted from pp. 56–57.

47. Of course, modern chemists are constantly referring to protons, neutrons, and electrons. I am making a more general point here that pertains more to the discovery of subnuclear structure, which can generally be ignored by the chemist.

48. G. Bachelard, *Le Pluralisme Coherent de la Chimie Modèrne*, Vrin, Paris, 1932 (translation of quotation provided by the present author).

49. For further discussion of these issues, see E.R. Scerri, Realism, Reduction and the Intermediate Position, in N. Bhushan, S. Rosenfeld, *Of Minds and Molecules*, Oxford University Press, New York, 2000, pp. 51–72.

50. E.R. Scerri, Response to Vollmer's Review of Minds and Molecules, *Philosophy of Science*, 70, 391–398, 2003.

Chapter 5

1. As previously mentioned, Mendeleev was not, in fact, the first to predict unknown elements. William Odling, John Newlands, and Julius Lothar Meyer all did so before him. E.g., Newlands left spaces for yttrium, indium, and germanium. For germanium, he predicted an atomic weight of 73, which compares very favorably with the current value of 72.59.

2. W. Brock, *The Norton History of Chemistry*, Norton, New York, 1992, pp. 324–325.

3. For general references to the prediction/accommodation debate, see chapter 2, note 73.

4. S.G. Brush, The Reception of Mendeleev's Periodic Law in America and Britain, *Isis*, 87, 595–628, 1996.

5. Posing such questions is complicated by the fact that, in some cases, the successful accommodation of an element relied on Mendeleev's correction of the atomic weights of some elements.

6. None of these comments should be taken to contradict what has already been said regarding Mendeleev putting more emphasis on the elements as unobservable basic substances.

7. These days, chemists can be associated with a single element. When I taught chemistry at Purdue University, I worked in the Brown building, named after H.C. Brown, who won a Nobel Prize in 1979 for his pioneering work in boron chemistry, which he continued to do until his death in December 2004 (at age 92). Similarly, the chair of chemistry during my days at Purdue was Richard Walton, who is the world's expert on the chemistry of rhenium. Few people, if any, are still being considered experts on the chemistry of all the elements.

8. The myths are explored and debunked in some detail in M. Gordin, *A Well-Ordered Thing*, Basic Books, New York, 2004.

9. This is not to say that chemical aspects were absent from Lothar Meyer's system. E.g., the headings for the various groups in his system consisted of different valences, as mentioned in chapter 3.

10. Not to be confused with group VIII in the modern table, the noble gases, which had not yet been discovered when Mendeleev discovered his periodic system. It was these elements, e.g., iron, cobalt, and nickel, that were termed transition metals in the sense of providing a transition between successive periods in the short-form table.

11. These elements were originally called the transition elements, whereas the term has now come to mean the elements in the central block of the medium-long form table as well as the elements placed under the table as a footnote. In modern terminology these elements from the d and f blocks respectively.

12. He may have reached this conclusion after seeing Lothar Meyer's published system of 1870, which already incorporated this improvement.

13. In fact, uranium is now known to form compounds with hydrogen, although they tend to nonstoichiometric, their formulas being written as UH_x, where x varies.

14. As is the case of Mendeleev's claimed interpolations to obtain the atomic weights of gallium, scandium, and germanium, his predicted values were not exactly as prescribed in his explanations.

15. Unlike in chapter 2, where the original version was given, relative to the O = 1 standard, the form given here is the more common one based on O = 16, hence, the difference in the constant cited in each chapter.

16. L.F. Nilson, O. Petterson, *Berichte*, 11, 381–386, 1878.

17. T.S. Humpidge, On the Atomic Weight of Glucinum (Beryllium) [Abstract], *Proceedings of the Royal Society*, 38, 188–191, 1884; Humpidge, On the Atomic Weight of Glucinum (Beryllium) Second Paper, *Proceedings of the Royal Society*, 39, 1–19, 1885.

18. More generally, the change in valence across the second and third periods shows a smooth increase from 1 to 4, followed by an equally uniform decrease down to 0 for the noble gases.

19. In 1946, Seaborg discovered that a major change was needed in the periodic table. Several elements that had been regarded as belonging to a fourth transition metal series were separated out from the main body of the periodic table to form the lanthanide and actinide series. Uranium is among these elements and is no longer regarded as a transition metal.

20. The prefix "eka-" is Sanskrit for the numeral one. Mendeleev also used "dvi-," or two, on some occasions to describe a second element that was like a particular known element, or a double of it.

21. Once again, this irregularity occurs because many elements consist of mixtures of various isotopes, and their atomic weights are thus averages of their values.

22. D.I. Mendeleev, Über die Beziehungen der Eigenschaften zu den Atomgewichten der Elemente, *Zeitschrift für Chemie*, 12, 405–406, 1869, quoted from p. 406.

23. Ibid., p. 42, note 16.

24. Strictly speaking, these provisional names were coined in the following year of 1871.

25. I thank Michael Gordin for this information. The article is D.I. Mendeleev, Die periodische Gesetzmässigkeit der chemischen Elements, *Annalen der Chemie und Pharmacie*, supplement 8, 133–229, 1872.

26. The first person to present such tables of comparison for Mendeleev's predictions was Per-Teodor Clève.

27. P. E. Lecoq De Boisbaudran, Charactères Chimiques et Spectroscopiques d'un Nouveau Métal, le Gallium, Decouvert Dans un Blende de la Mine de Pierrefitte, Valée d'Argèles (Pyrénées), *Comptes Rendus de l'Academies des Sciences, Paris*, 81, 493–495, 1875.

28. De Boisbaudran originally stated that the density of the element was 4.7 g/cm^3. Mendeleev had predicted a value of 5.9 g/cm^3 and wrote to De Boisbaudran asking him to redetermine the value, which he did, obtaining a value close to the present value of 5.904 g/cm^3.

29. Others maintain that he was really naming the element after himself in view of "Lecoq" contained in his own name. The Latin for cock is *gallus*.

30. D.I. Mendeleev, Remarques à Propos de la Découverte du Gallium, *Comptes Rendus de l'Academies des Sciences, Paris*, 81, 969–972, 1876.

31. D. Mendeleev, La Loi Périodique des Éléments Chimiques, *Moniteure Scientifique*, 21, 691–735, quoted from p. 692.

32. D.I. Mendeleev, The Periodic Law of the Chemical Elements, *Journal of the Chemical Society*, 55, 634–656, 1889.

33. The optical ether was a medium that had been invoked as the carrier of the electromagnetic force, although it had never been detected experimentally. Nevertheless, this concept served a mathematical purpose in the theory of electromagnetism until it was conclusively demolished by Albert Einstein's 1905 theory of special relativity.

34. Since the present book is primarily about the development of the periodic system and not about Mendeleev, very little has been said on the latter's view on the ether. Mendeleev's interest on this issue began early on in his career and was the motivation for his work on deviations from the gas laws carried out in the 1870s. Although he failed to uncover any evidence for the ether, he maintained a theoretical interest in it, which formed the basis for his speculations aimed at counteracting the discoveries of radioactivity and transmutation, which Mendeleev remained opposed to almost until his death in 1907. More information on Mendeleev and the ether can be found in Gordin's book, *A Well-Ordered Thing*, Basic Books, New York, 2004.

35. D.C. Rawson, The Process of Discovery: Mendeleef and the Periodic Law, *Annals of Science*, 31, 181–204, 1974.

36. D.I. Mendeleev, *Perioicheskii Zakon. Osnovye Stat'I, Compilation and Commentary of articles on the Periodic Law*, by B.M. Kedrov, Klassiki Nauk, Soyuz sovietskikh sotsial'sticheskikh respublik, Leningrad, 1958, p. 316, note 16 (emphasis added).

37. For fuller details of Mendeleev on magnesium and beryllium, see J.R. Smith, *Persistence and Periodicity*, unpublished Ph.D. thesis, University of London, 1975.

38. J. van Spronsen, *The Periodic System of the Chemical Elements, the First One Hundred Years*, Elsevier, Amsterdam, 1969, p. 215.

39. Although it must be acknowledged that the dramatic successes came first.

40. A similar table, although containing far fewer entries and only Mendeleev's later predictions, is given by W. Brock, *The Norton History of Chemistry*, Norton, New York, 1992, p. 325. It is interesting to also note that this author remarks, "[H]is predictions of eka-manganese, tri-manganese, dvi-tellurium, dvi-caesium and eka-tantalum were fortuitous guesses rather than predictions based upon a firm and accurate placing of their homologues in the table."

41. Entry under a section entitled Chemical Notices from Foreign Sources, *Chemical News*, 23, 1871, p. 252, mentioning Mendeleev's paper published in German in *Berichte der Deutschen Chemischen Gesellschaft zu Berlin*, 6, 1871.

42. P. Lipton, Prediction and Prejudice, *International Studies in Philosophy of Science*, 4, 51–60, 1990.

43. P. Lipton, *Inference to the Best Explanation*, Routledge, London, 1991.

44. See P. Lipton, Prediction and Prejudice, *International Studies in Philosophy of Science*, 4, 51–60, 1990.

45. Contemporary historians have begun to redress this imbalance. For example, S.J. Brush, The Reception of Mendeleev's Periodic Law in America and Britain, *Isis*, 87, 595–628, 1996.

46. W. Spottiswode, "President's Address," *Proceedings of Royal Society of London*, 34, 303–329, 1883, quoted from p. 392.

47. On this issue, unlike the alleged time lag, as I have called it, Maher cannot be exonerated on the basis of any mistake reported by his cited source on the history of chemistry, namely, the book by Ihde, since the latter does not mention the award of the Davy Medal. A.J. Ihde, The Development of Modern Chemistry, Dover Publications, New York, 1984, as cited in chapter 6, note 37.

48. P.T. Clève, Sur le Scandium, *Comptes Rendus des Seances de l'Académie des Sciences*, 89, 419–422, 1879, quoted from p. 421.

49. Editorial by W. Crookes, *Chemical News*, 1879.

50. D.I. Mendeleev, On the History of the Periodic Law, *Chemical News*, 43, 15, 1881.

51. Nevertheless, I note that Gordin argues that Mendeleev regarded predictions very seriously. M. Gordin, *A Well-Ordered Thing*, Basic Books, New York, 2004.

52. A. Wurtz, *The Atomic Theory*, translated by E. Cleminshaw, Appleton, New York, 1881.

53. This latter turned out to be another case of "pair reversal."

54. A. Wurtz, *The Atomic Theory*, translated by E. Cleminshaw, Appleton, New York, 1881, p. 16.

55. Marcellin Berthelot, *Les Origines de L'Alchemie*, Steinheil, Paris, 1885, p. 311.

56. D. Mendeleev, The Periodic Law of the Chemical Elements, *Journal of the Chemical Society*, 55, 634–656, 1889. This is Mendeleev's Faraday lecture. See the comments on p. 644.

57. W. Brock, *The Norton History of Chemistry*, Norton, New York, 1992, p. 324–325.

58. But to be more accurate, at least two authors, William Sedgwick and Jörgen Thomsen, had independently predicted the possibility of a group of completely unreactive elements. W. Sedgwick, The Existence of an Atom Without Valency of the Atomic Weight of "Argon" Anticipated Before the Discovery of "Argon" by Lord Rayleigh and Prof. Ramsay, *Chemical News*, 71, 139–140, 1895; J. Thomsen, *Anorganische Chemie*, 9, 282, 1895.

59. Recall that at the time the argon problem arose only one pair reversal had yet come to light, that of iodine and tellurium, and that its explanation would not be obtained until nearly 20 years later with the discovery of isotopes and the work of Moseley.

60. W. Ramsay and Lord Rayleigh, Argon a New Constituent of the Atmosphere, *Chemical News*, 71(1836), 51–63, 1895.

61. Since rotational motion can occur only about the common center of mass of a polyatomic system, its absence is an indication that the molecules in the gas are not polyatomic. Isolated atoms are perfectly spherical, so any rotational motion they might exhibit is undetectable. Similarly, vibrational motion can exist only between any two or more atoms in a polyatomic molecule, so its absence would also accord with monoatomicity.

62. This is the same Fitzgerald who anticipated, to some extent, Einstein's special relativistic length contraction.

63. W. Ramsay and Lord Rayleigh, Argon a New Constituent of the Atmosphere, *Chemical News*, 71, 51–63, 1895, quoted from p. 62.

64. Lord Kelvin, *Chemical News*, Argon a New Constituent of the Atmosphere *Chemical News*, 71, 51–63, 1895, quoted from p. 63.

65. Professor Mendeleev on Argon, (Report of the Russian Chemical Society Meeting), March 14, 1895, *Nature*, 51, 543, 1895.

66. Professor Mendeleev on Argon, (Report of the Russian Chemical Society Meeting), March 14, 1895, *Nature*, 51, 543, 1895.

67. As cited by J.R. Smith, *Persistence and Periodicity*, unpublished Ph.D. thesis, University of London, 1975, p. 456.

68. It has since been discovered, in work beginning in the 1960s, that the noble gases do in fact form stable compounds, with the exception of helium and neon, which still appear to be completely unreactive.

69. It should be remembered that even Newlands had anticipated this possibility, which would in no way destroy the periodicity of the remaining elements in the table.

70. As cited by J.R. Smith, *Persistence and Periodicity*, unpublished Ph.D. thesis, University of London, 1975, p. 460. Also see D. Mendeleev, *An Attempt Towards a Chemical Conception of the Ether*. This statement appears in the Russian edition of 1902 as a footnote.

71. S.J. Brush, Prediction and Theory Evaluation, *Science*, 246, 1124–1129, 1989.

Chapter 6

1. An interesting semipopular book on the life and work of Boltzmann is D. Lindley, *Boltzmann's Atom*, Free Press, New York, 2001.

2. Whether or not, or to what extent, Thomson discovered the electron has been the focus of much historical research. See various articles in J. Buchwald, A. Warwick (eds.), *Histories of the Electron: The Birth of Microphysics*, MIT Press, Cambridge, MA, 2001.

3. This is not exactly case. Isotopes of hydrogen, e.g., give rise to compounds that do show chemical differences. Nevertheless, for most purposes, chemical differences between the isotopes of an element may be taken to be insignificant.

4. It is by no means clear that Becquerel was the first to discover radioactivity, contrary to the most accounts and, indeed, the one given here. See T. Rothman, *Everything's Relative*, Wiley, Hoboken, NJ, 2003, pp. 46–52. Rothman makes a very good case for the prior discovery by Abel Niepce de Saint-Victor, who was the brother of Joseph-Nicéphore Niepce, who made the first ever photographic image.

5. The story of her early education, which has been told many times, is truly heroic, especially given the difficulties experienced by women wishing to study in universities at the turn of the nineteenth century. Curie was forced to go to Paris because Polish universities simply did not admit women at that time. After working for about six years as a governess and teacher she had saved enough money to undertake a trip to Paris to enroll at the Sorbonne in 1891. While living under very meager conditions she began by attending physics lectures and succeeded in graduating first in her class only two years later. She was immediately taken on to do some research on the magnetic properties of steels in the nearby laboratory of Pierre Curie, who was already a prominent French physicist. Eventually they would marry. During this period she also undertook another undergraduate degree in mathematics, finishing second in her class. She registered for a doctoral degree, which would be the first such degree awarded to a woman anywhere in Europe. The research she did for this degree would win her the first of her two Nobel Prizes. There are a number of detailed historical studies of Madame Curie, including S. Quinn, *Marie Curie, A Life*, Simon & Schuster, New York, 1995.

6. B. Bensaude, I. Stengers, *A History of Chemistry*, Harvard University Press, Cambridge, MA, 1996, quoted from p. 227.

7. In any case, the same authors are surely mistaken when they write, "In each place in Mendeleev's table there was no longer just an element, but a certain number of distinct atoms, all having the same chemical properties, but distinguished by their atomic weights

and the instability of their nuclei. . . ." B. Bensaude, I. Stengers, *A History of Chemistry*, Harvard University Press, Cambridge, MA, 1996, p. 230.

8. J. Perrin, Le Mouvement Brownien de Rotation, *Comptes Rendus*, 149, 549–551, 1909; H. Nagaoka, Motion of Particles in an Ideal Atom Illustrating the Line and Band Spectra and the Phenomena of Radioactivity, *Bulletin of the Mathematical and Physical Society of Tokyo*, 2, 140–141, 1904.

9. C.G. Barkla, Note on the Energy of Scattered X-radiation, *Philosophical Magazine*, 21, 648–652, 1911. This relationship held true for elements with atomic weights equal to or less than 32 ($A \leq 32$). In terms of atomic number, this is equivalent to the first 16 elements, or hydrogen to sulfur. It should also be noted that by this time it was understood that X-rays were produced by electrons.

10. A.J. van den Broek, The α Particle and the Periodic System of the Elements, *Annalen der Physik*, 23, 199–203, 1907.

11. In fact, the α particle is a helium atom that has been stripped of both of its orbiting electrons. It has a mass of 4 and a charge of +2.

12. A.J. van den Broek, Das Mendelejeffsche 'kubische" periodische System der Elemente und die Einordnuung der Radioelemente in dieses System, *Physikalische Zeitschrift*, 12, 490–497, 1911.

13. A.J. van den Broek, The Number of Possible Elements and Mendeléeff's "Cubic" Periodic System, *Nature*, 87, 78, 1911.

14. Van den Broek did not make this connection explicit, with the result that most writers on the periodic table and in history of science generally have failed to notice it. They merely state that van den Broek drew on the work of Rutherford and Barkla and went on to hint at the concept of atomic number. The point is that he had prior grounds for latching onto the work of Rutherford and Barkla.

15. A. Pais, *Inward Bound*, Oxford University Press, New York, 1986, p. 227.

16. A. van den Broek, *Physikalische Zeitschrift*, 14, 32–41, 1913.

17. N. Bohr, On the Constitution of Atoms and Molecules, *Philosophical Magazine*, 26, 1–25, 476–502, 857–875, 1913 (known as the trilogy paper). Van den Broek is cited on p. 14.

18. A.J. Van den Broek, Intra-atomic Charge, *Nature*, 92, 372–373, 1913.

19. E. Rutherford, The Structure of the Atom, *Nature*, 92, 423, 1913.

20. Many accounts incorrectly state or imply that Moseley achieved this feat himself.

21. Moseley's ancestors had all been prominent scientists. His father was a professor of comparative anatomy at Oxford. His grandfather, also called Henry, had been a famous mathematician and physicist at King's College, London. Henry Moseley the younger followed a rather typical aristocratic academic career in attending the public school Eton and going on to undergraduate studies at Trinity College Oxford.

Moseley's appetite for hard work is shown by the following anecdote: Charles Darwin the younger, the grandson of Charles Darwin of evolution fame and a good friend of Moseley's in Manchester, was later quoted as saying that one of Moseley's many talents was the knowledge of where one could find a meal at three o'clock in the morning in the streets of Manchester.

22. Fajans was visiting Rutherford's lab from Heidelberg at the time.

23. Strictly speaking, the planes in such substances as sodium chloride consist of ions and not atoms.

24. Barkla had actually distinguished two types of characteristic emissions, one more penetrating than the other. These he called K and L series, respectively.

25. As discussed in chapter 7, Bohr was to take the physical explanation of the periodic system to new levels when he began to use quantum theory to write electronic configurations for atoms and to relate these to the periodic table.

26. Moseley conducted the final stages of these experiments in Oxford in the laboratory of his former undergraduate professor John Sealy Townsend, who was able to provide him with space.

27. Three elements were missing between the first and last in the sequence that Moseley examined: phosphorus, sulfur, and scandium.

28. H.G.J. Moseley, Atomic Models and X-Ray Spectra, *Nature*, 92, 554, 1913.

29. Eka-manganese was eventually discovered and named technetium.

30. For further accounts of the discovery of elements and those that turned out to be spurious, see V. Karpenko, The Discovery of Supposed new Elements, *Ambix*, 27, 77–102 1980; and E. Rancke-Madsen, The Discovery of an Element, *Centaurus*, 19, 299–313, 1976.

31. As cited in B. Jaffe, *Crucibles: The Story of Chemistry from Ancient Alchemy to Nuclear Fission*, Simon & Schuster, New York, 1948.

32. As emphasized further below, Moseley himself did not conclude that seven gaps remained. In fact, his first estimate, based on the available evidence, was that there were just three.

33. Hahn would go on to discover the enormously important process of nuclear fission, along with colleagues Hans Strassman and Lise Meitner, thus paving the way to the development of atomic weapons and the later peaceful use of radioactivity in the generation of nuclear power.

34. According to M.E. Weeks and H.M. Leicester, *Discovery of the Elements* (7th ed., Journal of Chemical Education, Easton, PA, 1968), perhaps the most authoritative book on the discovery of the elements, protactinium was also independently discovered by Fajans and Soddy and independently by John Arnold Cranston and Alexander Fleck in the same year.

35. E.R. Scerri, Prediction of the Nature of Hafnium from Chemistry, Bohr's Theory and Quantum Theory, *Annals of Science*, 51, 137–150, 1994.

36. P.H.M. van Assche, The Ignored Discovery of Element Z = 43, *Nuclear Physics*, A480, 205–214, 1988.

37. Several earlier claims for the detection of naturally occurring element 61 were eventually refuted. Among the names in the course of the early claims were illinium (after Illinois), florentium (after Florence, Italy), and cyclonium (after the use of the cyclotron accelerator). Even the last claim was incorrect, although the eventual discovery was indeed made in a particle accelerator experiment.

38. W. Crookes, Address to the Chemical Section of the British Association, *Chemical News*, 56, 115–126, 1886.

39. A complete list of these radio-elements, including their eventual classification as isotopes of existing elements, can be found in the appendices of A.J. Ihde, *The Development of Modern Chemistry*, Dover Publications, New York, 1984.

40. J.W. Van Spronsen, *The Periodic System of the Chemical Elements, the First One Hundred Years*, Elsevier, Amsterdam, 1969, p. 309.

41. The symbol X was used in the mathematical sense of "unknown" since it was not known whether in fact these were new elements.

42. H.N. McCoy, W.H. Ross, The Specific Radioactivity of Thorium and the Variation of the Activity with Chemical Treatment, *Journal of the American Chemical Society*, 29, 1709–1718, 1907, quoted from p. 1711.

43. F. Soddy, *Annual Reports to the London Chemical Society* 285, 1910.

44. According to Alexander Fleck, a former student of Soddy's, the term "isotope" was suggested to Soddy by a family friend, Margaret Todd.

45. F. Soddy, Intra-atomic Charge, *Nature*, 92, 399–400, 1913.

46. This explanation is somewhat ahistorical in the use of mass numbers to characterize particular isotopes, e.g., uranium-235 or thorium-231.

47. To anticipate our current knowledge, chemical properties are governed by the number of electrons in an atom and not by its atomic weight. Two or more isotopes of the same element can therefore differ in mass, while having the same atomic number (number of protons) as well as the same number of electrons. The different weights that two or more isotopes of the same element display are due to their having different numbers of neutrons while sharing exactly the same number of protons. In approximate terms, the weight of an atom is given by the sum of the protons, neutrons, and electrons. The neutron was not actually discovered until 1930.

48. I. Lakatos, *The Methodology of Scientific Research Programmes*, edited by J. Worrall, G. Currie, Cambridge University Press, Cambridge, 1978. The pages dealing with Prout's hypothesis are 43, 53–55, 118–119, and 223.

Chapter 7

1. As mentioned in chapter 6, the notion that Thomson alone discovered the electron is hotly debated among historians of science.

2. M. Chayut, J.J Thomson, The Discovery of the Electron and the Chemists, *Annals of Science*, 48, 527–544, 1991.

3. There has been debate in the literature regarding the extent to which chemical or physical atomism was supported by various developments starting from John Dalton's theory. See, e.g., A. Chalmers, forthcoming.

4. As cited in A. Pais, *Inward Bound*, Clarendon Press, Oxford, 1986, p. 82.

5. Contrary to Thomson's original finding of a charge-to-mass ratio of 770 for cathode rays emanating from hydrogen ions, the electron was found to have a mass of 1,836 times less than that of the hydrogen atom.

6. J. Perrin, Les Hypothèses Moléculaires, *Revue Scientifique*, 15, 449–461, 1901.

7. Ibid., quoted from p. 460.

8. The concept of electron orbitals, or the earlier notion of electron orbits, began with Niels Bohr's theory of the atom, which is examined further below. Electronic orbitals have become perhaps the most important concept in the modern explanation of the periodic system, as discussed in chapters 8–10.

9. H. Nagaoka, Motion of Particles in an Ideal Atom Illustrating the Line and Band Spectra and the Phenomena of Radioactivity, *Bulletin of the Mathematics and Physics Society of Tokyo*, 2, 140–141, 1904.

10. H. Nagaoka, Kinetics of a System of Particles Illustrating the Line and the Band Spectrum and the Phenomena of Radioactivity, *Philosophical Magazine*, 7, 445–455, 1905.

11. J.J. Thomson, On the Structure of the Atom: an Investigation of the Stability and Periods of Oscillation of a number of Corpuscles arranged at equal intervals around the Circumference of a Circle; with Application of the Results to the Theory of Atomic Structure, *Philosophical Magazine*, 7, 237–265, 1904.

12. A.M. Meyer, A Note on Experiments With Floating Magnets, *American Journal of Physics*, 15, 276–277, 1878.

13. Rutherford described Thomson's plum pudding model as being like old lumber fit only for a museum of scientific curiosities. Source unknown to present author.

14. Optical spectra result from outer or valence electrons and should not be confused with the spectra obtained by Moseley using an X-ray source rather than visible light. X-ray spectra involve the excitation of inner electrons.

15. Another underrated contributor to this project was the British physicist John Nicholson. See T. Rothman, *Everything's Relative*, Wiley, Hoboken, NJ, 2003.

16. Experiments on incandescent objects inside a perfectly absorbing cavity produced a set of spectral distributions depending on the temperature of the heated object and the wavelength of light emitted by the object. The classical theory available to describe these spectral distributions appeared to be applicable only at high wavelengths; at lower wavelengths the theory predicted infinite emission intensities for the same heated objects, which did not agree with the experimental facts. Planck succeeded in explaining the experimental curves by assuming that, contrary to previous thinking, the energy of particles in the heated object, and consequently the energy emitted in such experiments, is not continuous but in the form of discrete units.

17. Strictly speaking, Planck's work revealed the quantization of "action," that is to say, energy divided by frequency. This quantity is now of historical interest only, and it is more common to refer to the quantization of energy, which is given by the action of a particular system multiplied by its frequency.

18. Conversely, electrons could undergo transitions to less stable orbits following the absorption of specific quanta of energy.

19. As many authors note, the quantization of angular momentum assumed by Bohr as well as the notion that electrons in stationary states do not radiate was somewhat ad hoc and only justified later by Erwin Schrödinger's approach to calculating the energy of the hydrogen atom.

20. N. Bohr, On the Constitution of Atoms and Molecules, Part III. Systems Containing Several Nuceli, *Philosophical Magazine*, 26, 857–875, 1913. Quoted from p. 874

20. Ibid, quoted from p. 875.

22. Bohr's atomic theory also provided an approximate explanation for the spectra of alkali metals, which have one unpaired outer-shell electron.

23. J.L. Heilbron, T.S. Kuhn, The Genesis of the Bohr Atom, *Historical Studies in the Physical Sciences*, 1, 211–290, 1969.

24. N. Bohr, On the Constitution of Atoms and Molecules, *Philosophical Magazine*, 26, 476–502, 1913 (table on p. 497).

25. E.R. Scerri, Prediction of the Nature of Hafnium from Chemistry, Bohr's Theory and Quantum Theory, *Annals of Science*, 51, 137–150, 1994.

26. For discussions of how Bohr argued as a chemist, see H. Kragh, Chemical Aspects of Bohr's 1913 Theory, *Journal of Chemical Education*, 54, 208–210, 1977.

27. This statement is a simplification and is only correct for the main-group, or representative, elements in the periodic table. In the case of the transition elements, the members of a group of elements have the same number of electrons in the same penultimate shell. In the rare earths, the elements in the same group have the same number of electrons in a shell located two shells from the outer shell. And there are further deviations given that about 20 elements have "anomalous" configurations, as discussed in chapter 9.

28. I. Langmuir, Arrangement of Electrons in Atoms and Molecules, *Journal of the American Chemical Society*, 41, 868–934, 1919; C.R. Bury, Langmuir's Theory of the Arrangement of Electrons in Atoms and Molecules, *Journal of the American Chemical Society*, 43, 1602–1609, 1921.

29. N. Bohr, Über die Anwendung der Quantumtheorie auf den Atombau. I. Die Grundpostulate der Quantumtheorie, *Zeitschrift für Physik*, 13, 117–165, 1923, English trans-

lation in *Collected Papers of Niels Bohr*, edited by J. Rud Nielsen, vol. 3, North-Holland Publishing, Amsterdam, 1981.

30. Quoted in H. Kragh, The Theory of the Periodic System, in A.P. French, P.J. Kennedy (eds.), *Niels Bohr, A Centenary Volume*, Harvard University Press, Cambridge, MA, 1985, 50–67, quote from p. 61.

31. Quoted in J. Mehra, H. Rechenberg, *The Discovery of Quantum Mechanics, 1925*, vol. 2 of *Historical Development of Quantum Theory*, Springer-Verlag, New York, 1982.

32. Quoted in Victor F. Weisskopf, *The Privilege of Being a Physicist*, W.H. Freeman, New York, 1970. quoted from p. 164.

33. N. Bohr, Uber die Anwendung der Quantumtherie auf den Atombau I, *Zeitschrift für Physik*, 13, 117–165, 1923.

34. P. Ehrenfest, Adiabatic Invariants and the Theory of Quanta, *Philosophical Magazine*, 33, 500–513, 1917, quoted from p. 501.

35. The term "adiabatic" has a different sense in thermodynamics than it does in quantum mechanics. In thermodynamics, it refers to a change carried out very quickly so that the system in question does not undergo any heat change. In quantum mechanics, an adiabatic change must be gradual so that the quantum states of the system are maintained following the change.

36. P. Ehrenfest, Adiabatic Invariants and the Theory of Quanta, *Philosophical Magazine*, 33, 500–513, 1917.

37. J.M. Burgers, Adiabatic Invariants of of Mechanical Systems, *Philosophical Magazine*, 33, 514–520, 1917.

38. H. Goldstein, *Classical Mechanics*, 2nd ed., Addison-Wesley, Reading, MA, 1980.

39. N. Bohr, Über die Anwendung der Quantumtheorie auf den Atombau I, *Zeitschrift für Physik*, 13, 117–165, 1923, quoted from p.129.

40. Ibid. Quoted from p. 130.

41. J. Honner, Niels Bohr and the Mysticism of Nature, *Zygon*, 17, 243–253, 1982.

42. A. Landé, Termstruktur und Zeemaneffekt der Multipletts, *Zeitschrift für Physik*, 15, 189–205, 1923.

43. A. Sommerfedl, *Wave Mechanics*, E.P. Dutton & Co., New York, 1930.

44. E. Stoner, The Distribution of Electrons Among Atomic Levels, *Philosophical Magazine*, 48, 719–736, 1924.

45. In modern notation, the third quantum number is labeled m_ℓ and the second quantum number is ℓ. Thus, if $\ell = 2$, m_ℓ can take values of $-2, -1, 0, +1$, and $+2$. The second or ℓ quantum number, in turn, is related to the first quantum number n as $\ell = n - 1, \ldots 0$. E.g., if $n = 3$, m_ℓ can assume values of 2, 1, or 0.

46. For a philosophical discussion of the nature of atomic orbitals, see E.R. Scerri, *British Journal for the Philosophy of Science*, 42, 309–325, 1991.

47. W. Pauli, letter to N. Bohr, February 21, 1924, quoted in Bohr-Pauli Correspondence, *Collected Papers of Niels Bohr*, edited by J. Rud Nielsen, vol. 5, North-Holland Publishing, Amsterdam, 1981. Translation on p. 412–414. This quotation is all the more remarkable because, as argued below, it was Pauli's own exclusion principle, formulated a few months later, that seemed to reinstate the notion of individual electrons in stationary states. The notion of individual electrons in individual stationary states was finally refuted with the advent of quantum mechanics. Only the atom as a whole possesses stationary states. The distinction is rather important for the physics of many-electron systems.

48. Complications included the occurrence of half quantum numbers, the problem of the anomalous Zeeman effect, and the doublet riddle. P. Forman, The Doublet Riddle and Atomic Physics *circa* 1924, *Isis*, 59, 156–174, 1968.

49. W. Pauli, Uber den Einfluss der Geschwindigkeitsabhängigkeit der Elektronmasse auf den Zeemaneffekt, *Zeitschrift für Physik*, 31, 373–385, 1925.

50. A. Landé, Termstruktur und Zeemaneffekt der Multipletts, *Zeitschrift für Physik*, 15, 189–205, 1923.

51. W. Pauli, Uber den Zusammenhang des Abschlusses der Elektrongruppen im Atom mit Complexstruktur der Spektren, *Zeitschrift für Physik,* 31, 765–783, 1925, quoted from p. 765.

52. Ibid. Quoted from p. 778.

Chapter 8

1. The following is a quotation from Lewis:

> I went from the Middle-west to study at Harvard, believing that at the time it represented the highest scientific ideals. But now I very much doubt whether either the physics or the chemistry department at that time furnished real incentive to research. . . . A few years later [1902] I had very much the same ideas of atomic and molecular structure as I now hold, and I had a much greater desire to expound them, but I could not find a soul sufficiently interested to hear the theory. There was a great deal of research being done at the university, but as I see it now the spirit of research was dead.

G.N. Lewis to R. Millikan, October 28, 1919, Lewis Archive, University of California, Berkeley, as cited in R.E. Kohler, The Origin of G.N. Lewis's Theory of the Shared Pair Bond, *Historical Studies in the Physical Sciences*, 3, 343–376, 1971, which is the definitive study on Lewis.

2. In addition, Thomson had not been able to deduce the correct number of electrons even in atoms as simple as oxygen.

3. The notion of a shared pair of electrons survives, to some extent, in the quantum mechanical treatment of the atom. In the current model, a bond consists of two antiparallel electrons within the same molecular orbital.

4. G.N. Lewis, The Atom and the Molecule, *Journal of the American Chemical Society*, 38, 762–785, 1916.

5. The term "amphoteric" is now taken to mean an oxide or hydroxide that can dissolve in acids to give salts and can also dissolve in alkalis to give metallates; i.e., it can show both basic and acidic properties. Examples include the oxides and hydroxides of boron and aluminum.

6. A. Stranges, *Electrons and Valence: Development of the Theory 1900–1925*, Texas A&M University Press, College Station, 1982. p. 204

7. G.N. Lewis, The Atom and the Molecule, *Journal of the American Chemical Society*, 38, 762–785, 1916, quoted from p. 768.

8. The notion of magnetic properties in electrons did not originate with Lewis but had been previously discussed by Alfred Parsons and William Ramsey. A. Parsons, A Magneton Theory of the Atom, *Smithsonian Miscellaneous Publications*, 65, 1–80, 1915; W. Ramsey, A Hypothesis of Molecular Configuration, *Proceedings of the Royal Society A.*, 92, 451–462, 1916.

9. See, e.g., J. Servos, *Physical Chemistry from Ostwald to Pauling*, Princeton University Press, Princeton, NJ, 1990.

10. This forms the basis of numerous examples used to this day to torment the lives of chemistry students attempting to write Lewis structures for any given number of molecules.

11. As discussed in chapter 6, Moseley provided a method of determining the nuclear charge and consequently the number of electrons in any atom.

12. This is true of the octet rule, e.g., which Lewis himself was not so keen on.

13. I. Langmuir, The Arrangement of Electrons in Atoms and Molecules" in the *Journal of the American Chemical Society*, 43, 969–934, 1921, quoted from p. 868.

14. Kossel was a physicist who published a theory of ionic bonding in 1916. W. Kossel, Uber Molekulbildung als Frage des Atombaus, *Annalen der Physik*, 49, 229–362, 1916.

15. Niton is now called radon.

16. The inclusion of Niβ and other elements followed by Greek letters is somewhat mysterious and not explained in Langmuir's text. It would appear that he is attempting to avoid any gaps in the periodic table, although gaps do occur in later parts of his table.

17. A cell can contain up to two electrons and, as such, can be thought of as analogous to the modern concept of an orbital, which can also accommodate a maximum of two electrons.

18. In the modern version of electronic configurations, there are several places in the periodic table where a new shell begins to fill even though the previous shell is not yet completely full. The attempts to explain such features are resumed in chapters 9.

19. This analogy depends on identifying one of Langmuir's cells, which can hold two electrons, with the modern notion of an atomic orbital.

20. This time, only some of the closed shell atoms appear twice in the table, namely, nickel and palladium, but not the noble gases.

21. Actually, Saul Dushman, one of Langmuir's colleagues at the General Electric Company in Schenectady, New York, had been the first to publish ideas on incomplete shells in transition metal atoms at the suggestion of Langmuir himself. S. Dushman, The Structure of the Atom, *General Electric Review*, 20, 186–196, 397–411, 1917.

22. C.R. Bury, Langmuir's Theory on the Arrangement of Electrons in Atoms and Molecules, Journal of the American Chemical Society, 43, 1602–1609, 1921.

23. Modern configurations differ slightly in that the transition metals are supposed to begin one element earlier, namely at scandium which is the first element where the filling of the third shell is resumed.

24. Superscripts denote the number of electrons in each quantum level.

25. C.R. Bury, Langmuir's Theory on the Arrangement of Electrons in Atoms and Molecules, *Journal of the American Chemical Society*, 43, 1602–1609, 1921.

26. M.E. Weeks, *Discovery of the Elements*, 6th ed., Easton, PA, 1956.

27. G. Urbain, Sur un Nouvel Élément qui Accompagne le lutécium et le scandium dans les terres de la gadolinite: le celtium, *Comptes Rendus*, 152, 141–143, 1911.

28. N. Bohr, Atomic Structure, *Nature*, 107, 104–107, 1921.

29. E. Rutherford, letter to N. Bohr, September 26, 1921, in *Collected Papers of Niels Bohr*, edited by J. Rud Nielsen, vol. 4, North-Holland Publishing Company, Amsterdam, 1981.

30. P. Ehrenfest, letter to N. Bohr, September 27, 1921, in *Collected Papers of Niels Bohr*, edited by J. Rud Nielsen, vol. 4, North-Holland Publishing Company, Amsterdam, 1981.

31. H. Kramers, quoted in H. Kragh, The Theory of the Periodic System, in A.P. French, P.J. Kennedy (eds.), *Niels Bohr, A Centenary Volume*, Harvard University Press, Cambridge, MA, 1985, 50–67, quote from p. 60.

32. N. Bohr, *The Theory of Spectra and Atomic Constitution*, 2nd ed., Cambridge University Press, Cambridge, 1924, p. 110

33. Ibid., p. 114.

34. Ibid., p. 114.

35. Ibid., p. 114.

36. Lecture Six of Seven Lectures on the Theory of Atomic Structure (Göttingen, 1922), in *Collected Papers of Niels Bohr*, edited by J. Rud Nielsen, vol. 4, North-Holland Publishing Company, Amsterdam, 1981 p. 397–406, quoted from p. 404.

37. N. Bohr, *The Theory of Spectra and Atomic Constitution*, 2nd ed., Cambridge University Press, Cambridge, 1924, appendix.

38. N. Bohr, letter to J. Franck, July 15, 1922, in *Collected Papers of Niels Bohr*, edited by J. Rud Nielsen, vol. 4, North-Holland Publishing Company, Amsterdam, 1981, p. 675, translation on p. 676.

39. N. Bohr, letter to D. Coster, July 3, 1922, in *Collected Papers of Niels Bohr*, edited by J. Rud Nielsen, vol. 4, North-Holland Publishing Company, Amsterdam, 1981, as quoted in H. Kragh, Niels Bohr's Second Atomic Theory, *Historical Studies in the Physical Sciences*, 10, 123–186, 1979.

40. N. Bohr, *The Theory of Spectra and Atomic Constitution*, 2nd ed., Cambridge University Press, Cambridge, 1924, quoted from p.114.

41. D. Coster, G. von Hevesy, On the Missing Element of Atomic Number 72, *Nature*, 111, 79, 1923.

42. J.D. Main Smith, Atomic Structure, *Chemistry and Industry*, 43, 323–325, 1924. Main Smith had previously published an article citing numerous criticisms of Bohr's electronic configurations but without suggesting any alternatives himself. J.D. Main Smith, The Bohr Atom, *Chemistry and Industry*, 42, 1073–1078, 1923.

43. J.D. Main Smith, *Chemistry and Atomic Structure*, Ernest Benn Ltd., London, 1924, p. 189.

44. Ibid, p. 190.

45. Ibid., p. 192.

46. In the sense of s, p, d, and f subshells, respectively.

47. J.D. Main Smith, The Distribution of Electrons in Atoms [letter dated September 8], *Philosophical Magazine*, 50(6), 878–879, 1925, quoted from p. 878.

48. E. Stoner, *Magnetism and Atomic Structure*, Methuen & Co., London, 1925, p. 1296 footnote.

49. H. Kragh, Niels Bohr's Second Atomic Theory, *Historical Studies in the Physical Sciences*, 10, 123–186, 1979; H. Kragh, Chemical Aspects of Bohr's 1913 Theory, *Journal of Chemical Education*, 54, 208–210, 1977.

50. Mansel Davies, Charles Rugeley Bury and his Contributions to Physical Chemistry, *Archives for the History of the Exact Sciences*, 36, 75–90, 1986.

51. K.J. Laidler, *The World of Physical Chemistry*, Oxford University Press, New York, 1993.

52. The question of the reduction of chemistry to quantum mechanics has been a central issue in the renewed interest in philosophical aspects of chemistry. L. McIntyre, The Emergence of the Philosophy of Chemistry, *Foundations of Chemistry*, 1, 57–63, 1999; J. van Brakel, On the Neglect of the Philosophy of Chemistry *Foundations of Chemistry*, 1, 111–174, 1999; E.R. Scerri, L. McIntyre, The Case for the Philosophy of Chemistry, *Synthese*, 111, 305–324, 1997.

Chapter 9

1. Sir Karl Popper has claimed that Bohr's prediction of the chemical nature of hafnium was "the great moment when chemistry had been reduced to atomic physics." K.R. Pop-

per, Scientific Reduction and the Essential Incompleteness of All Science, in F.L. Ayala, T. Dobzhansky (eds.), *Studies in the Philosophy of Biology*, Berkeley University Press, Berkeley, CA, 1974, pp. 259–284.

2. A. Sommerfeld, *Atombau und Spektrallinien*, Vieweg & Sohn, Braunschweig, 1919, p. 70.

3. It is also sometimes confusingly called the exchange force, although it does not constitute a physical force.

4. Since this book is about the periodic table of the elements, rather than compounds, the quantum theory of chemical bonding is not discussed. For a historical account of developments in molecular quantum chemistry, interested readers may consult J. Servos, *Physical Chemistry from Ostwald to Pauling*, Princeton University Press, Princeton, NJ, 1990.

5. This motivation, among others, has led to the widespread view that quantum mechanics supports an antirealistic interpretation. Such a conclusion is disputed by many philosophers, including Ernan McMullin, The Case for Scientific Realism, in J. Leplin (ed.), *Scientific Realism*, University of California Press, Berkeley, CA, 1984, pp. 8–40.

6. E.R. Scerri, Have Orbitals Really Been Observed? *Journal of Chemical Education*, 77, 1492–1494, 2000; E.R. Scerri, The Recently Claimed Observation of Atomic Orbitals and Some Related Philosophical Issues, *Philosophy of Science*, 68 Suppl., S76–S88, 2001. See also S. Zumdahl, *Chemical Principles*, 5th ed., Houghton-Mifflin, Boston, 2005, pp. 679–680, and W.H.E. Schwarz, Measuring Orbitals: Reality of Provocation? *Angewandte Chemie International Edition* 45, 1508–1517, 2006.

7. E.g., if two sets of concentric waveforms collide with each other, the result is a series of augmentations and reductions of the intensity of the waves. If two waves find themselves differing by a whole number of wavelengths, they produce constructive interference, leading to an additive effect. Conversely, two waves that are out of phase, differing by half a wavelength, will cancel each other out. The net result of these two effects is a series of so-called fringes consisting of alternating additions and cancellations of waves or, in the jargon, constructive and destructive interference.

8. C.J. Davisson, L.H. Germer, The Scattering of Electrons by a Single Crystal of Nickel, *Nature*, 119, 558–560, 1927.

9. Although Schrödinger did not wait for any experimental support for the wave nature of electrons.

10. The first to prove the equivalence of matrix mechanics and wave mechanics was Schrödinger himself. E. Schrödinger, Über das Verhältnis der Heisenberg-Born-Jordanschen Quantenmechanik zu der meinen, *Annalen der Physik*, 79(4), 734–756, 1926; English translation in *Collected Papers on Wave Mechanics*, translated by J.F. Shearer, W.M. Deans, Chelsea, New York, 1984. A more elaborate proof was later given by J. von Neumann, *Mathematische Grundlagen der Quantenmechanik*, Springer, Berlin, 1932.

11. Just like the square of the square root of -1, which is the real number -1.

12. More technically, it is the integral of the square of the wavefunction over a finite volume element that is observable, or $\int \Psi \Psi \delta \tau$.

13. As described above, even this step had not been possible within the old quantum theory.

14. They did not publish their work together. First Hartree established the basis of the method, and later Fock made it relativistically invariant.

15. Not everybody agrees with this claim, however. See B. Friedrich. . . Hasn't It? A commentary on Eric Scerri's Paper, Has Quantum Mechanics Explained the Periodic Table? *Foundations of Chemistry*, 6, 117–132, 2004; V.N. Ostrovsky, What and How Physics Contributes to Understanding the Periodic Law, *Foundations of Chemistry*, 3, 145–181, 2001.

16. E.R. Scerri, Have Orbitals Really Been Observed? *Journal of Chemical Education* 77, 1492–1494, 2000.

17. More correctly, the principle is stated by saying that the wavefunction for a system of fermions is antisymmetric on the interchange of any two fermions. This version correctly avoids the assignment of quantum numbers to each individual electron in a many-electron system.

18. Löwdin has expressed his views on the $n + \ell$ rule in P.-O. Löwdin, Some Comments on the Periodic System of the Elements, *International Journal of Quantum Chemistry*, 3 Suppl., 331–334, 1969.

19. This fact is also frequently downplayed in textbook accounts of the rules for obtaining electronic configurations.

20. E.R. Scerri, The Exclusion Principle, Chemistry and Hidden Variables, *Synthese*, 102, 165–169, 1995.

21. Moving to the many-electron case involves the use of analogous quantum numbers for which the usual one-electron-atom labels are retained.

22. Using the equation for the maximum capacity of any main shell, namely, $2n^2$.

23. E.R. Scerri, Transition Metal Configurations and Limitations of the Orbital Approximation, *Journal of Chemical Education*, 66(6), 481–483, 1989; L.G. Vanquickenborne, K. Pierloot, D. Devoghel, Transition Metals and the Aufbau Principle, *Journal of Chemical Education*, 71, 469, 1994.

24. Although the 3d orbital is lower than 4s in energy, as shown in figure 9.5. The resulting total energy of the atom when a 3d orbital is preferentially occupied can still be higher than the case of preferentially occupying 4s.

25. In fact, according to the more accurate treatment, it is incorrect to assume that the energy of an orbital is a fixed quantity. In effect, the energy of both the 4s and 3d orbitals depends on the relative occupation of these two orbitals. The energy of 4s, e.g., is different depending on whether it contains no electrons or one or two. A full analysis of the problem requires the comparison of five, not just two, orbital energies.

26. The number of neutrons added is, of course, variable and accounts for the formation of different isotopes of the same element.

27. These results were obtained using the Internet web pages designed by Charlotte Froese-Fischer, one of the leading pioneers in the field of Hartree-Fock calculations. http://atoms.vuse.vanderbilt.edu/.

28. This choice is made by convention. The energy corresponding to ionization is taken to be zero. All bound states have lower energies, and so the more negative, the more stable an energy level.

29. Of course, I am talking loosely since electrons are indistinguishable according to quantum mechanics.

30. E.g., very accurate calculations on the nickel atom include the use of basis sets that extend up to 14s, 9p, and 5d as well as f orbitals. K. Raghavachari, G.W. Trucks, Highly Correlated Systems. Ionization Energies of First Row Transition Metals Sc-Zn, *Journal of Chemical Physics*, 91, 2457–2460, 1989.

31. As I argued, nickel also has a configuration of $4s^1$ and not $4s^2$, as generally stated.

32. The possession of a half-filled subshell by any atom is neither necessary nor sufficient to ensure that an s^1 configuration is adopted.

33. A theoretical analysis of Hund's rule is given in J. Katriel, R. Pauncz, *Advances in Quantum Chemistry*, 10, 143–185, 1977.

34. R.L. Snow, J.L. Bills, The Pauli Principle and Electronic Repulsion in Helium, *Journal of Chemical Education*, 51, 585–586, 1974.

35. With the exception of some recent work on so-called Universal basis sets. E. V. R. de Castro, F.E. Gorge, Accurate universal Gaussian basis set for all atoms of the Periodic Table, *Journal of Chemical Physics*, 108, 5225–5229, 1998; Some considerations about Dirac–Fock calculations, A. Canal Neto, P. R. Librelon, E. P. Muniz, F. E. Jorge, and R. Colistete Júnior, *Theochem*, 539, 11–15, 2001.

36. However, there are several arguments that can be made in favor of the placement of helium among the alkaline earths. This is carried out in the left-step periodic table, e.g., G. Katz, The Periodic Table: An Eight Period Table For The 21st Centrury, *The Chemical Educator*, 6, 324–332, 2001.

37. Although, as noted above, the configuration of nickel is actually $4s^1 3d^9$, contrary to what is stated in most textbooks. Even if one considers the total number of electrons in the two most energetic orbitals, they do not all show the same value.

38. This point is disputed by V. Ostrovsky, What and How Physics Contributes to Understanding the Periodic Law, *Foundations of Chemistry*, 3, 145–182, 2001; see p. 175.

39. Of course, there are methods used in ab initio work that even go beyond the orbital approximation altogether, but this is discussed further below.

40. This expression is due to philosopher of physics Michael Redhead; see M. Redhead, Models in Physics, *British Journal for the Philosophy of Science*, 31, 154–163, 1980. A similar point is made by V. Ostrovsky, What and How Physics Contributes to Understanding the Periodic Law, *Foundations of Chemistry*, 3, 145–182, 2001.

41. There is also redundancy of information in the ab initio approach, in view of the fact that it operates in $3N$ dimensions rather than the familiar three-dimensional space in which density functional theory operates.

42. The recent reports, starting in *Nature* magazine in September 1999, that atomic orbitals had been directly observed are incorrect. J. Zuo, M. Kim, M. O'Keefe, J. Spence, Direct observation of d-orbital holes and Cu–Cu bonding in Cu_2O, *Nature*, 401, 49–52, 1999; P. Coppens, *X-Ray Charge Densities and Chemical Bonding*, Oxford University Press, Oxford, 1997.

43. The educational implications of the claims for the observation of orbitals are addressed in other articles, and I do not dwell on the issue here.

44. This is an advantage only for a realist. The antirealist is not unduly perturbed by the fact that central scientific terms such as atomic orbitals are non referring.

45. P.M.W. Gill, Density Functional Theory (DF), Hartree-Fock (HF), and the Self-consistent Field, in P. von Ragué Schlyer (ed.), *Encyclopedia of Computational Chemistry*, vol. 1, Wiley, Chichester, 1998, pp. 678–689.

46. The promise derives from some theorems proved by P.C. Hohenberg, L.J. Sham, and Walter Kohn, Inhomogeneous Electron Gas, *Physical Review B*, 136, 864–71, 1964; W. Kohn, L.J. Sham, Self-Consistent Equations Including Exchange and Correlation Effects, *Physical Review A*, 140, 1133–38, 1965.

47. An excellent account of ab initio and density functional quantum chemistry calculations is provided in the Nobel Prize acceptance address by J. Pople, *Reviews of Modern Physics*, 71, 1267–1274, 1999.

48. Dirac's famous statement concerning the reduction of chemistry was "The underlying physical laws necessary for the mathematical theory of a large part of physics and the whole of chemistry are thus completely known, and the difficulty is only that the exact application of these laws leads to equations much too complicated to be soluble. P.A.M. Dirac, Quantum Mechanics of Many-Electron Systems, *Proceedings of the Royal Society of London, Series A*, 123, 714–733, 1929, quoted from p. 714.

49. Other universal approaches to the problem of the periodic table have been pursued by Dudley Herschbach and colleagues; see S. Kais, S.M. Sung, D.R. Herschbach, Large-Z

and -N Dependence of Atomic Energies of the Large-Dimension Limit. *International Journal of Quantum Chemistry*, 49, 657–674, 1994.

50. As mentioned in note 18, this problem has been recognized by some leading quantum chemists, such as Löwdin. Several attempts to solve the problem have been published. See, e.g., V. Ostrovsky, What and How Physics Contributes to Understanding the Periodic Law, *Foundations of Chemistry*, 3, 145–182, 2001. Readers may also be interested in the present author's chapter in a book based on a recent international conference on the Periodic Table: E.R. Scerri, The Best Representation of the Periodic System: The Role of the n + ℓ Rule and the Concept of an Element as a Basic Substance, in D. Rouvray, R.B. King (eds.), *The Periodic Table: Into the 21st Century*, Science Studies Press, Bristol, 2004, 143–160.

51. J. Dupré, *Human Nature and the Limits of Science*, Clarendon Press, Oxford, 2001; J. van Brakel, *The Philosophy of Chemistry*, Leuven University Press, Leuven, Belgium, 2000 (see chapter 5 in particular).

52. B. Bensaude, I. Stengers, *A History of Chemistry*, Harvard University Press, Cambridge, MA, 1996 (see chapter 5 in particular); D. Knight, *Ideas in Chemistry*, Rutgers University Press, New Brunswick, NJ, 1992 (see chapter 12).

Chapter 10

1. The reader is referred to Helge Kragh, *Cosmology and Controversy*, Princeton University Press, Princeton, NJ, 1996, an excellent historical account of cosmology from which I have drawn liberally in the course of writing this section. Other good sources on nucleosynthesis are E.B. Norman, Stellar Alchemy: The Origin of the Chemical Elements, *Journal of Chemical Education*, 71, 813–820, 1994; P.A. Cox, *The Elements*, Oxford University Press, Oxford, 1989; and S.F. Mason, *Chemical Evolution*, Clarendon Press, Oxford, 1991.

2. One of the triumphs of the big bang theory is the successful prediction of the relative abundance of the two main isotopes of hydrogen: protium and deuterium. See chapter 20 of J.S. Rigden, *Hydrogen, The Essential Element*, Harvard University Press, Cambridge, MA, 2002.

3. Herman Bondi, one of the three founders of the steady-state theory, admitted defeat after learning of the hydrogen:helium ratio, which was interpreted as a "fossil" of the big bang. This is contrary to the popular story that it was the observation of the 3K cosmic background radiation in the 1960s that caused the steady-state theorists to throw in the towel. I am grateful to George Gale for pointing this out to me. He learned this through a series of interviews which he carried out with Bondi. Not all cosmologists have given up the steady-state theory, however. The husband and wife team of Geoffrey and Margaret Burbidge continue to support that theory, as reported in a recent article. R. Panek, Two Against the Big Bang, *Discover*, 26, 48–53, 2005.

4. This represents an example of simultaneous discovery, as thallium was independently isolated by C.A. Lamy working in France in the same year.

5. W. Crookes, The Genesis of the Elements, *Chemical News,* 55, 83–99, 1887; quoted from p. 83.

6. For a fuller account of Crooke's periodic system, see S.F. Mason, *Chemical Evolution*, Clarendon Press, Oxford, 1991. This is also an excellent source for the history of nucleosynthesis and the study of the origin of life.

7. D.I. Mendeleev, The Periodic Law of the Chemical Elements, *Journal of the Chemical Society*, 55, 634–656, 1889 (Faraday lecture), quoted from p. 641.

8. Whether Crookes can rightly be said to have been a chemist is debatable given his numerous interests, including spectroscopy, physics, and paranormal phenomena. The fact remains that he was initially trained as a chemist and retained a strong interest in chemistry throughout his life, including acting as the editor of *Chemical News* from when he founded it in 1859 up to the time of his death in 1919. See W. Brock, William Crookes, in C. Gillespie (ed.), *Dictionary of Scientific Biography*, vol. 3, Charles Scribner's, New York, 1981, pp. 474–482. Brock is currently writing a scientific biography of the life of Crookes.

9. The heat death had become widely accepted on the basis of the second law of thermodynamics and the associated increase in entropy of the universe.

10. A. Eddington, The Internal Constitution of the Stars, *Nature*, 106, 14–20, 1920.

11. Lemaître was more careful and went to some length to separate his scientific beliefs from his religious ones. J.D. North, Cosmology, Creation, and the Force of History, *Interdisciplinary Science Reviews*, 25, 261–266, 2000.

12. A further early contribution to the big bang theory was provided by the work of A. Friedmann, Über die Krummung des Raumes, *Zeitschrift für Physik*, 10, 377–386, 1922.

13. Bethe had not really participated in this research, but Gamov asked him to join the authors because the names of Alpher, Bethe, and Gamov would make for a nice prank. The paper has indeed become rather famous as the $\alpha\beta\gamma$ paper. R.A. Alpher, H. Bethe, G. Gamov, The Origin of the Chemical Elements, Physical review, 73, 803–804, 1948.

14. Two interesting biographies of the life of Hoyle have recently been published: Simon Mitton, *Conflict in the Cosmos: Fred Hoyle's Life in Science*, Joseph Henry, Washington D.C., NJ, 2005; Jane Gregory, *Fred Hoyle's Universe*, Oxford University Press, Oxford, 2005.

15. More recently, some direct evidence for hydrogen burning in the sun has become available through the study of solar neutrinos. K.S. Hirata et al., Observation of Neutrino Burst from the Supernova SN1987A, *Physical Review D*, 44, 2241–2260, 1991.

16. F. Hoyle, The Synthesis of the Elements from Hydrogen, *Monthly Notices of the Royal Astronomical Society*, 106, 343–383, 1946.

17. The term "black hole" was coined some time later by the physicist J.A. Wheeler.

18. F. Hoyle, D.N.F. Dunbar, W.A. Wenzel, W. Whaling, A State in C^{12} Predicted from Astrophysical Evidence, *Physical Review*, 92, 1095, 1953. This much-cited paper was in fact a brief announcement made at a conference and took up a mere 16 lines in a page arranged in two columns, the equivalent therefore of just eight lines of text. Hoyle's prediction is widely regarded as the only successful application of the anthropic principle, the notion that nature is the way that it is because this allows us to exist. Hoyle had reasoned that the resonant state of carbon had to exist since beings like us are made largely of carbon and are able to pose the question as to the formation of the element carbon.

19. Only Fowler was awarded the Nobel Prize for his work in nucleosynthesis, although it is generally agreed that the core of the discovery belonged to Hoyle. Many observers believe that Hoyle's combative style cost him a share in the Nobel Prize.

20. C. Seife, What is the Universe Made Of?, *Science*, 309, 78, 2005.

21. Strong nuclear forces are due to nearest-neighbor interactions. In the case of the light nuclei, a greater proportion of nucleons are at the surface, and so the overall force is of greater magnitude. In the case of heavy nuclei, however, the contribution is almost constant since most nucleons can be considered as being in the "interior" of the nucleus, and so they experience the maximum number of 12 nearest-neighbor interactions.

22. The value of 126 is only experimentally realized for neutrons given that the largest stable nuclei formed thus far have Z-values in the 110s.

23. In addition four of these elements have "doubly magic" nuclei since they have magic numbers with respect to both protons and neutrons. The doubly magic nuclei are ^{4}He, ^{16}O, ^{40}Ca, ^{208}Pb.

24. In the case of electrons in an atom, the force can be considered to be literally centrally directed, since it results from the attraction due to the central nucleus. In the nuclear analogue, each nucleon acts on every other one, and yet one can usefully assume a centrally directed field even though this is physically not the case.

25. Spin-orbit coupling also occurs in atoms but to a far lesser extent and is significant only for heavy atoms.

26. There are several more sophisticated approaches than nuclear-shell theory, but the empirical nature of the ordering of levels is a common feature among them.

27. Readers may wish to consult any of a number of excellent sources on inorganic chemistry, including F.A. Cotton, G. Wilkinson, C.A. Murillo, M. Bochmann, *Advanced Inorganic Chemistry*, 6th ed., Wiley, New York, 1999; N.N. Greenwood, A. Earnshaw, *Chemistry of the Elements*, Pergamon Press, Oxford, 1984.

28. Another unusual relationship, which is not discussed here, is the inert pair effect, whereby the lower members of many groups form stable compounds with a lower oxidation state than do the higher members. E.g., tin and lead form stable dichlorides compared to carbon and silicon, which produce only tetrachlorides. The electrons in the outermost s orbital of the lower members of these groups are said to be inert since they typically do not participate in bonding. A fuller explanation of the inert pair effect requires the application of relativistic quantum mechanics.

29. For further information on the diagonal relationship, see T.P. Hanusa, Reexamining the diagonal relationships, *Journal of Chemical Education*, 64, 686–687, 1987.

30. F. Habashi, A New Look at the Periodic Table, *Interdisciplinary Science Reviews*, 22, 53–60, 1997. Also note that IUPAC numbering for groups is used here and in the remainder of this chapter.

31. Ga^{3+} nonetheless shows a closed shell configuration given that the 3d subshell is filled.

32. Nor can the anomalous values for aluminum be attributed to the phenomenon of first-member anomaly for the simple reason that aluminum is the second member of group 13 as matters currently stand.

33. For a fuller discussion of these cases, the reader is referred the writings of Geoffrey Rayner Canham, which have been drawn on extensively in the writing of this section: The Richness of Periodic Patterns, G. Rayner Canham, *The Periodic Table: Into the 21st Century*, D. Rouvray, R.B. King, Research Studies Press, Bristol, UK, 2004, pp. 161–187; G. Rayner Canham, T. Overton, *Descriptive Inorganic Chemistry*, 3rd ed., W.H. Freeman, New York, 2003.

34. G.H. Lander, J. Fuger, Actinides: The Unusual World of the 5f Electrons, *Endeavour*, 13, 8–14, 1989.

35. AsCl$_5$ has now actually been prepared, but the difficulty in doing so is shown by the fact that this feat was achieved only in 1977, compared with PCl$_5$ and SbCl$_5$, which have been known since 1834. On a separate point, the screening explanations have been confirmed by relativistic quantum mechanical calculations carried out by P. Pyykkö, On the Interpretation of 'Secondary Periodicity' in the Periodic System, *Journal of Chemical Research* (Sweden), (S), 380–381, 1979.

36. One explanation is that, as the group is descended, the effective nuclear charge (protons minus inner-shell electrons) remains constant whereas the distance of the outermost electron increases, thus producing an overall decrease in ionization energy.

37. R.T. Sanderson, *Chemical Periodicity*, Reinhold, New York, 1960.

38. E.R. Scerri, Chemistry, Spectroscopy and the Question of Reduction, *Journal of Chemical Education*, 68, 122–126, 1991.

39. H. Obadasi, Some Evidence About the Dynamical Group SO (4,2): Symmetries of the Periodic Table of the Elements, *International Journal of Quantum Chemistry, Symposium* 7, 23–33, 1973; V. Ostrovsky, What and How Physics Contributes to Understanding the Periodic Law, *Foundations of Chemistry*, 3, 145–182, 2001.

40. Of the eight possible knight's moves available in the game of chess, it appears that the chemical knight's move represents just one of these. In all cases, it involves a movement of one step down in the periodic table, followed by two steps to the right.

41. Laing made the discovery of the knight's move relationship in the course of teaching a chemistry course for engineers that caused him to emphasize the similarity between zinc and tin (M. Laing, personal communication).

42. M. Laing, The Knight's Move in the Periodic Table, *Education in Chemistry*, 36, 160–161, November 1999; M. Laing, chapter 4, in D. Rouvray, R.B. King, Patterns in the Periodic Table, *The Periodic Table: Into the 21st Century*, Research Studies Press, Bristol, UK, 2004, pp. 123–141

43. E.g., cadmium, which lies directly below zinc, and lead, which lies directly below tin, are both highly toxic. However, cadmium appears to be an essential element for at least one organism, a marine diatom that produces a cadmium-specific enzyme that catalyzes the conversion of carbon dioxide and carbonic acid, as discovered in the year 2000. For further biological information on this element, see J. Emsley, *Nature's Building Blocks*, Oxford University Press, Oxford, 2001, pp. 74–76. This book is the standard reference for the detailed properties of all the elements, including their human, medical, economic, historical, environmental aspects.

44. Further information on the toxicity of organotin compounds is also to be found in Emsley, ibid.

45. For brass, the earliest regular production appears to date from the fourth century B.C. or perhaps earlier. From textual sources and actual artifacts, Taxila, in modern Pakistan, has produced the earliest brass, dated from that time. Bronze is far earlier. Copper began to be smelted in the late fifth millennium B.C. in the Near East. The first copper "alloy" was arsenical copper, where the arsenic had the same useful effect as did tin: hardening the metal, improving the casting quality, and lowering the melting temperature. P.T. Craddock (ed.), *2000 Years of Zinc and Brass*, British Museum Occasional paper No. 50, London, 1990.

46. M. Laing, The Knight's Move in the Periodic Table, *Education in Chemistry*, 36, 160–161, 1999.

47. More recently, Laing has noted even further possible knight's move connections between technetium and iridium (M. Laing, personal communication).

48. L. Suidan, J. Badenhoop, E.D. Glendening, F. Weinhold, Common Textbook and Teaching Misrepresentations of Lewis Structures, *Journal of Chemical Education*, 72, 583–586, 1995.

49. W.B. Jensen, Classification, Symmetry and the Periodic Table, *Computation and Mathematics with Applications*, 12B, 487–510, 1986; H. Bent, The Left-Step Periodic Table, *Journal of Chemical Education*, forthcoming.

50. I am assuming that hydrogen is indeed placed among the alkali metals, something that has been disputed, for example, by P. Atkins, H. Kaesz, The Placement of Hydrogen in Periodic System, *Chemistry International*, 25, 14–14, 2003. See also a response to this paper,

E.R. Scerri, The Placement of Hydrogen in Periodic System, *Chemistry International*, 26, 21–22, 2004.

51. In any case, there are only two elements in each f block group, which makes it more difficult to establish whether there is any anomaly whatsoever.

52. An excellent and detailed account of boron–nitrogen compounds and their similarities with carbon is given in N.N. Greenwood and A. Earnshaw, *The Chemistry of the Elements*, Pergamon Press, Oxford, 1997 pp. 234–240. The question of whether borazine displays aromatic characteristics is a matter of some debate in the literature. A.K. Phukan, E.D. Jemmis, Is Borazine Aromatic? *Inorganic Chemistry*, 40, 3615–3618, 2001.

53. Another example is the cyanide ion, CN^-, which behaves like a halide ion. See G. Rayner Canham, The Richness of Periodic Patterns, in D. Rouvray, R.B. King (eds.), *The Periodic Table: Into the 21st Century*, Research Studies Press, Bristol, UK, 2004, pp. 161–187. This author goes as far as to include NH_4^+ and CN^- in a newly designed periodic table, a tendency that seems to recapitulate some of the early periodic systems and earlier lists of elements, which also included ions and radicals.

54. D.E. Bergeron, A.W. Casteman, T. Morisato, S.N. Khanna, The Formation of $Al_{13}I$: Evidence for the Superhalogen Character of Al_{13}, *Science*, 231–235, 2004.

55. Anonymous, Evidence that Superatoms Exist Could Unsettle the Periodic Table, *The Economist*, 37, 475, 2005.

56. E.G. Mazurs, Graphical Representations of the Periodic System During One Hundred Years, University of Alabama Press, Tuscaloosa, AL, 1974.

57. The most authoritative website is by the inorganic chemist Mark Winter, WebElements™ Periodic Table, University of Sheffield and WebElements Ltd., 2006, available at http://www.webelements.com/.

58. Lithophiles are "rock-loving" elements found predominantly in oxide minerals or as halides. Siderophiles or "iron-loving" elements are found mostly in the earth's core, and chalcophiles are elements found in the earth's crust in combination with nonmetals, including sulfur, selenium, and arsenic.

59. B. Railsback, An Earth Scientist's Periodic Table of the Elements and their Ions, *Geology*, 31, 737–740, 2003.

60. F. Habashi, A New Look at the Periodic Table, *Interdisciplinary Science Reviews*, 22, 53–60, 1997.

61. In the 1960s and 1970s, philosophers of language, including Saul Kripke and Hilary Putnam, reanalyzed the question of sense and reference. They argued that reference was not fixed by the description of a natural kind term, such as "tiger," "quark," or "element," but by what they called its "essence," for which they appeal to the latest scientific knowledge. On this view, the reference of a particular element such as gold is given by its atomic number of 79 rather than by describing the properties of gold. I have argued that there are some parallels between the chemist's discussion of elements as basic and simple substances and the analogous discussion by philosophers in terms of sense and reference. Whereas the sense of an element is provided broadly speaking by the properties of the elements, its reference is defined through just one criterion: atomic number. E.R. Scerri, Some Aspects of the Metaphysics of Chemistry and the Nature of the Elements, *Hyle*, 11, 127–145, 2005.

62. Another radiochemist, Kasimir Fajans, was Paneth's leading opponent in believing that the periodic system would *not* survive the discovery of isotopes.

63. F.A. Paneth, The Epistemological Status of the Concept of Element, *British Journal for the Philosophy of Science*, 13, 1–14, 144–160, 1962, reprinted in *Foundations of Chemistry*, 5, 113–145, 2003.

64. Paneth and Hevesy showed that the electrochemical potential from two cells made from different isotopes of the metal bismuth was the same as far as experimental techniques of the day could distinguish. E.R. Scerri, Realism, Reduction and the Intermediate Position, in N. Bhushan, S. Rosenfeld (eds.), *Minds and Molecules*, Oxford University Press, New York, 2000, pp. 51–72.

65. The fact that more recent research has revealed some differences even in the chemical properties of isotopes does not alter the central issue under discussion.

66. S. Kripke, Naming and Necessity, in D. Davidson, G. Harman (eds.), *Semantics of Natural Language,* Reidel, Dordrecht, 1972, pp. 253–355; H. Putnam, The Meaning of Meaning, in his *Philosophical Papers*, vol. 2, Cambridge University Press, Cambridge, 1975, pp. 215–271.

67. If the possession of a particular electronic configuration were a necessary condition for membership to a particular group, this would imply that all members of a group have the same outer-shell configuration. This is violated by many groups of transition elements. If the possession of a particular electronic configuration were a sufficient condition, this would imply that elements with the same outer-shell configuration must be grouped together. This is violated by the case of helium, at least in the conventional representations of the periodic system. E.R. Scerri, How Ab Initio is Ab Initio Quantum Chemistry? *Foundations of Chemistry*, 6, 93–116, 2004.

68. Electronic configurations are known to be approximations, unlike atomic number, which can be given a clear realistic interpretation in terms of the number of protons in the nucleus of any atom. E.R. Scerri, How Ab Inito is Ab Initio Quantum Chemistry?, *Foundations of Chemistry*, 6, 93–116, 2004.

69. E.g., M.W. Cronyn, The Proper Place for Hydrogen in the Periodic Table, *Journal of Chemical Education*, 80, 947–951, 2003.

70. P.W. Atkins, H. Kaesz, The Placement of Hydrogen in Periodic System, *Chemistry International*, 25, 14, 2003.

71. Once again, unobservable apart from their possessing an atomic number.

72. C. Janet, The Helicoidal Classification of the Elements, *Chemical News*, 138, 372–374, 388–393, 1929; L.M. Simmons, The Display of Electronic Configuration by a Periodic Table, *Journal of Chemical Education*, 25, 658, 1948; R.T. Sanderson, A Rational Periodic Table, *Journal of Chemical Education*, 41, 187–189, 1964; G. Katz, The Periodic Table: An Eight Period Table For The 21st Centrury, *The Chemical Educator*, 6, 324–332, 2001; E.R. Scerri, Presenting the Left-step Periodic Table, *Education in Chemistry*, 42, 135–136, 2005.

73. From this point onward, I concentrate on the left-step table, although both representations are equally viable for what follows.

74. D. Neubert, Double Shell Structure of the Periodic System of the Elements, *Zeitschrift für Naturforschung,* 25A, 210–217, 1970.

75. Of course, it may be that future chemistry might reveal that helium does indeed belong in the alkaline earths. The notions of elements as basic substances and as simple substances are complementary, not contradictory.

76. As in the use of atomic number, the use of the $n + \ell$ rule appeals to elements as basic substances and not as simple substances. This rule represents a generalization concerning all the elements, although it is violated in some instances and is not concerned with any directly observable properties of the elements.

77. Charles Janet seems to have been the first author to publish this form of table. C. Janet, The Helicoidal Classification of Elements, *Chemical News*, 138, 372–374, 388–393, 1929. 78. On the subject of beauty and elegance, there are now a number of three-dimen-

sional periodic systems of which one of the finest is due to Fernando Dufour as shown in figure 10.15. See G.B. Kauffman, ElemenTree: A 3-D Periodic Table by Fernando Dufour, *The Chemical Educator*, 4, 121–122, 1999.

79. E.R. Scerri, The Tyranny of the Chemist, *Chemistry International*, 28, 11–12, May–June, 2006.

INDEX

Abegg, Richard, 208
ab initio calculations, 237, 243–245, 321
abstract elements, 289, 291, 294
accommodation, 123, 150, 280
    in Mendeleev's periodic system, 143–149
    of noble gases, 151–156
    versus prediction, 156
acid-base chemistry, 125
acids
    formation of, 18–19
    measurement of, 31
actinides, 275, 307
    discovery of, 129
actinium, 143, 170, 176
    placement of, 21
actinoid relationships, 269–270
ad hoc arguments, 150
"adiabatic," 315
adiabatic principle, 194–195, 200–201
affinity, chemical, xiii–xiv
air
    Dalton's research on, 34
    as element, 3
alchemy, xix, xvi–xvii, 3, 4, 34, 53
alkali metals, 265, 268–269, 275
    formulas for, 59–60
    Mendeleev's classification of, 104, 106
    Newlands's grouping of, 74
    similarity of hydrogen to, 281

alkaline earth elements, 265
    emergence of, 69
    helium as, 281
    Mendeleev's classification of, 116
alloys, 273
alpha particles, 159, 164, 165, 178
alpha rays, 163, 175
Alpher, Ralph, 254, 323
alphon particles, 165–166
aluminum, 174, 271, 324
    association with beryllium, 128
    Mendeleev's prediction of, 106
    placement of, 278
    similarity to iron, 276
    similarity to scandium, 267–268
    splitting of hydrogen, 277
    superatom clusters, 277
americium, 21
ammonia
    composition of, 35
    formula of, 294
    valence, 209
ammonium ion, 276
"amphoteric," 316
amphoteric elements, 208
Ampère, André, 63
angular momentum, 192, 199, 228, 245, 265, 314. See also spin angular momentum
anthropic principle, 323

antimony, 271, 272
  equivalent weights, 51
  Mendeleev's placement of, 129
  similarity to gallium, 273
aperiodic systems, 195
approximations, xvii
argon, 16, 117, 309
  discovery of, 151–156, 291
  electronic configuration, 234
Aristotle, xix
  nature of elements, 113
  philosophy of substance and matter, xv
Armstrong, Henry, 153
aromatic hydrocarbons, 25, 26f
Arrhenius, Svante, 251
arsenic, 271
  electronic configuration, 190
  equivalent weights, 51
astatine, 143, 291
  discovery of, 174
asterium, 172
astrology, 3
astronomy
  application to chemistry, 87
  impact on nucleosynthesis theories,
    253–254
  names of elements from, 8
astrophysics, 258
Atkins, Peter, xv, 281
Atomechanik (Hinrichs), 86
atomic combinations, 36
atomicity, of a gas, 151
atomic mass, 64
atomic number, xviii, 20, 58, 61, 131, 278,
    311, 327
  anticipation of, 76
  in density functional theories, 246
  discovery of, 165–169
  of isotopes, 313
  Moseley's experiments on, 170–173
  relation to charge, 171, 175
  relativistic effects and, 270
  stability and, 263
  triads and, 179–180
atomic physics. *See* physics
atomic radii
  among transition metals, 272t
  within periodic table, 266
atomic states, number of, 201

atomic theory, 37–38, 159
  of Dalton, 33–37, 57, 294
atomic volume, 98
  plotted against atomic weight, 97f
atomic weight, xviii, 38–39, 179, 291, 296,
    311
  versus atomic volume, 97f
  basic unit of measurement, 41–42
  Dalton's research, 34–36
  in De Chancourtois's tables, 71
  determination of, 57–61
  in development of periodic system,
    66–67
  versus equivalent weight, 44, 62–64
  gaps of Pettenkofer, 51t
  Gmelin's use of, 45, 46
  history of, 19–20
  interpolation, 110
  in Lothar Meyer's tables, 96
  Mendeleev's use of, 104, 115–116, 120,
    125–127
  Newland's use of, 72
  nonintegral, 176
  numerical relationships among, 53–54
  prediction of, 124
  Prout's hypothesis and, 40
  relation to atomic dimension, 89
  relation to charge, 164,167
  relation to chemical properties, 68, 109,
    313
  relation to specific heat, 127
  spectral line frequencies and, 88
atomism, Greek, 32–33
atoms, 120, 302. *See also* atomic radii
  concept of, xix, 64
  dimensions, 88
  formation of, 254
  Lewis's postulates of structure, 208–209
  models of, 183–187, 214f
aufbau principle, 190, 192, 233, 242, 243,
    265, 282
Austin, William, 35
Avogrado, Amedeo, 7, 38, 63, 302

Bachelhard, Gaston, 121
Balard, Antoine, 48
barium, 6
  atomic weight of, 301
  placement of, 50

spectral frequencies, 88
triad relationships, 42, 44
Barkla, Charles, 164, 170
bases
  formation of, 18–19
  measurement of, 31
basic substances, 281–283, 285, 304
basis sets, choice of, 241–242
Becquerel, Henri, 161–162, 310
Becquerel rays, discovery of, 160–162
Bent, Henry, 275
benzene rings, 276
benzenoid aromatic hydrocarbons, 25, 26f
Bertholet, Marcellin, 148–149
beryllium, 298
  association with aluminum, 128
  atomic weight of, 150
  formation of carbon from, 257
  placement of, 127–128, 156
Berzelius, Jacob, 40
  determination of atomic weights, 59, 104, 295
  element symbols of, 9
beta decay, 178
  formation of elements by, 254
beta particles, 178, 254
beta rays, discovery of, 163
Bethe, Hans, 254, 323
big bang cosmology, 249, 253–254, 258, 322
  stages in, 259t
binding energy, 259, 260f, 261
biology, philosophy of, xxii
Biron, Evgenii, 270
bismuth, 271
  electrochemical potential, 327
  placement of, 45
black-body radiation, 189
black holes, 256
Bohr, Niels, xix, xx, xxi, 7, 24, 199, 201, 217, 229, 312
  collaboration with Moseley, 171
  lecture quality and style, 193–194
  prediction of hafnium, 218–220, 319
  quantum theory, 188–197
  second theory of periodic system, 192–197
  trilogy paper, 168
  work method, 224

Bohrfest, 193
bohrium, 9
boiling points, knight's move relationships and, 274t
Boisbaudran, Emile Lecoq, 135–136
Boltwood, Bertram, 177
Boltzmann, Ludwig, 159
Bondi, Herman, 322
bonding, xiii, xvi
bonds. *See also* binding energy
  covalent, 207–208, 228
  interatomic, 153
  ionic, 207–208
  notation, 211
  single, double, and triple, 209–210
Bonifatii, Kedrov, 150
borazine, 276
boron, 271
  bonding with nitrogen, 276, 326
  equivalent weights, 51
  placement of, 45, 98
boundary conditions, 228–229
Boyle, Robert, xvi–xvii, 4, 34
brass, 325
Brauner, Bohuslav, 117, 129, 130
British Association for the Advancement of Science, 251
Brock, William, 123
Broglie, Louis de, 230
bromine, 15
  discovery of, 48
  placement of, 148
  reactions of, 49
  triad relationships, 43
bronze, 325
Brush, Stephen, 124, 150
buckyballs. *See* fullerenes
Bunsen, Robert, 87, 301
Burbidge, Geoffrey, 257
Burbidge, Margaret, 257
Burgers, J.M., 195
Bury, Charles, 173, 192, 213, 215, 224

cadmium, 85, 273, 325
  placement of, 98
calcium, 6
  atomic weight, 297, 301
  placement of, 50
  prediction of analogue, 142

calcium (*continued*)
  spectral frequencies, 88
  triad relationships, 42, 44
calorique, 5
Cannizzaro, Stanislao, 7, 38, 62, 66–67, 300,
  302
  atomic weights of, 104, 127
  influence on Lothar Meyer, 93
  influence on Odling, 82
  revival of Avogrado's work, 64–65
carbon, 271
  atomic weight, 65t, 296
  equivalent weight, 51
  formation of, 255, 257
  resonant state, 323
carbon-carbon bonds, 276
Cassebaum, H., 87
casseopeium, 172
cathode rays, 184–185, 313
cells, 212, 215, 317
central-field potential, 263
cerium
  electronic configuration, 216
  in table of Mendeleev, 110
cesium, 291, 298
  derivation of name, 8
Chadwick, James, 175
chalcophiles, 278
charge, 325
  atomic number and, 171, 175
  atomic weight and, 164, 167
charge density, 267, 276
Chemical News, 78, 147, 148, 250
chemical properties
  elucidation using triads, 43
  in left-step table, 286
  number of electrons and, 313
  similar, 109, 267–269
  in table of Lothar Meyer, 96
chemistry
  education, xx. *See also* textbooks
  as inductive science, 142
  international, 105
  numerical aspects of, 57
  philosophy of, xiii, xx, 113, 117–121
  versus physics, xvii–xviii, 211–212
  reduction to quantum mechanics, xiv,
    xxii, 247–248, 318
  role of alchemy in, xvi–xvii

Chemistry and Atomic Structure (Main
  Smith), 220
Chistyakov, V.M., 272
chlorine, 6, 15, 297
  atomic weight, 41, 176
  derivation of name, 8
  discovery of, 48
  electronic configuration, 191, 206
  reactions of, 49
  triad relationships, 43
chromium
  electronic configuration, 240
  orbital energies, 237–238
classification, xix, 65–66, 285
  based on atomic weight, 125
  early, of elements, xv–xvi
  Gmelin's, 48–50
  numerical relationships and, 57
  of periodic classifications, xiv
  primary versus secondary, 21–22
  Prout's hypothesis and, 40
Clausius, Rudolf, 151
Clève, Pierre, 137, 147
cobalt, 302
  placement of, 148, 171
colors, names of elements from, 8
columns. *See* groups
combustion, 30
composition, Lavoisier's analyses of, 31
compounds
  formation, xvi
  formulas for, 36–37, 59, 63–64
  tables of, 25, 27
constant differences, discovery of, 50–51
continuous recycling, heat death and, 252
contrapredictions, 124
Cooper, D.G., xv
copper, 6, 290
  alloys, 325
  electronic configuration, 240
  orbital energies, 238–239
  symbol for, 9, 10f
coronium, 172
Coryell, Charles, 174
cosmology, 249, 258. *See also* big bang cosmology
Coster, Dirk, 173, 219
Coulomb forces, 209, 256, 259
Courtois, Bernard, 48

covalent bonds, 207–208
  quantum mechanical calculation of, 228
Crookes, William, 153, 172, 176, 184, 250–251, 323
Crookes tubes, 161
crystal planes, angles between, 60
cubic particles, 3
cults, alchemy and, xvi
cultural anthropology, xvii
Curie, Marie, 162, 310
curium, 21
cyclonium, 312

Dalton, John, 7, 31, 63, 294
  atomic theory of, 33–38
  atomic weights of, 19, 41t
  element symbols used by, 9
  incorrect conclusions of, 57
  rule of simplicity, 58, 295
dark energy, 258
dark matter, 258
darmstadtium, 290
Dauvillier, Alexandre, 217, 219
Davies, Mansel, 224
Davisson, Clinton, 230
Davy, Humphry, 7, 39, 48
Davy Medal, 144–146
Döbereiner, Johann, xix, 7, 25, 42–43, 94, 179, 296
  triads of, 58t
De Boisbaudran, Paul Emile Le Coq, 68, 71
De Chancourtois, Alexandre Emile Béguyer, xix, 99
  role in development of periodic system, 68–72
  role in discovery of periodic law, 77
De Lapparent, Albert Auguste, 68, 71
Demidov Prize, 103
density
  of gases, 59, 64
  Lothar Meyer's use of, 98
density functional theories, 245–247
deuterium, 322
diagonal behavior, 265–267
Dias, Jerry, 25
diatomic molecules, 63, 64
didymium, 91, 302
Die Modernen Theorie der Chemie, 93–94

diffraction, by electrons, 230
diffraction techniques, 246
diffusion, Dalton's hypotheses, 35
Dirac, Paul, 247
Dobbs, Betty Jo, xvi
dodecahedral particles, 3
double bonds, 209, 210f
dubnium, 9
Dufour periodic tree, 286f
Dulong, Pierre-Louis, 59–60, 127, 299
Dumas, André, xix, 52–53, 62
  atomic weights of, 40–41, 60–61
dysprosium, 174

earth, as element, 3
Eddington, Arthur, 253, 255
education, chemistry, xx. *See also* textbooks
Ehrenfest, Paul, 194, 217
"eka-," 307
eka-aluminum, 132. *See also* gallium
eka-boron, 132. *See also* scandium
eka-lead, 275
eka-manganese, 172, 174
eka-silicon, 132. *See also* germanium
eka-stibium, 138
eka-tantalum, 173
eka-tellurium, 130
electrical oscillations, formation of elements through, 251
electric forces, 209
electricity
  isolation of elements and, 6
  use to form compounds, 37
electrochemical potential, 327
electrode potential, 267, 268t
electrolysis, 6
electromagnetism, 188
electron clouds, 24
electron density, 231, 246, 247
electronegativity, 297
  within periodic table, 266
  use in classification, 48
electron gas, 246
electronic configurations, 24, 185–187, 281, 283, 317
  among groups, 269, 270, 327
  of Bury, 213, 214
  by chemists, 205–211, 224–225, 227
  concept of, 21–22

electronic configurations (*continued*)
    ground state, 242, 244
    of Langmuir, 211–213
    of Main Smith, 220–223
    periodicity and, 190
    rearrangements, 190–191
    rules for, 232
    third quantum number and, 197–198,
        199t
    writing, 233–237
electronmagnetism, 308
electron rings, of Thomson, 187t
electrons, xix, 160, 293, 324. *See also* elec-
        tron density; electronic configuration;
        electron shells
    addition of, 196
    attachment sites, 208
    diffraction and interference by, 230
    discovery of, 159, 183–185, 205
    field of movement, 231–232
    magnetic properties, 209, 316
    number and chemical properties, 313
    repulsion, 241
    shared, 209, 316
    velocity, 24
    waves, 230–231
electron screening, 271
electron shells, xx, 190, 198, 221–222
    explanation for closing, 237–242
    filling, 212, 282
    number of electrons in, 198, 201, 208,
        215, 229, 232, 234, 242–243
    outer, 14, 24
    Pauli's scheme, 200t
    theories, 265
electron spin, 209
electron waves, 293
element 72. *See* hafnium
elements. *See also names of specific elements*
    abstract, 113–115, 117–118, 289
    abundance of, 258–265
    as basic substances, 278, 280
    chemical similarities, 29, 85, 125, 267–
        269
    classification, xv–xvi, 22, 40
    comparison of dissimilar, 54
    concept of, xv–xvi, xxii, 4, 280–281
    definitions of, xv–xvi

discovery of, 6, 7, 67, 172, 176
formation of, 249, 250–258
in Greek philosophy, 3
macroscopic behavior, 205
Mendeleev's evaluation of, 105–106,
    112–117
names and symbols of, 6–10, 290
number of, 22, 170, 174, 293
prediction of, 74, 306, 309
reactivity and ordering, 18–20
as simple substances, 281
synthesis of, 6, 21
unknown, 96, 106, 124, 131
empiricism, xv, 4, 31, 34
energy
    radiation of, 188–189
    released upon nucleus formation, 259
    zero-point, 252
energy states, 189
enneads, 55
equal volumes equal numbers. *See* EVEN
        hypothesis
equivalent weights, 19–20, 31–32, 38–39,
        291, 295
    versus atomic weight, 36–37, 44,
        62–64
    Pettenkofer's use of, 50–51
    in triads, 42–44
erbia, 217
erbium, 172
    derivation of name, 8
    placement by Mendeleev, 126
    prediction of atomic weight, 56
essentialism, 118
ether, 3, 252, 308
ether theory, 140
European periodic system, 13f, 14
europium, gadolinium and, 21
euxenite, 137
EVEN hypothesis, 38, 59, 61, 62, 295
exchange terms, 228, 241
experimentation
    beginnings of, 4
    philosophy of, xvii

failures, historical recording of, xviii
Fajans, Kasimir, 170, 178, 326
Faraday, Michael, 52, 82, 250

feldspars, 72
Fermi, Enrico, 246
fire, 289
   as element, 3
first-member anomaly, 275–276, 282
Fitzgerald, George, 153
florentium, 312
fluorine, 15, 297
   electronic configuration, 191
   placement of, 48
   prediction of analogue, 141
   reactions of, 49
   triad relationships, 43
Fock, Vladimir, 231, 319
Foster, George Carey, 78
Foundations of Chemistry, 287
Fowler, William, 257, 323
francium, discovery of, 174
Frankland, Edward, 37
Friedel-Crafts catalysts, 276
Friederich, B., 293
fullerenes, 291
fundamentalism, of scientific disciplines, xvii–xviii
fusibility, Lothar Meyer's use of, 98
fusion reactions, 256

gadolinium, europium and, 21
gallium, 267, 271, 291
   discovery of, 135–137, 156
   predicted and observed properties, 133–134t
   prediction of, 106, 123, 132
   similarity to antimony, 273
Gamov, George, 254, 323
gases
   atomic weight measurement, 62
   composition of, 251
   relative density, 59, 64
Gaudin, M.A.A., 299
Gay-Lussac, Joseph Louis, 37–38, 63
Geiger, Hans, 164
geography, names of elements from, 8
geology
   influence on atomic weight, 179
   periodic table for, 278
Gerhardt, Charles, 63, 72
   atomic weights of, 104

germanium, 15, 271
   discovery of, 137–140, 156
   predicted and observed properties, 139t
   prediction of, 96, 106, 123, 132
Germer, Lester, 230
Gladstone, John Hall, 155, 297
Glendenin, Lawrence, 174
Gmelin, Leopold, xix, 7, 44–50, 297
   atomic weights of, 40
   chemistry textbooks of, 46–50
   groupings of, 65
Goeppert-Mayer, Maria, 264
gold, 6, 174, 175
   alchemy, 3
Gordin, Michael, xv, 118
graphite, 276
gravitational force, on stars, 256
Greek mythology, names of elements from, 8
Greek philosophy. *See also* Aristotle
   view of elements, 3
Grignard reagent, 266
Grosse, Aristide, 173
ground state
   of carbon, 257
   configurations, 242, 244
   energy, 227
   of nickel, 239
   prediction of, 242
group displacement laws, 178
groups, 11, 14–15, 66. *See also* periodicity
   discovery of, 44–48
   electronic configuration and, 190–192, 242–243, 327
   of Hinrichs, 89–92
   of Mendeleev, 112, 125–126
   as natural kinds, 280–281
   of Newlands, 72–73, 80
   of Odling, 82–85
   similarities between, 267–269

Habashi, F., 278
hafnium, 175, 227, 283
   discovery of, 173
   electronic configuration, 215–217
   prediction of, 218–220, 224, 319
Hahn, Otto, 9, 173, 312
hahnium, 9

halogens, 297, 305
    classification by Mendeleev, 116
    equivalent weights, 51
    ordering of, 48
    properties of, 118
    reactions of, 49
Handbuch der Chemie (Gmelin), 44
harmonic interaction, 218
Hartog, Philip, 68, 71
Hartree, Douglas, 231, 319
Hartree-Fock method, 231–232
heat, as an element, 5
heat death, of universe, 251–252, 323
heat envelope, 57
Hefferlin, Ray, 25, 293
Heilbron, John, 190
Heisenberg, Werner, xix, 227–228, 230–231
Heitler, Walter, 228
helium, 16
    abundance of, 258
    calculation of properties, 230
    calculation of spectrum, 227
    derivation of name, 8
    discovery of, 155, 291
    electronic configuration, 211, 242
    formation from hydrogen, 253
    formation of carbon from, 257
    placement of, xxi, 22, 275–276, 281–
        283, 293, 321
    ratio to hydrogen, 249, 322
    stablity, 241
Hentschel, Klaus, 92
Herz, Heinrich, 184
Hevesy, György, 177, 219, 280, 327
high-temperature superconductivity, 10–11
Hilbert space, 246
Hinrichs, Gustavus, xix, 86–92, 99, 292
    periodic table of, 302
history of science, xviii
Hittorf, Johann, 184
Hoffmann, A.W., 250
holmium, 217
    derivation of name, 8
homogeneity, 177
homologous series, 94
horizontal relationships, 53–54, 302
    Mendeleev's analysis of, 105–106, 116
    in tables of Lothar Meyer, 94
Hoyle, Fred, 255–257, 323

Humboldt, Alexander von, 37
Hund's rule, 212–213, 233, 240–241
hydrides, 299
hydrofluoric acid, 48
hydrogen, 59
    abundance of, 258
    first-member anomaly, 275
    formation of helium from, 253
    as foundational element, 250
    Lavoisier's view of, 31
    placement of, 149, 281–282, 313, 326
    quantum theory of atom, 188
    ratio to helium, 249, 322
    splitting, 277
hydrogen burning, 256, 323
hydrogen peroxide, 295
hydroxides, 18
    reactions, 127
hypotheses, testing, 42

icosahedral particles, 3
illinium, 312
incandescent objects, 314
independent-electron approximation, 243
indium, 132, 290
    in table of Mendeleev, 110, 126
inert gases. *See* noble gases
inertness, 281
inert pair effect, 324
infinity, 32–33
inorganic atoms, 94
integral weights, 38–42
interatomic bonding, 153
interatomic distances, 25
interference effects, of electrons, 230
International Union of Pure and Applied
        Chemistry, 9
    naming conventions, 290
    periodic system, 13f, 14
    periodic table, 290
interpolation, 132, 142
    of Mendeleev, 131
iodine, 300
    discovery of, 48
    placement of, 96, 125–126, 130–131
    reactions of, 49
    reversal with tellurium, 76, 109, 117,
        148, 173, 178, 297–298, 309
    triad relationships, 43

ionic bonds, 207–208
ionic compounds, formation of, 206
ionium, 177
ionization, 236, 281
ionization energy, 25, 244, 271, 320, 325
  within periodic table, 266
ions
  inclusion in periodic table, 278
  that imitate elements, 276
iron, 6, 302
  formation of, 256
  similarity to aluminum, 276
  stability of, 259
isomorphism, law of, 59–60
isotopes, 58, 292
  abundance of, 249, 322
  atomic weights and, 73, 296
  discovery of, 41–42, 160, 176–179, 250,
    326
  elements with single, 299
  mass, 178
  periodic table and, 280
  triads and, 179–182
IUPAC. *See* International Union of Pure
  and Applied Chemistry

Janet, Charles, 327
Janet table. *See* left-step table
Jensen, Hans, 264
Jensen, William, 87, 275, 293
Jupiter, 258

Kaesz, Herbert, 281
Kaji, Masanori, 115
Karlsruhe conference, 63, 64, 72, 82, 93,
  103, 291, 299–300
Kauffman, George, 87
Kaufmann, Walter, 184
Kekulé, August, 37, 64
keltium, 175
Kelvin, William Thomson, 154
Khodnev, Alexei Ivanovich, 105
kinetic energy, 151, 194
Kirchoff, Gustav Robert, 87
Klassiker der Wissenschaften (Cannizzaro), 93
knight's move relationships, 272–275, 325
knowledge
  development of system of, xviii
  validity of, xvii

Kossel, Walther, 212
Kramers, Hendrik, 217
Kremers, Peter, 53–54
Kripke, Saul, 326
Kripke-Putnam view, xxii
krypton, 16
  discovery of, 155, 291
Kuhn, Thomas, xviii, 67, 190, 288, 298

Laidler, Keith, 224
Laing, Michael, 273, 274
Lakatos, Imre, 182
Landé, Alfred, 199
Langmuir, Irving, 211–213
lanthanides, 74, 270, 275, 307
lanthanum, 21, 272
  placement by Mendeleev, 126
Latin names, 290
Lavoisier, Antoine, xv, 7, 34, 288, 294
  nature of elements, 113
  quantitative analyses of, 29–31
  simple substances, 289
  view of elements, 4
law of chemical combination. *See* law of
  multiple proportions
law of conservation of matter, 114
law of constant proportion, 35–36
law of definite proportions by volumes,
  37–38, 59
law of integers, 63
law of multiple proportions, 36
law of octaves, 76–82, 85
lawrencium, 21, 272
lead, 15, 273
  atomic weight, 176
  placement by Lothar Meyer, 98
  radioactive decay, 178–179
  separation of radio-lead, 177
  transmutation, 3
  valence, 291
left-step table, 283, 284f, 285–286
Lemaître, Georges, 253
Lembert, Max, 179
Lenssen, Ernst, 55–56
Levi, Primo, 6, 290
Lewis, Gilbert Newton, 7, 205–211
light
  as an element, 5
  dispersion from atoms, 188

Lipton, Peter, 144–145
liquid drop model, 259, 261
literary criticism, xvii
lithium, 291, 301
    placement of, 50, 149
    similarity to magnesium, 265–266
    triad relationships, 43
lithophiles, 278, 326
logic, xxi
London, Fritz, 228
London Chemical Society, 78, 80, 82
Lothar Meyer, Julius, xix, 80, 92–98, 99
    Davy Medal, 145
    dispute with Mendeleev, 98, 147
    prediction of germanium, 139
    unity of matter, 294
    view of reductionism, 183
    work contrasted with Mendeleev's,
        124–125
Lothar Meyer, Oskar, 92
lumière, 5
lutetium, 21, 172, 174–175, 217, 272
    electronic configuration, 216, 218
Löwdin, Per-Olav, 233

macroscopic properties, 3
Madelung rule, 233
magic numbers, 263–264
magnesium, 6, 85
    placement of, 44, 50
    similarity to lithium, 265–266
    similarity to zinc, 267
magnetic fields
    effect on orbital motion, 197
    effect on spectral lines, 196, 199–201
Maher, Patrick, 144–145
Main Smith, John David, 225
    electronic configuration schemes,
        220–223
Marignac, Charles, 41, 217
Marinsky, Jacob, 174
Marsden, Ernest, 164
masses, ratio of, 36
mass number, 259, 292
masurium, 174
material ingredient, 113, 114, 118
matrix mechanics, 230
matter
    behavior of, xx

composition of, 258
dark, 258
fourth state, 251
subdivision of, 32–33
maximum simplicity, 62, 295
Maxwell, James Clerk, 188
Mayer, Alfred, 186
Mazurs, Edward, xiv, 277, 287
McCoy, Herbert, 177
measurable attributes, 115
Meitner, Lise, 9, 173, 312
melting points, 267, 268t
    knight's move relationships and, 274t
    prediction of, 136
Mendeleev, Dimitri, xiii, xix, xv–xvi, 80,
        99, 101, 278, 301
    on accommodation of argon, 154
    on accommodation of noble gases,
        155–156
    criticism of De Chancourtois, 71
    criticism of nucleosynthesis, 251
    development of periodic system, 67,
        105–112
    dispute with Lothar Meyer, 93, 98,
        147
    early life and work, 102–105
    influence of Pettenkofer on, 51–52
    misplacement of elements in table,
        303
    nature of elements, 112–117, 294
    predictions of, 117–118, 123–124,
        131–135, 140–149, 300
    publications of, 143
    reaction to discovery of gallium, 136
    style of work, 150
    triads, 179
    view of reductionism, 183
    views on ether, 308
mercury, 6, 15, 275
    placement of, 98, 126
    symbol for, 9, 10f
mesothorium, 177
metal oxides, 18
    reactions, 127
metals
    classification of, 50, 59, 90
    knight's move relationships, 273–275
    measurement of, 31
    properties of, 11

metaphysical explanations, 304
  of chemical phenomena, 113–114
  of elements, xv, 281
meterology, 86
Middle Ages, xix
  concept of elements, 3
Mitscherlich, Eilhard, 60
models, use in science, xvii
molecular tables, 25, 27
molecular weight, 66, 302
molecules, concept of, 64
molybdenum, 98
Monaghan, P.K., xv
monoatomicity, 154, 309
Morris, R., xv
Moscow Congress of the Russian Scientists
  and Physicians, 132
Moseley, Henry, 7, 131, 159–160, 169–175,
  311
motion
  frequency of, 194
  repetition of, 195
multiples of constant differences, 51

Nagaoka, Hantaro, 164, 185
naphthalene, 25
Naquet, Alfred, chemistry textbook of,
  65–66
natural kinds, xxii
  elements as, 280–281
nearest-neighbor interactions, 323
nebullium, 172
neodymium, 302
neon, 16
  discovery of, 155, 291
neo-ytterbium, 217
Neptune, 258
Nernst, Walther, 252
neutralization, 19
neutrinos, 323
neutron absorption, 254, 258
neutrons, 61, 320
  beta decay of, 178
  formation of, 256
Newlands, John, xix, 67, 72–82, 292, 300,
  301
  law of octaves, 76–82
  periodic table of, 91–92, 302
  suggestion of atomic number, 167

Newton, Isaac, xvi, 301
  atomism, 33–34
newtonium, 140–141
nickel, 302
  electronic configuration, 239–240, 320
  placement of, 148, 171
Nilsen, Lars Frederick, 137
niobium, 55, 290
nipponium, 172
nitrogen, 153, 298
  atomic weights, 296
  bonding with boron, 276
  electronic configuration, 190, 192, 196,
    198
  equivalent weights, 51
  first-member anomaly, 275
  placement of, 46, 49
noble gases, 150, 307, 310
  accommodation of, 156–157
  discovery of, xix
  effect on law of octaves, 78, 80
  electronic configurations, 224
  helium as, 281
  Mendeleev's analysis of, 140–141
  number of electrons in, 212
  prediction and accommodation, 151–
    156
  properties of, 16
noble metals, 269
  in tables of Mendeleev, 106
Noddack, Ida and Walter, 173, 289
nomenclature, 8
nonmetal oxides, reactions, 127
nonmetals
  Hinrichs's classification, 90
  placement of, 18
  properties of, 11
nonpolar organic compounds, formation
  of, 206–207
nuclear fission, 6, 312
nuclear forces, 259, 323
nuclear-shell model, 261, 263
nuclear-shell theories, 265, 324
nuclei, 120, 160
  discovery of, 164
  fusion of, 256
  stability of, 259–265
nucleons, 259, 263, 324
  spin-orbit coupling, 264–265

nucleosynthesis, 21, 249, 251–258
  conditions for, 257t
  in stars, 255–256, 259
numerical data, 296–297
numerical relationships, xix, xxi, 29, 44
  Mendeleev's use of, 141
  Pettenkofer's use of, 50

observable properties, 4
observers, versus theorists, 150
octets, 275
Odling, William, xix, 72, 82–86, 99, 300
  periodic table of, 74, 302
  role in acceptance of Newlands's work,
    78, 80
  role in discovery of periodic law, 77
Ogawa, Masataka, 172
orbitals, xx, 24, 185, 231, 246, 317
  energy levels, 233–238, 320
  filling of, 212–213, 233, 283
  versus orbits, 230, 293, 313
  reliance of quantum mechanical calcula-
    tions on, 247
orbits, 188
  versus orbitals, 230, 293, 313
  reality of, 198
  shape of, 192
organic compounds, classification of, 94
organometallic compounds, 266
organotin compounds, 273
osmium, similarity to xenon, 268
Ostwald, Wilhelm, 93
outer-shell electrons
  denotation of, 14
  similarities between elements and, 24
oxidation states, 270, 275, 291, 324
oxides
  chemical characteristics, 127
  Gmelin's formulas for, 50
  Lavoisier's view of, 31
  triad relationships, 42
oxygen, 59, 294
  atomic weights, 296
  discovery of importance in combustion,
    30–31
  electronic configuration, 190–191
  first-member anomaly, 275
  placement of, 46

pair reversals
  iodine and tellurium, 76, 109, 117, 130–
    131, 148, 173, 178, 297–298, 309
  nickel and cobalt, 171
  understanding of, 178
palladium, 8
Paneth, Fritz, 163, 220, 304, 326, 327
  atomic numbers and basic substances,
    278, 280
  separation of radio-lead, 177
  view of Mendeleev's work, 120
Pauli, Wolfgang, xix, xx, 7, 198, 199–201
Pauli exclusion principle, xx, 199–203, 227,
    229, 232, 233, 243, 315
Peligot, Eugène, 128
Perey, Marguerite, 174
periodicity, xiii, 84, 109, 149, 184, 229, 281,
    283
  Bohr's theory, 222
  concept of, 67
  discovery of, 99
  electronic configuration and, 187,
    190–192
  explanation of, 232, 242
  of Hinrichs's system, 92
  nature of, 16, 18, 71
  Pauli exclusion principle and, 202
  secondary, 270–272
  in tables of De Chancourtois, 69
  van den Broek's views, 166
periodic law, 16–18, 21–22, 76, 77, 281
  conserved in tables of Mendeleev, 118
periodic systems, xiii
  competing, 104–105
  of Crookes, 251, 252f, 253f
  development of, xv–xvi, xviii–xxii, 66–
    68, 99–100, 123
  discovery of, 299
  effect of physics on, 24–25, 229, 244
  first, xix
  of Mendeleev, 94, 96, 105–112,
    143–145
  modern, 248
  numbers of, 277
  versus periodic tables, 18
  philosophy of, xvii–xviii
  role of qualitative chemistry, 48–50
  symmetrical representation, 287

periodic tables, 20–21, 277–282
   best form, 282–283
   books on, xiv–xv
   circular, 291
   conferences on, 287
   contemporary reactions to, 146–149
   of De Chancourtois, 68–72
   format of modern, 10–16
   gaps in, 84, 96, 131–132, 172–174
   of Hinrich, 89–92
   horizontal series, 53–54
   influence of, 25, 27
   inorganic chemist's, 279f
   isotopes and, 160
   IUPAC, U.S., and European systems,
      13f
   long form, 17f
   of Lothar Meyer, 94
   medium-long form, 12f
   of Mendeleev, 106–112
   modified long form, 181f
   of Newlands, 91–92
   versus periodic systems, 18
   recent changes in, 21–24
   reduction of, 242–246
   short form, 14, 15f
   as a teaching tool, xx
   of van den Broek, 166–168
periods, 11. *See also* horizontal relationships
   beginning of, 282–283
   discovery of, 44
   length of, 85–86, 201–203
   number of elements in, 229
Perrier, Carlo, 174
Perrin, Jean, 185, 251
Petit, Alexis-Thérèse, 59–60, 127
Pettenkofer, Max, xix, 50–52, 94
philosophy. *See also under* chemistry
   of periodic system, xvii–xviii
   of science, xx–xxii
   of substance and matter, xv
phlogiston, xix, 30–31
phosphorescence, 161–162
phosphorus, 6, 271, 272
   electronic configuration, 190
   equivalent weights, 51
   Lavoisier's view of, 31
   oxidation states, 275

physical properties, 120
   in left-step table, 286
   Lothar Meyer's use of, 98
   preeminence of chemical properties
      over, 125
   used for classification, 22
Physical Society, 154
physicists, atomic theory and, 159
physics, xxi, 195
   versus chemistry, xiii, xvii–xviii, 120,
      211–212
   chemistry as subdiscipline of, xx
   effect on periodic system, xiv,
      24–25
   periodic table and, xix
   Taoism and, 291
pitchblende, 162
Planck, Max, 188, 314
Planck's constant, 189
planetary distances, 88t
planetary model, of atom, 87, 185–186
planets
   chemical composition of, 258
   names of elements from, 8
   symbols for elements and, 9, 10f
plasma state, 251
Platonic solids, 3
pleiads, 178
plum pudding model, 186, 314
plutonium, 6, 290
Pode, J.S.F., xv
Poincaré, Henri, 161
polonium, 176
   discovery of, 162
polyatomic molecules, 276, 309
Popper, Karl, 42, 319
positivism, xxi, 294
potassium
   electronic configuration, 234–235
   placement of, 50
   triad relationships, 43
potassium hydroxide, valence, 209
potentiality, 4
praseodymium, 302
precession, 192, 197
predictions, 74, 309
   versus accommodations, 156
   importance of, 56–57

predictions (*continued*)
  of Mendeleev, 106, 117–119, 123–124,
    131–135, 143–149
  of molecular properties, 25, 27
  of Odling, 83
  of trends in properties, 137
Priestley, Joseph, 31
primary and secondary kinship, 293
primary classification, 21–22
primary matter, 89
primary substance, Mendeleev's view of,
  103
Principe, Lawrence, xvi
Principles of Chemistry, The (Mendeleev),
  103, 117, 143
prism, 89
proactinium, 173
promethium, discovery of, 174
protium, 322
protons, 61, 160
  formation, 178
  number of, 20, 42
  stability and, 263
protyle, 39
Prout, William, xix, 39
Prout's hypothesis, xxi, 38–42, 61, 67, 119,
  149, 253, 296
  in light of isotopy, 160
  Mendeleev's view of, 103
  Moseley's work and, 175–176, 250
  triads and, 179, 182
Puddephatt, R.J., xv
Putnam, Hilary, 326
pyramidal periodic table, 283, 285f
pyroxenes, 72
Pythagoreanism, xxi, 87, 88

qualititative chemistry, role in periodic system, 48–50
quanta, 188
quantitative analysis, 29–31
quantization, 188, 192, 228
quantum mechanical nuclear-shell model,
  261
quantum mechanical tunneling, 256
quantum mechanics, xiii, xx, xxi, 120,
  228–231
  basis sets, 242
  effect on periodic system, 24–25

failure to explain periodic table, 237
  versus quantum theory, 216
  realism and, 319
  reductionism of, xiv, xxii, 247–248,
    318
  solving equations, 231–232
quantum numbers, xix, 232, 234, 320
  first, 282
  fourth, 199–201, 229
  principal and azimuthal, 192
  third, 197–198, 315
  values of, 201–202
quantum states, number of, 196, 198
quantum theory, 24, 224, 230, 312
  atomic, 188–192
  versus quantum mechanics, 216
  second, 192–197
  shortcomings of, 227
quintessence, 3

radiant light energy, 256
radiation, loss of energy through, 188
radicals, inclusion in tables of De Chancourtois, 69
radioactive atoms, creation of, 252
radioactive decay, 178–179
radioactive tracers, development of, 177
radioactivity, 117, 308
  discovery of, xix, 6, 159–163, 310
radiochemistry, 6
radiographs, first, 161
radiothorium, 176–177
radium, 117, 176, 298
  discovery of, 162
radon, 176
  discovery of, 291
Railsback, Bruce, 278
Ramsay, William, 7, 151–153
rare earth elements, 16, 150
  classification by Mendeleev, 117
  electronic configuration, 218–219
  placement of, 172
ratio of masses, 36
Rayleigh, John William, 7, 151–153
Rayner Canham, Geoffrey, 269
Rayner Canham table. *See* pyramidal periodic table
reactivity
  of metals, 50

ordering of elements and, 18–20
patterns, 49
realism, xxi–xxii, 21, 120, 278, 292
red giants, 258
reductionism, xviii, xx–xxi, xxii, 183, 247–248, 265, 283, 318
approaches to, 242–246
of Mendeleev, 119–121
refractive index, 155
relative density, 59, 64
relativism, xvii
relativistic effects, 237, 270, 293
relativity, xiii
effect on periodic system, 24–25
Remelé, Adolf, 98–99
repetition, 76, 77. *See also* periodicity
in chemical properties, 16
repulsion, 57, 241, 259
rhenium, discovery of, 173
rhodium, derivation of name, 8
Richards, Theodore W., 178–179
Richter, Jeremias Benjamin, 31, 36
Röntgen, Wilhelm Conrad, 161, 184
Rocke, Alan, 19
roentgenium, 290
Roscoe, Henry, 129
rotational energy, 151
rotational motion, 309
Royal Institution, 82
Royal Society of London, 152
r-process, 258
rubidium, 290, 298
Rücker, William Arthur, 154
rule of simplicity, 36, 295
Russian Chemical Society, 106
Russian Physico-Chemical Society, 155
ruthenium, 277
Rutherford, Ernest, xix, 159, 175–176, 188, 217, 314
assessment of van den Broek's work, 169
research on atomic nuclei, 164
research on Becquerel rays, 162–163

Sacks, Oliver, xv, 7
Saltpeter, Edwin, 254–255
Sanderson, Ralph, xv, 271
Saturn, 258
scandium, 217, 272
discovery of, 137, 156

electronic configuration, 235–236
orbital energies, 235, 237–238
predicted and observed properties, 138t
prediction of, 106, 123, 132, 147
similarity to aluminum, 267–268
scattering experiments, 164
Sceptical Chymist, The (Boyle), xvi
Scheele, Carl, 31, 48
Schrödinger, Erwin, xix, 7, 227, 230–231, 314
Schrödinger equation, xxi, 234, 244, 263
science
philosophy of, xx–xxii
postmodern critiques of, xvii
revolutions in development of, xviii
unity of disciplines, xxi
Seaborg, Glenn, 6, 7, 9, 21, 129, 269, 307
seaborgium, 9
secondary classification, 21–22
secondary periodicity, 270–272
Sedgwick, William, 309
Segrè, Emilio, 174
selenium, 300
atomic weight of, 299
deduction of atomic weight, 60
Gmelin's study of, 49
triad relationships, 43
separation, 176–177
discovery of isotopy and, 177–178
of rare earth elements, 172
separation energy plots, 261–263
Seubert, Carl, 98–99
siderophiles, 278, 326
Siegbahn, Manne, 175
silicon, 15, 271
equivalent weights, 51
Mendeleev's prediction of, 106
similarity to titanium, 268
silver, similarity to thallium, 273
simple substances, xv–xvi, 4, 5f, 117–118, 121, 289, 305
concept of, 113–114
periodic system and, 281–283
simplicity, rule of, 36, 295
single bonds, 209, 210f
single occupancy, 125
size ratios, of atomic dimensions, 88
slats, 18
social construction, science as, xvii

Soddy, Frederick, 163, 176, 177–179
sodium, 6, 301
    atomic structure, 206
    placement of, 50
    properties of, 11
    triad relationships, 43
sodium chloride, properties of, 113
Sommerfeld, Arnold, 192
specific heat, 59, 127, 151, 153–154
spectral frequencies, 25, 230
    atomic structure and, 185
    understanding of, 188
spectral lines
    effect of magnetic fields on, 196, 199–
        201
spectroscopy, 5, 71, 314
    role in discovery of elements, 67
    use by Hinrichs, 87–88, 92
spin angular momentum, 200, 233, 234
spin multiplicity, 240
spin-orbit coupling, 264–265, 324
s-process, 258
stability
    atomic, 188
    of isotopes, 249
    of nuclei, 259–265
    of orbitals, 236, 238
Stahl, Georg, 31, 288
stars
    contraction and collapse, 256
    nucleosynthesis in, 251, 253–255,
        259
    quantum mechanical tunneling in,
        256
Stas, Jean-Servais, 41, 103
stationary states, 195–198, 201, 316
steady-state models of the universe, 249,
    255
stoichiometry, 42, 305
Stoner, Edmund, 197–198, 222–223
Stoney, Johnston, 183
Strassman, Hans, 312
Strathern, P., xv
Strömholm, Daniel, 176
strontium
    placement of, 50
    triad relationships, 42
Strutt, R.J., 41
subshells, 222

substance, 304
    philosophy of, xv
Suess, Hans, 264
sulfur, 6, 300
    electronic configuration, 191
    Gmelin's study of, 49
    Lavoisier's view of, 31
    oxidation states, 275
    triad relationships, 43
sun
    age of, 258
    heavy elements in, 258
    hydrogen burning, 323
    temperature of, 255
superatom clusters, 277
superconductivity, 10–11
superheavy elements, 22
supernovae, 256, 258
supertriads, 55–57
Svedberg, Theodor, 176
symbols, for elements, 9–10
synthesis, of elements. *See* nucleosynthesis

tantalum, 175, 219, 290
teaching. *See* education
technetium, 6, 289
    discovery of, 174
Telluric Screw, 70f
tellurium, 300
    Gmelin's study of, 49
    placement by De Chancourtois, 69
    placement by Lothar Meyer, 96
    placement by Mendeleev, 125–126,
        130–131
    reversal with iodine, 76, 109, 117, 148,
        173, 178, 297–298, 309
    triad relationships, 43
temperature
    effect on triad relationships, 54
    needed for element formation, 255
    of superconductivity, 10–11
terbium, 174
    prediction of atomic weight, 56
tetrahedral particles, 3
textbooks
    explanations of electronic configura-
        tion, 235
    of Gmelin, 46–50
    of Lothar Meyer, 93–94

of Mendeleev, 103, 115–117
of Naquet, 65–66
thallium, 290
  discovery of, 250
  placement of, 98, 110
  similarity to silver, 273
theories, acceptance of, 123–124
theorists, versus observers, 150
theory of relativity, 153
  effect on periodic system, 24–25
thermodynamics, 200
Thomas, Llewellyn, 245
Thomas-Fermi method. *See* density functional theories
Thomsen, Jörgen, 309
Thomson, J.J., xix, xx, 7, 159, 164, 175, 205, 293, 313, 316
  discovery of electron, 183–188
Thomson, Thomas, 39–40
thorium, 176, 177, 270
  formation from uranium, 178
  radioactivity experiments, 162
thulium, 172, 174–175
tin, 15
  placement of, 98, 129
  similarity to zinc, 273
titanium
  placement of, 127
  similarity to silicon, 268
Tolman, Richard, 251
toxicity of elements, 273, 325
transformation. *See* transmutation
transition elements, 21, 307
  electronic configuration, 212, 215, 240–241
  electron shells of, 224
  occupation of orbitals, 236
  in tables of De Chancourtois, 69
  in tables of Lothar Meyer, 96, 98
transition metals, 302, 317
  atomic radii, 272t
  electronic configuration, 243
  in periodic tables, 14, 85–86, 300
  placement by Newlands, 77
translational energy, 151
translational motion, 153
transmutation, 43, 117, 163, 176, 253, 296, 304, 308
  hypothesis of Dumas, 52

refutation of, 34
trans-uranium elements, 6, 21
  naming, 8–9
triads, xxi, 53, 61, 94, 296
  conjugated, 54
  discovery of, 42–44
  expansion of, 45
  isotopy and, 179–182
  Mendeleev's view of, 103–104
  in modern periodic system, 57–58
  Pettenkofer's view of, 50
  super, 55–57
  use by Dumas, 52
triple alpha mechanism, 254
triple bonds, 209–210
truth, objective, xvii
tungsten, 270

units of measurement, for atomic weight, 41–42
unity of matter, 179, 294
universe
  age of, 258
  heat death, 251–252, 323
  steady-state models of, 249, 253
Upton, Thomas, 277
uranium, 21, 175, 270, 300, 307
  alpha decay of, 178
  correction of atomic weight, 128–129
  determination of valence, 126
  phosphorescence experiments, 161–162
  placement of, 110, 304
Urbain, Georges, 172, 173, 219, 220
  work with ytterbium, 217
U.S. periodic system, 13f, 14
utility, 286

valence, 19–20, 37, 190, 209, 291
  change across periods, 307
  determining, 126
  of groups, 221
  of metals, 128
  relationships, 302
  in tables of Lothar Meyer, 96
  use by Mendeleev, 103, 109, 116
vanadium, 98, 290
  placement of, 148
Van Assche, Pieter, 174
van den Broek, Anton, 7, 165–169

van Helmont, Johann, 294
van Spronsen, Jan, 47, 87, 142, 176, 297, 299
vaporization, 61
velocity, of electrons, 24
Venable, F.P., xiv, 296
vibrational energy, 151
vibrational motion, 154, 309
Vincent, Bernadette Bensaude, 305
von Hevesy, György, 173
von Laue, Max, 170
von Richter, Victor, 139
von Welsbach, Auer, 217

water
    data on composition of, 35
    as element, 3
    formula of, 57, 63–64, 294
    Lavoisier's view of, 31
    valence, 209
water vapor, formation experiments, 37
wavefunction, 228–234, 244, 246, 319–320
wavelength, 314
weight changes, of substances in chemical reactions, 30
Whiggism, xviii
Wiechert, Emil, 183
Williamson, Alexander, 74

Winkler, Clemens, 137–139
Wollaston, William, 292
Würzburg Physical-Medical Society, 161
Wurtz, Charles-Adolphe, 148

xenon
    discovery of, 155, 291
    similarity to osmium, 268
X-rays
    discovery of, 159–162
    link with luminescence, 161–162
    types, 170–171

ytterbia, 217
ytterbium, 172, 174–175, 217
    derivation of name, 8
yttrium, 272
    derivation of name, 8
    superconductivity of, 11

Zapffe, Karl, 86
Zeno of Elea, 32
zero-point energy, 252
zinc, 85, 290
    beta decay of, 258
    similarity to magnesium, 267
    similarity to tin, 273
zirconium, 219
zodiac, 3

BARBARA DELINSKY

Looking for Peyton Place

A NOVEL

Doubleday Large Print
Home Library Edition

Scribner

NEW YORK   LONDON   TORONTO   SYDNEY

This Large Print Edition, prepared especially for Double-day Large Print Home Library, contains the complete, unabridged text of the original Publisher's Edition.

SCRIBNER
1230 Avenue of the Americas
New York, NY 10020

SCRIBNER and design are trademarks of Macmillan Library Reference USA, Inc., used under license by Simon & Schuster, the publisher of this work.

ISBN 0-7394-5664-4

Manufactured in the United States of America

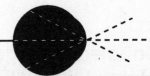

**This Large Print Book carries the
Seal of Approval of N.A.V.H.**

To my family,
with thanks and love

Acknowledgments

Gathering information on Grace Metalious was a challenge. So long after her death, precious little about her is in print. Two books, in particular, were a help: *The Girl From "Peyton Place,"* by George Metalious and June O'Shea (Dell Publishing Company, 1965) and *Inside Peyton Place: The Life of Grace Metalious,* by Emily Toth (Doubleday & Company, Inc., 1981). I found additional insight on Grace and her times in Ardis Cameron's introduction to the most recent edition of *Peyton Place* (Northeastern University Press, 1999). Finally, for information on Grace's childhood years in Manchester, New Hampshire, I thank Robert Perrault.

Having absorbed what I could from these sources, I tried to imagine what Grace would think, feel, and say. If my imaginings differ from what those who knew her saw, the fault is mine and mine alone.

For information on the current state of mercury regulation, I thank Stephanie D'Agostino, Co-Chair of the Mercury Reduction Task Force, New Hampshire Department of Environmental Services. For insight into what mercury poisoning actually *feels* like, I thank Claire Marino. At her behest, I advise readers that the treatment for mercury poisoning outlined in *Looking for Peyton Place* is still considered "alternative medicine." Though individuals do swear by its success, there is, as yet, no scientific study to suggest that it be the treatment of choice.

Thanks to the entire tireless Scribner publishing team. And again, always, I thank my agent, Amy Berkower.

Looking for Peyton Place

Prologue

I AM A WRITER. My third and most recent novel won critical acclaim and a lengthy stay on the best-seller lists, a fact that nearly a year later I'm still trying to grasp. Rarely does a day pass when I don't feel deep gratitude. I'm only thirty-three. Not many writers attain the success I have in a lifetime, much less at my age, much less with the inauspicious start I had.

By rights, given how my earliest work was ridiculed, I should have given up. That I didn't spoke either of an irrepressible creative drive or of stubbornness. I suspect it was a bit of both.

It was also Grace.

Let me explain.

I am from Middle River. Middle River is a small town in northern New Hampshire that, true to its name, sits on a river midway between two others, the Connecticut and the Androscoggin. I was born and raised there. That meant living not only in the shadow of the White Mountains, but in that of Grace Metalious.

Grace *who*? you ask.

Had I not been from Middle River, I probably wouldn't have known who she was, either. I'm too young. Her provocative best-seller, *Peyton Place,* was published in 1956, sixteen years before I was born. Likewise, I missed the movie and the television show, both of which followed the book in close succession. By the time I arrived in 1972, the movie had been mothballed and the evening TV show canceled. An afternoon show was in the works, but by then Grace had been dead seven years, and her name was largely forgotten.

I am always amazed by how quickly her fame faded. To hear tell, when *Peyton Place* first came out, Grace Metalious made head-lines all over the country. She was an unknown who penned an explosive novel, a New Hampshire schoolteacher's wife who wrote about sex, a young woman in sneakers and blue jeans who dared tell the truth about small-town life and—even more unheard-of—about the yearnings of women. Though by today's standards *Peyton Place* is tame, in 1956 the book was a shocker. It was banned in a handful of American counties, in many more libraries than that, and in Canada, Italy, and Australia; Grace was shunned by neighbors and received threatening mail; her husband lost his job, her children were harassed by classmates. And all the while millions of people, men and women alike, were reading *Peyton Place* on the sly. To this day, take any copy from a library shelf, and it falls open to the racy parts.

But memory is as fickle as the woman Grace claimed Indian summer to be in the opening lines of her book. Within a decade of its publication and her own consequent notoriety, people mentioning *Peyton Place* were more apt to think of Mia Farrow and Ryan O'Neal on TV, or of Betty Anderson teasing Rodney Harrington in the backseat of John Pillsbury's car, or of Constance

MacKenzie and Tomas Makris petting on the lakeshore at night, than of Grace Metalious. *Peyton Place* had taken on a life of its own, synonymous with small-town secrets, scandals, and sex. Grace had become irrelevant.

Grace was never irrelevant in Middle River, though. Long after *Peyton Place* was eclipsed by more graphic novels, she was alternately adored and reviled—because Middle River knew what the rest of the world did not, and whether the town was right didn't matter. All that mattered was the depth of our conviction. We knew that *Peyton Place* wasn't modeled after Gilmanton or Belmont, as was popularly believed. It's *us,* Middle River said when the book first appeared, and that conviction never died.

This I knew firsthand. Even all those years after *Peyton Place*'s publication, when I was old enough to read, old enough to spend hours in the library, old enough to lock myself in the bathroom to write in my journal and to have the sense that I was following in a famous someone's footsteps, the town talked. There were too many parallels between Peyton Place and Middle River to ignore, starting with the physical layout of the town, proceeding to characters like the wealthy owner of the newspaper, the feisty but good-hearted doctor, the adored spinster teacher and the town drunk, and ending, in a major way, with the paper mill. In *Peyton Place,* the mill was owned and, hence, the town controlled by Leslie Harrington; in Middle River, that family name was Meade. Benjamin Meade was the patriarch then, and he wielded the same arrogant power as Leslie Harrington. And like Leslie's son Rodney, Benjamin's son, Sandy, was cocksure and wild.

Naysayers called these parallels mere coincidence. After all, Middle River sat farther north than the towns in which Grace Metalious had lived. Moreover, we had no proof that Grace had ever actually driven down our Oak Street, seen the red brick of Benjamin Meade's Northwood Mill, or eavesdropped on town gossip from a booth at Omie's Diner.

Mere coincidence, those naysayers repeated.

But *this* much coincidence? Middle River asked.

There were other similarities between the fictitious Peyton Place and our very real Middle River—scandals, notably—some of which I'll recount later. The only one I need to mention now is a personal one. Two major characters in *Peyton Place* were Constance MacKenzie and her daughter, Allison. With frightening accuracy, they corresponded to Middle River's own Connie McCall and her daughter, Alyssa. Like their fictitious counterparts, Connie and Alyssa lived manless in Connie's childhood home. Connie ran a dress shop, as did Constance MacKenzie. Likewise, Connie's Alyssa was born in New York, returned to Middle River with her mother and no father, and grew up an introverted child who always felt different from her peers.

The personal part? Connie McCall was my grandmother, Alyssa my mother.

My name is Annie Barnes. Anne, actually. But Anne was too serious a name for a very serious child, which apparently I was from the start. My mother often said that within days of my birth, she would have named me Joy, Daisy, or Gaye, if she hadn't already registered Anne with the state. Calling me Annie was her attempt to soften that up. It worked particularly well, since my middle initial was *E*. I was Anne E. Annie. The *E* was for Ellen—another serious name—but my sisters considered me lucky. They were named Phoebe and Sabina, after ancient goddesses, something they felt was pretentious, albeit characteristic of our mother, whose whimsy often gravitated toward myth. By the time I was born, though, our father was sick, finances were tight, and Mom was in a down-to-earth stage.

If that sounds critical, I don't mean it to be. I respected my mother tremendously. She was a woman caught between generations, torn between wanting to make a name for herself and wanting to make a family. She had to choose. Middle River wouldn't let her do both.

That's one of the things I resent about the town. Another is the way my mother and grandmother were treated when *Peyton Place* first appeared. Prior to that time, Middle River had bought into the story that my grandmother was duly married and living in New York with her husband when my mother was conceived, but that the man died shortly thereafter. When *Peyton Place* suggested another scenario, people began snooping into birth and death records, and the truth emerged.

If you're thinking that my grandmother

might have sued Grace Metalious for libel, think again. Even if she could prove malicious intent on Grace's part—which she surely couldn't—people didn't jump to litigate in the 1950s the way they do now. Besides, the last thing my grandmother would have wanted was to draw attention to herself. Grace's fictional Constance MacKenzie had it easy; the only person to learn her secret was Tomas Makris, who loved her enough to accept what she had done. My real-life grandmother had no Tomas Makris. Outed to the entire town as an unmarried woman with a bastard child, she was the butt of sly whispers and scornful looks for years to come. This took its toll. No extrovert to begin with, she withdrew into herself all the more. If it hadn't been for the dress shop, which she relied on as her only source of support and ran with quiet dignity—and skill enough to attract even reluctant customers—she would have become a recluse.

So I did hold a grudge against the town. I found Middle River stifling, stagnant, and cruel. I looked at my sisters, and saw intelligent women in their thirties whose lives were wasted in a town that discouraged free expression and honest thought. I looked at my mother, and saw a woman who had died at sixty-five—*too young*—following Middle River rules. I looked at myself and saw someone so hurt by her childhood experiences that she'd had to leave town.

I faulted Middle River for much of that.

Grace Metalious was to blame for the rest. Her book changed all of our lives—mine, perhaps, more than some. Since Middle River considered my mother and grandmother an intricate part of *Peyton Place,* when I took to writing myself, comparisons with Grace were inevitable. Aside from those by a local bookseller—analogous in support to Allison MacKenzie's teacher in *Peyton Place*—the comparisons were always derogatory. I was a homely child with my nose in books, then a lonely teenager writing what I thought to be made-up stories about people in town, and I stepped on a number of toes. I had no idea that I was telling secrets, had no idea that what I said was true. I didn't know what instinctive insight was, much less that I had it.

Too smart for her own good, huffed one peeved subject. *There's a bad seed in that child,* declared another. *If she isn't careful,* warned a third, *she'll end up in the same mess Grace did.*

Intrigued, albeit perversely, I learned all I could about Grace. As I grew, I identified with her on many levels, from the isolation she felt as a child, to her appreciation of strong men, to her approach to being a novelist. She became part of my psyche, my alter ego at times. In my loneliness I talked with her, carried on actual conversations with her right up into my college years. More than once I dreamed we were related—and it wasn't a bad thought at all, because I loved her spirit. She often said she wrote for the money, but my reading suggests it went deeper than that. She was driven to write. She wanted to do it well. And she wanted her work to be taken seriously.

So did I. In that sense alone, Grace was an inspiration to me, because *Peyton Place* was about far more than sex. Move past titillation, and you have the story of women coming into their own. This was what I wrote about, myself.

But I saw what had happened to Grace. Initial perceptions stick; once seen as a writer of backseat sex, always seen as a writer of backseat sex. So I avoided backseat sex. I chose my publisher with care. Rather than being manipulated by publicity

as Grace had been, I manipulated the publicity myself. Image was crucial. My bio didn't
mention Middle River, but struck a more sophisticated pose. It helped that I lived in
Washington, a hub of urbanity even with its
political hot air—helped that Greg Steele,
my roommate, was a national correspondent
for network television and that I was his date
at numerous events of state—helped that I
had grown into a passably stylish adult who
could wear Armani with an ease that made
my dark hair, pale skin, and overly wide-set
eyes look exotic.

Unfortunately, Middle River didn't see any
of this—because yes, initial perceptions
stuck. The town was fixated on my being its
own Grace. It didn't matter that I had been
gone for fifteen years, during which time I
had built a national name for myself. When I
showed up there last August, they were convinced I had returned to write about them.

The irony, of course, was that I didn't seriously consider it until they started asking.
They put the bug in my ear. But I didn't deny
it. I was angry enough to let them worry. My
mother was dead. I wanted to know why. My
sisters were content to say that she died after a fall down the stairs, in turn caused by a

loss of balance. I agreed that the fall had killed her, but the balance part bothered me. I wanted to know *why* her balance had been so bad.

Something was going on in Middle River. It wasn't documented—God help us if anything there was forthright—but the *Middle River Times,* which I received weekly, was always reporting about someone or other who was sick. Granted, I was a novelist; if I hadn't been born with a vivid imagination, I would have developed one in the course of my work, which meant that I could dream up scenarios with ease. But wouldn't *you* think something was fishy if people in a small town of five thousand, max, were increasingly, chronically ill?

As with any good plot, dreaming it up took a while. I was too numb to do much of anything at first. My sisters hadn't painted a picture anywhere near as bleak as they might have, so my mother's death came at me almost out of the blue. I'd like to say Phoebe and Sabina were sparing me worry—but we three knew better. There was an established protocol. What I didn't ask, they didn't tell. We weren't very close.

The funeral was in June. I was in Middle

River for three days, and left with no plans ever to return.

Then the numbness wore off, and a niggling began. It had to do with my sister Phoebe, who was so grief-stricken calling me about Mom's death that she didn't know my voice on the phone, so distracted when I reached Middle River that carrying on a conversation with her was difficult. It was only natural that Mom's death would hit her the hardest, Sabina argued dismissively when I asked her about it. Not only had they lived and worked together, but Phoebe was the one who had found Mom at the foot of the stairs.

Still, I had seen things in Phoebe during those three days in Middle River that, in hindsight and with a clearer mind myself now, were eerily reminiscent of Mom's unsteadiness, and I was haunted. How to explain Mom being sick? How to explain *Phoebe* being sick? Naturally, my imagination kicked into high gear. I thought of recessive genes, of pharmacological complications, of medical incompetency. I thought of the TCE used to clean printing presses down the street from the store. I thought of *poison,* though had no idea why anyone would have cause to poison my mother and sister. Of all the scenarios I dreamed up, the one I liked best had to do with the release of toxicity into the air by Northwood Mill. I detested the Meades. They had been responsible for the greatest humiliation of my life. As villains went, they were ideal.

That said, I had been in enough discussions with Greg and his colleagues about the importance of impartiality to know not to point every finger at Northwood. During those warm July weeks, I divided my time between finishing the revisions of my new book and exploring those other possibilities.

Actually, I spent more time on the latter. It wasn't an obsession. But the more I read, the more into it I was.

I ruled out TCE, because it caused cancer, not the Parkinsonian symptoms Mom had had. I ruled out pharmacologic complications, because neither Mom nor Phoebe took much beyond vitamins. Mercury poisoning would have been perfect, and the mill did produce mercury. Or it had. Unfortunately for me, state records showed that Northwood had stopped using mercury years before.

I finally came across lead. Mom's store, Miss Lissy's Closet, had been rehabbed four years ago, largely for decorating purposes, but also for the sake of scraping down and removing old layers of paint that contained lead. My research told me that lead poisoning could cause neurological disorders as well as memory lapses and concentration problems. If Mom and Phoebe had been in the store while the work was being done, and ventilation had been poor, they might have inhaled significant amounts of lead-laden dust. Mom was older, hence weakened sooner.

Lead poisoning made sense. The clincher, for me, was that the Meades owned the building, had suggested doing the work, and had hired the man who carried it out. I would like nothing more than to have the Meades found liable for the result.

Facts were needed, of course. I tried to ask Phoebe whether she and Mom had been in the store while the work was being done, what precautions had been taken, whether the date of the work had preceded Mom's first symptoms. But my questions confused her.

Sabina wasn't confused. She said—unequivocally—that I was making things worse.

Worse was the operative word. Phoebe wasn't recovering from Mom's death, and my imagination wouldn't let go.

By the end of July, I made the decision. August promised to be brutal in the nation's capital—hot, humid, and highly deserted. Most of my friends would be gone through Labor Day. Greg had been given a month's leave by the network and was bound for Alaska to climb Mount McKinley, which was a three-week trek even without travel to and from. There was little reason for me to hang around in the District and good reason for me to leave. I had done all I could from afar. I had to be with Phoebe again to see if what I imagined seeing in her were really symptoms. I had to talk with people to learn how much of the paint removed had contained lead and how it had been removed. A phone call wouldn't do it. A *dozen* phone calls wouldn't do it.

I hadn't spent more than a weekend in Middle River in fifteen years. That I was willing to do so now vouched for my concern.

By the way, if you're thinking I never saw my mother and sisters during those years of exile, you're wrong. I saw them. Every winter, we met somewhere warm. The destination varied, but not the deal. I paid for us all, including Phoebe's husband while they were married and Sabina's longer-lasting husband and kids. Same with the summers we met in Bar Harbor. I had the money and was glad to spend it on our annual reunion. Unspoken but understood was my aversion to the tongue-wagging that would take place if I showed up in town.

I was right to expect it. Sure enough, this August, though I pulled up at the house on Willow Street at night, by noon the next day, word of my arrival had spread. During a quick stop at the post office, I was approached by six—*six*—people asking if I had come home to write about them.

I didn't answer, simply smiled, but the question kept coming. It came with such frequency over the next few days that my imagination went into overdrive. Middle River was nervous. I wondered what dirty little secrets the locals had to hide.

But dirty little secrets of the personal variety didn't interest me. I had no intention of being cast in the Grace Metalious role. She and I hadn't talked in years—as "talking" with a dead person went. I had earned my own name. I had my own life, my own friends, my own career. The only reason I was in Middle River was to find an explanation for my family's illness. Could be it was lead. Could be it wasn't. Either way, I had to know.

Then came the photo. Several days before I left Washington, I was at the kitchen table with my laptop, finishing the revisions of my next book. Morning sun burned across the wood floor with the promise of another scorcher. The central air was off, the windows open. I knew that I had barely an hour before that would have to change, but I loved the sound of birds in our lone backyard tree.

I wore denim shorts, a skimpy T-shirt, and my barest Mephisto slides. My hair was in a ponytail, my face without makeup. I hadn't been working fifteen minutes when I kicked off the slides. Even my iced latte was sweating.

Studying the laptop screen, I sat back, put the heels of my feet on the edge of the seat, braced my elbows on my knees and rested my mouth on my fists.

Click.

"The writer at work," Greg declared with a grin as he approached from the door.

Greg was usually the handsome face of the news, not a filmer of it. But he was a digital junkie. He had researched for days before deciding which camera to buy for his trip. "That the new one?" I asked.

"Sure is," he replied, fiddling with buttons. "Eight megapixels, ten times optical zoom, five-area autofocus. It's a beauty." He held out the camera so that I could see the monitor, and the picture he had taken.

My first thought was that I looked very un-Washington-like, very naive, very much the country girl I didn't want to be. My second thought was that I looked a lot like Grace Metalious had in her famed photograph.

Oh, there were differences. I was slim, she was heavier-set. My hair was straight and in a high ponytail, while hers was caught at the nape of her neck and had waves. In the photograph, she wore a plaid flannel shirt, jeans rolled to midcalf, and sneakers; I wore shorts and no shoes. But she sat at her typewriter with her feet up as I did, with her elbows on her knees and her mouth propped on her hands. Eyes dark as mine, she was focused on the words she had written.

Pandora in Blue Jeans, the shot was

called. It was Grace's official author photo, the one that had appeared on the original edition of *Peyton Place* and been reproduced thousands of times since. That Greg had inadvertently taken a similar shot of me so soon before my return home struck me as eerie.

I pushed it out of my mind at the time. Weeks later, though, I would remember. By then, Grace would be driving me nuts.

Her story had no happy ending. As successful as *Peyton Place* was, Grace saw only a small part of the money it made, and that she spent largely on hangers-on who were only too eager to take. Distraught over reviews that reduced *Peyton Place* to trash, she set the bar so high for her subsequent work that she was destined to fail. She turned to booze. She married three times— twice to the same man—and had numerous affairs. Feeling unattractive, untalented, and unloved, she drank herself to death at the age of thirty-nine.

I had no intention of doing that. I had a home; I had friends. I had a successful career, with a new book coming out the next spring and a contract for more. I didn't need money or adulation, as Grace had. I wasn't desperate for a father figure, as she was, and I didn't have a husband to lose his job or children to be taunted by classmates.

All I wanted was the truth about why my sister was sick and my mother was dead.

Chapter 1

I APPROACHED Middle River at midnight—pure cowardice on my part. Had I chosen to, I might have left Washington at seven in the morning and reached town in time to cruise down Oak Street in broad daylight. But then I would have been seen. My little BMW convertible, bought used but adored, would have stood out among the pickups and vans, and my D.C. plates would have clinched it. Middle River had expected me back in June for the funeral, but it wasn't expecting me now. For that reason, my face alone would have drawn stares.

But I wasn't in the mood to be stared at,

much less to be the night's gossip. As confident as my Washington self was, that confidence had gradually slipped as I had driven
north. I drank Evian; I nibbled a grilled
salmon wrap from Sutton Place and
snacked on milk chocolate Toblerone. I
rolled my white jeans into capris, raised the
collar of my imported knit shirt, caught my
hair up in a careless twist held by bamboo
sticks—anything to play up sophistication, to
no avail. By the time I reached Middle River,
I was feeling like the dorky misfit I had been
when I left town fifteen years before.

Focus, I told myself for the umpteenth time
since leaving Washington. *You're not dorky
anymore. You've found your niche. You're a
successful woman, a talented writer. Critics
say it; the reading public says it. The opinion
of Middle River doesn't matter. You're here
for one reason, and one reason alone.*

Indeed, I was. All I had to do was to remember that Mom wouldn't be at the house
when I arrived, and my anger was stoked. I
wrapped myself in that anger and in the
warm night air when, in an act of defiance
just south of town, I lowered the convertible
top. When Middle River came into view, I
was able to see every sleepy inch.

To the naive eye, especially under a clear moon, the setting was quaint. In *Peyton Place,* the main street was Elm. In ours, it was Oak. Running through the center of town, it was wide enough to allow for sidewalks, trees, and diagonal parking. Shops on either side were softly lit for the night in a way that gave a brief inner glimpse of the purpose of each: a lineup of lawn mowers in Farnum Hardware, shelves of magazines in News 'n Chews, vitamin displays at The Apothecary. Around the corner was the local pub, the Sheep Pen, dark except for the frothy stein that hung high outside.

On my left as I crossed the intersection of Oak and Pine, a barbershop pole marked the corner where Jimmy Sacco had cut hair for years before passing his scissors to Jimmy the younger. The pole gleamed in my headlights, tossing an aura of light across the benches on either side of the corner. In good weather those benches were filled, every bit as much the site of gossip-mongering as the nail shop over on Willow. At night they were empty.

Or usually so. Something moved on one of them now, small and low to the seat, and I was instantly taken back. Barnaby? Could it be? He had been just a kitten when I left town. Cats often lived longer than fifteen years.

Unable to resist, I pulled over to the curb and shifted into park. Leaving my door open, I went up the single step and, with care now, across the boardwalk to the bench. I used to love Barnaby. More to the point, Barnaby used to love me.

But this wasn't Barnaby. Up close, I could see that. This cat, sitting up now, was a tabby. It was orange, not gray, and more fuzzy than Barnaby had been. A child of Barnaby's? Possibly. The old coot had sired a slew of babies over the years. My mother, who knew of my fondness for Barnaby, had kept me apprised.

Soothed by the faint whiff of hair tonic that clung to the clapboards behind the bench, I extended a hand to the new guard. The cat sniffed it front and back, then pushed its head against my thumb. Smiling, I scratched its ears until, with a put-put-putter, it began to purr. There is nothing like a cat's purr. I had missed this.

I was straightening when I heard a murmur. *Cats have claws,* it might have said, but when I looked around, there were no shadows, no human forms.

The cat continued to purr.

I listened for a minute, but the only sound here on the barbershop porch was that purr. Again, I looked around. Still, nothing.

Chalking it up to fatigue, I returned to my car and drove on—and again the town's charm hit me. Across the street was the bank and, set back from the sidewalk, the town hall. The Catholic church was behind me and the Congregational church ahead, white spires gently lit. Each was surrounded by its woodsy flock, a generous congregation of trees casting moon shadows on the land. It was a poet's dream.

But I was no poet. Nor was I naive. I knew the ugly little secrets the darkness concealed, and it went far beyond those men who, like Barnaby, sowed their seed about town. I knew that there was a place on the sign between *Farnum* and *Hardware* where *and Son* had been, until that son was arrested for molesting a nine-year-old neighbor and given a lengthy jail term. I knew that a bitter family feud had erupted when old man Harriman died, resulting in the splitting of Harriman's General Store into a grocery and a bakery, two separate entities, each with its own door, its own space, and its own sign, and a solid brick wall between. I knew that there were scorch marks, scrubbed and faded but visible nonetheless, on the stone front of the newspaper office, where Gunnar Szlewitchenz, the onetime town drunk, had lit a fire in anger at the editor for misspelling his name in a piece. I knew that there was a patched part of the curb in front of the bank, a reminder of Karl Holt's attempt to use his truck as a lethal weapon against his cheating wife, who had worked inside.

These things were legend in Middle River, stories that every native knew but was loathe to share with outsiders. Middle River was insular, its face carefully made up to hide warts.

Holding this thought, I managed to avoid nostalgia until I passed the roses at Road's End Inn. Then it hit in a visceral way. Though I couldn't see the blooms in the dark, the smell was as familiar to me as any childhood memory, as evocative of summer in Middle River as the ripe oak, the pungent hemlock, the moist earth.

Succumbing for an instant, I was a child returning with chocolate pennies from News 'n Chews, shopping alone for the first time, using those rosebushes as a marker pointing me home. I could taste the chocolate, could feel the excitement of being alone, the sense of being grown-up but just a tad afraid, could smell the roses—unbelievably fragrant and sweet—and know I was on the right path.

Now as then, I made a left on Cedar, but I had no sooner completed the turn when I stepped on the brake. On the road half a block ahead, spotlit in the dark, was a tangle of bare flesh, one body, no, two bodies entwined a second too long in that telltale way. By the time they were up and streaking for the trees amid gales of laughter, I had my head down, eyes closed, cheeks red. When I looked up, they were gone. My blush lingered.

"Pulling a backabehind," it was called by the kids in town, and it had been a daredevil antic for years. Middle River's answer to the mile-high club, pulling a backabehind entailed making love in the center of town at midnight. This couple would get points off for being on Cedar, rather than on Oak, and points off if the coupling failed to end in, uh, release. Whether they would tell the truth about either was doubtful, but the retelling would perpetuate the rite.

Growing up here, I had thought pulling a backabehind was the epitome of evil. Now, seeing two people clearly enjoying themselves, doing something they would have done elsewhere anyway, I was amused. Grace would have loved this. She would have written it into one of her books. Heck, she would have done it herself, likely with George, the tall, sexy Greek who was her first and third husband and, often, her partner in rebellion.

Still smiling, I approached the river. The air was suddenly warmer and more humid, barely moving past my flushed face as I drove. The sound of night frogs and crickets rose above the hum of my engine, but the river flowed silently, seeming this night unwilling to compete. And yet I knew it was there. It was always there, in both name and fact. Easily 70 percent of the town's workforce drew a weekly paycheck from the Northwood Mill account, and the river was the lifeblood of the mill.

A short block up, I turned right onto Willow. It wasn't the fanciest street in town; that would be Birch, where the elite lived in their grand brick-and-ivy Colonials. But what Willow lacked in grandeur it made up for in charm. The houses here were Victorian, no two exactly alike. The moon picked out assorted gables, crossbars, and decorative trim; my headlights bounced off picket fences of various heights and styles. The front yards were nowhere near as meticulously manicured as those on Birch, but they were lush. Maples rose high and spread wide; rhododendron, mountain laurel, lilac, and forsythia, though well past bloom, were all richly leafed. And the street's namesake willows? They stood on the riverbank, as tall and stately as anything weeping could be, their fountainous forms graceful enough for us to forgive them the mess their leaves made of our lawns.

Quaint downtown, quintessentially New England homes, historic mill—I understood how a visitor could fall in love with Middle River. Its visual appeal was strong. But I wasn't being taken in. This rose had a thorn; I had been pricked too many times to forget it. I wasn't here to be charmed, only to find an answer or two.

Naturally, I was more diplomatic in the voice message I had left earlier for Phoebe. The few questions I had asked since I'd been here last hadn't been well received. Phoebe was unsettled and Sabina defensive. I didn't want to get off on the wrong foot now.

With Mom gone, Phoebe, who was the oldest of us three, lived alone in the house where we grew up. If I still had a home in Middle River, it was here. I didn't consider staying anywhere else.

"Hey, Phoebe," I had said after the beep, "it's me. Believe it or not, I'm on my way up there. Mom's death keeps nagging at me. I think I just need to be with you guys a little. Sabina doesn't know I'm coming. I'll surprise her tomorrow. But I didn't want to frighten you by showing up unannounced in the middle of the night. Don't wait up. I'll see you in the morning."

The house was the fifth on the left, yellow with white trim that glowed in my headlights as I turned into the driveway. Pulling around Phoebe's van, I parked way back by the garage, where my car wouldn't be seen from the street. I checked the sky; not a cloud. Leaving the top down, I climbed out and took my bags from the trunk. Looping straps over shoulders and juggling the rest, I started up the side stairs, then stopped, burdened not so much by the weight of luggage as by memory. No anger came with it now, only grief. Mom wouldn't be inside. She would never be inside again.

And yet I pictured her there, just beyond the kitchen door, sitting at the table waiting for me to come home. Her face would be scrubbed clean, her short, wavy blonde hair tucked behind her ears, and her eyes concerned. Oh, and she would be wearing pajamas. I smiled at the memory of that. She claimed it was about warmth, and perhaps it was, though I do remember her wearing nightgowns when I was young. The change came when I was a teenager. She would have been in her forties then and slimmer than at any time in her life. With less fat to pad her, she might have been chilled. So maybe it really was about warmth.

I suspect something else, though. My grandmother, who had always worn pajamas, died when I was fourteen. The switch came soon after. Mom became her mother then, and not only in bedtime wear. With Connie gone, Mom became the family's moral watchdog, waiting up until the last of us was home without incident—*without incident,* that was key, because incident led to disgrace. Public drunkenness, lewd behavior, unwanted pregnancy—these were the things Middle River talked about in the tonic-scented cloud that hovered over the barber-shop benches, through the lacquer smell in the nail shop, over hash at Omie's. Being the butt of gossip was Mom's greatest fear.

Grace Metalious had hit the nail on the head with that one. As frightened as the ficti-tious Constance MacKenzie was of her secret leaking out in *Peyton Place,* she was more terrified of being talked about when it did.

Mind you, other than the fiasco with Aidan Meade, I never gave Mom cause for worry. I didn't date. What I did, starting soon after I could drive, was to go down to Plymouth on a Saturday night, stake out a table at a cof-fee shop, and read. Being alone in a place where I knew no one was better than being alone on a Saturday night in Middle River. And Mom would be waiting up for me when I returned, which eased the loneliness.

Feeling the full weight of that loneliness now, I went on up the stairs, opened the door, and slipped inside. Mom wasn't there, but neither was the kitchen I recalled. It had been totally renovated two years before, with Mom already ill, but determined. I had seen the changes when I was here for the funeral, but, climbing those side steps, my mind's eye still had pictured the old one, with its aged Formica countertops, its vintage appliances and linoleum floor.

In the warm glow of under-cabinet halogens, this new kitchen was vibrant. Its walls were painted burgundy, its counters were beige granite, its floor a brick-hued adobe tile. The appliances were stainless steel, right down to a trash compactor.

I didn't have a trash compactor. Nor did I have an ice dispenser on the refrigerator door. This kitchen was far more modern than mine in Washington. I was duly impressed, as I had been too preoccupied to be in June.

The kitchen table was round, with a maple top, wrought-iron legs, and ladderback chairs of antique white. Setting my computer bag on one of the latter, I turned off the lights, went through the hall and into the front parlor to turn off its lamp. There I found my sister Phoebe, under a crocheted afghan on the settee. Her eyes were closed; she was very still.

Of the three of us, she resembled Mom the most. She had the same high forehead and bright green eyes, the same wavy blonde hair, the same thin McCall mouth. She looked older than when I had seen her last month, perhaps pale from the strain of carrying on. Totally aside from any physical problems she might be having, I could only begin to imagine what the past month had been like for her. My own loneliness in coming home to a house without Mom was nothing compared to Phoebe's feeling it all the time. Not only had she lived with Mom for all but the short span of her own marriage, but she worked with Mom too. The loss had to be in her face day in and day out.

Lowering my bags, I slipped down onto the edge of the settee and lightly touched the part of her under the afghan that would have been an arm. "Phoebe?" I whispered. When she didn't move, I gave her a little shake.

Her eyes slowly opened. She stared at me for a blank moment, before blankness became confusion. "Annie?" Her voice was uncharacteristically nasal.

"You got my message, didn't you?"

"Message," she repeated, seeming muddled.

My heart sank. "On your voice mail? Saying I was coming?" I had assumed the lights were left on for me.

"I don't . . . on my voice mail? I think so . . .

I must have." Her eyes cleared a little. "I'm just groggy from medicine. I have a cold." That explained the nasal voice. "What time is it?"

I checked my watch. "Twelve-ten." Deciding that grogginess from cold medicine was reasonable, I tried to lighten things with a smile. "The kitchen startled me. I keep expecting to see the old one."

I wasn't sure she even heard my remark. She was frowning. "Why're you here?"

"I felt a need to visit."

After only the briefest pause, she asked, "Why now?"

"It's August. Washington's hot. I finished the revisions of my book. Mom's gone."

Phoebe didn't move, but she grew more awake. "Then it is about Mom. Sabina said it would be."

"You told Sabina I was coming?" I asked in dismay. I would rather have called Sabina myself.

"I was over there for dinner. I couldn't not say."

Fine. I didn't want to fight. Sabina would have found out soon, anyway. "I miss Mom," I said. "I haven't taken the time to mourn. I want to know more about those last days, what was wrong with her, y'know?"

"What about the house?"

I frowned. "What about it?"

"Sabina said you'd want it."

"*This* house?" I asked in surprise. "Why would I want it? I have my own place. This is yours."

"Sabina said you'd want it anyway. She said you'd know all the little legal twists, and that it would be about money."

"Money? Excuse me? I have plenty of money." But I wasn't surprised Sabina would think I wanted more. She was always expecting the worst of me, which was why I would have preferred to phone her myself to let her know I was here. Then I might have nipped her suspicions in the bud.

"Have I asked for anything from Mom's estate?" I asked now.

Phoebe didn't reply. She looked like she was trying to remember.

"Mom's been dead barely six weeks," I went on. "Has Sabina been stewing about this the whole time?"

"No. Just . . . just once she found out you were coming, I guess."

That quickly, I was back in the midst of childhood spats. Sabina was the middle child, which should have made her the peacemaker of the family, but that had never been the case. The eleven months between Phoebe and her had left her craving attention, a situation I had aggravated with my arrival when Sabina was barely two.

Now I said, "This is why I called you and not her. I knew she wouldn't be happy about my visit, and that's really sad, Phoebe. Middle River's where I grew up. My family is here. Why does she have to feel threatened?"

Phoebe still hadn't stirred on the settee, but her eyes were as sharp as Mom's could be when she was worried we had done something wrong. "She doesn't know you anymore. I don't either."

"I'm your *sister*."

"You're a writer. You live in the city and you travel all over. You eat out more than you eat in. You know *celebrities*." Her eyes rounded when she recalled something else. "And your *significant other* is on TV all the time."

"He isn't my significant other," I reminded her.

"Roommate, then," she conceded and took a stuffy breath. "But even that's totally different from Middle River. Single women don't buy condos here with single men."

"Greg and I protect each other—but that's getting off the subject. I'm your *sister,* Phoebe," I repeated, pleading now, because the discussion was making me feel even more alone than I had felt entering the kitchen and finding no Mom. "I've tried to *give* in the last few years. Isn't that what our vacations were about? And the money for the new van? And even the new kitchen," I added, though my part was the appliances alone. "Why do you think I would try to take something you have?"

Seeming suddenly groggy again, Phoebe lifted an arm from under the afghan. She squeezed her eyes shut, rubbed them with thumb and forefinger. "I don't know. I don't keep lists."

"Phoebe," I chided.

"I guess not, but Sabina says—"

"Not Sabina," I cut in. "*You.* Do *you* distrust me, too?"

In a reedy voice, she said, "I'm confused sometimes." Her hand fell away. She opened her eyes, looking pitiful, and again, my heart sank. Something was definitely wrong.

"It's your cold," I reasoned, but was suddenly distracted. With the afghan lowered, I could see what she was wearing. Smiling, I teased, "Are those pajamas?"

She was instantly defensive. "What's wrong with pajamas?"

"Nothing. It's just that it was a Mom thing to do."

"She was cold. Now I'm cold."

The room was not cold. If anything, it was hot. I had entered town with my top down, while Phoebe had her windows shut tight. The house had no AC. Even the warmth outside would have stirred the air in here. Surely the moisture of the night would have helped Phoebe's cold.

Again, I thought how wan she looked. "Have you been sick long?" I asked.

She sighed. "I'm not sick. It's just a cold. They're a fact of life. Customers bring them into the shop all the time. It's late. I'd better go to bed."

I rose from the settee and shouldered my bags, then glanced back. Phoebe was holding the arm of the settee with one hand while she pushed herself up with the other. She reminded me of Mom the last time I had seen her. That wasn't good.

"Seriously, Phoebe, are you okay?"

On her feet now, she held up both hands.
"I. Am. Fine. Go on. I'll get the light."

I was in the hall when the parlor went
dark, leaving the stair lit by a lamp at the top.
I went on up, then down the hall to the room
that had always been mine. Dropping my
bags inside, I turned back to wait for
Phoebe. She walked slowly, seeming a tad
unsteady. Middle-of-the-night grogginess?
Possibly. But the niggling I had felt after being here last time was now a bona fide burr.

She came alongside, very much Mom's
height, which was several inches shy of
mine, and said, "Your room's just the same. I
haven't touched anything."

"I wasn't worried. What about your room?
Are you still sleeping there?"

"Where else would I sleep?"

"Mom's room. It's the biggest. This is your
house now. You have a right to that room.
Didn't Sabina suggest it last time I was
here?"

"I guess," Phoebe said, confused again.
"But it'd mean moving all my things, and I've
been in my room for so long." Her eyes grew
plaintive. "Do I really have the *energy* for
that?"

She should have it. She was only thirty-six. Clearly, though, she was depleted both physically and emotionally. I wondered how she managed to handle the store.

Rather than express doubt when she seemed so vulnerable, I said, "So, what time will you be up in the morning?"

"Seven. We open at nine."

"Can you stay home if your cold is worse?"

"It won't be worse."

"Okay then. I'll see you for breakfast?"

She nodded, frowned, added, "Unless I sleep later. I've been exhausted. Maybe it's the cold. Maybe it's missing Mom." With an oddly apologetic smile, she went on past me, down the hall.

"Want me to turn out the lamp?" I asked.

She looked back. "Lamp?"

I indicated the one at the top of the stairs.

She stared in surprise. "No. Leave it on. If it had been on that night, Mom wouldn't have tripped. It was dark. If she'd been able to see, she wouldn't have fallen, and if she hadn't fallen, she'd still be alive."

"She was ill," I reminded her. "It wasn't so much the dark as her balance."

"It was the dark," Phoebe declared and disappeared into her room.

* * *

I didn't sleep well. Once I opened the windows, pulled back the covers, and removed
every stitch of clothing, I could deal with the
heat, but the city girl I had become wasn't
used to the noise. Traffic, yes. Sirens, yes.
Garbage trucks, yes. Peepers and crickets,
no. Naturally, lying awake, I thought about
Mom, about whether Phoebe was sick, too,
and, if so, whether it was from lead or worse,
and each time I woke up, I thought of those
things. Dawn came, and the night noises
died, which meant that the river emerged.
Our house sat on its banks. The waters
rushed past, carrying aquatic creatures from
upstream, leaves and grasses from its
banks, all hurrying past the stones that lined
its bank.

Seven came, and I listened for Phoebe,
but it wasn't until seven-twenty that I heard
signs of life in the house. Wearing a night-
shirt now, I was sitting on the edge of the
bed, about to stand, when Sabina slipped
into the room.

Sabina and I were Barneses, with
Daddy's midnight hair, pale skin, and full
mouth. When we were kids, these features
had come together on her far better than

they had on me. Sabina was pretty and popular. I was neither. We were both five-eight, though Sabina had always insisted she was half an inch taller than I was. I didn't fight her on it. There was plenty else to fight about. Even as she approached the bed now, I felt it coming.

I tried to diffuse things with a smile. "Hi. I was going to call you. Is Phoebe awake?"

"No," Sabina replied in a low voice. She folded her arms and held them close. "I wanted to talk with you first. This has been really tough on her, Annie. I don't want you riling her up."

Dismayed by the abruptness of her attack, I said in a conversational tone, "I'm doing okay, thanks for asking. How are you?"

She didn't blink. "We could spend five minutes on niceties, but this is really important. Phoebe is having trouble accepting that Mom's gone. I don't know why you're here, but if you're thinking of doing anything to stir up trouble, please don't."

I was annoyed enough to lash back. "Phoebe is doing more than 'having trouble accepting that Mom's gone.' She looks physically ill. She says it's a cold. I'm wondering if it's something else."

"Oh, it is," Sabina confirmed, "but it's nothing you can fix. The way she's acting—like Mom did? It's a natural thing that sometimes happens when a loved one dies. I talked with Marian Stein about it."

"Who's Marian Stein?"

"A therapist here in town. I'm on top of this, Annie."

"Is Phoebe seeing her?"

"Of course, not. Phoebe doesn't need therapy, just time. This'll pass."

"Has she seen her doctor?"

"No need. Colds disappear. Symptoms pass."

I knew not to mention lead. It wouldn't be well received. So I said, "Mom was diagnosed with Parkinson's. It can run in families."

Sabina's eyes hardened. "And *that*," she said, still in a low voice but laced now with venom, "is why you shouldn't be here. She needs encouragement. You're so negative, you'll set her back."

"Oh, come on, Sabina," I scoffed. "I have enough sense not to mention this to her. But I Googled Parkinson's after Mom was diagnosed. If Mom had it, and if Phoebe has it, you or I may stand a greater risk. Aren't *you* worried about that?"

"*If* Mom had it?" Sabina charged. Her arms were knotted across her middle.

"There are other causes for the symptoms she had," I blurted out and regretted it instantly.

"I knew it! I knew you'd stick your nose in! Well, where were you last year or the year before that or the year before that? Fine and dandy for you to criticize us now—"

"I'm not criticizing."

"—but *you* weren't around. We were, Annie. Phoebe and I took Mom to the doctor, got her medicines, made sure she took them when she would have forgotten. Phoebe has been running the store for the last five years—"

"Five?"

"Yes, five. It's been *that* long since Mom was functioning *well.*"

Five years put a crimp in the lead theory. It would mean Mom had become ill long before the store was awash in lead-paint dust. There might yet be a connection, certainly with regard to Phoebe, but it would take some looking.

I was annoyed. "Why wasn't I told back then?"

"Because you weren't here!" Sabina

shouted and immediately lowered her voice. "And because the symptoms were so mild we thought it was age at first, and because Phoebe was there to cover at work, and because Mom would have been horrified if she'd known we were talking behind her back. You know how she was. She *hated* being talked about. So we didn't tell you—didn't tell *anyone*—until the symptoms made it obvious, and even *then* you stayed away. So don't criticize us, Annie," she warned. "You have no idea what it's been like. We did the best we could."

I was quiet. What could I say to that? Yes, I felt guilty. I had from the moment I learned Mom had died. I kept telling myself I was here on a mission; that was the initial premise. But maybe my mission was broader than I had allowed. So mentally I amended that premise with **TRUTH #1: Yes, I had come to Middle River to learn whether Mom had died of something that was now affecting Phoebe, but I was also here out of guilt.** I owed my sisters something. I wanted to make it up to them that I hadn't helped when Mom was sick.

Not that I could say that to Sabina. The words would positively stick in my throat.

Instead I asked, "How are Lisa and Timmy?" They were Sabina's kids, aged twelve and ten respectively. I actually knew how they were; I had an active e-mail relationship with them, and had been in touch with them a lot in the last month, though I don't think Sabina knew it. Her kids were astute; they knew there was tension between Sabina and me. My relationship with her kids was a little secret we kept. None of us was risking Sabrina's wrath by rubbing her nose in it.

She did relax a bit at mention of the kids. "They're fine. Excited that you're here. They'll probably ride their bikes over later. They want to know how long you're staying."

"They want to know," the devil made me ask, "or you do?"

She didn't deny it. "Me. Phoebe, too. This is her house."

It was a definite reminder. "For the record," I said, "I don't want the house. I don't want the store. I don't want Mom's money. The only thing I want, which I told you in June, is to have this room to use when I come."

Sabina looked dismissive, clearly doubtful I was telling the truth. "How long are you staying?"

"Assuming Phoebe has no problem with it, until Labor Day."

"A whole month?" she asked, seeming alarmed. "What'll you *do* all that time?"

"I have some work. Mostly I want to relax. Give Phoebe a hand. Help her get better. Talk with people around town."

"Who?"

The sharpness of the question put us right back in the boxing ring. "I haven't really thought that far."

"Are you kidding? Annie Barnes hasn't thought that far? I know what you're here for, Annie. You're here to cause trouble. You'll walk innocently around town like you did when we were kids, asking questions you have no business asking, pissing people off right and left, and then you'll go back to Washington, leaving us to mend fences. And then there's the thing about writing. You have some work. What work?"

"Whatever final edits my publisher wants on next spring's book. Written interviews that they'll need. Plotting a new book."

Sabina's mouth tightened. "Are you planning to write about us now that Mom is gone?"

"No."

"I think you are. You'll ask your questions and piss us off, and then when you're back in Washington and we're cleaning up the mess, you'll write something that'll make the mess even worse." She held up her hands, palms out. "I'm asking you. *Begging* you. Please, Annie. Mind your own business." The plea was barely out when she turned, strode to the door, and pulled it open.

Phoebe stood there. Seeming wholly oblivious, simply surprised to see Sabina, she said, "I didn't know you were here. But I'm making, um, I think, what was it I was thinking, well, I think I'll make eggs for breakfast. Should I make enough for three?"

Chapter 2

THE MIDDLE River Clinic was on Cedar Street, half a mile from Oak. Like the town itself—in that very euphemistic way—the clinic building was deeper than it appeared from the front, stretching an entire block to School Street over grassy dips and swells that were liberally strewn with, yes, cedars and oaks. Appropriately, the School Street entrance fed directly into a small emergency room that, over the years, had been a staging area for the treatment of countless playground wounds, strep throats, and allergy attacks. The oldest generation of Middle Riverites had been born and given birth here too, back when the clinic was a bona fide hospital. Nowadays, except for the most sudden of cases, childbirth and such took place in Plymouth.

The first floor housed offices for the doctors who serviced the town as part of the Middle River Medical Group. The second floor was rented out to independent practitioners, currently including a large team of physical therapists, a pair of chiropractors, and an acupuncturist. Psychologists were scattered throughout—small towns always had those—along with lawyers, investment counselors, and computer people. Sandy Meade wasn't picky; he owned the building and wanted every space filled. He might have balked at renting to a video store stocking adult entertainment—he did have standards—but he was fine with most else, as long as the rent money arrived in full each month.

The building was a handsome brick structure, two stories high, with white shutters, gutters, and portico. The last time I was inside was to visit the emergency room the Thanksgiving before I left town. Sam Winchell, who owned the newspaper and whose family lived out of state, had been joining us for Thanksgiving dinner since Daddy died; that last year, he cut himself sharpening the turkey knife. I was the person he was closest to and, hence, was his designated driver, while the others stayed home keeping the food warm.

By the time I arrived this August day, I had already dropped Phoebe at work, stopped at the post office and run into all those people who asked if I was going to write about them, visited Harriman's Grocery, and returned to the house to fill the refrigerator with food. It had been pathetically empty. Eggs for three? Phoebe hadn't had eggs enough for one. With Sabina gone—and no arguments there—we two had settled for stale bran flakes. Dry.

I was no fool. While in town, I had also stopped at News 'n Chews for a bag of chocolate pennies. A pack of M&Ms would have satisfied my chocolate craving, but I could get M&Ms anywhere, anytime. Chocolate pennies, hand-dropped by the Walkers for three generations and counting, were something else. They were well worth the risk of my being seen by even more people.

But then, word was already spreading. I suspected my own sisters had told friends, who had told friends, who had told friends. I wasn't about to drive my convertible down Oak Street—that would be a distasteful show on my first day back—but there was no way I could be invisible in a place desperately in need of food for talk.

Thomas Martin, the doctor who had treated my mother, was the director of the Middle River Clinic. He was new to town by Middle River standards, brought in three years ago when old Doc Wessler retired. The *Middle River Times* described Dr. Martin as not only a respected general practitioner, but a man with a business degree that would stand him in good stead for the demands of running a modern clinic.

I didn't call ahead for an appointment; the Middle River Clinic wasn't Memorial Sloan-Kettering. Nor did I tell either of my sisters what I planned to do. I didn't want an audience here, didn't want idle minds speculating. For the record, I simply wanted to thank the doctor for seeing my mother through her final days. Though I had met him after the funeral, I hadn't had the presence of mind to be overly gracious. Granted, my gratitude would be misplaced in the event that he had bungled her case and misdiagnosed her illness. But he was an outsider, like me; for that reason alone, I gave him the benefit of the doubt.

He was with a patient when I arrived, but his secretary knew me. Growing up, we had been classmates, my bookworm to her cheerleader. Her eyes widened and she smiled, which was more than she had ever done in response to me back then.

"You look great, Annie," she said in surprise, then added in immediate sympathy, "I'm sorry about your mom. She was a nice woman, always kind, whether you ended up buying something at the shop or not. And coming in here, she was always respectful. Some aren't, you know. They hate the paperwork and blame us for it. Well, we hate the paperwork too. Wow, Annie, you really *do* look great. Being famous must agree with you."

"I'm not really famous," I said, because success as a writer was different. There was no face recognition, no entourage, no advance man. My name was known in reading circles, but that was it.

Not that I said any of that now. I was socially tongue-tied and defensive, both of which were conditioned reflexes where Middle River and I were concerned.

Graceless, I forged on. "I was hoping to see the doctor to thank him for taking care of my mother. Does he ever take a break?"

"It's coming right up," she replied cheerily, standing. "I'll let him know you're here."

Five minutes later, the man appeared. He wore the obligatory lab coat, over an open-necked blue shirt and khaki slacks. His hair was short and dark, his eyes clear and blue, his body lean. When I stood to meet him, I saw that he was just about my height, perhaps the half inch taller than me that my sister always claimed to have.

An incompetent quack? If so, he showed no guilt, but rather approached with an easy smile and a friendly hand. "You're the lady of the hour."

I was immediately disarmed by his warmth, though, in truth, I might have liked him simply because he wasn't a native of Middle River. Moreover, since he wasn't from here, there was nothing emotional to hinder my speech. With a comfort I had developed in the last fifteen years, I said, "Sounds like you've had your ear twisted by someone other than Linda here."

"Three out of four patients this morning," he confirmed, eyes twinkling.

"Please don't believe what they say. There's a whole other side to me."

"As there is to us all," he remarked and, without ceremony, took my arm. "I'm buying coffee," he said, guiding me out.

The coffee shop was actually a small restaurant run by two of Omie's grandchildren. Younger than the diner by that many generations, it was called Burgers & Beans, the beans meaning the coffee kind, to judge from the rich smell that greeted us when we walked in the door. In addition to burgers, there were sandwiches and salads. We stuck to coffee.

"Want to chance it outside?" Tom asked, darting a glance at the patio, with its round wrought iron tables and umbrellas in the sun.

I smiled appreciatively. The fact that he understood my reluctance to be seen made me like him even more.

But the cat was already out of the bag. It would have been absurd for me to hide now. "Sure," I said lightly. Putting on a courageous face, I led him past several people who were definitely familiar and openly staring, to a table at the patio's edge. Whiskey barrels filled to overflowing with impatiens marked the border between flagstone and grass. I sat in the sun, just beyond the umbrella's shade. It was still early enough in the day for the warmth to feel good.

After settling himself, Tom said, "We have something in common, you and I. I spent ten years in Washington—college, med school, and residency."

"Did you?" I asked, our connection deepening. "Where?"

"Georgetown all the way."

"We're fellow alums. Did you live near Wisconsin?"

"For a while. Then Dupont Circle. It's a fabulous city. Hot in summer, but fabulous."

"Hot in summer," I confirmed and took a drink of coffee. Middle River could be hot— might well be hot this very afternoon—but not with the relentlessly steamy heat that seared Washington day after day.

"Is that why you're here," Tom asked, "to escape the heat?"

I was amused. "What did your patients say?"

"That you were here to write. That you're Middle River's version of Grace Metalious, and that of *course*," his eyes twinkled again, "I knew the Peyton Place link."

"You did know it," I surmised.

"How not to? It's the basis of tourism here. The inn offers Peyton Place weekends, the town historian gives tours of the 'real' Peyton Place, the newspaper puts out a *Peyton Place Times* issue on April Fools' Day each year. And then there's the bookstore. You can't walk in there without seeing it. *Peyton Place* is still prominently displayed, along with a *Reader's Companion* that outlines the parallels between Peyton Place and us."

"Do you remember *Peyton Place*?"

"I'm forty-two. It came out before I was born, but it was on my mother's nightstand for years after, all dog-eared and worn. I read it when I first came here."

"Did it make you rethink your decision?"

"Nah. *Peyton Place* is fiction. Middle River has its characters, but it isn't a bad place."

I might have argued and said that Middle River was a *vile* place for one who didn't conform to its standards. But that felt like sour grapes on my part. So I simply cleared my throat and said, "Sandy Meade must have made you one sweet offer to get you to leave wherever you were after Washington to come here."

"Atlanta," he said easily, "and I wanted a quieter life. I bought a house over on East

Meadow. It's three times the size of anything I could have bought elsewhere, and I had money to spare for renovations."

"Annie?" came a curious voice.

I looked around at a woman I instantly recognized, though I hadn't seen her in years. Her name was Pamela Farrow. She had been a year behind me in school, but her reputation for being fast and loose made the leap to my class and beyond. Back then, she had been a looker with shiny black hair, warm green eyes, and curves in the right places well before the rest of us had any curves at all. Now her hair wasn't as shiny or her eyes as green, and her curves were larger.

She hadn't aged as well as I had.

I'm sorry. That thought was unkind.

I atoned with a smile. "Hi, Pamela. It's been a while."

"You've been busy. We hear about the incredible things that are happening to you. Do you really live with Greg Steele?"

I wouldn't have guessed Pamela knew who Greg was; she didn't strike me as the type to watch the news, evening or otherwise.

And that was another unkind thought, which was why I hated Middle River. It brought out the worst in me. Back here for less than twelve hours, and I was being snide again. It was defensiveness, of course. I was lashing out with words, or in this case thoughts, to compensate for feeling socially inept. Which I wasn't now. At least, not in Washington. Here, I regressed.

I nodded in reply and drank more of my coffee.

"Well, we love hearing news of you," Pamela gushed. "You're our most famous native." She turned to the man with her. Unfamiliar to me, he wore glasses and a shirt and tie, had neatly combed hair, and exuded a style of conservatism. I was floored when Pamela said, "Annie, this is my husband, Hal Healy."

I would never have guessed it. Never. Either Pamela had changed, or this marriage was a mismatch.

I held out a hand; he shook it a bit too firmly.

"Hal was brought to town to be principal of the high school," Pamela said with pride, and here, too, I was surprised. The high school principal in my day was a sexy guy not unlike Tomas Makris in *Peyton Place*. Hal Healy had the look of a marginal nerd. "We've been married six years," Pamela went on. "We have two little girls. You and Greg don't have kids yet, do you?"

Not wanting to explain that Greg and I weren't sexually involved, I simply said, "No. We don't. I assume you both know Tom?"

"Oh yes," Pamela said with barely a glance at the doctor. I might been flattered by her attention to me, if I hadn't thought her profoundly rude to Tom. "So tell me," she went on in something of a confidential tone. "We're all wondering. Are you really here to write about us?"

I smiled, bowed my head, and rubbed my temple. Still smiling, I looked back up. "No. I'm not here to write."

Her face fell. "Why not? I mean, there's still plenty to write about. I could tell you stories . . ." She stopped when her husband gave her shoulder a squeeze.

"Honey, they're having coffee. This isn't the time."

"Well, I could," Pamela insisted, "so if you change your mind, Annie, please call me. You look so different. Very successful. I always knew you'd go places." She grinned, waved, and let her husband steer her away.

Dismayed, I turned to Tom. "She didn't

know anything of the sort. I was a pain in the butt back then—a gangly runt with an attitude. And I wrote lousy stories."

"Well, you don't now," the doctor said, suddenly seeming almost shy. "I read *East of Lonely* and liked it so much that I bought and read your two earlier books. You're an amazing writer. You capture emotion with such a sparseness of words." He was actually blushing. "I say your success is well earned. So when it comes to Middle River, you have the last laugh. People like Pamela, they really are proud of you. For what it's worth, I've never heard anything bad."

"You're too new to town," I said, shaking my head, but his words were a comfort. He reminded me of Greg, not as much in looks as manner. Both were easygoing and could blush. *Honest* was the word that came to my mind.

Yes. I know. I wanted to like him because I needed a friend, and because I needed a friend, I didn't want to consider the fact that Tom might be smooth and glib and not honest at all. My mother was dead. I had to remember that.

But then Tom said the one thing that could most easily make me forget. "What do you know about Grace Metalious?" he asked.

I grinned. "Most everything." I could be smug about this as I wasn't about much else in life. "What do you want to know?"

"Did she really drink herself to death?"

"She drank heavily, some sources say a fifth of vodka a day for five years. And she died of cirrhosis of the liver. *A* plus *B* . . ."

"Gotcha. What was so bad about her life that she had to escape into booze? Was it not being able to write a follow-up to *Peyton Place*?"

"Oh, she did a follow-up. Her publisher insisted on it. But she hadn't wanted to do it, so she dashed it off in a month. They had to hire a ghostwriter to revise it and make it publishable. In any event, it was widely panned. There were two other books after that, one of which was her favorite. Neither book was well received."

Smoothly, he returned to me. "So here you are with a major success, like Grace with her *Peyton Place*. Do you worry about matching the success of *East of Lonely*?"

"Of course I do," I said baldly. "There's ego involved, and professional pride, even survival as a writer. It's a mean world out there. But I do love the process of writing."

"Didn't Grace?"

"Yes. But my success didn't come with the first effort. Hers did. When you score a home run the first time up at bat, it's hard to top yourself the next time. Besides, there were other things that got her down—like her agent. He cheated her out of a lot of money. The little she did get, she spent. She had expensive tastes."

"For a Manchester girl?" Tom asked with a smile, and at my look of surprise said, "Hey, I took the Peyton Place tour. She grew up in Manchester with her mother and grandmother."

"Uh-huh. In a house full of women. Like me."

"Her father left," he cautioned. "Yours died. There's a difference."

"But we were both tenish when it happened, and the end result was the same. There was no father in the house to hold the reins. The women did it. They were strong, because they had to be. Which brings me to the reason I'm here." *About time, Annie.* "I want to talk about my mom."

He grew serious. "I'm sorry about her death. I wish I could have done more."

"The fall killed her. I know that. She broke her neck and died from asphyxiation. But before that—was it really Parkinson's?"

He didn't seem surprised by the question. "Hard to say," he admitted. "She had an assortment of symptoms. The tremor in her hand, the balance issue, the trouble walking—these were consistent with Parkinson's. The memory problem suggested Alzheimer's."

"But you choose Parkinson's."

"No. The most treatable of her symptoms were the ones associated with Parkinson's, so those were the ones I addressed."

"Did you suggest that she see a specialist?" I asked with something of an edge, because Tom Martin was, after all, only a general practitioner, and while I had the utmost respect for GPs and the way they juggled many different things, we weren't talking about a common cold or the flu.

"Yes," he said calmly. "I gave her the name of someone at Dartmouth-Hitchcock, but she never went. She felt it was too much of an effort. And that was okay. I consulted colleagues there and in Boston. They studied her file via computer and agreed with my diagnosis. There's nothing more that an actual trip to Dartmouth-Hitchcock would have accomplished. There are no tests to conclusively diagnose either Parkinson's or Alzheimer's. It's strictly a clinical diagnosis, a judgment call made by the physician. The medication I gave her helped as much as any could."

"There's no history of either disease in our family."

"Neither has to be hereditary."

"Is it possible she had a little bit of both?"

"It's possible."

"What would the likelihood have been of that at her age?"

"Slim."

Layperson that I was, I nonetheless agreed. "Could her symptoms have come from something else?"

He frowned slightly. "What did you have in mind?"

"I don't know." I was deliberately vague. "Something in the air, y'know?"

"Like acid rain?"

"Could be," I granted. "Air currents being what they are, New England has become a receptacle for toxins from plants in the Midwest. But I was actually thinking about something more local, like lead."

"Lead?"

"From lead paint. All the old paint was sanded off when Miss Lissy's was repainted. The air had to have been loaded with lead." Yes. I know. Sabina said Mom had been symptomatic for five years, and this work was done only four years ago. But what if Mom had been simply *aging* prior to then? Losing interest in daily chores? Experiencing a major postmenopausal funk? What if lead poisoning had taken over where the other had left off?

Tom smiled sadly. "Nope. Sorry. I tested for that at the start. There was no lead in her blood."

I felt a stab of dismay. I had been *sure* it was lead. "Could it have been there and gone?"

He shook his head.

"My sister seems to have some of the same symptoms."

Tom frowned at that. "Sabina?"

"Phoebe."

"I never saw any symptoms. What are they?"

"Poor balance. Bad memory. They're

pretty new. I only saw the bare beginnings of them when I was here in June. Sabina says it's part of mourning Mom. Sympathetic symptoms. But what if it isn't?"

"If it isn't, she ought to see me," he advised. "Think you can get her in?"

"I don't know. It might be hard with Sabina standing guard. If Phoebe were to see you, would you be able to tell whether the problem is real or psychosomatic?"

"Possibly. Your mother had no control over the symptoms. Your sister may. That would be significant. Plus, I can test for lead."

I finished my coffee and set down the cup. "Getting back to the other thing. You know, acid rain or whatever. Do you think there's an abnormal amount of illness in this town?"

He drank his own coffee, seeming momentarily lost in thought. Then, with a blink, he put the cup down. "There may be."

"What do you think is the cause?"

"I've been trying to figure that out since I arrived."

"How?"

"Watching. Asking questions."

"And what are your thoughts?"

He was quiet for a minute, turning the cup round and round on the table, watching it turn. Then he stopped, paused, finally raised his eyes. "Acid rain, or whatever. But you didn't hear that from me."

"Why not?"

"Because I have no scientific proof."

"You diagnose Parkinson's and Alzheimer's with no scientific proof," I said.

"The other is something else. It implies a very large problem, one that has become extremely political. I can state my opinion, and even point to a pattern, but there are those who'll say I'm crazy, and they might just be powerful enough to destroy my credibility in this town."

"Would you sacrifice the health of the people for the sake of your credibility?" I asked, edgy again.

He sat forward, more intense now. "I lived in Washington long enough to know people there. I say things—believe me, I do—and maybe my arguments have so far fallen on deaf ears, but to some extent I'm hog-tied. I've had to make a choice—politician or doctor. Yes, I'm worried about my credibility. I think I'm the best doctor this town has, in part because I ask every question I know how to ask and treat my patients accordingly. If I give up my practice to crusade for the cause, who will take care of the people here?"

"A replacement," I said, seeing his point, "but one who might be so beholden to the Meades as to disavow everything you would be down there in Washington trying to do. Have you talked to the Meades about this?"

"About what?"

"Acid rain, or whatever."

Cautiously, he asked, "Why would I talk with the Meades?"

"Because they run the only industry in town."

"I thought we were talking about pollution from the Midwest."

"We were."

Tom was silent. After a minute, he said, "They say I'm barking up the wrong tree."

"About which, acid rain, or whatever?"

His eyes held mine. "Whatever."

"What is whatever?"

"I'm not sure."

"Have you done any research?"

"Some."

"Can you give me a lead?"

"If I did that and certain people found out, I could lose my job."

"Like, you could lose your credibility?"

"You ask a lot of questions," he said with some irritation.

I eased up, actually smiled. "I'm more diplomatic about it back in Washington. Here, I revert to form. I always did ask questions. It didn't go over big when I was little."

"It may not now, either," he cautioned. "There are some who'll say you just need a scapegoat for your mother's death."

"I do. I half wanted it to be you."

"I half wanted it to be me, too. When a patient dies, I agonize over what I might have done differently. But I'm stumped in this case. I charted Alyssa's symptoms, consulted with colleagues, encouraged her to go elsewhere for a second opinion. I asked her every question I knew to ask to determine if she'd been exposed to something bad. Short of sending her off on a wild goose chase for a cure for a nebulous ailment, I did what I could to ease her symptoms."

"I know," I said. And I did. "But now I have to find someone else."

If he was relieved to be off the hook, he didn't let on. "Be careful, Annie. The Meades own this town. You know that better than me."

I paused. "You heard that story then?"

"The one about you and Aidan Meade?"
He nodded.

"Huh." I took a deep breath. "Well, the thing is, I'm a different person now from the one I was then. Middle River power isn't the only power in the world. You may be hog-tied, but I'm not. I know how to get things done. Look," I said calmly, "it could be that Mom's symptoms truly were from Parkinson's or Alzheimer's, but now Phoebe's showing the same symptoms, and you say it's not from lead. I have a month, I have energy, and I have an incentive to use both. So think of it as me doing your dirty work. I don't care about my credibility here, and I certainly don't have a job to worry about. You took care of my mother, so I'm here to thank you for that. That's what I told your secretary. It's probably well on its way around town by now." I leaned over the table toward him and half whispered an urgent, "Point me in the right direction, Tom. I'll be forever in your debt."

He finished his coffee and rose, took my empty cup, and deposited both in a nearby receptacle. I was beginning to think that was the end of it, when he returned, slipped back into his seat, and said with deadly quiet, "I'm wondering about mercury."

I sighed. "It can't be. The mill stopped using mercury long before my mother got sick."

"But mercury is unique," he said. "It enters the body, settles in an organ, and waits. Symptoms may not surface until years after exposure. That's what makes the possibility of mercury poisoning in Middle River an intriguing one."

I was feeling a rush of adrenaline. "Intriguing is an understatement. Are you sure that's what it does?"

"Lies dormant? Very sure. The problem is that I've never been able to connect one of my patients with an incident of actual exposure. Do that, Annie, and you'd top Grace."

Chapter 3

THAT QUICKLY, I shifted gears. Not lead. Mercury—and not the planet Mercury or the cute winged god of Greek mythology. This mercury was the slinky silver stuff found in thermometers. It was a metal. I was no expert on the subject, having explored it and quickly (if erroneously) dismissed it. But it was virtually impossible to live in Washington and not know of the controversy. Rarely did a week pass without an article on the mercury problem appearing in the *Post,* and a problem it was. Mercury emissions were a serious health hazard, environmentalists said. The other side, whose profits depended on the output of mercury-emitting plants, adamantly fought regulations.

As villains went, mercury would be perfect and, if the mill was the polluter, all the better for me. True, Northwood didn't currently use mercury. That had put me off the track. But Tom put me right back on with his claim that mercury could lie dormant in the human body for years before manifesting itself in symptoms. So what if something had happened way back when? The Meades had to know the potential for harm. If the mill had contaminated the town without the Meades' knowledge, shame on them. If they did know it and had either done nothing or, worse, covered it up, they could be criminally charged.

With deliberate effort, I reined in my thoughts. Mercury was a more explosive issue than lead. For that reason alone, caution was the way to go—not that I would have always taken that route. Once upon a time, I would have barged ahead with accusations. But I was a grown-up now and—I liked to think—responsible. Credibility rested on being cool-headed and deliberate. I knew not to draw conclusions before I had information, because there were, indeed, immediate questions. If mercury was the problem and the mill was its source, it remained to be seen why my mother, and perhaps my sister, was affected. Neither of them worked at the mill. Downriver from it, yes. But not at a place of immediate exposure. And why my mother and sister, and not my grandmother before them? After a life of consistently good health, Connie had died of an aneurysm.

Surely, there were answers to be had. I didn't want to get excited until I made a few connections. That meant going back to the drawing board to catalog the symptoms of mercury poisoning, the methods of exposure to mercury, and the kinds of plants that emitted it. Tom had been unable to connect any of his patients with exposure to the metal. I had to try.

First, though, I had to drop by the dress shop to see Phoebe. I had promised I would, since I was driving her van, and she might need it. I had also promised to bring lunch.

Armed with salads from Burgers & Beans, I went back down Cedar, crossed Oak, and drove on to Willow. This time, rather than making a right toward our house, I made a left and passed more Victorian homes. Quickly, though, residences ended and stores began.

If Oak Street offered necessities such as food, health products, and town services, Willow filled needs that were one step removed. Here were shops selling books, computers, and antiques. There were two beauty salons, a day spa, and the nail shop. There were stores that sold used furniture, office supplies, and candles and lamps. There was a workout place. And a picture-framing store. And an autobody shop. And there was Miss Lissy's Closet.

Miss Lissy was my great-grandmother, Elizabeth, who had first opened the shop. My grandmother, who ran it after Lissy died, might have changed the name to Connie's Closet, which had a certain cachet, had it not been for her obsession with keeping a low profile. By the time Connie passed on and my mother took charge, Miss Lissy's had become too much of a fixture in town to risk a change of name. Besides, with my mother being Alyssa, Lissy's Closet fit.

So, surprisingly, did her stewardship of the shop. She had wanted to be a writer, not a shopkeeper. Her move to the latter came after we girls started school, when it was clear that she could earn more money working for Connie than selling stories to regional magazines. And we did need the money. As a sales rep for the mill—yes, the very same Northwood Mill—Daddy earned a reasonable income, but not enough to save for our education. Mom's income was earmarked for that. Then Daddy died, and Mom's income was all we had.

Miss Lissy's occupied both floors of a small frame house in a row of other small frame houses, each with a tiny front yard and a porch, and just enough room between houses to allow for parking, loading, and fire prevention. The Meade family, which owned this entire end of the street and leased it to shopkeepers, took pride in this foresight. Having lost part of the mill to fire in the fifties before brick replaced wood, they knew what could happen when August turned hot and dry and the town grew parched. This summer was moist enough, but Middle River never forgot the acrid smell of smoke that lingered in the air for days after the inferno.

The shop was on the river side of Willow. The rents were higher on this side of the street, but the ambience was worth the cost. Each house had a back deck perfect for lunches, featured displays, and, in the instance of Miss Lissy's Closet, trunk sales.

Moreover, these back decks were connected to the river walk, the scene of the fabled October stroll, when fall foliage was at its height and Christmas shoppers first set out to buy gifts. Those on the landlocked side of Willow, in true sour-grapes fashion, reasoned that they were spared having to look across the water at "th'other side," as Middle River's poorest neighborhood was called. The Meades owned that land as well, and dressed up the riverbank with flowers and trees. Once leaves fell, though, nothing could hide the shabbiness of the homes there.

This August, the earth was moist, the willows grand, and the river world green.

As for Miss Lissy's, it was green every month of the year, inside and out, every shade of green imaginable, all of it blending surprisingly well. Prior to the painting done four years ago, it had been green, but monochromatically so. Now it was variegated in the way of a woodland glen. The shingles on the outer walls were painted celadon, with shutters and doors forest green. The inside walls were painted by room—one room sunshine green, one celery, one sea green, one olive. Wrapping tissue was sage, shopping

bags were teal, dress and suit bags a distinctive apple green, and no purchase at all
left the shop without a cascade of assorted
green ribbons attached. Over and around it
all was the store's logo, also redone during
my mother's watch when marketing and
modernizing came to the fore. It was a closet
door done in brushstrokes, with a zippy font
presenting the store's name in grass green.

I parked the van in the driveway and
walked around to the front. The look was
handsome, I had to say. As fearful as I had
been that so many shades of green would
be overkill, here on the outside, the celadon,
grass, and forest greens worked well. Likewise the hydrangeas that grew in profusion
against the front of the shop. They were the
palest green imaginable.

Climbing two steps, I crossed the porch.
The front door was already open. With the
tinkle of a bell, I pushed the screen and
went inside. For a minute, I just stood and
looked around. I couldn't remember when I
had been here last, surely not at the time of
the funeral, since the store had been closed
out of respect for my mother. I was in the
front room; doors straight ahead and to my
right led to other rooms. A staircase, farther

back on the right, led to the second floor, where there were yet other rooms. Each was a department, so to speak—underwear, outerwear, partywear, even makeup, to name a few.

Marketing and modernization? Oh yes. This front room was a perfect example of that. When I was a child, this room had held racks of the kinds of day dresses that the women of Middle River wore at the time. Now the room held blue jeans and slacks, T-shirts, polo shirts, and sweaters, all popular brands. It was definitely my kind of room.

"Annie, hi," said Phoebe, coming in from the room on the right. She seemed less groggy than she had earlier and sounded less stuffed. Dressed all in ivory—camisole, slacks, and slides—with her hair brushed and her makeup fresh, she looked elegant and in control. She also looked so much like Mom that I felt a catch in my throat.

As she headed for the checkout desk, she was trailed by a pair of customers who had to be mother and daughter. Though the mother was shorter and slimmer than her daughter, their features were nearly identical. Likewise their focus on me—even the daughter, who couldn't have been more than fifteen and therefore wouldn't remember me from before. But she knew who I was now. No doubt about that. Despite the long fall of hair that half covered her eyes, I saw recognition there.

I winked at the girl—why not give her a thrill? my perverse self thought—and nodded at her mother. "How's it goin'?" I said in passing as I slipped into the room from which they had come. This room held shoes. Again, though, the difference between these shoes and the ones that had been for sale the last time I was here was pronounced. There wasn't a leather pump in sight. Here were sneakers and casual sandals, flats and sling-backs and strappy things for evening wear, though surely those evenings were spent in places other than Middle River. I couldn't imagine anything dressy being worn here. Church events, weddings and anniversary parties, birthday bashes—they weren't terribly dressy unless they were held out of town, which they were now more often than not. From what I read, even the school proms were held elsewhere.

"May I help you?" asked a gentle voice.

I turned to face a woman in her late twenties. Dark-haired, she was tall and slim, wore a silk camisole, slim black pants, and black sandals. Classy and composed, she showed no sign of recognizing me.

Nor did I recognize her, for which reason I was able to smile easily. "I'm Phoebe's sister, Annie."

Her eyes did widen then. "Phoebe said you were in town, but she didn't say you'd be coming to the store." She put out a hand. "I'm Joanne. I work here."

"Full-time?" I asked as we shook hands.

"Yes. I started working weekends and summers when I was in high school, but I've been full-time since I graduated from college. Phoebe made me manager last year."

"Then I owe you thanks," I said. "The past month has probably been difficult here."

Joanne nodded. "I loved your mom. She was always good to me, and she had a great sense of style. Watching her go downhill was heartbreaking. She kept wanting to be here, but it got harder for her."

"Because of the stairs?" I was thinking of the balance problem.

"And illness. If it wasn't a cold, it was the flu. She was in bed half the time, and she hated that. She also hated forgetting people's names. She hated losing her train of thought in the middle of a sentence. The computer was a whole other issue. We had to make sure we were with her when she was entering sales. Poor Phoebe. The strain has taken a toll."

I might have asked about that toll—to wit, my sister's behavior and health. But Joanne was a stranger to me, which made prying seem like a betrayal of Phoebe. So I simply nodded and said, "The store looks wonderful. Very upmarket. How much of it is your doing?"

"Not much," she confessed. "Phoebe does the buying." She looked past me as Phoebe joined us, and said softly, "You wanted me to remind you about New York."

"All done," Phoebe said with flair. "I made the reservations."

"Reservations for a buying trip?" I asked.

"Yes. Do you want to go somewhere for lunch?"

By way of reminder, I held up the bag with the salads—to which, making light of forgetfulness, she tapped her head, rolled her eyes, and led the way out back.

The deck was warm, but the sun had already shifted enough to bring shade to the wide, weathered planks. A slight breeze came off the river, carrying the scent of wild-flowers—yellow wood sorrel and musk flower, blue vervain, purple loosestrife. I spotted a patch of forget-me-nots; it had been growing there forever.

Sitting at a small table, we opened our salads and bottled teas. "How do you feel?" I asked.

"I'm fine. Why do you ask?"

"Your cold seems better."

"It was only a cold." She lifted her fork. It hovered over the salad for a minute—hovered or shook, I wasn't sure which—before she set it back down, put her hand in her lap, and asked, "Where did you go after you dropped me here?"

That hand had shaken, I decided. She was covering it up.

Nonchalantly, between bites, I listed my stops in town, concluding with the one at the house to put groceries in the fridge.

"Then where?" Phoebe asked with neither an acknowledgment nor a thank-you.

"The clinic. I wanted to talk with Tom Martin."

"Why?"

"To thank him for taking care of Mom. Aren't you going to eat?"

"I will." She glanced at the discarded bag. "Any napkins in there?"

I checked. "No. Are there some inside?"

"In the office." She started to rise, but seemed to struggle holding the arms of the chair at the same time that she was moving it back.

"I'll go," I put in quickly to curtail that struggle. "I have to use the bathroom anyway." I left before she could argue.

"They're on top of the fridge," she called after me.

The office was on the second floor. I used the adjoining bathroom, which had been renovated since I had been here last, and quickly glanced around the office on my way out. It had been renovated too, and was neatly organized, with two desks facing opposite walls, built-in cabinets and shelves, appropriate spaces for computer, fax, and phone. One of the desks was clean, either Joanne's or a spare. The other, marked by a framed picture of our parents, was Phoebe's. It held a mess of papers and wasn't organized at all. A corkboard covered the wall behind it, and was covered with Post-its. There seemed to be duplicates. I saw variations of MAKE RESERVATIONS FOR NEW

YORK written on four separate sheets. Apparently, New York was something Phoebe didn't want to forget.

From the looks of those Post-its, New York was also something she didn't trust herself to remember. I wondered how many other of these notes served the same purpose, and whether this was compensation for a faltering memory.

Phoebe used Post-its; Mom had used notebooks. Long before she was ill, she had jotted down thoughts, notes, and reminders in a spiral bound journal, one journal per year. And there they were, in order, occupying a place of their own on the bookshelves to the right of Phoebe's desk.

Another time I would read through them. After all, Mom was a writer. I was willing to bet that there was something of her to be gleaned from all those entries.

For now, though, knowing Phoebe was waiting, I took a pair of napkins from the top of the waist-high refrigerator and returned to the deck. The warmth hit me first, and pleasurably, then the smell of newly cut grass wafting from the mower that droned several houses up. Birds were at a feeder by the riverwalk now—goldfinches, the male a bright yellow and the female a bland yellow-ish green. They came and went, vanishing into the willows at times before flitting back, at other times flying all the way across the river. On this sunny day, th'other side was at its innocuous best. I couldn't make out homes through the trees, though the river-bank was littered with fishermen. Both sexes, all ages. I counted a dozen, and those, in the small patch of bank I could see through our own willows. As I watched, a thin young man caught something, unhooked it, and dropped it in a pail. This fishing wasn't just for sport. It was for dinner.

The setting was a throwback, yet not—old-fashioned, but very real. Warmth, smells, birds, fish, and leisure—it was idyllic, I had to say.

A Trojan horse, came a whisper. I quickly turned my head, thinking that someone was being very funny. But the buzz had come from a fly.

Swatting at it, I turned back to Phoebe. She had eaten some of her salad while I was gone, but the fork was now idle against the rim of the plastic tray. She was slouched lower in the chair, her head resting against its back and her eyes closed. I let her rest while I ate. When I dragged my chair deeper into the shade, though, the small sound made her bolt upright. Her eyes flew to mine. She looked blank one minute, irritated the next.

"You must be tired," I said, and passed a napkin across.

Relaxing in her seat again, she took the napkin, studied it, then set it down. With her hands on the arms of the chair, she eyed me with caution. "Tell me where you went this morning."

I shot her a curious look. "Didn't I already do that?"

She didn't speak for a minute, seeming to process my question and regroup. "Well, tell me how it felt. Was it embarrassing?"

"Embarrassing?"

"Seeing Middle River. Knowing you come from here. It must be a pretty ridiculous place compared to Washington."

"I don't compare the two. Middle River is what it is. It doesn't embarrass me."

"But you pretend it doesn't exist."

"I don't. It's my hometown. I've discussed it in interviews. It isn't a secret."

Phoebe furrowed her brow. "Did you say

you went to the clinic?" I nodded. "To see Tom?" I nodded again. "What did he say?"

"We have Washington in common, so we talked about that. He's an impressive man. I think it's remarkable the Meades were able to lure him here."

"He came because of his sister," Phoebe said. She picked up her fork, then put it down and picked up her iced tea instead. The bottle went to her mouth with passable steadiness.

"He didn't mention a sister. Does she live here?" I asked.

Phoebe's eyes lit. She was obviously pleased to be the informant. "She moved here with Tom. She's retarded and was institutionalized for a while, until Tom took her out. He can take care of her here."

I was touched. "That is the *kindest* thing."

"So don't expect anything," Phoebe warned.

I didn't follow. "Expect?"

"By way of a date. The two of you might have lots in common, but he doesn't date. He spends all his free time with his sister. She's his mission in life. Some people wonder about that."

"Wonder what?"

Phoebe gave a sly smile. "Whether they have something going, y'know?"

It was a minute before I realized what she meant, the idea was so contrary to my impression of Tom. "That's disgusting, Phoebe."

"I'm not saying *I* think it," she argued, "just that some do. You know how they are. They go looking for anything remotely perverse. I like Tom. He's very attractive, don't you think?"

"I do, and we'll be friends. But anything more?" I shook my head no. There was no chemistry at all.

"How's Greg?" Phoebe asked.

"Fine. He'll be leaving for his trip in another day or two."

"Why aren't you going with him?"

"Because I'm here." But that wasn't really what Phoebe was asking. I sighed. "I've told you before. We're not an item."

"Why not? What're you waiting for?"

"Nothing," I said calmly. "I'm not in a rush."

"You should be. Look what happened to me. Miscarriage after miscarriage, because I waited too long."

This was the first I had heard about

Phoebe's age being a factor in those miscar-riages—but then, it was something we didn't often discuss. Painful subjects were gener-ally off-limits during those weekends we spent together. But I was curious now that she had raised the issue. "I wouldn't call early thirties old," I said. "It's very much par for the course in Washington."

Phoebe was suddenly irritated. "Well, Washington is *not* Middle River. Maybe those women are different."

"Did your doctor really blame it on age?"

"He didn't have to. It was obvious."

I didn't think it was obvious at all. Very few of my friends had children yet; they were busy building careers. But I knew plenty of women in their late thirties, even in their for-ties, who had successfully carried babies to term. Conversely, I had heard of plenty of younger ones with a history of miscarriage.

I wondered if exposure to mercury could cause miscarriage. Nearly every article on the subject mentioned the vulnerability of young children. Pregnant women were ad-vised not to eat fish that contained high con-centrations of mercury, because even a small amount could affect a fetus.

It struck me that as a Washingtonian I

knew far more about the politics of mercury regulation than I did about what mercury actually did to the body.

I had to learn more.

Back at the house, I turned on my laptop and used Phoebe's dial-up to access the Web—and how not to check e-mail first? I told myself I had to do it for work; that was the premise. But of course there was more to it than that. So here was **TRUTH #2: Ego is involved; we want to be wanted.** In my case, given my childhood isolation, I view e-mail as a party. Granted, it's one I throw for myself. But how nice to enter my little cyber-home from wherever I happen to be, and find all sorts of greetings from friends.

I wasn't disappointed now, though the notes in my in box weren't only from friends. I quickly deleted the spam and filed a handful of work-related messages that needed no reply—one from my editor acknowledging receipt of my revisions, one from my agent wishing me a good trip. My publicist wanted to know if I was interested in speaking at a fund-raiser in Idaho the following spring, but since that was my serious writing time, I respectfully declined.

Having saved the best for last, I read notes from my three closest friends—Amanda the graphic artist, Jocelyn the college professor, Berri the professional volunteer—each asking her version of how the drive had gone and how my reception in Middle River had been.

Greg had sent a quickie. His were always quickies, often just a note in the subject line. This one said, *Did you arrive?* I clicked on "reply," typed in, *Yup. Are you packed?* and sent it back.

Done playing, I pulled up Google, but I had barely typed in *mercury poisoning* when I heard something on the stone drive. Recognizing the sound, I smiled and rose from the kitchen table. I was on the side stairs when my niece and nephew, Lisa and Timmy, dropped their bicycles and ran toward me.

I hugged both of them at the same time. This was my homecoming. If there was one single reason for me to spend time in Middle River, these two were it.

I held them back and looked from one tanned, beaming face to the other. "I hope you noticed," I said, "that I'm not even leaning over. You two are getting so tall. What is *happening* here?"

"I'm turning thirteen in three months," Lisa pointed out.

"And I'm gonna be tall like my dad," said her brother, not to be outdone.

"You are both gorgeous," I decided, "and so sweet. You're just what I need my first day back."

"Mom says you're staying the month?" Lisa asked.

When I nodded, Timmy complained, "But that's not fair. We have to go back to school in three weeks."

"We can do plenty in three weeks," I told them. "First, though, I'm making limeade," which I knew they loved but their mother hated, "and then I want the details of Girl Scout camp," this to Lisa, "and I want to know what you've been doing that kept you from e-mailing those details sooner."

"Robert Volker," Timmy said with displeasure.

"Oh, you're so smart," Lisa sang in insult, but she was blushing.

"Robert Volker." I was testing the name, gently teasing, when Timmy broke away. "And you have to tell me about baseball," I called after him. "Did you win the series?" He was jogging deeper into the driveway. "Come back here, I'm not *done* with you." Suddenly, though, I understood.

"This car is *cool,*" came the boy's awe-filled cry as he circled the BMW. "Oh *wow.*"

Keeping an arm around Lisa's shoulder—I wasn't ready to let her go—I drew her with me toward the convertible.

From the opposite side, Timmy looked over at me, eyes wide. "Take us for a ride, Auntie Anne?"

Lisa had started the Auntie Anne stuff. She had been two, suddenly talking, but consistently putting an *ee* sound on the wrong word. No one ever corrected her, least of all me. I like being called Auntie Anne. To this day, it makes me feel warm and special and loved—all of which I wanted to wallow in for a while. I figured that if I could distract them long enough, maybe with a batch of sugar cookies to go with the limeade, they might forget. I wasn't yet up for driving topless through town.

"Later," I told him. "Let's visit first."

But Lisa freed herself and ran to the car. "I *love* this, Auntie Anne. It must be *awe*some to drive. Three years from October, I can do that. Will you bring it back then?" With barely a breath, she said to her brother, "Wouldn't it be *awesome* if we had a car like this?"

"Can we go for a ride—*please,* Auntie Anne?" Timmy begged.

With his close-cropped sandy hair, his broad shoulders, and the promise of height, he was very much his father's son. Conversely, Lisa was a Barnes, with her long black hair—worn in a ponytail now—and her slender build. Both children had their dad's personality. I had always liked Ron Mattain. He was a decent, good-natured sort, far easier to take than my sister.

But I'm being unfair. People always like Sabina. She can be kind and funny and sweet, simply not to me. The same is true of Phoebe, if on a lesser scale, and part of me knows I deserve their hostility. I was awful to them growing up. Totally aside from my gawking at them around the house, they were the butt of more than one of the scathing little editorial pieces I wrote when I was the high school correspondent for the *Middle River Times*.

I had tried to make amends with the vacations I planned, but if Sabina's remarks this morning meant anything, I missed the mark. She was right. This definitely was **TRUTH #1: I should have been here while Mom**

was sick. They had shouldered the full responsibility—and it was all fine and good to say that they hadn't let on how bad things were, or that we sisters weren't close, or that I hadn't felt welcome. Alyssa was my mother. I should have helped out.

Our family was shrinking, the older generation now gone. This was no doubt why having sisters suddenly seemed important to me, and I don't just mean *having* sisters. I mean being close to them. The ill will between us had never bothered me before. Now it did.

Of the truths I acknowledged while I was in Middle River, that would be another, but not just then.

"*Please,*" Timmy pleaded, standing on the far side of the car.

Lisa joined his entreaty. "We could go for a ride and *then* talk."

"And where will you sit?" I asked pedantically. "This car only holds two."

"We'll share the passenger's seat," Timmy said.

"We're very small," added Lisa.

"Take us for a ride now, and I'll wash your car."

"I'll make you dinner. I am *the* best cook."

"Is that so?" I asked Lisa, thinking that at not-quite-thirteen, she had remarkable confidence. "What do you make?"

"Chili."

"Ah. That's tempting."

"Wait'll the guys see *this,*" Timmy said as he ran a worshipful hand along the rim of the open car. He raised expectant eyes and waited.

Had anyone else asked, I might have said no. But I adored these two children and loved spoiling them. If Middle River saw me driving through town in my vintage BMW with the ragtop down, surely it would know I was doing it for the kids. I certainly wasn't doing it to impress the natives. I didn't care what Middle River thought of me.

Besides, I was suddenly in the mood for a ride. "Okay," I said in surrender. "You twisted my arm."

In short order, with both children belted into the passenger's seat, we set off. It was a perfect afternoon for a drive, just warm enough so that the breeze coming at us felt good. Warmth and breeze were nothing new for me; I experienced them all the time in

D.C. What was different here was the lack of urgency. Time was ours for the taking. We were tooling around, with nowhere special to go. I could hear my engine (beautifully smooth, I might add), the occasional lawn mower, the hiss and click of a sprinkler, the kids shrieking at friends. It was fun. Novel. Restful.

I cruised down Willow all the way to the end and turned this way and that under a dappled canopy of leaves, along back roads on the outskirts of town, the kids waving at everyone they saw. We drove along the river for a while, then turned inland and headed back toward the center of town, which was about when I realized I needed gas. So I swung into the filling station at the very bottom of Oak.

It was a tactical error. For starters, nothing in Middle River was self-serve; after all, gossip couldn't spread unless people had the opportunity to talk. I had no sooner pulled up to the gas pump when Normie Zwibble emerged from the mechanics' bay. Normie had barely graduated high school and, ever since, had run the station with his father. The Normie I remembered had more hair, but he had always been pudgy. He had also always been friendly and sweet, neither of which had changed, which was why it took twenty minutes to fill the car with gas. Before the nozzle was even in place, Normie had to properly *oooh* and *ahhh* over the car. Then, while he filled the tank, he kept up a running commentary about all of the people I might remember from Middle River High. He continued talking long after the nozzle had quit on its own. I paid in cash, simply to speed up our escape.

By then, though, Timmy and Lisa were asking about the time, exchanging worried looks, saying that they were supposed to pick up a prescription for their mother at The Apothecary, but that the pharmacy closed at four and it was nearly that time now.

I had no choice. Driving straight down Oak, I pulled into a diagonal space outside the drugstore, and sat in the car while the kids went inside. They hadn't been gone more than a minute when a monster of a black Cadillac SUV with darkened windows whipped around the corner. It passed me and abruptly stopped, then backed right up until it was blocking my tail. The driver rolled down his window. There was no mistaking that square jaw, that beak of a nose, or that full head of auburn hair.

I had known I would run into Aidan Meade if I was in town for a month. I had just prayed it would be later, rather than sooner.

Chapter 4

KAITLIN DUPUIS was dying. She hated shopping with her mother in the first place, hated the clothes her mother made her buy (clothes she wouldn't be caught *dead* in when she was with friends) and the noise that went with it—*this one's more slimming, that one's more flattering, the other makes you look smaller, for God's sake get that hair out of your eyes.* Her mom never told her outright that she was ugly and fat—she was too PC for that—but those little digs did the same thing. Being with Nicole was demoralizing for Kaitlin, but there was no way around it. She might hate the docility she had to show, but one thing was tied to the next—docility to trust, trust to privileges, privileges to independence, and independence to Kevin Stark, though of course her mother didn't know about Kevin. Kaitlin had to keep it that way, but she couldn't think straight with her mother in her face, and after Miss Lissy's Closet, there had been lunch and then a mother-daughter tennis match at the club (wearing the *totally* dorky fat-farm whites the club required), and driving back and forth, all with her mother *there,* even in the ladies' room, because this was Wednesday, and during the summer Nicole DuPuis had Wednesdays off.

Nicole was Aidan Meade's executive assistant, meaning that she was his liaison to the outside world. She handled his phone calls and opened his letters, but the crux of her day was spent on e-mail. Aidan liked e-mail. He felt that it effectively countered the image of backwoods New Hampshire, and though little of the e-mail he received was crucial, since his father ran the show, he was forever being cc'd on matters, so there was volume indeed. Nicole's job was to give Aidan the *perception* of importance by replying to each and every e-mail in a manner that suggested he was at his desk, hard at work, and on top of every part of the business.

Basically, as Nicole had told her friends often enough for Kaitlin to know, it was a sell job, and that was right down Nicole's alley. She was good at selling, marketing, packaging. She had built a life doing it for herself, transforming the poor girl she had once been into a woman with just enough savvy and skill to snag a wealthy husband. The fact that Anton DuPuis didn't have anywhere near the money Nicole had thought he did was only a temporary setback. So she wouldn't have a cushy nest egg for her old age or millions to pass on when she died. Nicole only cared about the here and now anyway. She had enough money to create the *perception* of wealth, and that was all that mattered. The DuPuises lived on Birch Street. They drove late-model cars, belonged to the country club, and shopped to their hearts' content.

So then how to explain the fact that Nicole worked? Power. She didn't have to work, she told friends. She *chose* to work, because, after all, she had only one child, who was in school and didn't need her, and charity luncheons weren't really her thing. She needed an intellectual outlet (Kaitlin nearly gagged each time she heard *that* one, because *she* knew that her mother had lunch at her desk not to get more work done or to read Jane Austen but to watch *All My Children* on the portable TV she kept hidden in the drawer), and Aidan Meade had come to depend on her, so wasn't that nice? What would she *do* with her time and her mind if she didn't have this? If the job paid well—which this one did—so much the better. And she did love those Wednesdays off.

"This gives me quality time with my daughter," she said—another PC sound bite—and planned the entire day to make up for all the time she was at work—which was why only now, *hours* after the fact, Kaitlin was finally alone in her bedroom and able to make the call.

Kevin answered after barely a ring. "Hey, cutie."

"She knows," Kaitlin said in quiet panic. She had her head bowed, hair hiding her face in a way that was usually a comfort, but there was no comfort now. "We're in trouble."

"Your *mother*?"

"Annie Barnes. She saw us last night."

There was a silent heartbeat, then an incredulous, "No way," from Kevin. "It was dark. It was like five seconds in the headlights. She doesn't even know who we are."

"I keep telling myself that, but she walked into Miss Lissy's Closet while I was there with my mom, and she winked at me. I mean, it was deliberate. If you'd seen the look on her face, you'd know it was true."

"How? How could she tell?"

"I don't know," Kaitlin said. "Maybe my hair?" Or my fat butt, she was thinking, though she didn't say it. Kevin claimed that he didn't think her butt was fat, but that was because he knew it would upset her if he told her the truth.

"Like, none of your other friends has long blonde hair?" he asked. "She *can't* know who you are. She doesn't *live* here."

"Kevin, she *winked*. I'm telling you, I don't know how, but she knows."

Kevin was quiet for a minute. "Did she say anything?"

"Of course, she didn't! There was no time for that—I mean, like she went off into the other room, and we didn't see her again— but I'm sure she told her sister, and by now Phoebe's probably told Joanne, and Joanne will tell her mother, and her mother works for

James Meade and sees *my* mother at the office all the time. What are we going to do, Kevin? If she finds out, we're sunk! She'll be *furious*. She'll dock me for months, she'll interrogate me every time I go out, she'll send me *away* if she has to. She'll threaten your parents with God-only-knows-what. She'll accuse you of *rape*."

"I never raped you."

"It's a *legal* thing, Kevin. I'm underage!"

"And your mother wasn't doing what we are when she was seventeen?"

"Of *course* she was, which is why she's so distrustful of *me*. She's terrified I'll do what she did."

"She didn't do so bad."

"She hates my father. They barely talk. They don't even make love anymore."

"How do you know?"

"I heard her tell a friend on the phone."

"Well, that's not the situation with you and me. We love each other. Besides, we were just having fun."

Kaitlin might have strangled him for missing the point. "Kevin. Think. Having fun is even worse, because having fun leads to babies, and that's how I came to be. She'll go ballistic if she knows I'm on the pill, and she won't *hear* the part about love. She'll just think about your daddy working at the mill." Kaitlin rubbed her temple. She might have known she would be caught. She was *always* caught. "What were we thinking?"

"That it was nice."

Oh, it had been nice. It always was with Kevin, right from the start. He had been her first. And gentle? Omigod. So sweet touching her breasts, so excited taking off her clothes that his hands shook, and then so careful and apologetic and *upset* when she bled. She had expected to feel grateful that he wanted to make love with her. What she hadn't expected was to feel the fire. But it was there every time. This wasn't the first backabehind they had pulled. Lacking the privacy of a bedroom, they had made love in the woods, behind the school, even out at Cooper's Point. They had made love on Oak Street, Willow Street, and now Cedar.

Kaitlin wouldn't find another guy like Kevin. If it ended, she would be back to being the only one without a date. If it ended, she would feel more ugly than ever. If it ended, she would absolutely, totally cease to exist.

"Annie Barnes will tell," she cried fearfully.

"She'll even do it out of spite, because *she* never pulled a backabehind. Like, who'd have slept with her? She had no friends. I mean, ze-ro. She was ugly, and she was odd."

"Yeah, that's what they always said, but I saw her today, and she wasn't ugly. She came into Harriman's while I was stocking the shelves."

"Omi*god. That's* how she knew. You must have looked totally guilty."

"She didn't see me."

"Not once?"

"Not once."

Kaitlin flipped her hair back from her face. "I can't believe Annie Barnes chose *that* instant to drive down the street."

"Maybe she won't tell."

Kaitlin felt despair. "Yeah. Right. Like she won't write it into something? Kevin, do you know who she *is*? She writes books, and she's come here to write about Middle River, because that's what she used to do, only now she's rich and famous, so the whole world reads what she writes. We're in *big* trouble," she cried. "What are we going to do?"

LOOKING FOR PEYTON PLACE 114

Chapter 5

HEART POUNDING, I watched through my rearview mirror as Aidan Meade studied first my car and then me, but that pounding heart had nothing to do with physical attraction. I'm not even sure I had felt that for Aidan back when I was eighteen, when we were meeting in the woods at Cooper's Point. What I felt then was awe; Aidan was the most sought-after twenty-one-year-old in Middle River, and he was interested in me— or so I thought at the time.

I knew differently now, which was why the pounding of my heart came from anger. I tried to get a grip on it during the time he spent studying my car, but it was unabated, even stoked when he opened his door, climbed out, and approached.

Fifteen years was a long time for anger to simmer. *Be cool, Annie,* I cautioned. *Be the deliberate woman you are in D.C.—the one who doesn't act on impulse and has more power for it.*

"That's some car, Annie Barnes," he said in a smooth voice, but the closer he came, the less sure he seemed. "Annie?"

"Uh-huh."

"You look like her, but not. Wow, you've changed."

Had he made it a compliment, I might have calmed, but as stated it annoyed me all the more. "Not you, Aidan," I observed. "You look the same. Older, but the same."

He smirked. "I was hoping you'd say 'older but wiser.'"

"Wiser?" I couldn't help it. "You're on what marriage now? Fourth?"

"Third. Four would be pitiful, don't you think?" I thought three at his age was pitiful, but before I could tell him, he said, "Been following my life, have you?"

"Not yours, per se. I read the *Middle River Times* cover to cover each week. It's better than *People*—more juicy. I'm always amazed. It reports each of your weddings like it was the first."

"*Dad-dy!*" came a shriek from the car.

Aidan held up a hand to still the voice.

I couldn't see in the windows, though there was truly no need. "And now you're a daddy, too," I observed politely. "How many kids do you have?"

"Five."

"In all?"

"Three with Judy and now two with Bev. Lindsey and I didn't have any kids."

I frowned, puzzled. "Lindsey was the first. Didn't you marry *because* she was pregnant?" Of course, he had. The town had been abuzz with the news. There had been a miscarriage and a speedy divorce. The marriage had lasted a mere six months.

"No," Aidan lied, then glanced at his car in response to a cry.

"*Daddy,* he's *kicking* me! *Stop* it, Micah!"

"Micah, keep your feet to yourself!" Aidan yelled. He faced me again. "You have an edge, Annie. But then, you always did. As I recall, we had something going, until you turned sour."

The barb wasn't worth answering. "As *I*

recall," I said with a smile, "we never *got* something going, because you were never there. You'd tell me to meet you at eight, and I'd sit alone in the woods until you showed up at ten or eleven. You'd give me some story about work holding you up and how exhausted you were and that you'd call me in a couple of days, and sure enough, you did. Then Michael Corey accused you of having an affair with his wife, and you said it couldn't possibly be so, because on the dates in question, you were with me. You said all that in a sworn affidavit. And I didn't deny it."

"Nope," he said smugly.

"Because to deny it," I went on, welcoming the catharsis, "would have meant admitting to the town that we hadn't been together, and you knew I wouldn't do that. You knew I had never dated anyone else and that I thought it was awesome to be picked by a Meade. You *knew* I would jump at the chance to be yours."

He grinned. "You did."

"And that I would lie, rather than say you had basically stood me up all those times."

"You did that, too."

"Lied, yes—but not under oath like you

had, never under oath." I stopped smiling. "I was your alibi right up to the night of my senior prom. You offered to take me."

"In gratitude," he said.

"And then stood me up."

He smirked. "Couldn't be helped. I was tied up."

"So I sat home alone that night, all dressed up in the prettiest dress from my mom's shop. I had told everyone at school that you were my date. When I never showed up, they decided I had made it up."

"So then you backtracked and said you lied about the other," Aidan picked up the story, "only no one believed you. After all, I was the one who had sworn to tell the truth. You were such a pitiful thing. People understood why I did what I did."

"Even though you lied."

"My lie was for the greater good. Wouldn't have done Mike Corey any good to know the truth about Kiki and me. They got back together again after she and I broke up."

I gave him my most serene smile. "For the greater good? Huh. What would you do if I told you I was wearing a wire right now?"

I saw a second's surprise in his eyes, but he was distracted when a blood-curdling screech came from the car. This one was higher pitched, if smaller.

"Leave the baby alone, the *both* of you," Aidan roared, "or you'll get it when we get home!"

Beat your kids, too? I wanted to ask, but it would have changed the subject. We weren't done with this one yet.

Aidan was less cocky now. "You're not wearing a wire. You had no idea I'd be coming round that corner."

"No, but, y'know, I'm glad you did," I said and meant it. I had been dreading this meeting. Now it had happened, and I hadn't crumbled. This was proof that I wasn't the lonely girl I had once been, the one Aidan Meade had left waiting in the woods, the one who had been stood up on the night of her only senior prom, the one who had been disappointed and humiliated and compromised. The woman I was now said in a voice that was forceful for its self-possession, "I've always wanted to tell you what a snake I think you are. You used me, Aidan. I won't ever forget that."

I glanced past him when another big SUV rounded the corner. This one was identical to Aidan's, dark windows and all, except that it had the Northwood Mill logo on the side of the door. It came to a stop as suddenly as Aidan had, idling in the street while the driver opened his door and walked over. Fortyish and intelligent-looking, he wore jeans and a jersey with the same mill logo on the breast pocket.

"Your father's on the warpath," he told Aidan. "The ad folks just showed up, and you haven't answered your cell."

Aidan glared at his own car and shouted, "Did that phone ring, Micah?"

I half turned in my seat to get a glimpse of the shadow of a child, but another shadow caught my eye first. This one was in the passenger seat of the Northwood Mill SUV. Judging from the size of it, it was a man and from the profile, a Meade. That would be James. He was the older brother, the heir apparent, the brains behind the mill's recent growth, and the father's right-hand man.

I don't know whether the child answered Aidan or not, but Aidan was arguing with the man from the mill. "They weren't supposed to be here until four." When he saw his man looking at me, he said with a snide edge, "This is Annie Barnes. She's come back to town to make trouble."

I wasn't bothered by the remark. I looked good, and I knew it. Offering the man a hand, I smiled pleasantly. "And you are?"

"Tony O'Roarke," he said.

"He's our VP for operations,"Aidan put in, "which means he's the hands-on guy at the mill *and* the one the old man calls when he has a gripe with one of us." To Tony, who had shaken my hand quite nicely, he said, "Those people said they'd be there at four, so that's when I'll be there."

"He wants you now."

Aidan's look was as cold as his tone. "I'm busy now."

Clearly feeling the chill, Tony raised a low hand and took a step back, then retreated fully. Aidan watched them drive off. Mouth and nostrils were tight when he faced me again, but he picked right up where we'd left off.

"So now you've come back to town to get revenge. Sorry, sweetheart, but there's nothing to avenge."

"Then you have nothing to worry about, do you?" I said offhandedly, and that *did* make him nervous. Everything about him seemed suddenly tighter.

Warily, he asked, "Are you here to write?"

I glanced at The Apothecary just as Lisa

and Timmy emerged. "Well, writing *is* what I do."

"How much do you make per book?"

"*East of Lonely* has just gone back to press again. I think the in-print total is approaching two million."

"How much do you make?"

I was amused. "I'm sorry. I don't discuss money."

"I could find out."

"I don't think so. Meades control Middle River, but they don't control New York—or Washington, for that matter." I smiled. "You don't scare me, Aidan." I turned the smile on my niece and nephew. "Get what you needed?"

By way of an answer, Lisa held up the prescription bag. "Hi, Mr. Meade," she said with a deference that was echoed by her brother as they cinched themselves into the single seat belt again.

Politely, I said to Aidan, "I'm afraid I'm going to have to ask you to move your vehicle."

He probably would have argued— Meades prided themselves on getting the last word—had one of his children not yelled, "*Dad-dy, I need to go potty bad!*"

With a grunt of displeasure, he gave my BMW a pat, and walked off.

I watched in my rearview mirror as he climbed into the big black SUV, slammed the door, and drove away. There was no pounding heart now. I might have prayed that my confrontation with Aidan would have taken place at a later time, but Someone upstairs had known better. It wasn't that my prayer hadn't been answered, simply that I had prayed for the wrong thing. I should have prayed to get it over and done. Now that it was, I was composed and content.

I couldn't have asked for a more satisfying meeting.

Back at the house, while Timmy washed my car with a ten-year-old's love and Lisa the almost-woman made chili from a mix, I logged onto the Web for a refresher on mercury poisoning.

There were two kinds—acute and chronic.

Acute mercury poisoning results from intense exposure over a short period of time, most notably either from eating the stuff or inhaling undiluted vapors. Symptoms start with a cough or tightness in the chest, and progress to breathing and stomach troubles. Fatalities occur in cases where pneumonia develops. In cases where mercury is actually swallowed, there can be intense nausea, vomiting, and diarrhea, not to mention permanent kidney damage.

I didn't see my mother or sister falling into this category. *Chronic* mercury poisoning, though, was less easily dismissed. It consists of repeated low-level exposure to contaminated materials, and since the symptoms are slow to appear, exposure can occur much earlier. Moreover, the symptoms of chronic mercury poisoning vary widely from victim to victim. One might suffer bleeding gums, another numbness in the hands and feet. Others suffer slurred speech and difficulty walking, or mood changes that include irritability, apathy, and hypersensitivity. In the later stages of chronic mercury poisoning, the central nervous system can be affected, as can kidney and liver functions. Birth defects are a serious risk, hence the warning that pregnant women not eat those kinds of fish that retain higher concentrations of the metal. The relationship between mercury poisoning and autism in young children is a whole other story.

I pulled up more sites in search of other symptoms. By the time I was done, I had found descriptions of all of the ones I had seen in my mother and was starting to see in my sister—the fine tremor of the hand, the trouble with balance, the memory loss and frequent inability to hold a train of thought for long.

Unfortunately, these malfunctions could also be attributed to Alzheimer's or Parkinson's, or to any number of other conditions. Worse, identifying chronic mercury poisoning with certainty is next to impossible. Acute mercury poisoning can be proven, since blood tests done within several days of exposure to high doses of mercury show elevated levels of the metal. After those few days, though, it moves into the nervous system and no longer shows up in the blood, at which point a blood test is useless. Urine tests are even worse; since mercury doesn't *ever* leave the body through urine, it never shows up in a test.

The bottom line? Chronic mercury poisoning can never be diagnosed for sure.

On the other hand, if it could be shown that a person with specific symptoms has been exposed to mercury, a circumstantial case can be made. This was where North-wood Mill came in. Yes, I was biased. I was looking to pin this on the mill. But what else could be responsible? If people in Middle River were getting sick in the numbers I thought they were, the source of mercury had to be big. The mill was big.

"What are you doing, Auntie Anne?" Lisa asked.

"Oh, some research," I said, knowing she would assume the research was for a book. Highlighting the latest batch of information, I copied it into a folder with the rest. "It's kind of a fishing expedition at this point." I clicked the folder shut and turned away from the laptop. "That chili smells really good, sweetie. Did your mom teach you how to make it?"

"Nope. My dad. He's the cook. Mom only gets home in time to eat." Her eyes flew to the door and widened in excitement. "Here she is now. She left work early to see *you*." She ran to the door and opened the screen. "Hi, Mom. Come on in here. I'm making din-ner for Aunt Phoebe and Auntie Anne."

Sabina came through the door looking tired and tense. "You are the best girl in the world," she said and kissed her daughter on the forehead. "Want to help Timmy finish the car, so I can have a minute alone with Auntie Anne?"

"I have to stir the chili first," the girl said and returned to the stove. With deliberate motions—carefully taught, I decided—she put on a mitt, lifted the lid of the big pot, stirred the chili in a way that avoided splatter, then returned the lid to the pot, removed the mitt, and smiled at us.

"Well done," I said, smiling back.

Looking proud, she went outside.

The screen door had barely slapped shut when Sabina slipped into the chair kitty-corner to mine at the table and said, simmering, "I just got a call from Aidan Meade. He wants to know why you're here. What did you say to him?"

"Nothing," I replied. I was annoyed by her annoyance. "He didn't ask why I was here." At least, not in as many words. "Why's he asking you?"

"Because *you* must have said something that set him off. I asked if you were writing a book, and you said you weren't. He thinks you are. He thinks you're still fixated on doing what Grace did."

"This has nothing to do with Grace."

"With you it *always* has to do with Grace. She had a grudge against small towns. So do you."

"She didn't have a grudge against small towns. She just never fit in."

"She *hated* them. Look at *Peyton Place*."

"She painted a realistic picture. There's as much to love about that town as there is to hate."

"See? You defend her."

"I understand her. I know what it's like not to fit in. I have good reason to hold a grudge against Middle River. But why would I want to write a book about it? I have ideas aplenty for future books that have nothing to do with this town. Aidan Meade must have one guilty conscience if all he can think is that I'm writing a book about him."

"Or about us. Why else would you be here?"

"This is my home."

"Correction," Sabina said. "Your home is in Washington. This is where you grew up, but you left. You rejected us."

"Correction," I shot back, "you rejected me. Middle River made my life so miserable that I had to leave. And I'm not moving back here. Trust me on that. I like my life. But my roots are here. I feel a need to connect."

"Why? Mom's gone."

"But you're here. Phoebe's here. You're the closest blood relatives I have." There it was, **TRUTH #3: My need for family,** bubbling to the surface before I could hold it down—and believe me, I would have liked to. Sharing emotions wasn't something we usually did. Tipping my hand to Sabina made me vulnerable.

But the words seemed to quiet her. She drew in a tired breath and sat back in the chair. "Then try to see it from my point of view, Annie. I'm in a difficult position. He's my employer."

Sabina was Northwood Mill's main computer person, very much the jack-of-all-trades where information technology was concerned. When a computer malfunctioned, she was there. When an employee needed instruction, she was there. She was there when computers were added to the network, there when the system had to be wiped clean of viruses and worms, and she had a major say when it was time to up-grade. The last, of course, meant that she had to train every last employee when up-grades were made.

Being only functionally computer-literate myself—i.e., I could do what I needed to do, namely e-mail, write, and search, but little else—I had the utmost respect for her. I had a feeling she worked longer hours than any of the Meades.

"They need you more than you need them," I said on that hunch.

"Not true," Sabina argued. "They pay me well, and the benefits are twice what I'd get elsewhere. We have two kids to take care of. But you wouldn't know about that."

No. I wouldn't. Steering clear of that dis-cussion, I said, "Sabina? Do you know any-thing about mercury poisoning?"

"Should I?"

"The symptoms Mom had are very similar."

"Mom had Parkinson's." She scowled. "Why can't you accept what she had? Are you so afraid you'll get it too?"

"It's not that. It's just that these symptoms bother me. Same with Phoebe's symptoms. And from what I've read in the *Middle River Times,* it seems like lots of people are suffering from one or another of the same symptoms."

"Are they?" she asked doubtfully.

"If what I've read is correct. You know the column—"

"I do, but the ailments go across the board. Face it, Annie. Middle River is getting older. Older people have more ailments."

"More frequent miscarriages in twenty-somethings? More autism in children?"

"What's your point?" my sister asked blandly.

"The mill. Do you think it could be a polluter?"

"No."

"Have you ever heard the word *mercury* spoken there?"

"No."

"Are you aware that mercury is a political hot button?"

"*No.* Where are you heading with this?"

I backed off. "I don't know. I'm just wondering about these things."

Sabina rose, tired and sad now. "I know you, Annie. Wondering leads to no good. Please. I'm begging you. *Please,* don't do anything to make life difficult for us here. My husband works for the mill. *I* work for the mill. This is our livelihood, our future, our *children's* future. We can't afford to have you meddling."

I wasn't meddling. All I was doing, I decided, going online again that evening after Phoebe went upstairs, was learning more about a problem that maybe, just maybe, was causing harm.

Basically, there are two sources of mercury emissions—natural and man-made. Natural mercury emissions come from volcanic eruptions, the eroding of soil and rocks, and oceanic vapors. These have been around forever, and can't be contained.

Man-made sources of mercury pollution are another story. There are two major villains on this end. The first are trash incinerators that dispose of mercury-laden goods such as thermometers, fluorescent lightbulbs, and dental fillings. The second are oil- and coal-powered plants. These plants are the nation's single largest mercury polluter; once emitted, their toxins remain in the air for centuries. Moreover, once airborne mercury drifts to earth and mixes with bacteria in the soil, it becomes methyl-mercury, and methyl-mercury is extraordinarily toxic. It is also bioaccumulative, meaning that it grows more potent as it climbs the food chain. Mercury in a minnow isn't as bad as it is in trout, which isn't as bad as it is in swordfish or tuna, which are larger. Man, being at the top of the food chain, suffers the harshest effects.

Although some airborne mercury comes from industrial plants in the Midwest, nearly 50 percent is generated in New England. New Hampshire has yet to prohibit mercury pollution and, hence, is one of the worst offenders.

That was something. But I was still in need of a crucial piece of information. I had to dig for it; clearly the industry didn't want the past rearing its ugly head. Beneath layers of links, though, I finally found what I sought: paper mills were indeed known mercury pollutants.

If Northwood Mill released mercury, its most probable victim of contamination was the river. My mother's shop was on the river, not far downstream of the mill. She had spent much of her working life there. Now my sister was taking over. And all those other people worked in all those other shops on the shores of the Middle River. And all those customers sat for hours on those back decks drinking in sunshine and potentially harmful vapors. And all those people on the other side fished for dinner and *ate* what they caught.

The possibilities were endless.

I didn't know whether to be excited or horrified. In need of grounding, not to mention direction, I turned off my laptop and phoned Greg. I half expected that he would be out somewhere celebrating his last night in Washington, but my luck held. He was home.

"Hey," I said with a relieved smile.

"Hey, yourself," he said back and answered my e-mail query. "All packed, thanks."

"Then why aren't you partying?"

"I am."

Ah. He wasn't alone. "Oh dear. Can we talk another time?"

But he was easygoing as ever. "We're waiting for take-out delivery, so this is a great time. What's up?"

It was all the encouragement I needed. "I might be onto something here," I said eagerly. "My mother's symptoms are identical to those of chronic mercury poisoning."

"Mercury? Huh. Any idea how she might have been exposed?"

"Our paper mill."

"Does it produce mercury waste?"

"Not now. But it did. Paper mills are known polluters. Their waste goes right into the river. My mom worked along the river for years. Now my sister's been there awhile, and she isn't well."

"Same symptoms?"

"Some. And she has a cold. One of the effects of chronic mercury poisoning is a weakened immune system. Lots of colds, the flu, pneumonia."

"I'm not sure you can get mercury poisoning just from working alongside a river. I think there has to be direct contact, like something eaten or touched."

"That's possible," I conceded. "But refresh my memory. What's the latest on mercury regulations?"

Greg didn't hesitate; recapping the news in layman's terms was his forte. "Everyone is in agreement that emissions need to be reduced. The disagreement comes in how much and when. The Clean Air Act suggested guidelines and deadlines, but the EPA has scaled back the guidelines and extended the deadlines."

"Under pressure from lobbyists?"

"You bet. And then there's the credit-trading business. All plants pollute. How much they're allowed to pollute is spelled out in pollution credits. A plant that cleans itself up so that it doesn't have to use its credits can sell them to a plant that doesn't want to clean up. The clean plant makes money to compensate for getting clean, and the dirty plant pays to stay dirty. Opponents say this will create hot spots where the pollution is worse than ever. Do we want to live in one of those spots?"

"No. But my mother and sister—much of Middle River—may be doing just that."

"Direct contact, Annie. Keep that in mind. Besides, I wouldn't accuse anyone until you know more. The issue is disposal. If it followed the rules, your mill may be on the up-and-up."

"How would I find out?"

"Ask."

"I can't ask at the mill. They'd go berserk."

"Then call the state environmental agency."

"Like they'd know if the mill has broken rules? Wouldn't Northwood cover it up?"

"They'd try. If they were caught, they'd pay a fine and cover *that* up. But it'd still be in the state records."

It was a good thought, precisely why I had called Greg. His suggestions were always intelligent, purposeful, and diplomatic. Prone to impulsiveness, I'd had, especially, to learn the last.

State environmental agency, I wrote down, then said, "I'm going to go back through the archives of the local paper to find out who is sick and with what. The health column is thorough."

"It would be better to talk with local doctors."

"But then the cat would be out of the bag."

"Would that be so bad?"

"Yes. The Meades own the mill, and they're evil. If they get wind of the fact that I'm looking into this, they'll take it out on my sisters. I did talk with my mother's doctor. He heads the clinic. He put the bug in my ear about mercury, but he's in a precarious position. The mill controls much of the town, including the clinic." I paused. "When you talk Middle River, you talk the mill. It always comes down to that."

"And you'd love to get back at the Meades," Greg said.

"Mom had something. Phoebe may, too."

"And you'd love to get back at the Meades," he repeated.

I was silent. He knew me too well. "You bet I would," I finally admitted.

"Thinking of doing the Grace thing?"

"Definitely not. Grace wrote a book. I'm not doing that."

"Even if you find a connection? Even if you follow this all the way and manage to bring about change?"

"If things change, there's no need for a book."

"What if you find a cover-up? It would make a terrific story."

"That's *your* field, Greg. You're the journalist. I write fiction."

"So did Grace."

"I'm not Grace."

"So you've said," he teased but with a gentle whiff of truth. "You've said it over and over again in the twelve years I've known you. But she is your model. You cut your teeth on her legacy."

"My career has been totally different from hers."

"And you've made it big now, which means you can write what you want and be guaranteed a large audience."

"My audience would hate it."

"But you'd have quite a platform." On a

genuinely curious note, he asked, "Did Grace do it on purpose—write *Peyton Place* with the explicit intention of upsetting the apple cart? Did she do it to create a scandal?"

"With malice aforethought?" I asked with a smile. It was a phrase Greg's lawyer friend Neil—our friend—used all the time. "I don't think so. She loved writing, and she knew small towns. She was simply writing about what she knew."

"But she did have gripes."

"Yes. She was never one to conform to the traditional expectations of women—just couldn't be the pretty, smiling, docile wife and mother that people wanted her to be. She wore jeans and men's shirts. She never cleaned house, except for the very corner where she kept her typewriter, which is such a telling thing," I mused. "She disdained the fifties image—couldn't function as the schoolteacher's wife and do committees and chaperoning and bake sales. She was a brilliant writer who antagonized her landlord with the constant banging of her typewriter keys. Writing was what she did. She just didn't care about the rest. So she was ostracized."

"And she lashed back."

"She told the truth."

"And got revenge."

"That's arguable, given how fast and far she fell after the book came out."

"But she did make a point."

"She did," I had to concede. "What's yours, Greg?"

"I love you. I don't want to see you hurt. I admire you for being there and wanting to know what's wrong with Phoebe, and if you can show a pattern of illness that suggests large-scale mercury poisoning *and* that the mill is the culprit, I'll *help* you write the book. I just don't want you to suffer in the process."

"How will I suffer? People in Middle River already hate me. What do I have to lose?"

He was a long minute in answering. Then he spoke with the kind of caring I was desperate to find. "You've come a long way, Annie. When I first met you, you were still working to distance yourself from Middle River, and you've succeeded. But you revert when you talk of the place. You're prickly, and you're defensive. I'm not sure that town is good for your health."

"Neither am I," I said with new meaning and deepened resolve. "I love you too, Greg. Travel safe."

Chapter 6

TOM MARTIN was late leaving the clinic again. Today, he'd had to referee a clash between two radiologists fighting over assignments, but more often, his lateness was patient-related. Medical emergencies were forever cropping up at the end of the day, and Tom was a softie. He didn't have the heart to make a patient worry through the night about what ailed him, when an extra thirty minutes of his own time would do the trick. Fortunately, the woman he employed to look after his sister understood.

That said, he hated to push his luck. Making a beeline for home, he whisked out the side door of the clinic and was striding toward the parking lot when he heard his name called. He didn't slow, simply looked around. Eliot Rollins had left the building and was trotting toward him.

Eliot was an orthopedist, a large teddy bear type with a close-cropped beard and a wide white smile. He was a nice guy, a genial sort with more social skill than intellectual curiosity. Thanks to partying in college, much of it with Aidan Meade, he'd had to take several shots at medical school before finally getting in. He had come to Middle River straight from his residency and had been at the clinic for three years before Tom was hired.

Tom might have blamed the distance between them on that; Eliot wouldn't have been the first to resent the new guy on top. But there was more. They had different manners, different tastes, different friends. As a matter of rule, each went his own way.

That was one of the reasons Tom was surprised to see Eliot now. Another was that Eliot was usually among the first to leave work, not the last. He had a rambling old home on the south end of town, where he kept a wife, two children, and a cooler of beer by the pool. Summers, he was usually gone from the clinic by four.

"Hold up," Eliot said in his usual genial way. "I have a question."

Tom slowed only marginally. "Sorry. I'm running late. Have to get home."

Eliot drew abreast and matched his pace. "How's Ruth?" he asked.

"She's doing well, thank you."

"It's working out with Marie Jenkins, then?"

Marie was the woman who looked after Tom's sister. She had been an ER nurse on the verge of burnout when Tom hired her, and the job suited her well. Ruth required custodial care, but there was no trauma in her day. To the contrary. Everything about Ruth was slow. She had to be helped with washing and dressing, and had to be driven to and from programs three times a week in Plymouth. She needed help with the DVD player; though, if allowed, would watch *Finding Nemo, Beauty and the Beast,* and *The Lion King* for hours on end. She loved being read to, even when she didn't understand the content, and often fell asleep before the first chapter was done.

Marie was so grateful for the respite from

the tension of the ER that she didn't mind the tedium. During quiet times, she read or cooked. Tom had no problem with that.

"It's working really well," he told Eliot. Having reached his pickup, he opened the door, tossed in his briefcase, rolled down the window. "What's up?" he asked as he climbed in.

"Julie and I are having a cookout Friday night. Want to drop by?"

Tom smiled. "Thanks, but I can't. Marie is taking off for the weekend, and Ruth has been nervous with strangers lately, so it could be more trouble than not." He pulled the door shut.

"Oh," said Eliot and curved a hand over the open window. "Well, maybe another time." He caught a quick breath. "Say, was that Alyssa Barnes's daughter who dropped by today?"

And *that,* thought Tom, was the question Eliot had really wanted to ask—and with some urgency, if the price of asking it was having to invite Tom to his cookout. Amused by that, and curious enough to set aside his own rush to get home, Tom said, "It was. She wanted to thank me for taking care of her mother."

"Well, that was nice. She didn't catch you at the funeral?"

"Not for that."

"Huh. So she caught you now. She's here for a month, y'know. Did she say what she'll be doing?"

"I don't think she knows for sure."

"Then she didn't mention doing some writing?"

"Not to me," Tom said with more innocence than was perhaps justified, but he sensed where Eliot was headed. "Are you worried she'll write about us?"

"That depends," Eliot mused casually. "People are saying she might. I don't know about you, man, but I wasn't in town ten minutes before someone was telling me about *Peyton Place*. I mean, Aidan used to talk about the connection, and, sure enough, there it was. It's an obsession."

"*Peyton Place* is history. That all happened nearly fifty years ago."

"Middle River says it could happen again. They're saying Annie Barnes could to it. I'm not sure we want her doing that."

"Why not?" Tom goaded. "We're good folk."

"Sure we are, but we help each other out in ways we don't always want strangers to know. You want her writing about Nathan Yancy?"

No. Tom didn't. Nathan had been a staff hematologist when Tom had first arrived at the clinic. It was a full year before Tom had a credentialist in place who discovered that Nathan had bogus degrees. He was quietly asked to leave, and he went without benefit of recommendation. It would be all well and good to say Tom wasn't responsible for a doctor who had been hired on someone else's watch. The fact remained, though, that if word got out about Nathan, confidence in the clinic would weaken and its reputation would suffer.

The same thing might happen if word got out about Eliot Rollins. Unfortunately, Eliot Rollins was in the Meade inner circle and hence felt a certain immunity. That said, Tom had no intention of taking the fall for the powers-that-be, if and when the issue erupted.

"What can I say, Eliot. I've warned you. If you're overmedicating certain patients—"

"I am not overmedicating anyone," Eliot cut in, suddenly less genial. "I've said it before, and I say it again. Pain management is a vital part of practicing medicine. It's the cutting edge."

"Sure it is, for geriatric patients. Our police chief doesn't fall into that category."

"You want to tell him he has to suffer?"

"He doesn't have to suffer," Tom argued. "He could lose forty pounds and start exercising. Add a little physical therapy, and his back would be fine."

"You're an expert on orthopedics now?"

"Nope. Don't have to be. It's common sense. You gave him a crutch, Eliot. He's addicted now, so the problem is compounded."

"You have no proof that he's addicted."

"Anyone who takes the painkillers he does with the frequency he does is addicted."

Eliot put his hands in his pockets. He was chewing on a corner of his mouth. "Was she asking you about that?"

"Who?" Tom couldn't resist asking.

"Annie Barnes. When she talked with you today."

"No. She was thanking me for taking care of her mother." He had already said that, but it bore repeating.

"That wouldn't take more than a minute or two. You were out there longer than that."

"Were you watching us?" Tom asked, but didn't wait for an answer. Eliot hadn't had to watch them to know they were there. Middle River was a town filled with spies. He had known that his having coffee with Annie would make the rounds. So, casually, he said, "We were talking about Washington. We both graduated from Georgetown."

Eliot didn't look convinced. "All that time you were talking about Washington?"

"Most of it. We talked some about her mom. That's foremost in Annie's mind." Tom fished out his key and started the engine. "Sorry, but I really do have to get home. Hey, tell you what. If I think of someone to stay with Ruth, I'll try to drop by Friday night. Fair enough?"

Eliot didn't answer. But then, Tom didn't give him the chance. Backing out of the space, he turned the wheel, stepped on the gas, and, with satisfaction, left the teddy bear of an orthopedist behind. Glancing in his rearview mirror, he saw that Eliot had put his hands on his hips. And Tom knew two things for sure—first, that he would opt to be with his sister over Eliot any Friday evening, and second, that Eliot wasn't done with the matter of Annie Barnes.

Chapter 7

I STARTED running when I was eighteen. Having just moved to Washington, I was erasing the past and reinventing myself— new attitude, new friends, new interests. Middle River did baseball, basketball, and football, but running? Not in *our* town. That alone would have been reason enough for me to take it up. There were, of course, other reasons, the prime one being Jay Riley, a junior who had helped freshmen move into my dorm and, yes, was a runner. I had a crush on him from day one.

As fate had it, my roommate, Tanya Frye, whom I adore to this day, had been running competitively since she was twelve, and though I had nowhere near the natural aptitude she did, I took to the sport with remarkable ease. Granted, she started me slowly. By the end of freshman year, though, I could do a 10K in forty-six minutes—which, FYI, is nothing to sneeze at.

I never did catch Jay's eye, but my new persona flourished. I dated other guys. Some were runners, some were not. All admired my legs in shorts. It was a whole new experience for me.

Over the years, I have come to race less and run more for the simple enjoyment of it. I always feel better after I run. I'm told it's a chemical thing having to do with endorphins, though I've never delved into the science of it. I do know that after a run my head feels more clear, my limbs more limber, my insides more in sync with the rest of me. After a run, I feel well oiled. Conversely, when several days pass without a run, I'm logy.

That's how I felt waking up Thursday. So I put on a singlet, shorts, and running shoes, stretched right there in the kitchen, and slipped out the door while Phoebe slowly emerged from her own lethargy over coffee and eggs.

Normally I run at six in the morning to beat the heat and then return for my prime hours of work, but Middle River was cooler than D.C. and I wasn't working. Running at eight here, though, raised a different problem. Middle River would be up and about.

So, as I had done in my convertible the day before, I stuck to the back roads, and it was really nice. There were no cars on the road; the area was strictly residential, houses spaced farther apart, and though I spotted the occasional native watering the lawn or weeding the beds, there was enough land between us so that I was barely seen.

Running north up Willow to the end, I turned east and cut across town on pavement that was buckled and cracked but forgiving in the way of a withering hag. These roads weren't named after trees; when the town's founding fathers had run out of those, what recourse did they have but to name streets after themselves? So there were Harriman, Farnum, and Rye streets. There were Coolidge, Clapper, and Haynes streets—and, of course, Meade Street. None of these namesake families lived in this neighborhood; they were closer to the center of town in homes that were larger and more imposing. The houses here were even more modest than ours on Willow. These were small Capes that, if enlarged, had been done so in the form of caboose-type additions tacked on. All were well-tended, though, with pretty walks and flower beds, and matching shutters and doors.

I had a favorite here. It was a bungalow at the very end of Hyde Street, tucked under a clan of hemlocks and looking like a gingerbread house, with its chocolate siding and pale pink trim.

Omie lived here. But more on Omie later.

Feeling good, I ran on at a steady pace. I turned right at the end of the road and was in the process of returning to the center of town when I spotted another runner.

Another *runner*? I mused wryly. What was Middle River *coming* to?

I might have asked him—yes, the runner was a guy—had he been closer, but he was on the cross street several blocks ahead. He glanced at me as he passed, then disappeared.

Oh boy, sweetie, he's a good-looking one, said a salacious little voice in my head.

Unable to see much this far away, I promptly dismissed the remark, but my commentator didn't leave. Neither a fly nor the purr of a cat this time, the words were crystal clear.

He's tall. I like them tall, she reminded me. ***Who do you think he is?***

Haven't a clue, I thought.

There's nothing like a bare-chested man to stir the juices.

He's a runner, I argued. It's a different kind of bare-chested. He's probably all matted and sweaty.

Sweat is sexy as hell. Follow him, sweetie.

I will not.

No? For God's sake, what good are you?

"Hey! Aren't you Annie Barnes?" asked a man who had emerged from his front walk during my brief distraction. He was short and beyond middle age, very much the opposite of the runner (okay, I did get a *general* impression), and he seemed to expect me to stop and talk.

Stopping and talking was something Middle Riverites did. Not me. Especially not while I was running.

I raised a hand as I ran past, but didn't slow down. I had hit my stride. I felt good.

Was I being rude? Probably. When in Rome, as the saying goes.

But my feelings of guilt only went so far. The town already thought the worst of me. What was a little rudeness but a validation of that?

By now, you may be thinking I was disliked by *everyone* in Middle River during the years I lived here. This is not so. I had a few friends. Granted, they weren't exactly my peers. But they were supportive.

One of those was Marsha Klausson. For as long as anyone could remember, she had owned The Bookshop. It was located on Willow, three doors down from Miss Lissy's Closet, and it had been something of a second home for me. As a child I spent hours cross-legged on the scuffed wood floor in front of the bookshelves, browsing through everything appropriate for my age before picking one to buy. That was the deal my father made—we could each buy one a month. Though at that time we had little money to spare, he wanted us to build our own little libraries. One book a month would give us a good start, he announced.

Then he died. I was ten at the time, and

my mother carried on the practice, but she never beamed over my purchases the way Daddy had. She also directed me toward what she felt to be the "proper" book, at which point I began spending more time at the town library. My tastes were mature. I had moved on to J. D. Salinger, Jack London, and Ursula K. Le Guin when most girls my age were still on Louisa May Alcott. I'm not sure my mother would have approved of my reading *Lord of the Flies* or *The Great Gatsby,* but these books intrigued me. They were my escape.

Mrs. Klausson seemed to understand that. I was her "staff reviewer" long before the practice became commonplace. By mutual agreement, we kept my reviews unsigned. We were co-conspirators in this. I got to read books without buying them; Mrs. Klausson got to talk books without reading them. I continued to send her reviews even after I left Middle River. Once in a while, I still did. We corresponded by traditional mail. As computerized as the store had become, she wasn't an e-mail person. Nor could I ever call her Marsha. Nearing eighty now, she would always be Mrs. Klausson, and not only in deference to her age. She had been natty before nattiness reached Middle River. I suspect I was drawn as much to this difference in her as I was to her books.

I was also drawn to her scent. Everything about her was honeysuckle, from the oil she dabbed behind her ears, to the sachets that scented her clothes, to the soap with which she washed her hands. She generated an aromatic little cloud that followed her around the store. Honeysuckle might not have been my favorite scent, any more than roses were. But just as the smell of roses conjured up Road's End Inn and told me I was on my way home, honeysuckle conjured up the haven of The Bookshop. Like one of Pavlov's dogs, I no longer needed the scent to capture the feeling; one step in the door of this shop and the welcome came to me all on its own.

Fifteen years had made little difference in Mrs. Klausson's appearance; she still wore a pressed blouse and neat linen slacks. I could see that she was shorter than she had been when I left town, and more wrinkled—neither of which I'd had the presence of mind to see when she had come to the house to pay her condolences in June. But her eyes lit up when she saw me enter, and, abandoning a pair of clients at none other than Grace's table, she was by my side in a flash.

"Well, well, well," she said with a twinkly smile and a fond once-over, "look at you. Our most famous writer."

"Second only to you-know-who," I reminded her. It was our long-standing joke.

"She's dead, but you're alive," the woman declared. Favoring her arthritic neck, she turned her whole body and, clutching my hand, pulled me toward the clients she had left. "This is Annie Barnes. Annie, you remember my old friend Carolee Haynes. And this fabulous woman with her is the daughter of an old friend of hers. Her name is Tyra Ann Moore, and she's in visiting us from Tucson with her husband and two girls."

Carolee smiled dutifully. She might have been a friend of Mrs. Klausson's, but she had never been a friend of mine. Nor, come to think of it, had she ever been a friend of my mother's. The Hayneses lived in a Georgian mansion on Birch; Carolee had always considered those of us in Victorians on Willow to be beneath her. She was tall, lean, and starched. Even the wrinkles around her mouth looked ironed in, no doubt from pursing those thin lips in disdain.

Tyra Ann Moore was another story. Blonde-haired and standing barely five-four, she exuded warmth and generosity. Her voice held both. "You're Annie Barnes? I don't believe it! You're one of my favorite writers."

"Then we're even," I said with a crooked grin. "Tucson is one of my favorite places."

"Annie grew up here," said Mrs. Klausson with pride, "which was what I was about to tell you right before she arrived. She was a fixture in this shop for more years than I can recall. That's why I keep her books on the shelf nearest to this table."

"In the shadow of Grace," Carolee intoned dramatically.

"*East of Lonely* was the best," Tyra said with a hand on her chest. "I can't believe it's really you! You look so normal. I mean, down to earth. Do you spend your summers here? Do you come back to write? Are you working on something new?"

"I'd like to know that too," Carolee remarked.

"I've already finished my next book," I told Tyra. "It's being published in the spring."

"Which didn't answer the question," Carolee said.

"Oh, Carolee," scolded Mrs. Klausson, "let her be. She's just lost her mother. She's back to be with her sisters, aren't you, dear?"

I barely had a chance to nod when a third customer joined us. She was close to my age, and had dark brown hair and deep olive skin. Her voice was vibrant. "This has to be the best place for a writer. There's a secret around every corner."

"Juanita," Carolee scolded.

Mrs. Klausson made the introductions. "Annie, this is Juanita Haynes. She's married to Carolee's youngest boy, Seth."

I had never had cause to care for Seth, who had never had cause to care for me. But my feelings changed in the instant. Knowing that he had fallen in love with a Latino woman and had dared bring her to Middle River endeared him to me. Likewise, knowing what this poor woman would face with a mother-in-law like Carolee in a town like Middle River endeared *her* to me.

"I'm glad to meet you, Juanita. Do you and Seth live in town, or are you visiting?"

"Just visiting," Carolee answered for her daughter-in-law. "They're both in hedge funds in New York. We rarely get to see them. A few days here, a few days there. Surely not enough to hear secrets."

Juanita grinned mischievously. "But I did hear about Father William."

Carolee shushed her, but Tyra was clearly interested. "What about him?"

"He has a sweetheart."

"Mary Barrett is his housekeeper," Carolee corrected primly. "She lives in, because there happens to be an extra room in the parish house, and Father William welcomes the rent."

"He doesn't charge her rent," said Mrs. Klausson.

"I know that, Marsha, but we're quibbling with words. He lets her use the room in exchange for cleaning the house and the church. It's a business arrangement, very much the *equivalent* of rent. There is nothing going on between them."

"That's not what Seth's friend Peter said," Juanita teased. "Peter does carpentry work for Father William. He saw them in bed together."

Carolee's mouth was as pinched as could be. "Peter Doohutton is a drunk, like his father was before him, and he's no friend of Seth's. They happened to have been in the same class in school. Truly, Juanita, I wouldn't trust what Peter says. Father William is a man of the cloth, and he's been here a *lot* longer than you."

I would have spoken up on Juanita's behalf, if Juanita hadn't been so clearly enjoying herself. A beautiful woman, she exuded confidence and smarts. I figured that Carolee had met her match with this one.

"Hey," Juanita said now, "I'm fine with Father William *whatever* he does. He's a great guy."

"Then let's not spread gossip."

"Spread?" Juanita echoed, barely containing a smile. "There's no spreading when it's already done—at least, that's the impression I got when we went to church."

"*We* are Congregationalists," Carolee remarked. "Father William is a Catholic. We got *no* impression of him, because we were not in his church."

"But plenty of people were," Juanita said. "Seth's friend John was telling us about it at brunch—and neither he nor his father are drunks. They're selectmen, both of them, here in Middle River. John says attendance at Our Lady is as high as ever. He says that most people don't really care what Father

William does at night, as long as it's with a woman."

Carolee glared at Mrs. Klausson. "Do you see what's happening here? They're condoning it."

Mrs. Klausson humored her friend. "No, no, Carolee. They're just leaving well enough alone."

"But the talk—"

"The talk does no harm. It dies at the town line."

"Unless," said Carolee, "someone chooses to write a book about it." Lips pursed, she looked at me.

I was annoyed enough by the woman to want to lash out, and probably would have disgraced myself by saying something like, *Write a book? That might be a good idea after all, because Father William isn't the only one having an affair with his housekeeper, and in fact, maybe he got the idea from your husband, who for years and years and years was slipping home every Friday while you were having your nails done, to settle up the bill with your longtime housekeeper, what was her name?* if Tyra Ann Moore, still starstruck, hadn't jumped in and saved us all.

"Will you tell us about your next book, Annie? What's the title? Will you carry any of the characters over from *East of Lonely*? When is it coming out—I mean, what month?" Before I could begin to answer, she said, "I just can't believe you're here. I'm buying a copy of each of your books for you to sign. Would you do that? No, I'll buy two copies, one for me and one for a friend back home who loves your work as much as I do. . . ."

Sam Winchell was another of the few I had called a friend during my Middle River years. Sam was younger than Mrs. Klausson, probably sixty-five now, which meant he had been in his forties when we had worked together and fifty when I left town. Like Seth Buswell in *Peyton Place,* Sam was the son of an influential man, in this instance a United States senator. Like Grace's Seth, Sam had inherited enough money to own and operate the local weekly and, at the same time, afford a house on Birch. Like the fictitious Seth, Sam was disdainful of bigots and impatient with those who threw their power around, meaning that he had limited patience not only for people like Carolee Haynes, but for Sandy Meade, as well.

That said, Sam was an avid golfer, part of a foursome that played eighteen holes at a club forty minutes away on all but the most rainy of Thursdays. He had been doing this when I had worked for him those high school summers, and judging from the reverence with which he still talked of golf in the paper, he hadn't changed.

Sam's foursome included Sandy Meade. Stories of their bickering on the greens were legendary, as were those few occasions when a game was played minus one of the men when a particularly strong disagreement cropped its ugly, if short-lived, head.

By the way, I had every bit the respect for Sam that I had for Mrs. Klausson. But Sam was always called Sam. The only people who called him Mr. Winchell were from away, and it lasted only until he could wave a hand and say a brusque, "Call me Sam."

I timed my visit so that I would arrive at the *Middle River Times* office on Thursday at eleven. Yes, I wanted to say hello to Sam, but even more, I wanted to pore through the archives to make a tally of the town's sick. Granted, it wouldn't be exact. Greg was right; doctors could give me better figures, but only if they chose to, and it was a big *if.*

They could easily stonewall, and then turn around and tell the Meades I was asking. Was it worth the risk? Not yet. I figured I could visit with Sam until he left to play golf, then peruse the paper to my heart's content. Sam's staff wouldn't be around. The paper came out Thursday mornings, and this was Thursday. Within an hour, the newspaper office would be deserted.

The place was front and center on Oak Street, its brick newly painted, mullioned windows washed, trim freshly painted. Sam had always been fastidious about things like these, and clearly that hadn't changed.

No sooner had I let myself in the door, though, when I was hit by the smell of cigar. Sam's fastidiousness had never included his lungs. The cigar in his mouth was as much a fixture of the man as his gray pants and bow tie. No one working for Sam ever complained about the smell; to do so would have meant instant dismissal. I figured that come the day Sam was gone, this place would have to be fumigated *big*-time to remove the last of that smell.

The front desk was deserted, the nearby leather chairs empty. Two offices branched off the main room. I went to the one on the right, led there as much by the strength of the scent as by habit.

Sam, who was tipped back in his chair reading the hot-off-the-press edition of the *Middle River Times,* spotted me over the paper and broke into a grin. His chair creaked as it straightened; the paper landed on the desk with a careless rustle. Rising, he came forward, put an arm around my shoulder, and gave me a fatherly squeeze.

"It's about time you got here," he complained gruffly. "You've been most everywhere else in town first."

"I have not been most everywhere else," I returned. "In case you've forgotten, there are not too many places where I'm welcome."

Sam made a sputtering sound. "That was history. You're a noted writer now. You can go anywhere you want. I'm real proud of you, Annie."

I smiled. "Thank you. That means a lot. The rest, well, you know, people jumping on the bandwagon, it's pretty empty."

"And that, right there, is why you'll continue to be a success. Nothing goes to your head, and that's just one *other* way you're different from our Grace," he added, picking up the thread of a conversation we had often had before. Sam, like Mrs. Klausson, shared my fascination with Grace.

I defended her as always. "I don't know that success went to her head. She was just unprepared for it and had no idea what to do. I'm not sure I would have known either, back then. *Peyton Place* was the first of a new breed. No one knew what to expect."

"Well, we do now," Sam concluded, arching a brow. "So I want you to make me a promise. Whatever you write while you're here, I want to serialize it in the *Times*."

I laughed. "I am not here to write."

"Sandy says you are. He called me earlier and ordered me to wheedle what I can from you." He shot a look at the wall clock. "I'm meeting him in forty minutes, and you can be sure he'll know you've been here and want to know what we said."

"Is he nervous?" I asked, enjoying myself. After growing up powerless in this town, I was feeling a definite satisfaction at their reaction to me now. I had felt it with Aidan in a very personal way. With Sandy it was broader, more general, and, given the extent of his power, even sweeter.

"Oh, you know," Sam said dismissively, but I wasn't letting go.

"No, I don't. What have the Meades done?"

Sam gave a grunt. "What haven't they done? But don't get me started on this. Sandy and I walk softly around each other. I could cite you half a dozen things he does that drive me crazy."

"Cite one," I invited.

Separating himself from me, Sam went to the desk and neatly folded the paper. "One? If I told you one and you proceeded to write about it, I might find my computers not working next Tuesday, and if that were to happen, there'd be no paper next Thursday. Not that my livelihood depends on one week of the paper, but it would be a royal pain, not to mention cause a big mess with advertisers who pay by the week."

"Sandy wouldn't sabotage your computers. My sister is his computer person, and she would never do something like that."

"Not if her job was on the line? Not if her husband's was, too? They both work for Meade, and he would use that for all it's worth. If they lost their jobs, you can be sure they wouldn't find others in this town, so how're they going to support their kids?"

"They could bring charges against Sandy."

"And who's gonna testify?" Sam smiled sadly. "So there's your one thing he does that drives me crazy. This is nothing new, Annie. Nothing's changed."

I was thinking that it *ought* to change, that the Meades were truly evil, that if I could prove the mill was contaminating the town with mercury waste and was breaking laws to do it, the family might finally get what it deserved, when Sam said, "I gotta run or I'll miss my tee time. Come on. I'll walk you to your car. Let word get back to old Sandy about *that*."

"Actually, I'd love to sit here awhile and read the paper. Catch up a little."

"Why? I send you the damn thing every week."

I slid him an apologetic look.

He sighed. "And you don't read them."

"I skim," I said, not quite a lie. I often did skim one article or another. Usually, though, my reading was thorough. Not so my memory, which was why I needed this time.

Sam, in his innocence, was flattered enough to leave me to my work.

I started with the current issue, the one he had been reading. It told me, in grand detail, that the library had been given another

Meade grant to buy new books, that a full-time ophthalmologist was now affiliated with the Middle River Clinic, that a fire on th'other side had gutted two houses, leaving eleven people homeless, that the state was cutting the money for our schools again, necessitating an emergency town meeting in September to discuss what to do, that a local plumber had been detained for driving under the influence, that for the third weekend in a row vandals had taken baseball bats to mailboxes on Birch and Pine, that the Hepplewaites, who owned Road's End Inn and sponsored the Peyton Place tours of the town, were celebrating their fiftieth wedding anniversary on Saturday, and that Omie was recovering from pneumonia.

The news about Omie concerned me. I liked Omie. She had been old since I could remember, everyone's grandmother and a great-grandmother in her own family many times over. Never a big talker, she was kind to the nth degree. When I used to sit by myself in a booth at the diner, she would bring over her tea and keep me company. There were times when that made things worse— like going to the movies with your mother on a Saturday night and seeing all your friends with dates. Other times, it was a godsend.

Omie's heart was kind. I made a vow to stop by and see her.

Setting aside the current issue of the *Times,* I relocated to the front room, where the most recent months' issues hung on tidy wood racks, and began to peruse the archives. The "Health Beat" column was always on the fourth page, top left. Sam was fastidious in this, too. He believed that people should know exactly where their favorite columns were, and didn't want phone calls coming to him when readers couldn't find what they wanted.

Working backward chronologically, I read column after column. It didn't take long to exhaust the hard copy, at which time I moved to the left inner office to access those on microfiche—and all the while, I kept notes on a pad beside me, marking dates, names, and illnesses.

Most newspapers probably wouldn't list enough sick people to make this search worth my while, but Sam was a clever marketer. He knew that people loved seeing their names in print. Middle River was small enough and the *Middle River Times* hungry enough for copy, to make details not only feasible but advisable. What another town might consider irrelevant was included—a chicken pox outbreak in the local Cub Scout den with the names of children stricken, or random cases of Lyme disease or asthma. Broken limbs were reported case by case, with names, cause of the break (e.g., auto accident, fall from bicycle, soccer game), and often even the color of the cast applied and by whom. The column also listed what doctors in town attended which conferences out of town. Our Tom Martin took the lead in those, both attending and conducting seminars. His specialty was general practice, or the revival thereof, which didn't tell me much about the possibility of mercury poisoning.

Other things did, though. I noted each mention of Alzheimer's and Parkinson's disease. I noted references to autism (e.g., Alice Le Claire's three-year-old son had started at a special school for autistic children), references to breathing difficulties and digestive problems, and I put asterisks beside those instances in which a history was mentioned (e.g., Susannah Alban was recovering from another miscarriage). I noted when babies were born either very early or with illnesses that kept them hospitalized.

In skimming the microfiche reel, I found an announcement of the wedding of Seth and Juanita Haynes. It was on the front page of the paper two years ago last June. The event was in New York City and sounded elegant, from the elaborate description. Interestingly, there was no picture, though I had no doubt pictures did exist. Carolee Haynes was not only a bigot, but a coward.

I went back through five years of the *Middle River Times*—not because I planned to go that far or spend so long at it, but because I kept munching on my chocolate pennies thinking that I would read just one more issue, and those "just one more"s added up. By the time I was done, I had many pages of scribbles.

I flipped slowly through, hoping a pattern would jump out. Oh, the potential was there. Lots of people were sick, many with symptoms like my mother's. But was it mercury poisoning? I had no idea.

I told myself that I needed more information, that I had to sit down with a map of the town and put dots in the spots where each of the afflicted lived to see if there were pockets of illness, that I had to learn whether Northwood did indeed produce mercury waste and, if so, I had to go to the next step, whatever that was. I told myself that I shouldn't worry about the lack of instant answers, that this was just the start.

But I was discouraged. Having exhausted my supply of chocolate pennies, I was also hungry. It was midafternoon, and I was craving protein. I could have gone home to Phoebe's, but by now everyone in town knew I was here, so keeping a low profile was needless. Besides, I kept thinking of Omie. She was the third and last of the friends I'd had in town. I had seen her briefly at the funeral but hadn't had the state of mind to talk with her, any more than with Mrs. Klausson or with Sam. Today's paper said she'd been sick. I really wanted to say hello.

So I returned the microfiche reels to their drawers—neatly and in order, so that no one would know I had been here—and left the newspaper office. Though Middle River was no humming metropolis on a hot Thursday afternoon in August, the center of town was far from deserted. My car was one of a dozen or so parked in those diagonal spaces that ran up the street. Granted, mine was the only convertible. Convertibles weren't practical in towns like Middle River, where winters required four-wheel-drive vehicles, chores involved carrying large loads, and family space was a necessity.

I tossed my purse into the car and slid in. Fortunately, I was wearing jeans. Even through the thick fabric, I could feel the heat of the sun-baked seat. Feeling the heat all over, actually, I slipped on sunglasses, started the engine, and looked behind to back out of my space. I saw no cars in the road. But I was being watched. Directly across the street, several people looked on from the sidewalk in front of Farnum Hardware, and down a bit, several more from just outside Harriman's Grocery. A pickup was idling nose-in at the curb in front of News 'n Chews, driver waiting—watching me too— until a trio of children ran out with small brown bags of candy and climbed up into the cab.

I didn't acknowledge anyone. My Middle River persona wasn't attuned to social niceties. Shifting, I drove on down Oak, crossing Pine, then Cedar, where the roses at Road's End Inn positively burst with sweetness in the midafternoon sun. At School Street, I made a left and passed several frame houses with daylilies growing orange and yellow and wild, before turning left again into Omie's lot. There were only a handful of other vehicles, which was good. Much as I wanted to see Omie, I wasn't in the mood to wait for a seat, and it wasn't only a reluctance to be stared at. My stomach was growling, and the food smells even out here were strong.

The diner was the real thing. Originally built for Omie's father in the early 1900s as a lunch wagon with (new at the time and quite the thing) wooden counters that folded out to make a walk-up window, it had been modified, enlarged, and renovated many times over. Through it all, though, it retained the look and feel of a diner.

Omie was similar, in a way. A small woman with white hair in a bun, a face filled with creases, and eyes that were eternally blue, she was either ancient or not, depending on what one wanted to think. Her actual age was irrelevant. Same with her last name. The few of us who knew it could neither spell it nor pronounce it, and it didn't matter. Her daughters had Americanized the original Armenian, and then had all married local men of French-Canadian descent, so the original name was basically gone. Though Omie had great-grandchildren, perhaps even great-greats by now, she remained a grandmother to everyone who opened the diner door and went inside.

Indeed, though she still had the look of a traditional Armenian grandmother, inside she was anything but. For one thing, she had an exquisite business sense. For another, she thrived on work. Long after other great-grandmothers in town had retired to their back porches to alternately rock and crochet, she was at the diner overseeing hiring, firing, redecoration, and menu changes. She had officially passed ownership to her eldest daughter years before, but the transfer was only on paper. The diner was Omie's.

Where wheels had been in the structure's infancy, now stood a base of solid brick. The body of the diner had sides of stainless steel with burnished circles, and panels of decorative tiles between wide windows. The tiles were blue, green, and white, as was the sign high on the roof. OMIE'S, it read in acknowledgment of the woman's force. It had been a

gift from her family several decades before,
prior to which there had been no sign at all.

I heard a little whispering, but this wasn't
Grace. It was my own inner someone telling
me that this place was special, and it was all
the urging I needed. Grabbing my purse, I
climbed from the car and went inside.

Chapter 8

SOMETHING WAS cooking—chicken fajitas, I guessed, from the smell and the siz-
zle—but with Omie I could never be sure. As
Americanized as she was, as wise to the
needs of Middle Riverites, Omie prized her
Armenian heritage. That meant she cooked
as often with lamb as with beef, baked as often with phyllo dough as with flour, seasoned as often with cinnamon, cardamom,
nutmeg, and clove as with salt, pepper,
parsley, and thyme. New items like fajitas,
pastas, and Caesar salads popped onto the
menu to keep the diner current, but there
were always old standbys—chicken pot pie,

macaroni and cheese, and Omie's hash—all of it cooked with an Armenian flair. For that reason, any food smell that wafted from the diner carried the uniqueness of Omie.

Moreover, the scents lingered, because Omie hated air-conditioning. She liked having the windows open, and kept her customers comfortable with old-fashioned fans. Since these never quite eliminated the humidity, smells clung.

Those fans whirred now, stirring the air. Simon and Garfunkel sang softly through ceiling speakers, while the cold drink case hummed an even softer refrain. These things I heard clearly, because whatever conversation had been ongoing when I walked in the door had come to a dead stop.

Feeling awkward, I retreated to my Washington mantra. I was successful in Washington. I had friends in Washington. I could walk into Galileo, Kinkead's, or Cashion and be just like everyone else. That was what I had wanted most when I left Middle River—not the walking into restaurants part, but the being just like everyone else part. Isn't that what most of us want, to be part of something, to belong? Here was **TRUTH #4: We can thumb our noses all we want at people who are different from us, but in the end we ache to belong.**

Standing now with the nearest fan stirring loose wisps of my hair, and all eyes turned my way, I ached for that in this town too. Middle River was where I was born. I had spent the most important eighteen years of my life here, many of those wishing, praying, *yearning* to be part of it all. Of course, I wouldn't have admitted it back then. I hadn't *realized* it back then. It was only now, in this very instant, that I saw how true it was.

I needed Omie.

Drawing myself up, I ignored stares and looked around. Over the years, the diner had grown deeper. To the right of the kitchen, through an archway, were tables that sat sixty, but they were empty now. This time of day, the action was up front, in the diner's heart. Five men sat at the counter—a trio along the front arm, another two at the rear of the L. Of a dozen booths hugging the front of the diner and its adjoining right side, three were taken.

I didn't see Omie. But she would find me. She always did.

My favorite booth was five away from the door, the one in the corner before the line of booths turned and, like the counter, shot toward the back of the diner. I had always felt enveloped in this booth, and hence safer. Unoccupied now, it beckoned.

Heading there, I passed a booth with two women I recognized as neighbors of Sabina's. Their eyes met mine. They didn't smile. Same with the family of four in the next booth.

The third occupied booth was midway among those running from my corner back along the side wall of the diner. Pamela Farrow and her husband sat there, snuggled side by side, holding hands. Hal was fully engrossed in his wife. Pamela's eyes flicked briefly to mine before returning to his—no smile, no hello, just a glance—quite different from the way she had fawned over me the day before.

So maybe she was just so in love with her husband that she couldn't think of anything else. More likely, someone had gotten to her.

Or, perhaps she was simply acutely aware that one of the two men at the far back end of the counter was James Meade. He couldn't be missed, with his straight back and dark, brooding look—not to mention striking salt-and-pepper hair. I could see the latter now as I hadn't seen it when he had been in the shadow of his car, and might have been surprised by it—James wasn't quite forty—had not the father been silver forever.

James didn't intimidate me. I wasn't in his employ and didn't owe him a thing. Naturally, Pamela would be thinking that if she gushed over me, James would tell his father, who chaired the School Committee, which determined the tenure of her husband as high school principal.

And that was fine. I didn't miss her fawning. I wasn't here to make friends and influence people. Class reunions had never been my thing. I had friends. I had *lots* of friends.

Thinking about them now, I slid into my booth. As James Taylor took over the airwaves, I set my purse by my hip on the wood bench and my elbows on the dark green Formica. In that moment, at least, I was confident.

A menu was wedged between the napkin holder and a jar of ketchup. I pulled it out, opened it, and began to peruse. Little by little, conversations that had been interrupted by my arrival resumed.

"Just when I needed a cup of tea," came a sweet, familiar, and very welcome voice.

I looked up as Omie lowered herself to the opposite bench. Those endless blue eyes had faded some, and her cheeks had the texture of washed crepe. She looked paler than I recalled, but her smile was as sincere as ever. On impulse, I rose and, bracing myself on the table, leaned across and kissed her cheek.

"You look better than last time," she remarked in a grandmotherly voice that held only the faintest trace of an accent. She had been a small child when her parents reached America, and had grown up speaking English with other children. What accent she had she chose to have. It gave her voice a distinctive ring.

"Time has passed," I said. "The shock has worn off. I hear you've been sick."

"It's nothing."

"Pneumonia isn't nothing."

"Well, it is if it's gone," she said with the dismissive wave of a hand. "These things happen when you get to be my age. Two, three times a year it's a cold or such. Too many people in here with their germs. I can't throw things off like I used to. I'm getting old."

I wondered if it was that, or if her immune system was being depleted by something else. It was definitely worth considering, but later. I gestured toward the menu. "From the looks of this, you could be forty. There are lots of new things mixed in with the old."

Omie smiled. "I have to please everyone if I want to stay in business."

I lowered my voice. "If the goal is staying in business, you shouldn't be having your tea with me. I sense that your other customers would rather I not be here." The women who lived near Sabina had already left their booth and were at the cash register.

Omie, too, spoke more quietly. "Ignore them. They're frightened of you."

"Frightened?"

"You speak the truth."

"And they don't?"

Omie didn't answer. We both knew that hypocrisy festered in towns like Middle River. Under the facade of beauty, there was a stratum of dirt. Grace had known this. It was a major theme running through *Peyton*

Place, a major theme running through her *life.* She'd had dreams and ideals that were all too often betrayed.

"What are they hiding?" I whispered now.

"What aren't they hiding?" Omie whispered back.

"Tell me the latest."

Omie seemed happy to do that. "Those three at the counter? Doug Hartz is the one in the middle, the one looking uneasy. See how he keeps glancing back at you?"

I did. "What's his problem?"

"He prides himself on not paying taxes. Tells enough people so the whole town knows. Seems to think that if he runs a cash business, he'll slip by under the radar."

"So why do I make him nervous?"

"You're from Washington. The IRS is in Washington. He thinks you'll tell."

"That's pretty funny," I said, because the connection was absurd. "And the two women who just left? What chased them off?"

"Your success. They meet here every afternoon at two to work on a book they're writing. They've been doing this for three years, and there's no sign of any book. And now here you are. For many people in this town, you're everything they can never be."

"Even James?" I asked, because his presence at the counter with his pal was so *there.*

"Even James," Omie said and gave a little snort of dismay. "Staying at that mill under the thumb of his father—can you imagine? It's a waste. James is the smartest of the bunch."

"Not smart enough to leave Middle River," I remarked.

But Omie drew the line there. "Why would he leave?" she asked with a softness that made its point. "Life here is better than it would be most anywhere else. He's tasted what it's like to live away. He was away for college and graduate school. There was a time when he was on the road four days a week doing work for the mill. He knows what it's like, and he chose to come back."

"Chose? Or was forced?" I could imagine Sandy telling James he was free to leave, but that if he chose to, it would be without a cent of Meade money. "Extortion can take different forms."

Omie didn't reply to that. Instead, she

beckoned her grandson over and asked me, "What will you have to eat?"

I ordered the balsamic chicken salad and an iced tea. The latter arrived before either of us could resume the discussion. By that time, the family of four had skedaddled, Van Morrison was singing, and I was studying the man with James. He sat with his shoulders hunched, his eyes on his hands, and his hands on a bottle of beer. He had the look of a man besieged. I might have guessed he didn't even know I was here.

I had a sudden start of recognition. "Is that Alfie Monroe?" I whispered. At Omie's nod, I looked at him, then away, then sipped my tea and studied him over the rim of the glass. Alfie Monroe had been at the mill all his life. Hardworking and decent, he had worked his way up until he was one step below the Meade sons, and well respected for it. The man I remember was good looking and proud. He would be in his early fifties now. This man looked older than that, and I wouldn't have noted either good looks or pride.

Had I read anything in the paper about him? I didn't think so.

"Is he not well?" I asked but my imagination had already taken off. If toxic waste had been released by the mill, someone who worked in its bowels stood an even greater risk of affliction. The Meades worked in offices far from the nitty-gritty. Not so the plant manager.

"He's well in body, but not in mind," Omie said. "He was passed over when Sandy hired Tony O'Roarke to manage the plant."

No. I hadn't read this. There had been nothing in the paper, and I knew why. Sam Winchell would have disapproved of an outsider displacing Alfie, but Sam was a pragmatist. He knew that nothing he said in his paper could either change what Sandy had done or help Alfie. His protest was to say nothing at all.

I pictured Tony O'Roarke driving James Meade around town in that big black SUV, and felt a stab of annoyance. "Why didn't Alfie get the job?"

"The Meades wanted someone more experienced."

"Who could be more experienced than Alfie? He's been with the company all his life."

Omie thought about that as she sipped her tea. Eyes sad, she set down the cup. "Sandy felt they needed a new approach."

"Approach to what?"

"Keeping things streamlined."

I interpreted that to mean cheap. "Sandy wants more productivity for less money."

Omie didn't argue. As she took another sip of tea, Pamela and Hal left their booth and passed by without a glance. My salad arrived. I began to eat. Doug Hartz and his friends slid off stools, dug into pockets, and dropped cash by plates, then followed Pamela and Hal out the door. In their wake, a group of teenagers entered, girls identical with their low-cut jeans, skinny tops, and ironed hair. Two of the five had cell phones to their ears. Several glanced my way as they passed. One did a double-take.

She looked familiar. It was a minute before I placed her as having been with her mother at Miss Lissy's Closet the day before. At the time, I would have guessed she was no more than fifteen. Now with friends, all of them long-haired and mascaraed, she looked eighteen.

I raised several fingers in a covert wave, a simple acknowledgment that we had seen each other before. After all, she was Phoebe's customer. I didn't want to be rude.

The poor thing seemed horrified. I could only begin to imagine what nonsense someone had fed her. Quickening her step, she followed her friends to the very last of the booths, where they all crowded in. Shania Twain began to sing.

"So," Omie said with a gentle smile, "they say you've come to write. Are you still Grace's girl?"

"No. I've made my own way."

"Do you two still talk?"

Had anyone else asked me that, I would have been mortified. Hearing voices—having conversations with people who no longer existed—wasn't something sane people normally did.

But Omie knew about these conversations. I had confided in her once, when I thought I was losing my mind. I was fifteen at the time. My body had finally—just barely and very late—begun to change and, while my family of women was relieved that I was "normal" in this sense at least, they were themselves too blasé about things like periods, breasts, and emotions to want to hear my qualms, even if I *had* been able to express them, which I hadn't. Me, I was feeling like a stranger in my own skin. Well aside from those physical changes, I was terrified of my relationship with Grace.

"You need a friend," Omie had offered by way of explanation back then. "Grace understands you."

"Grace is *dead*," I had replied with a touch of panic.

"Well, yes. But that doesn't mean she isn't here with us."

"Like a ghost?" I asked skeptically.

"That depends. Have you seen her?"

"No. I just hear her. But how can that be? I never heard her voice in real life. How do I know what it sounds like?"

"You know enough about her to imagine how it sounds."

"Grace is dead," I repeated.

"This is her spirit."

"Oh, Omie. I don't think so."

"Well, I do. You came to me to ask, and this is what I say. Grace Metalious was not a happy woman. She died alone and unfulfilled. You may be her vehicle for fulfillment." Before I could argue, Omie asked, "When she talks to you, what does she say?"

What had Grace said? She had given me encouragement, even goaded me on at times. We could argue—I said some awful things—but she always came back for more. She told me I had the makings of a good writer, and, back then, it was what my battered fifteen-year-old ego needed to hear.

We continued to talk, Grace and I, until I left Middle River. Once in Washington, I became a different person. I no longer needed Grace, so our conversations stopped.

Did Grace and I still talk? Omie asked.

There were those few whisperings I had heard, but I couldn't be sure they weren't just the purr of a cat or the buzz of a fly. And the woman goading me on when a well-formed man in running shorts and sneakers had crossed my path this morning? That was Grace all right, but just for fun—my alter ego, more than anything else.

"No," I told Omie in a version of the truth. "Grace doesn't do Washington. She was there once to promote the sequel to *Peyton Place.* The press wasn't kind."

"She was a small-town girl at heart, I think."

"Yes and no. She hated the scrutiny of small-town life, hated the expectations of small-minded people. Cities fascinated her—Paris, Los Angeles, Las Vegas, New

York. But she couldn't hold her own in those places. She didn't know what to wear or what to say. She never became savvy and was forever putting her foot in her mouth and then, after the fact being hit in the butt by her own words."

Omie beamed. "Not you. See how far you've come?" She leaned in. "So, about the writing. Is it true?"

My initial hunger sated, I put down my fork. "I'm already involved with a book. It's due to be published next year." I paused. "Omie?"

"What, sweetheart?"

"Do you think people here are unusually sick?"

She considered that. "I don't know what it's like anywhere else."

"The doctor says my mother had Parkinson's disease. She isn't the only one in this town who has it."

Omie studied me more closely now. "No."

"Do you think there's a reason why so many do?"

"What kind of reason?"

"I don't know. Has there been any talk— you know, maybe about something in the air making people sick?"

"There's always talk. Someone gets a cold, they blame it on airborne germs."

"That's not what I mean."

"I know."

"Alice LeClaire's three-year-old son is autistic," I said. "Would she talk with me?"

"I doubt it. In addition to that child, she has four other children to care for, and no man to help."

"No man? I thought there were at least two." My mother had maintained—with proper scorn—that pairing up fathers with Alice's children was a popular game in Middle River. Apparently, Alice never quite broke it off with her exes.

"The latest count has it at three," Omie confirmed, "but since the little one was diagnosed, they're all staying clear. They don't want to be blamed for it. They're convinced it has to do with genes."

A muffled phone rang. Instinctively, I went for my purse, but this ring came from somewhere back in the kitchen. Seconds later, a closer phone rang. James Meade flipped open a cell.

"Autism can also result from exposure to toxicity," I told Omie. "Wouldn't Alice want to know that?"

Omie smiled sadly. "And think she had let her baby be exposed to something bad?" The more distant phone rang again. "Besides, the Meades pay for medical insurance for their workers."

Ah. And Alice had a job at the mill. Considering the special needs of an autistic child, she would be lost without insurance, not to mention that job.

James Meade pocketed his phone. Alfie Monroe pushed himself off the stool and strode past.

"What about the Dahills?" I asked Omie when he was out of earshot. "How many in that house have kidney problems?" My reading told me they were on the third generation of it. All three generations worked at the mill, though I didn't recall in what capacity.

"Kidney problems are hereditary," Omie said.

"Possibly. Okay—probably. But what if they're job-related?"

Omie seemed about to caution me when her grandson called from the kitchen cut-through, "Omie, it's cousin Ara on the phone."

Her eyes lit. "I have to take this," she said. Rising, she leaned close, kissed my cheek, and whispered, "Try the McCreedys. They've suffered a long string of problems. They're looking for a reason."

I had barely taken that in when Omie was gone, and a separate movement caught my eye. James Meade had risen. He put money on the counter and approached me. There was nothing casual about his course. He knew I was here and meant to stop.

James was impressive. Not only was he the tallest of the Meades, but he was the best looking, and I don't say that to get back at Aidan. It was simply true. Like Sabina and me, James and Aidan had similar features— thick hair, deep brown eyes, pointed nose, square jaw. But with James, like Sabina, the whole was more than the sum of its parts. You might notice Aidan's hair, or his eyes, or his mouth. What you noticed with James was his authority. He was quiet. But when he spoke, you listened.

Now, eyes somber, he stood before me. He wore jeans and a pressed blue shirt, neck open, sleeves rolled to the forearm. Those threads of silver in his hair lent him greater bearing.

"My brother just called in a stew," he said in that quiet, authoritative voice. "He saw your car here. He thinks you're up to no good."

Something about James unsettled me. I wished he wasn't so tall, wished those eyes weren't so penetrating. Feeling a need for distance, I sat back. "I was hungry," I said. "I came here for food."

"You make him nervous."

"Why?"

James seemed to consider the question, but his eyes never lost their intensity. Imaginative on my part, yes, but I felt they were searching me for motive and thought. Finally, quiet still, he said, "I'd say Aidan was feeling guilty for what happened all those years ago, but I doubt that's it. More likely, he's afraid of your pen. He doesn't know what you're going to write."

I smiled. "If he's afraid of my pen, then he has something to hide. I wonder what it is."

James almost smiled back. "I wouldn't do that if I were you."

"Do what?"

"Wonder. It could create problems. Aidan doesn't want that."

"Should I care what Aidan wants?"

"Yes," James said in a factual way. "We employ Sabina and her husband, and we hold Phoebe's lease. There's a lot at stake here."

I felt an inkling of anger. "Is this a threat?"

"Not from me. I'm just passing on what my brother said. He also wanted to remind you that we were good to your father when he was sick. Northwood paid him long after he stopped being a productive employee. We did it out of loyalty."

The implication, of course, was that we owed loyalty in return—or rather, that I did, since I was the one threatening the status quo—and that angered me all the more. I didn't owe the Meades a thing.

"Was it loyalty?" I threw back at him, perhaps rashly, but I didn't like James Meade. "Or was it fear? My father worked at your plant all his life. Maybe he knew things you'd rather he hadn't known. You treat all your employees a little too well. Is that to instill loyalty, so that they won't blow the whistle if they see something wrong?"

I shouldn't have said it. I knew that the instant the words were out of my mouth—and if I hadn't realized it on my own, James's expression would have tipped me off. He seemed suddenly alert, suddenly personally invested in the discussion.

Eyes barely leaving mine, he slid onto the bench where Omie had been. He put his forearms on the table, large hands maddeningly relaxed. "Is something wrong at the mill?" he asked with a calm that could grate.

I nearly backed down. Then I caught myself. Call me impulsive to have blurted out what I had in the first place. But backing down was idiotic when something higher was at stake.

So I said, "I ask you. Is something wrong at the mill?"

He sat back. He didn't answer, just looked at me. At one point I saw his jaw move, but otherwise he was still. He seemed very intelligent. That bothered me the most.

He stared. I stared back. Elton John was singing. I liked Elton John, but took no pleasure in him now.

"What," I finally demanded.

"Are you writing about the mill?" he asked.

"Should I?"

"Are you?"

"It's amazing," I mused. "No one seems to want to talk with me here except to ask if I'm writing a book. The more they ask, the more intrigued I get. If one more person asks, I may do it."

He didn't speak, just continued to look at me.

"*What*," I said again, more irritably this time.

It was a minute before he responded, and then he seemed genuinely curious. He frowned. His voice dropped even lower. "Do you hate me?"

The question took me aback. "Not you. Who you are."

"My family."

"What it represents."

"Power."

"No single family should have that much. People shouldn't have to fear for their lives if they speak up."

"Their lives?"

"Their livelihood," I corrected. "You just threatened my two sisters and my brother-in-law."

"I didn't. Aidan did. I'm just the messenger."

"Same difference."

"No," he said with calm assurance. "It isn't. I'm not my brother. And I'm not my father."

I eyed him dead-on. "Then you wouldn't commit perjury like Aidan did, and then bribe enough public officials like your father did so that nothing would come of it?"

James didn't blink. "I have never committed perjury."

"What about bribing public officials? Ah, but you don't need to do that. Daddy does it for you."

I was hitting a nerve. I could see that in his dark eyes and tight jaw.

"You don't know me," he said. "I wouldn't burn bridges if I were you."

"Does that mean if I were to write a book about the mill, you'd cooperate?"

"It depends what you want to say in your book. If it's simply the diatribe of a woman who hates this town—"

"I don't hate Middle River. If I went to the effort of writing a book, it would be because I *care* about the town."

"Do you?"

"I grew up here. My family still lives here."

"Answer the question."

"I would *never* write a book out of spite. I don't think I'm capable of doing that."

Still his question hung in the air. He eyed me steadily, daring me to confess. And suddenly, what he asked wasn't so bad. I was a big girl. Spite didn't have to play a part in my response. I could rise above the past.

"Yes, I care," I admitted. "There's lots that's positive here."

"Like?"

"Four seasons. Trees and flowers. Good schools. Marsha Klausson and Sam Winchell. This diner. Omie." When he said nothing, I added, "The barbershop. The nail shop. Miss Lissy's. Brunch at Road's End Inn. Chocolate pennies." He remained silent. "The October Stroll," I tacked on, because it was a lovely tradition, and there were others. "Christmas Eve at the intersection of Oak and Pine. Fireworks in the stadium on the Fourth of July. Even the Backabehind Club." I smiled. "Can't believe the kids are still doing that, but I caught a pair of them at it on my way in the other night."

"Did you ever do it?"

My hackles shot back up. He was ridiculing me. He knew I hadn't dated. The whole town knew I hadn't dated.

But his expression wasn't so much smug as curious.

"No," I said flatly. "I spent most of my senior year at Cooper's Point waiting for your brother to show up."

"For what it's worth," James said, "he was wrong when he did that."

"Which part—carrying on with someone's wife, or using me to get away with it?"

"Both."

As apologies went, it was indirect. But it did seem genuine. And that struck me as odd. Reminding myself that James was a Meade, I put my elbows on the table and leaned in just a bit. "If you're trying to soften me up, don't bother. I'm here to keep an eye on my sister. She doesn't seem well."

"What's wrong with her?"

"I don't know."

"Did Tom have any thoughts?"

"Tom?" I asked in surprise, but quickly regrouped. "Tom isn't treating Phoebe. I was just thanking him for taking care of my mom."

"What about Sam? He have any thoughts?"

There was clearly a point here. James was giving me a message. It was a continuation of the one he had set out to deliver in the first place, namely that I was being watched.

"Sam's a friend," I said. "I dropped by to say hello."

"And stayed for three hours after he left," James reported.

"And that," I declared, "is one of the things

I *detest* about Middle River. Whose business is it but mine where I go and what I do? Do you Meades have a spy in every corner?"

James snorted. "No need for spies, and we're not talking corners here. That car of yours can't be missed. Leave it home next time."

"Next time, I will," I said with resentment, then added dryly, "Any more warnings? Words of advice? Messages?"

James slid out of the booth and rose to his full height. "Just a question. Aidan wants to know why you aren't in D.C. with loverboy."

Mention of Greg made me feel stronger. "Loverboy? Cute."

"The correspondent."

"I know who you mean. I'm not there with him, because I'm here," I said quite logically and tipped up my chin. "Does Aidan have any more questions?"

Chapter 9

KAITLIN DUPUIS didn't have much of a choice as to where she would sit in the booth. Totally distracted by the sight of Annie Barnes, she had just been carried along by the others. When the dust cleared and they were seated, she found herself on the bench that faced the front of the diner, and a lucky thing it was. If she had been sitting on the opposite side of the booth, she wouldn't have seen Annie talking with James Meade, and this was important. Annie talking with James was bad.

Kaitlin didn't care what Kevin thought; Annie *did* recognize her. First a wink at Miss

Lissy's Closet, now a wave. Annie wasn't waving at any of the other girls. She knew about Kaitlin and Kevin. Oh, she knew. And what if she was telling James at this very moment, just as part of the conversation, like, *You'll never guess what I saw the DuPuis girl doing the other night*? James would tell his brother, who would tell Kaitlin's mother, who would then have heard it from *two* sources, if Phoebe's manager's mom, who was James's secretary, was counted, and no *way* would Kaitlin be able to talk her way out of it then. Not that she would have had much of a chance of doing that even with one source. Nicole was *obsessed* with her daughter's virginity, even more than she was with her weight.

What to do? In Kaitlin's dreams, Kevin would have already taken the lead, sought Annie out, and sworn her to secrecy. But Kevin still thought she was imagining things. Besides, Kevin would be totally intimidated approaching someone like Annie. And why would he bother? His parents didn't care if he was with Kaitlin. His dad would cheer him on and reach for another beer. And his mom? His mom *loved* Kaitlin. It would be her *dream* to have her son marry up.

Kaitlin wasn't marrying Kevin until she was older. But she sure didn't want him in jail now.

"*Kaitlin.* Where *are* you?"

She blinked and looked around to find her friends staring at her. Embarrassed, she shot back at no one in particular, "Can't I think about something myself?"

"What was it?"

"Had to be Kevin. She's *always* thinking about Kevin."

"Not always," Kaitlin replied, even in that, though, feeling pride. She was the last of her friends to be with a guy. For the longest time, she thought it would never happen. Fearful that something would screw it up now, she had an idea. "I was actually thinking of Annie Barnes. Do you know that's her back there?"

"I saw," said Bethany.

Kristal nodded. "Me, too."

"Annie Barnes *here*?" Shawna asked, twisting around to look, and in the next instant, Jen was twisting around, too.

"No, no," Kaitlin cried in hushed alarm. "Don't turn around. She'll know we're looking."

"Why's she here?"

"That's what *I* want to know," Kaitlin declared. "Any ideas?"

"Look. She's with James Meade. Oh wow."

"Like, they're *dating,* Kristal? No way. He's too old."

"He isn't so old. My mom says the gray in his hair is from grief from his dad."

"I think it makes him look like Richard Gere."

"Talk about *old.*"

"Annie Barnes isn't so young herself. They could be dating."

"He's not interested. He has a little kid."

"What difference does *that* make?"

"Are we really *not* supposed to know about that baby?"

"Pu-leeze. Everyone knows. Sam Winchell didn't say anything in the paper, 'cause Sandy Meade told him not to. Sandy's annoyed that James went and adopted a baby. He doesn't understand why his son couldn't make one the usual way."

"Maybe he can't."

"Can't?"

"Like, is sterile?"

"Yeah. Right."

"Has anyone seen the baby?"

"I don't know, but it's a girl. My mother saw James at Harriman's buying Snow White Huggies. The baby was home with the nanny."

"The nanny's from East Windham. Can you believe? He couldn't hire someone from Middle River?"

"Sandy probably told him not to. He doesn't want people seeing or talking. The baby's from China."

"Vietnam."

"Whatever."

"Talk about raising a kid with zero of its heritage. Does James know what he's doing?"

"James Meade always knows what he's doing," declared Bethany.

Kaitlin agreed with that, all right. James did always know. So maybe what Kaitlin had to do was to talk with *him*. Maybe she could convince him not to tell. Now that he was a father himself, he might appreciate her dilemma.

Right, Kaitlin. And pigs could fly. She returned to her original idea. Annie was the one to corner. "Is Annie Barnes staying with her sister?"

"Yes."

"She isn't very friendly," Bethany complained. "She was running this morning and just ran right past Buzz Madigan after he said hello."

"Would *you* have stopped to talk with Buzz?" Shawna asked. "Like, I mean, he can go on and on and on. If I was running to get a workout, and I stopped to talk with him, the workout would be shot."

"Do you think she'll spend much time at the store?" Kaitlin asked casually.

"At Miss Lissy's Closet? No. She's going to write."

"Where'll she do that?" Kaitlin asked. "At the house?"

"Who knows?"

"Why does it matter?"

Kaitlin shrugged. "I was just thinking that if everyone is saying she's come here to write, maybe she's rented a place of her own."

"No. She's staying at Phoebe's."

Bethany leaned forward, eyes wide. "What if Greg Steele comes to visit?"

"That'd be *great*."

"Omigod. He's so hot."

"Could be trouble if she's hooking up with James."

"Does she have any friends in town?" Kaitlin tried.

"No."

"None."

"Absolutely not."

"What about the gym?" Kaitlin asked. "Your mom's a member, Jen. Think Annie Barnes might work out there?"

"Mom didn't mention it," Jen said and shot her a suspicious look. "Why are you asking these questions?"

"I'm *curious*."

"Don't tell me you've read her book."

"No. I'm just curious. Is there a law against that?"

Eliot Rollins left the clinic early, but not to go home. While his car sat back in the lot, he walked down a block, crossed the street, and entered the low brick building that housed the police station.

The police station. He still choked a little calling it that. Yes, the place had a holding cell—two, actually, but it was questionable as to whether either one could actually hold a person who wanted to escape. Eliot had grown up in New York, where police stations were the real thing. This one was mostly for show. Likewise the chief of police. Marshall Greenwood did what he could to give the illusion that Middle River abided by laws. The truth was that little crime occurred here that wasn't committed by someone directly or indirectly related to the mill, and the mill took care of its own.

Marshall was out and about three times a day, patrolling the streets in his cruiser. He stopped to talk with whomever he saw, so that what would otherwise take only half an hour sometimes stretched to two. The rest of the time he was at his desk.

He was there when Eliot entered, and yes, indeed, he was overweight. Tom Martin had that one right, but then, it wasn't something the man could hide. Winters, he wore a jacket to cover the gut hanging over his belt. Summers, the gut strained against the buttons of his shirt. That shirt was blue to match his jeans. He called this his uniform, which meant that he didn't have to waste time picking out clothes in the morning. Blue shirt, blue jeans, Demerol, OxyContin, Xanax.

He sat straight in his wide-armed, high-backed chair. It was an ergonomic one bought for him by his friend and benefactor Sandy Meade, and it probably cost more than the rest of the office furniture combined. He was doing a crossword puzzle in a fat book of crossword puzzles. A stack of others leaned against the file cabinet behind the desk. Marshall was addicted to these, too.

He spotted Eliot. "Good timing, Dr. Rollins," he said in his gravelly voice. "I need a word for a skin disease. Six letters, third one's a *z*."

Eliot felt a stab of annoyance. He didn't want to be here at all. The problem wasn't his. He was just the man in the middle, and he didn't do zits. "I'm an orthopedist, not a dermatologist."

"Come on," Marshall coaxed. "You studied everything in med school. It can't be real difficult if it's in my puzzle."

"Put the puzzle down, Marshall. We have to talk." He put his hands in the pockets of his slacks. "Annie Barnes is snooping around. She was at the clinic yesterday talking with Tom, and he says she didn't raise this issue, but it's only a matter of time. You and I could both be in trouble if she gets wind of certain prescriptions."

Marshall frowned. "Why in the world would she get wind of that?"

"Because she's here looking for dirt and this particular dirt is hot stuff. Think about it. 'Middle River police chief addicted to painkillers.' Makes a good headline, y'know?"

Marshall set his pencil down. "I take

painkillers for a back problem. I'm not addicted."

"Want to prove that by going cold turkey?"

The look on Marshall's face said that he did not. Then he frowned. "You're acting like this is my fault. I didn't ask for Jebby McGinnis to drive drunk right into my car. I was injured in the line of duty, and you were the orthopedist who headed my case. I'm just following doctor's orders. You're the one writing the prescriptions."

"You're the one demanding them," Eliot shot back, because he wasn't taking the fall here. "You're the one who suggested I write prescriptions in the name of your wife so that insurance will pay for the stuff you take above the normal allowance. That's health care fraud. There's another hot topic. Want to try for that one?"

"I don't want to try for any one," Marshall said in a gruff, law-and-order voice that told Eliot he had his attention. "Is there a point to this discussion?"

"It's already been made," Eliot said with satisfaction and headed for the door. "Sounds like you appreciate the problem. I'm assuming you'll do what you can to make sure nothing comes of it."

"What in the hell'm I supposed to do?" Marshall called, but Eliot was already on the threshold.

"You'll figure something out," he said. "And it's eczema. *E-c-z-e-m-a.* Don't you wish *that* was your only problem?"

Chapter 10

I HAD MY list of people who were sick in Middle River. And I knew that paper mills produced mercury waste. Now I had to connect the two. Specifically, I needed to know whether Northwood was one of the paper mills that had cleaned itself up, or whether it continued to pollute.

How to find out? That was the dilemma I grappled with as I stood at the kitchen window with my coffee early Friday morning. The window was a bay with a three-sided view of the yard. I saw a soft stretch of lawn, bordered by my mother's flowers—the purple, orange, and white of asters, daylilies, and hostas, respectively. Deeper into the yard, gangly black-eyed Susans swayed in the breeze along with the low-hanging willow limbs farther behind them. And farther yet, the river flowed, a rippling swath of gray-blue.

The day was overcast, seeming torn between rain and shine, and the indecision mirrored my mood. Yes, the state environmental agency would have information about the status of the cleanup at Northwood Mill, but going there wasn't my first choice. All it would take was one tattletale phone call by a person there with Meade ties, and the you-know-what could hit the fan.

An alternative was to talk quietly with a mill insider, but my choices were limited. Whoever it was had to be high enough up the mill ladder both to have legitimacy and to know the truth about what the mill did. I wouldn't put it past Sandy to tell rank-and-file workers that the mill was clean, when it might be anything but.

As candidates for mole went, I could think of four. One was Alfie Monroe. If he had an ax to grind against the Meades after being passed over for the position for which he had worked all his life, he might help me out. Another was Tony O'Roarke. He would have an insider's knowledge of the workings of the mill, without a history of Middle River ties to make him beholden to the Meades. Suppose, just suppose, he had found something amiss at the mill, had complained to Sandy Meade and been rebuffed. He might become my ally simply to save his own skin down the road.

Aidan Meade was another possibility. That's right. Aidan Meade. Aidan had a huge ego, always had, always would, and it was common knowledge that he was concerned about his rightful place at the mill. If he was fed up with playing second fiddle to James, I might goad him into staging a coup around the issue of mercury. Would he listen to me? Oddly, I think he would. I had the distinct impression that he was intrigued, perhaps even excited, by where I lived and what I had become. Perhaps I was deluding myself—or buying into my own hype—or simply thinking too Washington—but I half suspected that I could seduce him if I chose to.

I wouldn't, of course. Even aside from the ethical issue of his being married, I despised him too much to want his skin touching mine. But I could lead him on, right up to the brink. I could get what I wanted from him, then drop him when I had enough dirt on the mill. I could use him the way he had used me. That would be poetic, don't you think?

And then there was James. He had hinted that he stood for what was right. But if that was so, and mercury waste was a problem, he should have acted on it already. He was first in line for the throne. Did I really think he would risk that position by bucking his father? No, I did not.

A loud thud came from the hall, then two more fast thuds. Terrified, I set my mug on the table and ran. I found Phoebe halfway up the stairs, sitting upright, rubbing her elbow. Relieved that she hadn't fallen far, I quickly climbed the steps. I pushed up her pajama sleeve to see the elbow. There was nothing protruding.

"I don't think it's broken," I said. "Does anything else hurt?"

Phoebe looked awful, more washed-out and disheveled than ever. She stared at me fuzzy-eyed and asked in a feeble voice, "Why do I keep forgetting you're here?"

"Can you bend your elbow?"

She did it, albeit gingerly. "I banged it on the banister. I don't know what happened. My foot just slipped."

Had she been wearing slippers with a smooth sole, I might have bought her explanation. But her feet were bare, like mine, and the carpeted stair was textured. "You lost your balance. Were you feeling dizzy?"

"No. I'm just slow waking up."

"You never were when we were kids."

"It's my cold."

"The cold sounds better," I argued. Her voice was far less nasal than it had been. "How did you sleep?"

"Like a log," she replied.

It was not the truth. I had woken up to the sound of floorboards creaking not once, but twice in the night. I didn't think she was lying. She simply didn't remember.

"Here," I said, taking her other arm. "Let me help you down."

We started slowly; she was shaky. By the time we reached the kitchen, though, it was clear she was unhurt. I poured her a cup of coffee and sat with her.

Gently, I asked, "Does it worry you that you have some of the same symptoms Mom had?"

"*Mom's* symptoms? Oh, I don't have those. I have a *cold.*"

"With symptoms just like Mom's. Phoebe,

what if Mom didn't have Parkinson's disease at all? What if something else was making her sick—something environmental like, say, mercury?"

Phoebe eyed me as if I were crazy. "Where on earth would she have found that?"

"The mill."

"*Our* mill?"

I nodded.

"There's nothing wrong with our mill. Besides, Mom and I don't go *near* the mill."

"You don't have to go near it," I said, speaking hypothetically. "If the mill is polluting the river and you eat fish—"

"Not from the river. We buy our fish at Harriman's. Besides, Sabina works right *there* at the mill, and she's fine. They're *all* fine. I was at her office a few weeks ago, picking her up from work. We were going to plant flowers at Mom's grave."

Mom's grave. My heart lurched at the words. It was one thing to think about Mom being sick or Mom falling down the stairs or Mom visiting the mill. It was another to think of her buried in the graveyard on the hill behind the church. I felt a great yawning inside.

It grew worse when Phoebe's eyes drifted

to the window. "We need rain, it's been a *horribly* dry summer," she said in a voice so uncannily like Mom's that even she looked quickly back at the stove, where Mom should have been. When there was no Mom, she started to cry. She bowed her head and covered her eyes, and for the life of me I didn't know what to do. We weren't a touchy-feely family, and I was grieving, too. But Phoebe needed something.

Tissue, I decided and made a dash for the bathroom. I returned with several, which she instantly took. While she pressed them to her eyes, I pulled a chair close, sat down, and rubbed her arm. It seemed paltry comfort to offer. When her crying slowed, I said, "I'm sorry, Phoebe. The past few weeks have been hardest on you."

"It's just that I don't know what's wrong," she cried in a broken voice, holding the tissue to her eyes. "I keep tripping and losing my train of thought and doing all the things Mom did, but I'm too *young* to have what she had, aren't I?"

"Yes," I said. Suddenly I was thinking that it might not be mercury poisoning or anything else at all. My sister could well be clinically depressed. "I think you should see Dr. Martin."

Phoebe shook her head at that. Uncovering her face, she drew herself up and seemed to calm. "No. I'm really all right," she said as she wiped her eyes. "It's only natural that I worry sometimes. Summer colds hang on forever."

"Maybe you need to take time off from work."

She looked alarmed. "And do what?"

"Rest."

"No. I'd only sit here and worry more. Besides, the store needs me." She blew her nose and managed a wobbly smile. "There. I'm better now."

I was feeling helpless again. "Can I get you anything? Eggs? Cereal? . . . Pizza?"

Phoebe smiled more naturally then. "Mom hated it when we ate pizza for breakfast, didn't she?"

I smiled back. "She did."

"You were never into it as much as Sabina and me. We would deliberately leave slices from dinner so we could have it for breakfast the next day."

"I haven't seen any in the fridge."

"No. No pizza. But that's okay. Food tastes lousy to me lately." Her eyes met mine. "Even coffee doesn't do it the way it used to. What does that mean?"

"I don't know. Dr. Martin might."

"I'm not seeing Dr. Martin."

"If you're afraid of what he might find, that's self-defeating."

"Fine for you to say. You're not the one feeling lousy."

No. Not lousy. But in need of fresh air. I glanced at the clock. It was nearly eight. "Are you going to sit here a bit?" I asked.

Phoebe nodded.

"Mind if I go for a run?"

"Why ever would I mind?" she snapped.

I didn't answer, just drained the last of my coffee and went up and changed. When I returned to the kitchen, Phoebe hadn't moved. She was staring at her coffee mug, just as I had left her.

"Can I get you anything before I leave?" I asked.

She looked up and frowned. "Where are you going?"

"Running," I said, though I had already told her that. "I'll be gone forty minutes. Don't have breakfast. I'll make it when I get back." I went out the door before I could worry, stretched quickly, and set off.

The sky had grown darker, and the air was heavy. Slogging through it, I began slowly, then picked up the pace once my legs warmed. I followed the same route I had taken yesterday—at the same time, yes, I did notice that—and I was on the lookout. I don't know why. I certainly wasn't in the market for a love interest here. Maybe someone with whom to talk about running?

And hell is spiked with ice. Face it, sweetie. You liked the way he looked.

I didn't *see* the way he looked, I argued. He was too far away.

You saw enough. Why else would you be chasing him now? Hey, is that him?

No. It's a man mowing the grass at the edge of the street.

Oh. Too bad. I want to see our guy, too. Who is he?

How would I know? I replied, breathing harder as I increased my pace again. I haven't lived here in fifteen years. I don't know half the people in town.

But they know you. Now you know how it feels to be stared at. I kept telling you it was bad, but you didn't believe me.

It's not so bad if you know to expect it.

Not so bad? People eyeing you like you have horns, then asking why you can't be more like Harriet Nelson?

You're dating yourself, Grace. Besides, it was worse for you because you drank and swore and wore men's clothes.

By Jesus, that was my right.

True, but there are consequences to some behavior. You went out of your way to shock people.

And you didn't?

I did. But I was young. You were in your twenties and thirties. Why were you so contrary?

How else could I know who was friend and who was foe? People are two-faced, sweetie. They're all peachy-keen when they want something from you, then once they have it, they'll stab you in the back. I thought you learned that with Aidan Meade.

I was thinking that not all people were two-faced—certainly not Greg and others of my Washington friends—when I was suddenly distracted, and not by the light rain that had started to fall.

There! Oh boy. That's him.

Indeed it was. I had just turned a corner, and there he was, running a block ahead. My breathing seemed suddenly louder. I was amazed that he didn't hear, but I suppose the patter of rain hitting leaves overhead covered it up. He didn't slow, didn't turn. I had a fabulous view of his tapering back. He was shirtless again; his shorts were close-fitting and hit at the thigh.

That's an impressive backside.

Uh-huh.

Run faster.

I can't. I'll mess up my workout. Besides, I like watching him—even if he isn't the best runner I've ever seen. He's not terribly graceful. I think he's flat-footed.

But he's tall and broad-shouldered. And dark. I like them dark. My George had dark, curly hair. He was Greek. Did I ever tell you that?

Uh-huh. Many times.

Boy, did my mother hate that. She wanted me to marry a purebred American. We were French-Canadian ourselves, like so many of the others who came to Manchester to work in the mills. My mother wanted to be better than them. Nothing my father did was ever good enough for her. She drove him away. Shit, he's turning the corner! Don't let him get away!

What do you want me to do? I thought

with a crazy half laugh. I was running faster than I normally did and had gained on him, but I was breathing perilously hard. Flat-footed or not, he was fast.

Call him! Grace ordered. *Make him stop. He may be the one.*

What one? I asked.

The perfect one. The Adam.

I'm not looking for any Adam.

Sure you are, honey. We all are.

The perfect man doesn't exist. Isn't that the point you made in your books? Your men were all scum.

Not all of them. I liked Armand Ber-geron and Etienne deMontigny. They fell hard for their women in No Adam in Eden.

Etienne was an abuser.

His wife drove him to it. She turned into a real bitch. But what about Gino Donati? He's as good as they get. I would have loved a man like Gino. He had the potential for being perfect. So does that guy up ahead. Christ, sweetie. Move it. He's around the corner. Run faster.

I had no intention of doing that. Soothed by the coolness of rain on my skin, I settled back into a saner pace and ran straight ahead. I might never have known who the other runner was if that little voice inside hadn't prodded me into glancing to my right as I ran.

Well, well. I'll be damned.

The runner had stopped halfway down the block and was standing there in the rain, looking back at me with his hands low on his hips. I could see that he was breathing hard and that because he was dripping wet, his hair looked darker than it really was. In the seconds before a huge oak tree came between us, I saw the gray there. Then the next block began, and he was gone.

I ended up going to work with Phoebe—I mean, actually working at the store. It wasn't something I would have normally chosen; when I was a child, Miss Lissy's Closet had been my least favorite place. It represented everything I was not. Understanding this at some level, my mother put me to work in the back room. Naturally, I felt I was being hidden so that I didn't hurt sales.

It struck me that I still might hurt sales, albeit for entirely different reasons. But I was concerned enough about Phoebe to ignore my misgivings. She had warned me that mornings were bad, and this morning saw that out. She couldn't find her pocketbook, though it was in plain sight at the foot of the credenza in the hall, and once she found the pocketbook, she couldn't find the keys to the van, though they were right there inside. She searched for these things while she was still in her bra, because she couldn't find the blouse she wanted to wear, and when she finally located the blouse, it was on the laundry room floor, fallen from its hanger.

She lost it then—just fell apart. "Look at *this*," she cried, holding up the blouse. "I pressed it so carefully the other day, and now look. *I* can't wear *this*."

"It looks okay."

"It's *wrinkled*. I can't wear a wrinkled blouse."

"Then pick another," I suggested.

"But *this* blouse goes with this skirt, and now look at it. I hung it *so* carefully—I don't know why it fell down—but I can't show up at work this way, so *now* what should I do?"

I set up the ironing board, heated the iron, and ironed the blouse. Phoebe took it back upstairs and returned holding the banister and looking lovely, but she was clearly unhappy. She said it was the rain, then murmured something about buying the wrong things.

"Wrong things?" I asked.

She waved a hand in dismissal.

"What wrong things?" I asked, to which she replied a testy, "What are you talking about?"

Sensing a losing battle, I sat her at the table with a glass of juice, eggs and toast, and a cup of coffee, and while she picked at the food with a shaky hand, I showered and dressed. When she appeared to have finished, I took the keys, held an umbrella over us going out to the van, then drove her to the lower end of Willow.

It had been years since I had been at the store this early, and though computerization had streamlined things, certain chores remained the same. Tallies from the previous day had to be re-checked; messages left by West Coast suppliers after closing time the day before had to be accessed; bank deposits had to be prepared. Newly received shipments had to be checked, ticketed, and shelved; customers waiting for those clothes had to be called. Yesterday's coffee grinds had to be disposed of and a fresh pot put on to brew.

Like Mom, Phoebe had a crew that cleaned weekly, but vacuuming was done daily. Shelves had to be neatened and racks straightened; clothes that had been disturbed the day before were refolded and rearranged by size. Dressing rooms had to be checked, errant clothes put away, and the soft green privacy curtains hooked open.

Since Miss Lissy's had built a reputation for carrying lines of clothing that few other stores in central and northern New Hampshire carried, telephone orders were a constant; most such orders were filled during slow spots in the day, but others were done prior to opening to assure that they shipped with the morning mail. This day, there was only one order to box.

Phoebe assigned it to me. She made it clear that she and Joanne (who, to my relief, was already at the computer when we arrived) were the only ones who knew enough to handle the *complicated* stuff (Phoebe's italics), and I wasn't offended in the least, not then, nor when she directed me to refold T-shirts and jeans and put wayward shoe boxes in order of size. It was a lovely exercise in helping—not greatly taxing, but practical—and it enabled me to watch the give-and-take between Joanne and her.

This held no surprises. Having already sensed Joanne's competence, I figured that she would be doing most of the work. Yes, it discouraged me. It was further proof of Phoebe's impairment. That said, Joanne was wonderfully gentle and understanding and kind. She worked with Phoebe in a way that made my sister feel she had done her part.

Joanne was an enabler. So, for that matter, had I been that morning, helping Phoebe get ready for work. We were making it easy for her to blame the problem on a lingering cold or, worse, to ignore it entirely. But we did that for those we loved, didn't we? And we did it for ourselves. *We* wanted Phoebe to be well too.

And she did seem to be, once the store opened, but I suspected it was a triumph of will. I could see the mask of sweet competence take over her face when a customer appeared, then fall away as soon as the *ting-a-ling* announced that the customer had left. And there were plenty of *ting-a-lings*. While the store would never be a high traffic zone, à la Neiman Marcus the day before Christmas, the clientele came in a steady stream. Granted, it was Friday. If ever there was a time for the impulse buy of shorts, a T-shirt, or sandals for a weekend that promised sun despite the continuing rain, this was it. The mailman came and went, with *ting-a-lings* and wet footprints on the floor. Same with the UPS driver, who delivered three large boxes.

"August is busy," Joanne explained as I helped her check out what proved to be jeans, long-sleeve T-shirts, and sweat suits. "Hot as it is, people are starting to think about fall. School clothes have to be out on display. Same with fall sweaters and slacks. These were ordered back in March."

"Were you at the show with Phoebe?" I asked. I knew that Mom had been too sick to go.

"No. She did it alone."

That made me feel better. I could see from the boxes' contents that whatever she had done was done right. Colors, styles, and fabrics were all appropriate and smart.

But that was March. This was August. "Is she up for New York?" I asked.

Joanne met my gaze and said a quiet, "I've offered to go with her, but she insists she needs me here. What do you think?"

I thought no, she was not up for New York, and I became even more convinced of it during lunch. I had run out for salads, but since it continued to rain, we ate in the office—and was the place a mess? *Big*-time. Apparently Phoebe, in a frenzy while I was out, had decided that since she couldn't *find* anything, reorganization was needed. In the space of those minutes, she had unloaded bookshelves and drawers. Aghast at the mess, I offered to help put things back, but she insisted she had to do it herself. I suggested we eat somewhere else. She nixed that idea too.

So we sat there, but Phoebe did little eating. Seeming lost, she picked up her fork, put it down, ferreted through the clutter on her desk to pull out a note. Barely reading it, she put it down, picked up her fork and seemed about to eat, when she put the fork down and unearthed another note. Back and forth it went, with nothing accomplished on either end, and it struck me how closely she guarded herself when customers were in the store, how deliberately she spoke, how casually she touched whatever she passed—

wall, doorjamb, counter—for balance, how desperate she was to create the perception of well-being, when she wasn't well at all.

When the phone rang, it was even worse—clearly a supplier, I gathered from the one-sided conversation, but Phoebe was bewildered, couldn't remember the supplier or putting in an order. Muddling through, she was close to tears when she finally hung up the phone.

Seeming to forget I was there, she put a hand on her head and murmured, "I'm *so* losing it—oh God—what is the *matter*? I can't be sick—I'm too young—I have the store and without me the store isn't—isn't— won't run—and I don't have anyone to take over for me if I get worse—and it's bad enough here but now there's New York."

Aching for her, I asked, "What does New York entail?"

"Next week—will I be better? I just don't know—and if I'm not I won't be able to han- dle it—I mean, it's one thing waiting on cus- tomers here—I've done it so long I could do it in my sleep—but the other is hard—you have to be able to think straight and some- times now I am *so* far away from that, that it *terrifies* me."

"Phoebe," I shouted to get her attention. When her eyes flew to mine, I asked again, "What does New York entail?"

She blinked, swallowed, looked away and then back. "It's a day or two of shows. There are aisles of booths. My vendors will all be there showing the holiday line."

"Do you have to order things on the spot?"

"No. I can wait until later in the month. But there's a load of paperwork and it's hard keeping things straight—who's selling what—and seeing the trends and deciding what'll work here." She pushed a hand through her hair and said plaintively, "I just *need* to be better."

If she had Parkinson's disease, medication could help. Of course, if she had Parkinson's, the long-term prognosis wasn't good, in which case the buying trip might be a practice in futility, since the store would eventually have to close.

But I didn't believe she had Parkinson's, and if the symptoms were psychosomatic and related to depression, missing the trip would make them worse. Besides, if I could get her to New York, we could see a specialist. I wouldn't even tell her ahead of time. She might be furious with me, but the deed would be done, and with no one the wiser in Middle River.

"I could go with you," I said.

Phoebe was startled. "You? *You* don't know anything about buying clothes."

"I do it all the time," I said with a laugh.

"It's not the same."

"I know." I grew serious, because she was, very much so. "But I could handle the details and the paperwork. I'd be your assistant. I'd take care of peripheral things, like flights and hotels and what vendor is where. You'd just have to look at the clothes and decide which you like. It would be fun, Phoebe. We've never done anything like this, just the two of us."

She seemed interested, but cautious. "No. We haven't."

"I know some great restaurants. We could make it a mini-vacation, a real breather for you. You haven't had that since Mom died."

Her eyes lit. "I might finally kick this . . . whatever it is I have."

"Definitely," I said.

"Well, then . . ." Her voice trailed off. Her brow knit. In the silence, I first feared she had lost her train of thought. But she hadn't. To the contrary. We were in a lucid stretch, as was evidenced by the clarity in her eyes. With that clarity came concern, though, and it suddenly hit me that she might not want to be stuck in New York alone with me. As it happened, her concern was different. "What about Sabina?" she finally asked.

Sabina and Ron lived on Randolph Road, another of those quiet streets named after early Middle Riverites. Their house was a sweet Cape, older than some but kept in mint condition, thanks to Ron's skill with a hammer and nails, and painted a perfect pale blue with white trim. There was no lawn to speak of, no elaborate shrubbery, no cultivated flowers. Landscaping here was entirely natural, consisting of pine trees overhead, their needles underfoot, and clusters of wild ferns.

After a stop at Harriman's, I arrived midafternoon with the makings of a fabulous dinner. Sabina and Ron were at work, and Timmy was nowhere in sight. Lisa, who had been reading a book in the hammock out back, was delighted to see me. Excitedly, she peered into the grocery bag I put into her arms. "What do you *have,* Auntie Anne?"

"I have a tenderloin of beef," I said, taking the second bag and heading for the house. "And fresh green beans and little red potatoes and makings for a spinach salad. I have Brie and crackers. I have blueberries and raspberries and a recipe for shortbread. Want to help me surprise your parents?"

We had a ball. I set her to work washing things, then showed her how to trim the ends off the beans and halve the potatoes. We mixed the former with slivers of almonds and the latter with chopped fresh oregano. We mixed the shortbread batter, clumped it on a baking tin, and put it in the oven. We fixed a rub for the tenderloin, applied it, and put the meat aside. We washed the fruit. We washed the spinach. We set the table.

I didn't often cook, but when I did, no recipe was too complex—at least, not with a friend like mine. Berri Barry was into cooking. All I had to do was tell her what I wanted—as I had in an e-mail from the store, to which she had promptly replied—and I was given a menu plan with recipes, plus schemes for substitutions in the event that the ingredients she called for weren't at hand.

I went simple here, mainly because Ron was a meat-and-potatoes man. Nonetheless, by the time Timmy rode his bike up the drive, the kitchen smelled divine.

Phoebe joined us a short time later, and then Sabina. I handed both glasses of wine. Tired, Phoebe was content to retreat to a chair on the porch. Sabina was more watchful, clearly unsure of why I was there.

And why *was* I there? Officially, I wanted to make a dent in her distrust, so that she wouldn't object to my taking Phoebe to New York. But here too, there was a deeper truth, one that was tied to the last. I was reaching out to Sabina. I was acknowledging that I knew how hard she worked, and trying to make things easier for one night at least. I was thanking her for taking care of Mom when I hadn't. I was celebrating her two incredible children and her home.

One look at Ron rushing in late, though, and I forgot all those things. I don't know why it didn't occur to me sooner. My source was right under my nose. Ron worked in the maintenance department at the mill. If anyone would know about mercury at the plant, he would. Right?

*   *   *

Wrong. And wrong on *two* counts. But I had to wait until dinner was nearly done to learn that, because I couldn't ask him point-blank when he walked in the door, any more than I could ask in front of everyone at dinner. Sabina would have hit the roof. So I bided my time as we ate, and I made a point of being open and agreeable. I answered the kids' questions about Washington, Ron's questions about Greg, Sabina's questions about the herbs I had rubbed on the tenderloin. My opening came when we moved on to dessert. I had forgotten whipped cream for the shortcakes. Sabina offered to run out for some.

She took my convertible and the kids. Phoebe returned to the porch, while Ron and I set to washing pots and pans, and I didn't mince words. As soon as we were alone, I told him what I suspected and, outright, asked what he knew.

"Not much," he said, long arms up to the elbows in a sink filled with suds. "I work mostly in packing and shipping, over at the garage and the shipping docks. That's way on the other side of the mill."

"Have you heard talk?"

"We talk about cars and boats and sports."

"No one mentions the words *mercury* or *pollution* or *regulations*?"

He didn't answer, simply rinsed a serving dish and passed it to me.

"Is there a pattern of ill health at the mill?" I tried.

When he took a steel wool pad to the roasting pan, seeming content to let the matter ride, I asked if he would tell me if there was.

He looked at me then, eyes filled with affection, though serious indeed. "I like you, toots. I always have. I've fought for you sometimes when I thought Sabina was being too hard. But this is different. This is our lives. Even if I did know, which I don't, there's too much at stake."

"Something more than life and death?" I asked in amazement. I would have thought *nothing* was more important than life and death. But that was just another premise of mine, in rebuttal to which Ron calmly offered **TRUTH #5: More than life and death? You bet. My work and my wife. Without those,**

I'm dead anyway. You don't understand the power the Meades have. They can ruin us."

"I *do* understand. I felt it once."

"When you were eighteen. It's different for me, Annie."

"But what about *mercury*?" I asked, because that seemed the bottom line. "Yes, you have a family, but what of their *health*?"

"Look at them. They're fine. I'm a lucky man," he said, and returned to the pan.

Back at the house later that night, I was no more decided about where to turn next than I had been that morning. Discouraged and in need of a lift, I went online to commiserate with friends, but there was the usual spam to wade through first. I highlighted the immediate suspects—then paused. One stood out. The username was TrueBlue, and the domain was a common one. I didn't know any TrueBlue, but the subject line caught me cold.

I know about the mill.

Could be advertising a coffee mill, I thought. Or a pepper mill. Or, sickly, a people mill. Could be carrying a virus, I warned myself.

But if someone had information on *our* mill, I wanted it.

As determined as I've ever been in my life, I clicked on.

Chapter 11

To: Annie Barnes
From: TrueBlue
Subject: I know about the mill.

You're snooping. I have info. Ask.

There was no electronic signature, no information anywhere to suggest the identity of
TrueBlue. I clicked on every possible spot,
but found nothing beyond what I saw on my
screen. On the plus side, my computer didn't
self-destruct, which meant that TrueBlue
wasn't lethal, at least not instantly so.

I hit "reply," but even when I right-clicked

on TrueBlue in the address line, I learned nothing. So. What did I know? I knew that TrueBlue had my name and e-mail address, that he (or she—it could easily be a woman) knew I was interested in the mill, and that the e-mail had been sent at nine-thirty that night, roughly an hour ago.

To: TrueBlue
From: Annie Barnes
Subject: Re: I know about the mill.

Who are you?

My finger hovered over "send." I knew that in answering, I would confirm my own existence, which was a major no-no in fighting spam. But this wasn't spam, as in a mass e-mailing that advertised something I didn't want. TrueBlue knew I was curious; he (or she) had mentioned the mill. But *our* mill? Not necessarily. This sender could be an ecology buff responding to my research on mercury. I had visited enough sites; any one of them might be able to track its guests—and reply, if so inspired. Did that imply evil intent? I didn't see how. Besides, it wasn't like there were other volunteers lining up at my door.

I sent the e-mail, then sat for ten minutes, clicking on "send/receive" every thirty seconds, until I realized the absurdity of assuming that TrueBlue was sitting in front of a computer just waiting for a reply on the odd chance I was at *my* computer at this particular hour.

That said, the longer the minutes dragged on, the more I wanted to know who he (or she) was. On impulse, I picked up the phone and called my sister the computer expert.

It did dawn on me while her phone was ringing to wonder if, given the subject matter, calling Sabina was such a great idea. But it was already too late; she had caller ID and would know it was me, or Phoebe, and in either event, she would call back. Then she answered, and the point was moot.

"Hi," I said. "You weren't sleeping, were you?"

"No," she replied, audibly cautious.

"I have a question. I just got an e-mail from someone with a username I don't know. How can I track it down?"

"Is it spam?" she asked, seeming to relax some.

"No. It's about work." I clicked "send/receive" again. Nothing.

"A fan?"

"No."

"Is it threatening?"

"No. I think it's about some research I'm doing. I want to check the legitimacy of the source."

"You don't recognize the domain name?"

"Well, I do." I told her what it was.

"That's not good," she advised. "It's a Web-based e-mail service, which makes it next to impossible to trace. Have you tried a reverse search?"

"No. I just got the note."

"Well, try that. You could also Google the address. That might turn something up."

"On *Google*? A person who wants *anonymity*?"

Sabina ignored my challenge. She was calm, clearly knowing of what she spoke. "Whoever it is may have used the address in a chat room or on a bulletin board. If it's been picked up by a search engine, it could show up—not that you'd necessarily get a real name through that, but you'd have somewhere else to look. Try those things. If nothing works, I could snoop behind the scenes. I might be able to find code that would tell you something."

The offer would have pleased me had the circumstances been different. But I was suddenly feeling guilty for using Sabina to get information that might lead to my getting the kind of scoop she had warned me against. She would be furious if she knew.

"Let me try these," I said. "Thanks, Sabina. This is a help."

"Thanks for dinner," she replied. "Now we're even."

I hung up the phone thinking that tit-for-tat between sisters was sad, but I was instantly distracted when I clicked on "send/receive" again.

To: Annie Barnes
From: TrueBlue
Subject: Re: I know about the mill.

It doesn't matter who I am. I know about Northwood Mill.

I answered immediately.

It totally matters who you are. How can I trust anything you tell me, if you won't give me your name? For all I know, you're an ex-employee who holds a

*grudge and will say anything to cause
trouble. Only cowards hide behind
anonymity.*

Besides, who says I'm snooping?

*And how did you get my e-mail
address?*

*And if you're looking for money,
forget it.*

As soon as I sent the note, I imagined
Greg's dismay. *You have a potential source,
and you're calling him names? That's a no-
no, Annie. TrueBlue has something you
want. Until you have it, treat him with kid
gloves. Lose your source, and you're up a
creek without a paddle.*

Click as I might on "send/receive," no reply
came through. I was starting to think I *was* up
that proverbial creek, when TrueBlue replied.

Sorry, toots, but goading me won't work.
I already know I'm a coward. And I'm no
ex-employee. I've been at Northwood
awhile and plan to stay, so I don't need
your money. I know you're snooping,
because the whole town knows it. And
e-mail addresses are a lot easier to find
than to trace.

His answer was telling—and, yes, I say *his.* A woman wouldn't address me as toots. Not even Grace used that word, and she could be tough.

My brother-in-law Ron had used it, barely two hours ago, but if you're thinking Ron was TrueBlue, think again. Ron had meant what he said when we were doing the dishes. He wouldn't risk his job or his wife by telling me mill secrets. Besides, Ron didn't have a computer of his own, which meant that if he were TrueBlue, he would be using Sabina's, and he was way too smart for that. He was also way too de-voted to my sister to be out of the house using someone else's computer at this hour of night

No, TrueBlue wasn't Ron. He could be any of the four men I had already pegged as possible moles. Or he could be one of hun-dreds of others. The mill employed that many in various capacities.

Taking Sabina's suggestion, I Googled the e-mail address. There was no match. Same thing when I ran it in a reverse direc-tory. *E-mail addresses are a lot easier to find than to trace.* TrueBlue was right about that. So how had he gotten mine? Possibly, if he was hooked into the Northwood network and was computer savvy, from Sabina's machine. Possibly, totally apart from Northwood but certainly vulnerable to a hacker, from Phoebe's. Or from Sam's. Unfortunately, to ask any one of them would mean tipping my hand.

Actually, tipping my hand was also a problem with TrueBlue. For all I knew, he had no intention of giving me answers but was simply trying to find out what I was after. Aidan Meade, for one, wouldn't *hesitate* to approach me under false pretenses; Lord knew he had done it before. Once he found out I was after mercury, he could make sure I got nothing.

I typed my response this time with more care.

Snooping sounds awfully negative for what I'm doing. I'm just trying to find the truth about my mother's death. I'm not accusing anyone of anything. I'm just spending my summer vacation trying to explain her death.

I don't want to think she had Parkinson's disease. There's no villain there. I'd rather think she was exposed to

something toxic. She didn't work at the
mill, so that's a long shot.
You say you have info. What's it on?

It was nearly eleven-thirty. Granted, it was
Friday night, so TrueBlue wouldn't necessarily
have to be up the next morning for work. Still,
I wondered why he was alone at his computer. So I added a question to that effect.

And why are you awake and e-mailing so
late? Too much caffeine? Scary movie?
Guilty conscience?

As I sent off the note, I realized that I had
likely forgotten the most probable reason:
confidentiality. TrueBlue wouldn't want anyone
else knowing what he was doing. That would
include a wife, significant other, whomever.
A reply came within minutes.

Call it dedication to the cause. I ought to
be sleeping. I have to be up in a couple
of hours. What about you? Are you
always a night owl?

I typed quickly, the first of several rapid
exchanges.

Only when I get notes from strange people who say they have information for me. Why do you call yourself TrueBlue?

Now, if I told you that, I'd be giving something away. Why do you type in script?

I'm an artist. It's my prerogative. Do you share your computer with anyone?

No. What about you? Are you at your sister's computer?

Definitely not. For what it's worth, she does not condone this search. Nor does my sister Sabina, so if you're thinking of ratting on her, save your breath. She would be furious if she knew I was corresponding with you. Do you people have to sign a loyalty oath or something?

No loyalty oath. Not yet, at least. What have you come up with for possible toxins that might have killed your mother?

If I told you that, I'd be giving something away——to quote you. But you've offered your services. No one else has

approached me with that kind of offer.
Why you? Why now? What's your role
here? What's in it for you?

A length of time passed without a reply, but I wasn't sorry for the questions. For one thing, I felt they were important. For another, the pause gave me time to e-mail Greg. He would have flown from San Francisco to Anchorage earlier that day. He was carrying his Blackberry.

I explained what was happening and asked his advice. To trust or not to trust— that was the question.

It was nearly thirty minutes before True-Blue e-mailed again. He didn't answer any of my questions, simply asked one of his own.

Are you writing a book?

I'm always writing a book. But it isn't
about Middle River. It takes place in
Arizona and has to do with a family of
packrats.

Packrats? Is this a children's book?

No. Human packrats. They don't say
much of substance to each other, but

they stash away all the little clues to their lives. My main character is the oldest daughter, who is struggling to figure out who she is and what she wants. When her parents suddenly die, she uses all those little things her parents stashed away to figure out her past. As my publicist puts it, the things that were hidden away become clues in a personal journey of self-discovery.

Skeletons in the closet? There's a novel idea.

For your information, the only truly novel ideas that exist are those that have to do with new developments in technology or medicine. I write about human relationships. It's all about painting an emotional picture that gives readers pause.

Touchy, are we? But now you've told me your plot. Aren't you afraid I might steal it?

Are you kidding? Give the same plot premise to six different writers, and you'll get six entirely different books. Are you thinking of trying to write?

No. But if I was, I don't know if I'd be as cavalier as you. Wouldn't it bother you if someone was to up and write your plot?

Only if he did it better than me.

Ha ha. What are you doing while you wait for my replies?

It appears that I'm plotting. It helps to talk a plot aloud, or in this case to type it to a friend, not that you're a friend, but you know what I mean. I do a lot of my best thinking at night. Speaking of thinking, I have an idea. Want to call me on the phone? I could give you my cell number, so no one would be the wiser that you called. Or you could give me your number. This waiting for a few sentences to go back and forth is pretty silly. Do you IM?

No. I don't IM. I don't have time for that. And no. No phone calls. This is safer.

But you're anonymous and I am not. What fun is that?

I think it's great fun. It's nice to be free of who I am and what I do. Are you working at a desktop or a laptop?

I tried to decide which of my suspects might want to escape himself. Surely not Aidan Meade. He was too much of a narcissist to want distance. Any of the other three could want it, I supposed. Didn't most of us want to escape our identities from time to time?

Laptop. You?

Same. In bed?

No. In the kitchen. I need a phone jack. What about you?

In bed. I'm wireless.

Wireless? In Middle River? I'm impressed.

I had no sooner said that when I realized two things. First, if he was using his computer in bed, he was sleeping alone. And second, we were actually flirting.

Here too, that ruled out Aidan Meade.

Aidan didn't flirt, he took. And I ruled out his brother James; no sense of humor there. TrueBlue could type—not a mistake yet—and was more articulate than I imagined, which made me think he wasn't Alfie Monroe. Alfie was a large-motor, big-machine kind of guy.

Tony O'Roarke was a possibility. I knew nothing about Tony, least of all whether he slept alone.

And I would have included Tom Martin, except that he worked at the clinic, not the mill.

Flirting. Interesting. I had never cyber-flirted before. Greg might not approve at all. But would it hurt to milk the thing? It was in that spirit that I added a quick line to my e-mail before sending it off.

Are you male?

Last time I looked.

Ha ha. What do you do at the mill?

Enough to know what I'm talking about. What are you looking for?

Something to convince me that you're legit.

Try this. Paper mills like Northwood take wood and turn it into paper. Part of the process entails using bleach to make the paper white. Bleach, or chlorine, as it is more commonly called, is produced in a chlor-alkali plant. Northwood has a small one that fills its needs.

Did you know that?

I did not. Go on.

Chlor-alkali plants use salt and electricity to produce chlorine. Traditionally, mercury was used to stabilize the product when electricity was passed through the salt. I say "traditionally," because we now know that mercury is extremely toxic. When it is released from chlor-alkali plants as waste, it fouls air, ground, and water near the plant. Add flow patterns and wind currents to that, and you have serious pollution.

Do we have serious pollution here?

Northwood has complied with every state mandate. We no longer use mercury.

So what's the problem?
Your turn. You tell me.

He might be strong enough to resist goading, but I was all too human. I was also impatient. Flirting was all fine and dandy— *caution* was all fine and dandy—but he was the one who had mentioned mercury. If he had information, I wanted it.

Mercury. Either you still use it in some illicit capacity, or there is pollution from the past that was never cleaned up. How'm I doing?

Not bad. Forget the first; we don't use mercury now. But the past continues to be a problem.

In what sense?

If I tell you, what'll you do with the information?

That depends on whether I decide I can trust what you say. You still haven't told me why you're doing this. What's in it for

*you? What do you expect me to do? And
why don't you do it yourself?*
Why don't I do it myself? Oh, I've tried.
I've talked to people, but they refuse to
listen. And I'm in a difficult position. If I
go too far, I stand to lose my ties with
Northwood. Those ties mean a lot to me.
    What do I expect you to do? There's
some information I don't have. I need you
to be my legs for that. For the rest, I
need you to be my voice.

I thought of Tom. Hadn't he too talked with
people? Wasn't he too in a difficult position?
No, I didn't think Tom was TrueBlue. True-
Blue said he had been with Northwood
awhile. Not only was Tom relatively new to
town, but he wasn't with the mill at all.
    Still, he was using me. These two men
shared that. So now I replied to TrueBlue.

*You need me to be your sacrificial lamb,
you mean.*

I hit "send" with more force than was
needed, but I was amazed at the man's gall.
He wanted me to do his dirty work. So much
for flirting. He was *using* me.

I was tempted to turn off my computer—to hang up on the guy, so to speak—just turn around and walk away. That's what my passionately defensive Middle River persona would have me do.

But I was grown-up enough to pause, to think, to realize that if this man had information that would benefit me, the using was two-sided. If he had information that would explain my mother's illness, and if his information implicated Northwood Mill, wouldn't I be successful on two counts?

How would you be a sacrificial lamb? You'd be righting a wrong. You'd be finding out why your mother died. You'd be doing something good for all the people who live in Middle River. If nothing else, think of the children. Mercury is devastating in children. How can you turn your back on that?

You're doing it.

No. If you don't agree to help, I'll find someone else. I've been at this awhile. I have other coals in the fire. I may not be willing to be out there and in your face. I may not be willing to sacrifice my name and my place. But I'm committed. Think about it—you may not get a better offer. I'm logging off now.

I did think about it for a good long time and then, naturally, I overslept the next morning, which meant that since I had promised to help Phoebe out at the store, I didn't have time to go for a run. That put me in a bad mood, which probably explained my paranoia. I was convinced that Phoebe didn't want me around, because no sooner did we reach the store, she gave me a long list of menial errands to run—I was sure it was deliberate on her part—bringing trash to the dump, buying cleaning supplies, hand-delivering a new pair of panty hose to a local customer.

Wherever I went, people stared; I was convinced of that too. It was like they knew I was in touch with TrueBlue, like they knew we were conspiring to rock the boat.

In fact, I hadn't heard from TrueBlue again. I did hear from Greg, though, a truncated note Blackberried shortly before his helicopter touched down at a base camp on the Kahiltna Glacier. He said that he was cold but energized, that this would be his last e-mail until they cleared the 12,000-foot mark four days from now, weather permitting, and that I should corroborate whatever any informant said—and even *that* hit me the wrong way. Well, of course, I would check it all out. Did Greg think I was totally naive?

There were more errands to run as the day passed. I mailed bills, fetched lunch, gassed up the van—the last after waiting ten minutes behind the minibus from Road's End Inn carrying its load of weekenders taking the Peyton Place tour.

Phoebe had her good moments; she did clean up the mess she had made in the office the day before. But in addition to assigning me menial errands, she gave me a few that were absurd. For instance, at one point she sent me with a list of places to go in search of the carpenter who had promised to spend Sunday at the store rebuilding one room's worth of display shelves. She wanted to confirm that he was coming. Me, I wanted to know why the guy didn't have a cell phone she could call. Lord knew, most everyone in town seemed to have one.

I never did track him down, which made me think Phoebe had known I wouldn't—

and yes, I was being bitchy. She hadn't asked for my help; I was the one who had offered it. Did I really expect she would put me in charge of the store? Maybe that was what she was afraid of—that I *would* try to take over—which was as ridiculous as it got. I was a writer, not a shopkeeper. And I say that with the utmost respect for what Phoebe did. I would have gone out of my *mind* with the indecision of some of her customers— *the black skirt is best—no, the navy one will work better with my shoes—but the black one makes me look slimmer—of course, the beige one will be perfect for September.*

Writing was a solitary profession, meaning that I did what I wanted when I wanted to do it. Yes, I had to conform to a publisher's schedule, but my book was my book. I was in total control of its contents. My modus operandi didn't involve waiting for someone to decide whether to buy loafers or slides, V-neck or crew neck, wool or cashmere. I had no patience for that. No *way* would I want the store.

Maybe *Sabina* had put that bug in her ear.

More paranoia.

And that was *Phoebe's* fault. Had she given me chores that involved thinking, I wouldn't have had time to stew. But what she gave me was busywork, nothing more. Running mindless errands around Middle River? My worst nightmare. Talk about awkward. Not only did people look at me wherever I went, but when we had cause to exchange words, they were guarded.

And then there was my shadow. Someone was following me. I was convinced of it. I can't tell you how many times I felt a prickly sensation at the back of my neck and looked fast over my shoulder.

There was never anyone there, of course.

But someone had been. I didn't find out who it was until I visited my mom's grave the next day. Until then, I was just in a lousy mood, which might be juvenile, but was fact. By the time Saturday night came, I was annoyed enough at Middle River to agree to help TrueBlue with anything his little heart desired.

To: TrueBlue
From: Annie Barnes
Subject: Your offer

You have a point. Where do we go from here?

Chapter 12

MARSHALL GREENWOOD was on edge. He knew that if Annie Barnes was on his case, there could be trouble. One solution was to retire. He was sixty-six and would be doing it soon anyway. As chiefs of police went, he was old. But Middle River wasn't demanding of him, and he needed the money. Besides, his wife didn't want him around the house. She hadn't said it in as many words, but that was the general drift.

Barring retirement, the other obvious solution was to ditch all the pills. If he wasn't taking anything, he wasn't an addict. He could simply deny the allegations. Over and done.

He lasted three hours, at which point the pain in his back was so bad and he was so ready to crawl out of his skin, he gave up on that idea. Besides, even if he were able to kick the habit now, the past would be hard to deny. Among his doctor, his HMO, and the drugstore was a paper trail a mile long. No, there had to be another answer.

He spent most of Friday trying to find it. But he had never been good at finding answers. He did things that were obvious. When someone was drunk, Marshall either drove him home or hauled him in to sleep it off. When a car hit a tree, he called the ambulance first and then the tow truck. When a man beat his wife, he drove the woman to her mother's house, and if the guy did it again, he called the county sheriff.

The only clues Marshall dealt with were in his crossword puzzles. Friday night, he agonized over one that read "major backup." The word had six letters, the second an *a,* the last an *n.* He went through every possible version of *traffic tie-up* before squeezing his eyes shut, shaking his head hard, and taking a whole different approach. Bingo. The answer was *patron*.

Taking that as a message, Marshall picked up the phone early Saturday morning and called Sandy Meade. Sandy was his most respected friend in Middle River. They had gone to school together, and though Marshall was two years behind, they had been teammates in every sport. Sandy was inevitably the star, but Marshall was the better athlete. He made Sandy look good on enough occasions that, by way of reward, he had Sandy's undying loyalty. Sandy was responsible for his being hired for the security force at the mill years before, and when he won a spot in the Middle River Police Department, Sandy was responsible for that, too.

"Got a minute?" he asked now. He was sitting in the driveway, calling from his car so that Edna wouldn't hear, but he timed the call well. At this hour on a Saturday morning, Sandy would be out back on the stone patio of that big Birch Street home, reading the newspaper over a third or fourth cup of the strong black coffee his housekeeper kept hot in the carafe.

"Perfect timing," Sandy said. In the background was the muted rustle of paper, and in the foreground, that Meade voice filled with steel. "You're the person I want to talk with. Annie Barnes is back. You know that?"

"I do," Marshall said, delighted that Sandy had raised the subject first.

"Well, Aidan's in a stew. He thinks she's here to dredge up old business. You think so?"

The perjury business hadn't occurred to Marshall. But he was implicated in that, too. "Geez, I hope not. There has to be a statute of limitations on that. What does Lowell say?" Lowell Bunker was Sandy's attorney.

"He says what you do, that the statute has expired, so there can't be any prosecution. Personal assassination is something else. She could write that whole thing into a story, just barely disguised."

"But why would she have to come back here to do that? She knows all the facts."

"She wants more. She can't make a book out of that one incident, so she's back here fishing for dirt."

"On what?" Marshall asked. He wasn't stupid; he wasn't volunteering the business about the pills unless he had to.

"My guess is the mill. She'll accuse us of all sorts of things, and she won't be able to prove any of it, but she'll go on television shooting her mouth off about art mirroring life, and the talk alone can bring the EPA. She's a dangerous woman. There's a good reason they've always compared her to Grace."

Marshall fed off Sandy's strength. "What can I do?" he asked with crusty sternness, just like the stalwart police chief he most often imagined he was. *Stalwart*—as in "tough as nails"—had been in a puzzle of his not long before. He liked that word a lot.

"Hassle her," Sandy said. "The way I see it, the sooner we let her know she's not welcome here, the sooner she'll leave. If she keeps at it, I have an option or two on my end, but before that, let's see what you can do."

Marshall was not a mean person. He saw his job as being reactive rather than proactive—two more good words. But now he had a mandate from Sandy Meade, and Sandy was a man to please.

Finding Annie was easy. The pale green van with Miss Lissy's logo on the door was nearly as hard to miss as if she had been in the convertible, and with Annie popping in and out of stores, driving from one end of town to the other—Sandy was right, she was definitely fishing—Marshall had her pegged in no time flat.

For a time, he followed her, staying the proper distance behind, but watching every move she made. Given the slightest provocation, he would have pulled her over and given her a ticket. But she stayed within the speed limit, used her blinker at every turn, even stopped for two kids in the crosswalk in the center of town. He thought for sure he would be able to get her when she made a right at the corner of School and Oak; half the town slipped around that corner without coming to a complete stop, as the stop sign dictated. But she did.

Unable to hassle her this way, he tried another tack. It was a more backhanded one (*furtive* was a more subtle word), but again, given Sandy's mandate, Marshall felt justified. He became an investigator, visiting those places where she had made stops and fishing for what *she* was fishing for.

"So," he said to Jim Howard, the bottle sorter at the town dump, "I saw Annie Barnes here earlier. What'd she have to say?"

"Not much," said Jim. He was a sad-eyed

and silent man, not personable at all, which was why he was a bottle sorter, not a cop, Marshall mused.

"She didn't ask questions?"

"Nope."

"Didn't say anything?"

"She said 'hi.'" He returned to his work.

"Did she talk with anyone else?"

He shook his head. Then he spotted a brown bottle in the midst of the green ones in the large bin, and pulled it out.

"Well, let me know if she does, okay? She's clever, that one. Thinks if she asks people like you, you'll tell all for the sake of the glory. She's writing a book, you know— or trying to."

Marshall repeated the same line a bit later at the gas station, where he found Normie Zwibble to be just as naive as Jim Howard.

"She's writing a book?" the mechanic asked in disbelief. "That's cool."

Marshall knew admiration when he saw it. Needing to nip it in the bud, he was deliberately cruel. "Not if she writes about an overweight grease monkey who spends half his pay on lottery tickets. Who do you think people will point at?" They would point at Normie, and that would mean trouble. Everyone knew Normie played the lottery—everyone except his parents, who believed gambling was evil and who, moreover, believed that what money Normie didn't have in his pocket at week's end had gone into the collection box at church. No matter that Normie was over thirty. He still lived at home, still worked for his daddy, still went to every town dance with his hair slicked back and his face filled with hope.

That face had gone pale. "She wouldn't do that," he protested.

"Oh, she would," Marshall said. "If I were you, I'd steer clear of Annie Barnes. And I'd tell my buddies to do it too," he tacked on and was immediately pleased with himself. Getting other people to do his dirty work was a brainstorm.

So he tried it on Marylou Walker at News 'n Chews, but only after he bought some almond turtles while waiting for a pair of *Peyton Place* tourists to leave the shop. Marylou was in her late forties, the second generation of three running the store. "I saw Annie Barnes in here before," he remarked when he finished the last of his third turtle. "What'd she buy?"

"Chocolate pennies," Marylou answered

with pride. "She's been in here nearly every day. She says she's missed our pennies. Washington doesn't have anything like them."

"She walked out with more than a bag of pennies."

"Uh-huh. She bought the newspaper, several postcards, and a map of the town."

"Postcards? And a map? Why a map?"

"I guess to see what streets are new."

"I'd say she was doing research," Marshall advised. "You do know that she's writing a book. How would you feel about being in it?"

"Why would I be in a book?"

"Because you're part of a family that has, let's say, an interesting history."

Marylou was a minute in following. Then she frowned. "If you're talking about my cousins, that's *ancient* history."

"It was incest," Marshall reminded her. "Do you want it revived in a book?"

"No."

"Then I'd make sure you don't get too friendly with Annie Barnes. That's what writers do, y'know. They sweet-talk you into telling things you wouldn't normally tell. I'd warn your family, too. Be careful, Marylou. Your parents were around when *Peyton*

Place came out. Ask them. They remember what went on in this town. Grace Metalious did it then. This is Annie Barnes's chance for revenge." Seeing that Marylou was listening, he knew he had made his point.

Taking another turtle from his little bag, he left the store, but he was thinking about other people who needed to get the message. Even past the *Peyton Placers,* there had been kids in the shop. They were all eyes over Annie Barnes. Hadn't he seen the DuPuis girl several times that day, covering much of the same ground as Annie?

Naturally, when he thought of talking with Kaitlin, she was nowhere in sight. So he headed to Omie's for a late lunch, and there, just coming out with his wife, was Hal Healy. It was a stroke of luck. Hal was even better. He could reach more people in an hour than Kaitlin DuPuis could in two weeks.

"Got a minute?" Marshall asked. It was his standard opening.

Hal smiled at Pamela. "Wait for me in the car, hon?" He kissed her forehead and watched her walk off.

Marshall watched, too. Pamela was a good-looking woman. He never would have paired her up with a formal guy like Hal. Sour grapes? Maybe. But not because he wanted Pamela himself. The thing was, he remembered when he and Edna had been the way Pamela and Hal were. It was really quite sweet.

"How can I help you?" Hal asked in his quiet, no-nonsense voice.

Marshall refocused. "Annie Barnes. Do you know the name?"

"Of course. She and my Pam were friends back in school."

"Do you know the whole business about Annie and Grace?"

"Yes. Marsha Klausson told me. I've read *Peyton Place* several times."

"Then you know what's inside. Hot stuff, huh?"

Hal grew red. He gave half a shrug, clearly uncomfortable talking about it.

"Yup," Marshall said, letting him off the hook. "Hot stuff. From what I hear, Annie Barnes is of the same mold, and quite frankly, I'm worried. She's been all over town. I see the kids looking at her like she's their newest idol. She could be a bad influence. School isn't in session for another couple weeks, but you're already holding faculty meetings. Think you ought to alert the teachers to what's going on?"

Hal thought about that for a minute. He actually seemed torn. "I'd hate to stir something up that would otherwise rest in peace."

"Rest in peace?" Marshall asked and cleared his throat to keep the frogginess at bay. "The whole town's concerned. Annie Barnes has a national following, and the media are vultures. She says the right thing in the right place, and we're smeared all over the news. Is that what you want?"

"How will my talking with teachers prevent that?"

"It'll keep them from talking to Annie Barnes. You see, that's the key. If we stand together and stonewall, she can't get a thing." Stonewall. He *liked* that word.

Hal must have, too, because he nodded. "That makes sense. Good idea. Thanks, Marshall." Arching a mischievous brow, he cocked his head toward his car, tossed a thumb that way, and took off.

Kaitlin DuPuis followed Annie around town most of Saturday, but she didn't get her opening until Sunday morning, and that was purely a fluke. She rarely went to church—hated sitting between her parents listening to all that talk of love, when there was none to the right or the left. As far as she was concerned, this was the most hypocritical of the hypocrisies in her home.

But it was important to her parents that people see them together in church, and because Kaitlin didn't know what would come of the Annie Barnes thing, she figured that this week she'd better go too.

Then she saw Annie in the parking lot, climbing out of her convertible and running around to the passenger side to help her sister Phoebe. Once they connected with Sabina Mattain and her family, Phoebe went up the stone steps and inside with them. Annie separated and headed around behind the church.

Kaitlin followed her parents up the steps and inside, then stopped and said the one thing that was guaranteed to buy her time. "I want to go see Gramma. I'll be right back." She went down the side stairs and out the lower door, which put her one walkway, a small span of neatly mown grass, and a white picket fence away from the graveyard.

Hurrying through the gate in the fence,

she proceeded down the stone path that led to her grandmother's stone, but she didn't stop there. She continued up over the rise, then down to a hollow ringed by woods. Though this was technically the back end of the cemetery, Kaitlin had always thought it the prettiest part.

Annie was at her parents' grave, and suddenly that gave Kaitlin pause. The stone with BARNES on it wasn't new, but the sod on Alyssa's side was. This death was fresh. If Kaitlin was intruding on Annie's privacy, it wouldn't help her cause.

Then again, for all Kaitlin knew, Annie had been here before. Kaitlin came often to sit with her grandmother, just as she had done at the nursing home in the months prior to her death. Her grandmother had loved her— *really* loved her. When the rest of the world seemed unbearable to Kaitlin, visiting here helped.

So maybe Alyssa Barnes was helping Annie, in which case Annie would be softer, more relaxed, and receptive to Kaitlin's plea. Or so Kaitlin hoped. Not that she had other options. Catching Annie alone in a place where all of Middle River wouldn't see was proving to be next to impossible.

She slowed when she neared. Annie sat on the grass with her legs curled to the side. She was wearing sunglasses. Still, it was clear she was looking at that stone with its newly etched addition.

Kaitlin waited quietly, hoping Annie would look up and smile. When that didn't happen, she took one step forward. She waited, then took another. She was about to clear her throat when Annie finally did look. After a moment of nothing, Kaitlin thought she saw her eyebrows go up. Surprise? Recognition?

"Hello," she said. Her voice gave nothing away.

"Hi," Kaitlin replied in a voice that gave *everything* away. It was shaky. She was totally nervous. Given the slightest provocation, she would have turned and run. Lest she do that, she rushed out the words. "I have to ask you a big favor. It's about the other night. I know you know it was us, but like, I want you to know that it'll be really, *really* bad if you tell anyone about it, and it'll be even *worse* if you put it in your book. See, my parents don't know about Kevin. They would go *ballistic* if they did because he isn't the kind of guy they want me with, and it doesn't do any good for me to say we're in love, because love doesn't mean a *thing* to my parents." She pressed her heart, which was where she was feeling the ache. "Kevin is *so* special to me. He's the first guy who has ever, *ev*-er been interested in me, and he didn't just do it for the sex, because if he had, he'd be gone, because I don't think I'm very good at that, either. He doesn't care that I'm not pretty. I mean like, he loves me— is that awesome?—*loves* me—like nobody else ever has except my grandmother, and she's dead too."

"The other night?" Annie asked. Her brows were knit now.

Kaitlin felt herself blush. "You know. The other *night.*" When Annie didn't say anything, just sat there looking confused, Kaitlin felt her first inkling of doubt. "You saw us, I know you did. You knew it was me the minute I walked into the front room at your sister's store." The inkling grew. "Didn't you?" Still Annie seemed baffled. That came through, sunglasses and all. "Like, if you didn't, why did you wink?"

"I winked because you were staring at me."

"But then it happened again at Omie's—I mean, you waved at me then."

"I recognized you from the shop."

Annie Barnes was perfectly serious. And Kaitlin DuPuis felt like a *total jerk*.

"Omigod," she whispered, then did it again, because she didn't know what else to do. Should she stay? Run? Dig a big hole and crawl in beside Gramma?

She was looking back, then ahead, then back again when she heard a vague "hey" from somewhere. When it came a second time and louder, her eyes flew to the source. Annie had taken off her sunglasses and was gesturing her closer.

Kaitlin didn't budge. "I don't believe I did this," she said in dismay and put a hand on the top of her head. "You didn't know."

"How could I see? It was dark."

"That's what Kevin said, but we were there in the headlights, and I was sure you had. So now *I've* told you." She wrapped her arms around her middle, but that didn't stop her eyes from tearing. "This is so bad. Like, I am such a *loser*."

"You're not a loser."

"What do *you* know about it?" she asked. She didn't care if she was being rude. Annie Barnes would do what she wanted, regardless of how Kaitlin behaved.

"I've been there," Annie said. "Want to come sit?"

"What I *want* is for you to forget what I said, but that won't happen, will it." No question there. Kaitlin brushed at the tears and looked back in the direction of Gramma. Gramma would know what to do next. Standing here, though, Kaitlin didn't feel any vibes.

"I won't tell anyone," Annie repeated.

Kaitlin should have listened to Kevin. He had been right. Now, she had messed up *everything.* "I'm cooked. I'm *done* for. My parents will accuse him of rape. Do you know how *awful* that'll be?"

"I said, I won't tell."

Kaitlin was embarrassed just thinking about it. She put a hand back on her head, like that would somehow keep her grounded. It was only then that she heard Annie's words. She looked back. Annie seemed serious.

Actually, she looked like she might have been crying. Her eyes weren't exactly red. But the area around them was shiny. Like, wet. And still she said, "Why would I tell? What's the point?"

Kaitlin could think of several, but mentioned only the most obvious. "Uh, your book." Her voice rose at the end. It was a no-brainer.

"I'm not writing a book."

"Everyone says you are."

"They're wrong."

"You could still tell my parents."

"Why would I do that? I don't owe your parents anything. What you do is not my affair. How old are you?"

"Seventeen."

"And you don't think that's a little young for sex in the street?"

Kaitlin stiffened. "See? You agree with them."

"No." She gave an odd kind of smile. "But I'm an adult. I'm supposed to say that."

Kaitlin tried to interpret that smile. "Yeah, are you supposed to tell, too?"

"Why would I bother? No one would believe me anyway. They'd say I'm envious, because I never pulled a backabehind. And they're right. I didn't have many friends, much less of the male variety. Like I said, I've been there." She patted the ground. "Are you sure you don't want to sit?"

Kaitlin *did* want to—not because she

needed to sit, but because something about Annie pulled at her. It had to do with that smile, or whatever it was. It suggested Annie wasn't one of them. But Kaitlin already knew that. She wasn't supposed to be talking with Annie. Word around town said she was dangerous.

She didn't seem dangerous to Kaitlin, at least not sitting here at her mother's grave. She looked . . . sad.

Of course, it could be an act. It could be that she would leave here and tell the first person she met that it had been Kaitlin DuPuis and Kevin Stark making love in the middle of Cedar the night she had blown into town. But she wasn't getting up to leave and do that. And now that Kaitlin was here, she didn't see the harm of staying. It was better than sitting with her parents. It could even help. If she and Annie became friends, she might be able to convince Annie not to talk. Wasn't that one of Kaitlin's mother's pet thoughts—win 'em over, then call the shots?

She started forward, but stopped. "Are you sure you want me here? Aren't you, like, talking with your mom?"

"No. I'm just sitting. It's lonely. I'm not exactly Miss Popularity around town."

Kaitlin knew how *that* was. It had only been the last two or three years that she'd had friends—at least, friends that mattered. Coming the rest of the way, she lowered herself to the grass.

Annie gave her the smallest smile, before looking back at the gravestone.

"Are you sure I'm not intruding?" Kaitlin asked.

Annie nodded, seeming content with the silence. And so, for a short time, was Kaitlin. Birds were making noise in the woods, but not so loud as to drown out the distant strains of the church hymns. Kaitlin enjoyed the hymns from here. They helped clear her mind, helped her focus.

"And you really won't tell?" she asked Annie.

"I won't tell."

"You didn't already tell your sister Phoebe? Or James Meade? I saw you talking with him at Omie's."

"How could I tell either of them? I didn't know it was you."

"Will you tell them, now that you do?" Kaitlin asked, cursing herself *again* for her own stupidity.

Annie looked at her. "There are more important things. I don't even know your name."

Unfortunately, everyone else in Middle
River did, which meant that even if Kaitlin re-
fused to give it, it was easily found out.
"Kaitlin DuPuis."

"Kaitlin. That's pretty."

"Uh-huh. It's pretty and light and perky—
all the things my mother hoped I'd be that
I'm not."

"Why are you so down on yourself?"

"Because it's the truth. I'm my mother's
biggest disappointment. Well, next to my
dad."

"I'm not touching *that* one. About the
mother thing, though, maybe we're all destined to disappoint them. We just can't be
what they want."

"You weren't?" Kaitlin asked in surprise.
"Why not? I mean, like, look at you. You're
wicked successful."

"My books sell. That's not everything."

Kaitlin thought about it. No, selling books
wasn't everything. Still. "You don't seem so
bad."

Annie made a throaty noise. "Not as bad
as they say?"

"I'm sure they meant when you were little," Kaitlin put in. "You know, before you left here. I also thought you were ugly."

"I thought so, too."

"But you're not at all. I mean, you're really beautiful. I'd give anything to look like you. I got all of my dad's bad features—bad nose, bad hairline, bad skin." She pointed at her eyes. "These are contact lenses, and this nose is fixed. Same with my teeth and my jaw. I had *the* worst receding jaw, but they fixed that with a retainer too. We've done everything we can to make me look a little attractive."

"I think you're *very* attractive."

"I will *ne*-ver be very attractive. I mean, all we can do is patch things up and then watch for the bad stuff popping up again in my own kids."

"Who says that?"

"My mom. She tells me I'd better marry someone rich to pay for plastic surgery for my kids."

"Are you serious?"

"Oh yeah. My mom's into beauty. She's done everything she can for me. She's harping on low-carb now, but she's just about

given up on the weight thing, because packing it on is just like my dad, too. I don't know
if South Beach will work. Nothing else
seems to. I mean, if we're at Omie's and
everyone's ordering stuff with fries, am I
supposed *not* to order that? It's like I'd be
wearing this, this scarlet letter on my forehead—*F* for fat."

"You aren't fat."

"I am. Ask anyone."

"I wouldn't have singled you out from your
friends as being fat. Don't put yourself down.
You don't want to be as scrawny as some of
those others."

"See?" Kaitlin caught her there. "You did
notice. I am bigger."

"Not bigger. *Normal.* Much prettier and
softer for it."

Kaitlin didn't believe that for a minute. But
she did like hearing Annie say it. Prettier and
softer? She certainly *wanted* to be that.

Annie was after something. She had to
be. Why else would she be passing out com-
pliments? It struck Kaitlin that they could
make a deal. Kaitlin could give her something else for her book, in exchange for Annie keeping quiet about Kevin and her.

"I know things about this town that you'd probably like to hear," she said quietly. "I'd be willing to trade."

But the words weren't even fully out when Annie was shaking her head. "Your secret is safe with me. You don't have to pay to make it so."

"Are you sure? I mean, I really don't want them knowing about Kevin. And I will help you. I know a lot of stuff."

Another head shake. "Thanks. But I'm fine." She looked back at the grave.

"I'm sorry about your mom." When Annie simply nodded, she pushed herself up. "I should be getting back. I don't want my parents to come looking for me. They'll die if they know I've been talking with you."

A rustling came from the woods. Kaitlin looked there just as a man emerged. It was Mr. Healy, the high school principal. He seemed as startled to see them as they were to see him. Actually, Kaitlin was more than surprised. She was *appalled*.

"Omigod," she murmured. "Omigod, I'm outta here." Backing away from Annie, she raised her voice. "Hi, Mr. Healy. Just leaving. I was over there visiting with my gramma and I heard a noise down here, so I ran over. Just leaving. Gotta get back. Bye-bye."

Hal Healy watched her go. When she disappeared over the rise, he slicked back his hair with a hand. His tie was neat; he had checked that before. Running his thumb and forefinger along either side of his mouth, he turned to Annie. She was sitting on the grass, looking, frankly, provocative. He didn't doubt for a minute that she was a scandal waiting to happen.

"Why was Kaitlin here?" he asked.

"She was visiting her grandmother's grave."

"Yes, she did say that, but I'm not sure I believe her." He sighed. "I wasn't going to approach you about this, but finding you here says I'm meant to do it. I'm worried, Miss Barnes."

"Annie."

"You know what my job is. So much of what the kids in this town do is my responsibility. Now, here you come, back to town after fifteen years, and you're just about the age Grace Metalious was when she was shocking the living daylights out of folks living here."

"You don't need to tell me about Grace. I felt the effects of her book more than most."

"Then you'll understand what I'm saying. The kids here are impressionable. They're also apt to be a little in awe of your fame. If they see you snooping around, looking for titillating stories, they'll be fired up to give them to you. We have enough trouble trying to control them without that kind of instigation."

He had to hand it to her; she was tough. She stared at him with cold, hard eyes—and those, here at her mother's new grave. Her voice, too, was cold. "I'm not in the market for sex and titillation."

"Well, that's good," he said. "I'm pleased to hear you say it. These kids are our future. We have to make sure they're up to it. I wouldn't want our jobs to be made more difficult than they are."

She continued to stare, silent now. He figured he had made his point.

"That's all I wanted to say. Thank you for listening. I'm sure you'll keep this in mind. We only want the best for our children, agreed?"

"Totally," she said.

He nodded. Raising one hand in a half wave, he stepped around the Barnes burial plot and made his way up the hill, back toward the church. Once inside, he slipped into a free seat in the back row. He could see Pamela's head up near the front, her black hair distinct. He could also see Nicole DuPuis across the aisle, with Kaitlin dutifully returned to her side.

He was prepared to let it go at that. As the service went on, though, he began to worry that Kaitlin would tell her mother she had seen him coming out of the woods. He would rather Nicole hear his story than hers.

So he waited until the service was over and the congregants were milling around out front. Kaitlin went off with her friends, while her father went off with his friends. Hal positioned himself not far from Nicole's car. Unfortunately, though, before she could reach him, Pamela appeared, and the moment was lost.

Chapter 13

SOMETHING HAPPENED to me that Sunday in the cemetery, and it had nothing to do with what I learned about Kaitlin DuPuis or would later learn about Hal Healy. For starters, it had to do with having a good cry and accepting that my mother was gone and that nothing I did could bring her back. I could fight. I could rant and rave about what might have poisoned her. I could breathe hellfire and vengeance. But Mom had lost her balance, fallen down the stairs, and died of a broken neck. What good was vengeance? Would it put her back in the kitchen, waiting for me to come home?

Then Kaitlin came by with her fears, and I got to thinking about the ways in which we disappoint our parents. And Hal Healy had suggested I was a bad influence on Middle River's youth, so I got to thinking about that.

I was still at it when the service finished, and Phoebe found her way to where I sat. I watched her come toward me steadily, then not so, seeking preventive purchase on each gravestone she passed. I felt something then—an awakening—but it wasn't until she and I had spent another few minutes of quiet at our parents' grave, before the awakening took shape. It crystallized when Sabina arrived.

"I was worried," she complained, looking straight at me. "I thought maybe something had happened. You never made it inside at all."

Inside? Did I need to see the Hayneses, the Clappers, the Harrimans, or the Ryes? Did I need to see the *Meades*?

"No," I replied calmly. "I needed to be here more."

"I'd argue with that. The pastor's sermon was on respect within families."

"Sabina—" Phoebe warned.

"It was about understanding the needs of

others," Sabina went on, clearly not done with me, "even when those needs aren't your own. We saved you a seat. It was glaringly empty through the entire service. Okay, I know you're not into organized religion. Anything requiring conformity goes against your grain. But even *aside* from the pastor's message, which was really good, family is key. You sitting in church with your family would have gone a long way toward showing certain people in this town that you can fit in once in a while. Show them that, and *we* might not be on the hot seat quite so much. My friends are asking questions. My neighbors are asking questions. My boss is still asking questions. But you don't care. It just doesn't concern you." Face filled with disgust, she turned to Phoebe. "My family is waiting. Can I drop you home?"

Phoebe smiled. She might not have had the wherewithal to tell Sabina off. But she didn't cave in. "I'll stay a little."

Sabina strode off. And we didn't discuss it, Phoebe and I. She had made a small statement, and I was grateful.

I was not smug, however. Smugness was a luxury I couldn't afford where my sisters were concerned.

I had moved on—or back, however you choose to see it—and as we sat, my thoughts gelled. Seeking revenge might be noble if a Meade cover-up was involved. But it was also exhausting. It required the sustaining of anger, something I might have done at eighteen. At thirty-three? I couldn't.

Nor, though, could I let it all go. I couldn't bring my mother back, but, **TRUTH #6: She wasn't the only one involved.** I could fight this fight for Phoebe, who was clearly ill. I could fight it for Sabina, who talked of respect but didn't have a clue about other threats. I could fight it for Hal Healy, whose concern for the morals of the town's kids was almost laughable, when you thought about the possibility of poison in the air they breathed, the water they drank, and the fish they ate.

I'm not sure I can say that I heard my mother speak. I'm not even sure I can say that she would have wanted me to do this. Like Sabina, she was afraid of talk.

I did hear Grace, but only from afar. Grace didn't do graveyards. That said, she did like unearthing things that were buried. Oh yes, Grace was all for it.

Bottom line? Mom was gone. Phoebe was sick. And totally aside from Grace's need to shock people, I knew this was the right thing to do.

To: Annie Barnes
From: TrueBlue
Subject: Where do we go next?

That depends. What are your plans?

*To: TrueBlue
From: Annie Barnes
Subject: Re: Where do we go next?*

I haven't thought past finding out whether something in Middle River made my mother sick. No, I'm not planning a book. Isn't that what you're asking, again? And if not you, then everyone else, but the question is getting old. Are you willing to help me, or not?

That depends. There are ways besides a book to publicize a wrong. You could turn the information I give you over to *The Washington Post,* which would be no different from your writing a book. Same thing if you give it to your pal Greg Steele.

*I take it you don't like those options. Are
you getting cold feet?*

Cold feet? Not by a long shot.
Remember, I live here. I have even more
reason than you to want things fixed. But
here's my problem. If you do the wrong
thing with whatever you find, this town
will be transformed in ways your Grace
couldn't fathom. Book, newspaper,
nightly news—it doesn't matter how the
story breaks, but if you go for something
splashy, Middle River will be overrun not
only by the media, but by lawyers. Know
what happens then?

They swarm.

That's putting it mildly. Personal injury
lawyers come in droves, making wild
promises to every possible victim. They
organize their class-action suits and film
their front-page stories on the lawn of the
town hall, and they get their headlines
and their big trial and their hefty
settlement. Unfortunately, they're the only
ones who make anything on the deal.
Northwood loses a bundle paying

damages and expenses, and worst-case
scenario, goes bankrupt, in which case
the economy of the town goes to hell
right along with the jobs of the people
who live here. And the victims who are
supposedly receiving money for pain and
suffering? Once the lawyers take their
share and the court fees have been paid
and the rest has been divvied up
between all the people involved, any
single victim gets a pittance.

I take it you don't like lawyers.

Wrong. My college roommate is a lawyer.
I'd walk through fire for him. But he's the
first one to tell me to avoid litigation. So
that's what I'm doing. I want things fixed,
if not for me then for the kids around
here. I don't want the town destroyed,
and that's what will happen if you go for
headlines.

*I don't need headlines. I need answers. If
the answers warrant it, I want change.*

If that's all, then we're on the same page.
The thing is, once I give you information,

you can basically do what you want. Can
I trust you're telling me the truth now?

I ask you——can I trust you'll tell me the
truth? How do I know you aren't an
instrument of the Meades and won't
send me on a wild goose chase just to
keep me busy while I'm here?

Try this. Government regulations allow for
a certain amount of pollution. When a mill
like Northwood exceeds that amount, it is
required to load it into 55-gallon drums
and have it carted away to an authorized
toxic waste disposal area. This is costly. It
cuts into profits. On occasion, Northwood
used other methods.

What methods?

Your turn. Give me something. We're trying to establish a mutual trust here. Put
up or shut up.

I have just finished marking a map of the
town with dots. Each dot represents
someone who has been seriously ill dur-
ing the last five years. In some instances,

there's no pattern. In other instances, there are definite clusters. Like along the river. Hyperactivity, muscular dystrophy, autism——lots of trouble with kids on th'other side.

Could be genes. Could be coincidence. Could be toxicity.

What do you think?

I didn't hear back from TrueBlue, but that didn't alarm me. By the time of our last exchange, it was very late. I went to bed and again slept longer than I would have. But I was on vacation, wasn't I? What were vacations for if not to sleep late?

That set me up to go running at eight. Grace couldn't have been happier.

Good girl, playing with fire. You're very boring where men are concerned.

Excuse me, I demurred. You don't see me in Washington. I've dated some unusual men.

Unusual?

Impressive.

James Meade is something else. There's drama here. He's Archenemy Number One.

Actually, he's Archenemy Number Two. Aidan is Number One, based on past behavior, still unforgiven. But don't get excited—there's no drama. So I bump into him running. So what?

You know what. You like the way he looks.

Correction. I like the way he *runs*.

Same difference. Where is he?

We're not there yet. He has a certain route.

Why does he take that route? Is it near where he lives?

I don't know where he lives.

Haven't you asked?

No, I haven't asked. That would suggest I want to know, and I don't.

Is he married?

Not that I know of. There's never been a notice in the paper. There has never been mention of a wife, period—no picture of the two of them at a social event. Had there been one, Sam would have run it. He loves the visual.

What's James waiting for? What's wrong with him?

I don't know, and I don't care. I told you. I like the way he runs. That's it.

I had barely decided that, when I saw him emerge from the cross street ahead. I fully expected him to continue on as he had in the past. Instead, he looked directly at me and slowed. He ran in an oval until I reached him, then, without a word, started up again.

I fell into pace behind him. He must have wanted this, or he wouldn't have slowed, and I couldn't look a gift horse in the mouth. Flat-footed? He was. But he was good. When you run with someone better than you, you run better. It's that way with most sports, don't you think?

I wasn't let down. He ran at a speed that may or may not have been his usual but that was a challenge for me, and I kept up. I wasn't close enough to get the benefit of drafting, but he paced me in ways no one had since I had belonged to a running club years before. Okay, okay. There was an element of pride involved for me. He had issued a challenge; I was determined to meet it. But there was something poetic here. I was using him. I liked that idea.

Staying ten feet behind, I followed him through the back roads of town, and though there were few houses here, we were passed by several cars. Was I worried to be seen running with James? Not at all. My image in town was at rock bottom; I had nothing to lose. James's image was another story. It could be tarnished if he was seen running with me.

But this was his idea, wasn't it? At any time, he could have poured it on and left me in the dust.

I half expected he would, if only to put me in my place. It would have been a very Meade thing to do. Then again, he was a runner. I had come to think of runners as a notch above.

Indeed, he stayed with me—or let me stay with him—until we reached the intersection of Coolidge and Rye, where we had first connected. Then he pointed me off toward Willow, raised his hand in a wave, and continued on straight without looking back.

Tuesday we ran side by side. We didn't talk. He gestured when he wanted to turn, taking a slightly different route from the one the day before, and I was fine letting him choose. More so than me, he knew the roads that were best for running. That freed me to focus on striking the ground with the outside of my heel, on keeping my knees properly flexed, on modulating my breathing, and on keeping up with James. But I did it. Breaking off again at Coolidge and Rye, I felt proud.

He called the house that night. I don't know what he would have done if Phoebe had answered the phone. I didn't ask him that. Our conversation was brief.

"Are you running tomorrow?" he asked.

"Yes."

"Want to try off-road?"

I was game. My knee was bothering me a little. A dirt path would be more forgiving than pavement. "Sure."

"The varsity course at eight?"

"I'll be there."

We were the only two who were there, I realized as I drove through the parking lot to the very back and turned a corner to the strip at the edge of the woods, but that was no surprise. Varsity runners didn't do the cross-country course at eight in the morning, not during the school year, not during preseason, which this was. They would be here later. For now, the place was as empty as the path through the woods promised to be.

Not that I would have missed James, even if the lot had been full. He was driving the large black SUV I had seen him in once before. Tony O'Roarke had been at the wheel then, but there was no driver now. The windows were open—it was hot already at eight—and James was stretching on the grass not far from the start of the path.

I parked and joined him there, and I have to say, I felt a qualm then. Shy? I don't know. He had been bare-chested each of the previous days; today he wore a tank top. Somehow, that seemed more personal—as if, since he had known for sure today that we would be running together, he had given the matter of bareness some thought.

Okay. I know. He didn't want anyone getting the wrong idea—namely, me.

But covering up just that little bit didn't do a thing in terms of propriety, at least, not in *my* mind. His arms and legs remained bare—long and firm, with lean ankles and wrists—and the tank top didn't hide wisps of chest hair or the darker shadow under his arms. Or the shadow of his beard. Or his Adam's apple. James Meade was very male.

Then again, maybe he seemed more imposing simply because he was standing in place rather than running. Granted, he looked something like a heron, holding one foot back by his butt and balanced on the other. But he was impressive even on one foot.

Whatever, I felt vaguely intimidated. Greg and I had been at a media dinner once and I had been introduced to George Clooney. Okay, maybe George Clooney isn't *your* cup of tea, but he stirs something in me. James had the same effect now, possibly for a similar reason. He was a celebrity of sorts, certainly the star attraction of Middle River. Given the synergy between the town and the mill, he was the direction both would take once Sandy retired. In that sense, he was a powerful man.

Power was alluring. That was the intellectual me speaking.

The visceral me suddenly saw raw chemistry. I hadn't felt anything physical for Tom Martin, and with Aidan Meade, I had probably been too young and green to get past who he was. Not so with James. He was hot.

Shy? Intimidated? Baloney. I was *attracted* to him—which was the *stupidest* thing in the world. Was I a *masochist*? James was a Meade, of the same ilk as Aidan. Being attracted to James was just plain dumb.

I was that. I was also tongue-tied.

So I stretched. Make that—*we* stretched. I

went through my usual routine, all habit, which was good, because my mind wasn't on it. I didn't have to look at James to be aware of what his body was doing. Long legs stretching—torso bent over spread thighs—chest lifted by hands clasped higher than I could ever reach—head angling slowly from one side all the way to the other.

Foreplay? Oh boy. By the time we started to run, I was loaded with so much energy that I would have beat my own time even without James's pacing.

The path was narrow, so he went ahead, and I focused my thoughts on the run. Off-road was different from road running. It took more concentration simply because the terrain was less even.

By the way, the varsity course was on Cooper's Hill. *Cooper's* Hill. Ring a bell? If so, you are astute. Cooper's Hill is indeed home to Cooper's *Point,* the site of my humiliation at the hands of Aidan Meade. The point, as opposed to the hill, is a lookout over the town, reached by a path that ascends through the woods. It's an easy climb, even by flashlight at night—ten minutes, max. As for the hill itself, its only other attraction is a slope for sledders in winter.

The running course, on the other hand, is a favorite of cross-country skiers. It undulates gently around the lower part of the hill for a total of two miles. And if you're thinking that two miles isn't much of a run, keep in mind that two miles running off-road is the equivalent of three on a level surface in terms of time and physical drain.

That said, I was game for another go-round when we finished the first, and gestured this to James when he looked questioningly back. Yes, my knee was tired, but the rest of me was eager. Having gotten past the attraction thing—one cross over the path to Cooper's Point and I was cured—the varsity course was the perfect place to run on a hot, sunny day like this. Other than the length of grass that spanned the sledding slope, the path was generously shaded. Here we ran on a bed of leaves, pine needles, and dirt. Yes, there were exposed tree roots to navigate. I tripped over one early on, and barely managed to catch myself before James looked back. I didn't trip again.

The second round was more draining. I kept up, but was nonetheless grateful when we reached our starting point. James was covered with sweat—rivulets dripping down his face until he brushed them away with an arm, that arm and the other glistening, body hair plastered to skin—but I was no better myself. The hair that had escaped from my ponytail was glued to my slick neck, my face was gleaming, my singlet and shorts clung to sweat dripping beneath. We were both breathing hard, but I wasn't thinking about his body then. I was thinking that the run had been fun.

He must have been thinking it, too, because the look on his wet face was surprisingly pleasant. We stood panting for a minute, just looking at each other. And I grinned. Why in the devil not? If returning to the vicinity of Cooper's Point was a test, I had passed. I had kept up with James. He was staring at me, and I refused to look away.

After a minute, he gave a quick little head shake and went to his car. Pulling two bottles of water from a cooler in the backseat, he handed me one. I finished it off in no time, and gratefully took one of a second pair that he fetched. This one I held to my face; it felt delightful against my flushed and sweaty skin. In time, I closed my eyes, tipped my head back, and put the cool bottle against my neck.

When I finally righted my head and opened my eyes, he was watching me.

Actually, he was watching my breasts.

I cleared my throat. His eyes met mine. Was he embarrassed? No. But that was the power thing. The Meades oozed entitlement. They were shameless when it came to using people—and that was what this was. There was no way that James Meade really wanted Annie Barnes, unless it was tied to thwarting my mission here. But I wasn't falling for it. Like the saying goes—fool me once, shame on you; fool me twice, shame on me.

I wasn't being fooled. If anyone was a user, this time it would be me. That decided, I continued to look at him as I stretched. Oh yes, he was male. If I was drawn, what harm would it be to play awhile? If James could do it, so could I. Running was my Washington thing; as long as my contact with him related to that, I was strong.

Did I feel duplicitous? Not on your life. If anything, I took satisfaction in the knowledge that while I was running with James Meade, I was plotting to screw him.

Uh-oh. That was a poor choice of words. Clearly just a figure of speech.

But you get my point. I might have no longer been out for revenge. But if, as True-Blue implied, it proved to be the case that toxic waste had been recklessly disposed of, James Meade and his family had some answering to do.

We never did say much of anything that morning, James and I. He didn't look at my breasts again. He looked at my mouth, my eyes, my legs—and he seemed puzzled, like he hadn't expected that I *had* any of those things or that they would function like the same body parts on other women functioned. He looked confused, like he had never dreamed I could run, much less keep up with him.

Of course, with a Meade you never knew what a look meant.

But I wasn't virginal and naive, I wasn't employed by the mill, and I wasn't afraid of James. I thanked him for the water. That was all I said before I headed for my car.

Grace was the one who felt the need to talk. I was barely out of the parking lot when she lit into me.

What are you doing? she asked. She was clearly annoyed. *You don't just . . . toy . . . with a man like that. You go after him with all you have. You could have charmed him. You could have said something sweet. You could have told him he's a great runner. You could have batted your eyelashes, for God's sake.*

Batted my eyelashes? People don't do that nowadays.

If you want to play, play. Milk it, sweetie. This could be a central part of your book.

Book? What book? I'm not writing a book.

I think you should. But you need sex in it. Sex sells. Real sex. Down-home-and-dirty sex.

I'm selling just fine without that.

You'd sell better with it. Remember what happened when I finally sold Peyton Place? *My publisher made me add a sex scene between Constance and Tomas. I wrote it in her office in an hour, and I was not happy. But my readers loved it.* Peyton Place *sold twelve million copies. Have any of your books done that?*

No, I reasoned, accelerating as I turned left from School onto Oak, because times have changed. Virtually no book today sells twelve million copies. There's too much competition, too many other books, too many other diversions, like movies, DVDs, and ca-ble TV. Besides, in its day, *Peyton Place* was unique with its sex. Sex in books today is commonplace.

So what's your problem? Seduce James Meade, and you have a great plot.

I stopped where Cedar crossed Oak, let a car pass, then accelerated again. Grace was starting to annoy me. I am not seducing James Meade, I insisted. I'd only be burned. And I'm not writing a book.

You're a disappointment.

You're a pain in the butt.

I'm leaving.

Good. Go. I'm stopping in a second anyway. I want *The New York Times*. And chocolate pennies.

Okay, she tried. *Forget the book. Seduce James Meade, and you'll be able to get the dirt you need on the mill.*

That's disgusting, I mused.

It's the best way to get information. Done all the time in my day. You all are even more promiscuous, so what's the problem?

Leave.

I will. But when your search goes nowhere, remember what I've said.

Go!

I heard a siren and at first thought it was a warning—to me, to Grace, to us both. Then I realized it came from the car behind me. Car? Make that police cruiser, from the looks of the top bar with its lights popping and flashing.

Sirens and lights were rare in our town. Thinking something big must have happened, I pulled over by the barbershop to let him pass. To my dismay, he pulled over right behind me and killed the siren. The lights remained on.

I was trying to decide why that was so, when Marshall Greenwood approached. Leading by his middle with his back ramrod straight, he took his sweet time.

I looked up at him. "Hi."

"License and registration, please," he said in a voice that had grown more gravelly in the years I'd been gone.

I blinked. "Did I do something wrong?"

"You were speeding. License and registration, please." He hooked his hands on his belt, waiting.

"Speeding?" I echoed. "Here? I just stopped. How could I be speeding?"

"The limit's twenty. Sign's right up the block. You were going more than twenty."

I looked around. There had been a car

ahead of me, and there were others coming along now that were going at the same rate I had been going.

No. That's not right. They weren't going at the same rate. They had all slowed down to stare at me.

I cleared my throat. "*May*-be I was going twenty-five, but even that's pushing it."

"Twenty-five is above the limit."

"Are you using radar?"

"Don't need to. I know when someone's speeding. License and registration, please."

I had never gotten a speeding ticket in my life. Nor had I ever seen—ever *known* of anyone else to get one on this particular strip. "Is this a new crackdown or something?" I glanced around. There were men on the barbershop bench, men on the rockers outside Harriman's Grocery, men and women in the chairs on the lawn of the town hall. All were watching the goings-on—garish blinking lights were definite attention-getters—but not a one of them was crossing the street. "Was I endangering anyone?"

Marshall sighed. "That's not the point, Miss Barnes. The folks who live here, well, they know the ropes. It's you folks from away, come here and try to do things the way you want with no regard for the greater good. When you're in this town, you have to abide by our laws." He put out a hand, waiting.

Short of causing a scene, I took my license and registration from the glove box.

He was ten minutes in writing up my ticket. I'm sure most of the town knew I was getting it before I actually had the paper in my hand. Cars and trucks rolled past, drivers' heads turned my way. People went in and out of buildings, craning their necks to see. And those blinking lights remained on.

I sat, and I sweated—I was parked in direct sun. I weathered the heat of dozens of pairs of eyes.

Oh yes, he was doing this on purpose. Marshall Greenwood was Sandy Meade's puppet. But if Sandy thought to intimidate me, his scare tactics were laughable. Sitting there waiting for my speeding ticket under the eyes of every Middle Riverite who passed, I was suddenly inspired.

I did stop for the *Times* and some chocolate pennies, but if Marylou Walker was more chilly than she had been on previous days ringing me up, I didn't care. I was out of there in a flash, tossed the goodies in the car, and drove on home.

Fortuitously, Phoebe had already gone to the store. That meant I didn't have to talk low when I picked up the kitchen phone to call the New Hampshire Department of Environmental Services.

Yes, I know. I argued earlier that I didn't dare call the agency lest someone with Meade connections get wind of it. But that had been Friday; this was Wednesday. That was before TrueBlue appeared as an ally, before my epiphany at the graveyard gave me renewed incentive, before I had held my own side-by-side with none other than the powerful James Meade. It had been before Marshall Greenwood gave me a speeding ticket with *Get out of town* written all over it in invisible ink.

And it had been before my devious little brainstorm.

Chapter 14

I DIDN'T CALL direct. Rather, I called Greg's assistant in Washington, who knew all the little ins and outs of these things and who just happened to be a friend of mine, too. She patched me through on one of the lines that the network used when discretion was in order.

Didn't think that happened? Think again. Media people have all sorts of tricks up their sleeves. Making untraceable calls is the least of them.

I would far rather have taken the high road, of course, but the circumstances called for caution. How would you feel if you had been stopped in the middle of town for doing what everyone else *always* did, and then were made to sit there while everyone watched? Was Marshall Greenwood really checking my registration that whole time? Was he really checking to see if there were any outstanding warrants against me? No, he was not. He was making me wait for the humiliation of it.

For the record, I was not speeding. This was intimidation, pure and simple. Was *that* taking the high road? Marshall was the law, yet he made me feel helpless and harassed. Moreover, of all the people watching, not one came forward in my defense. They were turning against me, and no, this wasn't more paranoia on my part. Heretofore, their stares had been largely benign. Now they were censorious. Marylou Walker's coldness was a case in point.

What had I ever done to them? The injustice of it infuriated me.

If you've ever felt that, then you'll understand my resorting to subterfuge in making this call—and, as it happened, I didn't even have to lie about who I was. I introduced myself as a novelist who was exploring the issue of mercury pollution for a possible book, and in theory it was true. Though I wasn't planning a book on Middle River, who knew whether the book after the one I was now plotting wouldn't include a twist of this type?

The woman at NHDES never even asked me my name. Trusting and pure, she simply asked what I wanted to know.

"I'm interested in plants in this state that produce mercury waste," I began in a most general way.

"We do have those," she replied.

"Mostly paper plants?"

"That produce mercury? Not exclusively. But mostly."

"Do you monitor them?"

"Yes. Any plant that discharges waste of *any* sort into our waters is required to take out a permit. Then there's testing, monthly or quarterly. Water samples are taken from the point of disposal and are sent to the lab for analysis. We test the air, too, by taking samples of emissions from stacks."

"Who does the testing?"

"Water tests are done by the plant itself. When it comes to air tests, there's usually an observer from our department."

"You don't observe the water sampling?"

"Not usually."

This struck me as a case of the fox guarding the chicken coop. "What's to prevent someone from taking bottled water and claiming it's from the river?"

"The lab would see the difference. We know the chemical makeup of the river."

"What's to prevent someone from taking river water and simply filtering out the bad stuff?"

"If someone went to that extreme, it would suggest there was major pollution, and if that's the case, the air sample would show it. We also do annual inspections to see that plants are properly disposing of their hazardous waste. Disposal is key. Not only does on-site disposal need to be done to code, but everything exceeding the legal limit must be loaded into drums and sent to an authorized site. The mills pay for the disposal of this extra waste. They keeps records of how much goes where. They also pay the state up front for every pound of waste produced."

"What kinds of numbers are we talking about?"

"That depends on the plant."

"Say, Baxter Mills. Or Wentworth Paper. Or Northwood."

"Any of those could produce upward of six or seven hundred pounds in a given period.

At three cents per pound to the state, there would be an assessed fee of two thousand dollars."

Two thousand dollars didn't strike me as being prohibitive. Northwood could certainly handle that without folding. "Do you check the accuracy of what they report?"

"Only when there's a discrepancy from the norm."

"How do you know the norm?"

"We have records. There's a Web site, if you'd like to look." She gave me the URL and explained how it worked.

"And test results?" I asked. "Does the Web site show those, too?"

"No. Mills keep the results. So do we, for a while. Then they're archived."

Northwood wasn't about to show me its files. "Can someone like me access your archives?"

"Yes. But there's a process."

I knew how that went. Processes required applications. They also required time. Between the information I would have to give on the former, and the ability that the latter gave for someone from NHDES to notify Northwood, I would be cooked.

"Okay," I said, "in a slightly different vein,

exposure to mercury is known to create health problems. Does your department deal with those?"

"We know what the health problems are. Do we monitor them in specific cases? That would be difficult. Mercury poisoning is hard to diagnose. We hear about the occasional acute case, but that's it. A local health commissioner might know more. You could try contacting one of those."

"That's a thought," I said appreciatively. And it was a thought, albeit a moot one. Middle River didn't *have* a health commissioner. "You're very kind to be helping me this way."

"It's part of my job," she replied pleasantly.

"Then you don't mind another question or two?"

"Of course not."

I spoke hypothetically now. "If, say, a paper plant wanted to cover up the extent ot its production of mercury waste, what would it do?"

"It might bury waste improperly. Or falsify records."

"Could it cover up an actual spill?"

"Technically, yes. If the spill was a major one, though, there would be obvious repercussions, people suddenly very ill. In that case, a plant might try to cover up the extent of the spill, but they wouldn't be able to hide the fact of its occurrence."

"What about bribery?" I suggested.

She chuckled. "That would be hard, given the numbers of people who work at these plants."

Clearly, she didn't know the power of the Meades. "How could a person—my protagonist, for instance—uncover an attempted cover-up of a spill?"

"That's a little outside my jurisdiction."

"Take a guess. Pure speculation."

She was a minute thinking about it. "There would likely be memos, but they would be internal. He would need someone on the inside to provide him with those. Same thing if you want to show that records had been falsified."

TrueBlue. He was my man. He could give me this.

Knowing what I had to do next, I gave in to genuine curiosity. "How does one clean up a mercury spill?"

"Not easily," the woman replied. "A cleanup involves decontaminating everything that came into contact with the mercury—clothes, skin, flooring, machinery. Since mercury is heavy, it sinks into the ground, so in the case of a ground spill, a cleanup could entail digging up dirt and extracting whatever of the metal that seeped in. It's very costly, and it's time-consuming."

"Do you believe that there have been spills in New Hampshire?"

"I know there have been."

"Any plants I know?" I teased and added a quick, "I could interview people there for a touch of authenticity."

"Fortunately for us—unfortunately for you—they've closed. That's the thing. A mercury spill can devastate a company. You understand why they'd try to keep it under wraps."

I checked out the URL she had given me. Much of the information provided was technical, such as identification numbers and codes, but when it came to waste description, total quantity, and weight in pounds, the data were clear. Up until eight years ago, Northwood had produced mercury waste. When I compared the amount with that of other paper mills, it was similar. Not that I cared about other mills. My sister wasn't sick in those towns. Only in Middle River.

To: TrueBlue
From: Annie Barnes
Subject: Possibilities

It appears that Northwood is not on a DES watch list, which is consistent with the claim that it no longer produces mercury waste. So I'm guessing that one or more of the following is true:

(1) Northwood falsified past reports to hide the amount of waste produced.
(2) Northwood falsified past tests to hide the strength of toxicity in that waste.
(3) Northwood improperly disposed of waste.
(4) Northwood covered up the incidence of spills.

Do you have——or can you get—— information on any of this? Ideally, I'd like copies of internal memos.

I sent it off with a sense of satisfaction. Moving ahead on the mercury front was key.

Right behind it was Phoebe. She wasn't getting any better. I had made reservations to fly to New York with her, but she was still worried about Sabina.

I wanted Sabina in my corner. I really did. I figured she would continue to fight me on the mercury thing, but I was hoping to reach some kind of agreement on Phoebe.

And then there was Tom. I owed him an update on what I was doing. I also wanted him to give me the name of a doctor I might take Phoebe to see in New York.

Killing two birds with one stone, I phoned my friend Berri, the one in Washington who had helped me plan the dinner I made at Sabina's last Friday.

"Hey," I said, delighted to hear her voice at the other end of the line. "I'm so glad you're home." I meant that, and it had less to do with menu plans than with warmth. I had been in Middle River for a full week now. My reception was getting cooler by the minute.

"Annie, how *are* you, sweetie?"

"Better for having you pick up the phone. It's lonely up here."

"Lonely?" she teased. "With your sisters and Sam and Omie and all the other people you've warned me would be watching what you're doing?"

Berri had been an outlet for me during

moments of doubt prior to my leaving Washington. I had prepared her well—prophetically, actually, given the earlier scene on Oak. People were certainly watching me. *Loads* of them.

"It's not the same as meeting you and Amanda for coffee. I feel like I've been gone a month."

"Me, too," she said. "By the way, we're all reading *Peyton Place*."

"You are? Amanda and Jocelyn, too?"

"All three of us."

"That's fabulous," I said in excitement. I was also touched. My friends were busy women with limited time to read. One book a month was pretty much their limit, and it was usually a current best seller. The only reason they would have chosen *Peyton Place* was me. That expression of devotion couldn't have come at a better time. I was definitely suffering from a need to feel loved. "So, what do you think?"

"I *like* it. I mean, I'm only into the first fifty pages, but she's a good writer. I didn't think she would be."

"Thought it was just a trashy novel?"

"*Yes*. We're meeting next week to discuss it."

"Oh no," I cried. "Wait'll I get back!" I wanted to be in on that discussion.

"Preliminaries are next week. Trust me, we won't have finished it. Amanda's work schedule is light, but Jocelyn has to have syllabuses ready in two weeks. "Me, I have a diversion, too." Her voice lowered in excitement. "I met a guy."

This wasn't news. Berri was *always* meeting guys. When I laughed and said that, she said, "I mean, *met* a guy. His name is John. He is very smart, very handsome, and very cool. He was at the function I ran last night. I'm seeing him Friday night."

Berri was a professional volunteer. Last night was the Kidney Foundation, if memory stood me well. "What does he do?"

She snorted. "This is Washington, so he's a lawyer. But he's not like any other lawyer I've met. He has long hair, for one thing. And an earring. And a tattoo." She tacked on the last with something I can only describe as pride.

"Where's the tattoo?"

"I don't know. Obviously, I saw the earring and the hair, but when I went looking for the tattoo, he got all smug. He said he didn't *do* that on the first date. I mean, he's just adorable. And a good person. He does legal work for the foundation pro bono. And he wants kids."

"How do you know that?" I asked.

"He said it right off the bat, because there was this *impossible* child that someone brought to the event, and John was the only one who could get the kid to keep his hands off the hors d'oeuvres. I mean, the child touched every single one on the tray. When I said John was a natural, he said he adores his nieces and nephews and can't wait to have some of his own. He's twenty-nine."

I caught my breath. "A younger man." Berri was thirty-three, like me. "Oh, Berri, I hope it works for you. I'll keep my fingers crossed." And I would. Berri wanted to be married. She wanted the house, the kids, the cars. She wanted *love*.

But then, didn't we all?

"How was the tenderloin?" she asked.

"Incredible. The whole meal was incredible. They loved it. That's one of the reasons I'm calling. I need another meal plan."

"Who's this one for?"

"Some of the same people, some different. I want to make something here that I can bring to different houses ready to eat."

"A one-dish meal?"

"If possible."

"Give me five. I'll call you back."

The phone rang in three. "That was fast," I said without a hello. "You are such a sweetie. Do you know what it means to me to have someone like you?"

There was a split second's silence, then a low and amused, "Actually not. Tell me."

It was James, of course. I would have recognized his voice even if there were dozens of other men calling me here, which there weren't. Even kept down in volume, his voice was deeper and more resonant than that of most men I knew, and it did the same thing to me as the sight of him did. Chemistry? Big-time.

I would have been lying if I said I wasn't pleased that he'd called. I was glad, though, that he couldn't see the flush on my cheeks or hear the thud-thud-thudding in my chest.

What could I do?

I laughed. "Sorry about that. I was just talking with a friend in D.C. and she promised to call me back. Did you *hear* what happened to me on the way home this morning?"

"I did. Marshall takes his job too seriously sometimes. I assume it's just a fine."

"It's more than that. It's the principle of the thing. He's had his eye on me. He's been waiting. So what's next?"

"Next is you go down to town hall and pay the fine so that no one can find fault. Then you drive very, very carefully."

I had been thinking of what *Marshall* would do next. As for the other, of course, James was right.

He went on, still in that low voice, and it struck me that he was trying to be discreet about the call. "I have a meeting about to start and another one early tomorrow, so running before work is out. Can you make it after work? Say, seven?"

"Sure," I said with remarkable calm. "Same place as today?"

"If that works for you."

"It does."

James's call had made my day. It gave me something to look forward to, something to *plot* about as I waited for TrueBlue's reply. I logged on soon after Berri called back, but there was nothing then, nor when I checked at midday. I checked again when I returned from Harriman's, and again, later, when the chicken pot pie went into the oven.

Yes, chicken pot pie. Make that *pies,* plural; there were actually three biggies. But these were no ordinary chicken pot pies. Berri had merged several recipes to come up with what she called a *Tuscan* chicken pot pie, which contained, among other things, artichoke hearts, black olives, sun-dried tomatoes, and garlic. Had I ever doubted her skill, those doubts would have vanished when the kitchen filled with unbelievably good smells.

The crusts had puffed and turned a warm amber when the timer rang. I took the pies from the oven and put them on the top of the stove, and the smell was even better then.

I checked the time. It was nearly four. Yes, I could drop Sabina's at the house. But I really wanted to go to the mill. I hadn't seen it in fifteen years. I was curious.

I opened the lower cabinet, where Mom had always kept the insulated bags she used for toting hot food to church suppers. When they weren't there, I checked each of the other lower cabinets. I checked the higher cabinets. I checked the pantry.

I could have called Phoebe, but I was trying to leave her alone for the day. I had been in and out of the store Monday and Tuesday. I was tired of running menial errands. I was also tired of seeing Phoebe struggle to hide whatever it was that ailed her, which also raised the issue of whether, given her current state of mind, she would remember whether the insulated bags had been tossed out in the kitchen renovation, or if not, where they were.

I drove back into town. Harriman's had what I wanted. I bought two. Ignoring the pointed stares of the clerks, I paid for them, returned to my car, and drove home.

Five minutes later, with two of those chicken pot pies duly packed and riding shotgun with me, I approached the mill.

So here we are, at the mouth of the ogre's cave, I mused. It didn't look evil, that's for sure. The entrance was actually lovely. Indeed, the whole place was quite attractive, but I'm getting ahead of myself. Let me paint the picture of what I saw; it's what I do well.

The mill complex was located north of town, which, given what we now know about toxicity flowing downstream, was not the wisest choice. At the time of its founding, though, the roots of the town were already in place, and the land south of those roots lay on solid granite. Hence, by default, the mill was built to the north.

We thought of Benjamin Meade as the founder. That was the story taught in our schools and reinforced during the civics course, when third graders were given a tour of the mill, but it was actually Benjamin's father, Matthias, who got the thing going. In the early 1900s, it was strictly a lumbering operation. Forestry done farther north floated logs to Matthias's mill, where they were skinned and cut into wood planks for housing in the cities farther south. By the 1930s, Benjamin had taken over, and in a short time had broadened the scope of operations enough to genuinely merit being credited as the founder of the modern mill.

Even then, though, it began small, a single structure on the river at the end of a road cut through woods, and this same road was what I saw now, albeit with a new, large, and—I have to say—tasteful sign. As it had a century before, the woods that rose around it were filled with hemlock and pine, balsam and spruce—all evergreen, thick and picturesque throughout the year and deeply fragrant in ways the mill itself was not. A stone entry had been added, waist-high and crafted of rocks displaced when the more recent of the buildings was built. The stonemason was always an Arsenault; Arsenaults had been doing stonework around the mill since Benjamin's earliest days. They were actually kept on the payroll. This was one of the little gems we learned in third grade. Did we appreciate it then? Of course not. Now I understood that this fact was yet another nail in the coffin of dissension that might mar the face of the mill. No Arsenault would report suspicious activity. Those you pay, repay you with loyalty.

Arsenaults were artists; their stonework, done without mortar, was a sight to behold. Here at the entry to the mill, the mosaic of rock ranged in color from amber to slate and in size from small, narrow slabs to pieces the size of a desktop monitor. I have watched stone walls being built and know the artistry involved. Here, those rocks were arranged not only in a way to make the wall sturdy, but to make it an interesting tableau of local stone. The finished product curved gently on either side of the road in a pastoral invitation to enter.

Enter I did, and without pause. There was no guard's station here. The Meades didn't need a guard, but ensured their safety in more subtle ways. I guessed that there were cameras mounted on the trees, with a security guard monitoring the comings and goings. I wondered what he would make of my car.

The road was wider than I remembered it, a change made perhaps in recent years to accommodate the Meades' SUVs. The service road would be at the other end of the complex and wider yet to accommodate the gargantuan semis that carried the mill's goods from the shipping bays to the world.

From these beautiful stone walls, the road stole through scented woods for a quarter of a mile before the first of the buildings appeared. They were attractive redbrick Capes with a Colonial feel—mainly single-floor structures, they had tall doorways with pediments atop, dormer windows in the roof, and white shutters and columns out front. No single building was large, but there were many of them, added as the need arose. This was the Administrative Campus, the sign said—don't you *love* the word *campus*?—and beneath the heading were a flurry of arrows. One pointed to the building that housed the sales department, another the marketing department. Individual arrows pointed toward the executive offices, product development, and the Data Center.

Sabina would be in the last, but I didn't immediately drive there. Taking the scenic route, I followed the road around and between these buildings—and yes, I did give a special look to the one housing the executive offices. James's SUV sat out front. Make that Aidan's SUV, since there was a child seat in the back. There was also a large, dark sedan with tinted windows that had Sandy Meade's name written all over it.

The building that held the Meades' offices was similar to the others in its redbrick and Cape style, but the roof was higher and the dormer windows not only for show. This building had a spacious second floor that was largely glass from the back. Sandy Meade had his office here, beside the conference room where he ruled. He claimed that the second floor gave him a view of the river that he wouldn't have if he were downstairs. We all suspected he simply liked being raised above the rest.

With the exception of that second floor, the executive offices matched the other buildings in appearance. Trees and shrubs were well tended, a testament to Northwood's full-time grounds crew. Lawns, newly mown and still bearing the mark of the mower and a sweet, summer-warm smell, were dotted with white lawn chairs. Lilac and rhododendron, their blooms long gone, softened the expanse of brick between windows. There wasn't a weed in sight.

The guts of the mill lay ahead, but they were hidden behind trees. First came a trio of buildings that were as charming in purpose as they were in style, starting with the Clubhouse. Wholly subsidized by the mill, the Clubhouse was a meeting place for town events. I remembered my mother spending time there in the formative days of a group called the Middle River WIBs—Women in Business. Meetings were held over dinner, the latter provided by the mill in the form of a bribe. No one particularly wanted to drive all the way out there to meet, but the promise of a free meal—and a gourmet one at that—brought the WIBs. Other groups followed suit, but only after the place was rebuilt after one of those gourmet meals caused a fire in the kitchen that quickly spread to the rest. No one was injured in the fire, and the rebuilt Clubhouse had every safety feature imaginable. According to the *Middle River Times,* most every major civic group held its meetings here now. Northwood catered them all.

Stifling dissent? Oh yes.

Opposite the Clubhouse was the Gazebo. Embedded in trees, it overlooked the river and was totally charming. Many a Middle River mill couple had married there. The Meades donated flowers and champagne. Wasn't that nice?

And finally, ahead just a bit more, was the Children's Center. Northwood was in the forefront here, opening an on-site day care center before it was the thing to do. Parents paid a small fee, with the rest of the costs picked up by the mill. Can you imagine the gratitude those parents felt for the Meades?

The working plant was so close it was frightening. On certain days there was a faint sulfur smell, but I didn't catch that today. Perhaps I was too focused on the sudden fall-off of the woods and the emergence of those redbrick buildings, risen like a phoenix from the ashes of the big fire. That red brick notwithstanding, there was no charm here. These buildings were functional—large square and rectangular things that housed the machines that produced paper from logs. As a third grader viewing them, I had thought they were cavernous. Looking at them from the outside now, parked at a guard station, I still found them huge. There were also more now than back then, a tribute to Northwood's growth. Looking down the main drag, I saw one redbrick building after another. I also saw people on foot and in cars. The day shift was finishing its work, leaving the plant on the service road.

"May I help you?" asked the guard.

I took in a deep breath. "Actually, I think not. I'm going to the Data Center." Shooting him my most brilliant smile, I shifted into reverse, backed around, and cruised toward the forest again. Was I afraid of getting close to the plant? Probably, and it had nothing to do with exposure to mercury. The Meades didn't like trespassers—and, yes, there was that little sign.

The Administrative Campus was far more welcoming. Pulling in beside Sabina's ageless Chevy in the small parking strip at the Data Center, I climbed from the car, lifted one of the insulated bags from the seat, and approached the door.

The Data Center was unique. For one thing, the air here was cool to accommodate the needs of the machines. For another, rather than four lavish offices, each with a smiling face near its door, there was a large open room on one side with three desks and, on the other side behind a glass wall, the server.

Two of the three desks were taken, the one farthest from the door by Sabina. She wasn't surprised to see me; someone had warned her I was here. I smiled at the other worker, a man, and went over to Sabina.

"This should probably go in your car, so that it stays warm."

Sabina rose quickly and led the way back outside. If I didn't know better, I'd have thought she wanted me gone. Either that, or not seen. Or not *heard*. Who knew how trustworthy her coworker was?

Funny, though, she didn't stalk ahead. Once I cleared the door, she walked by my side. "What's with all the cooking? Is it the old 'way to the heart through the stomach' theory?"

"Actually," I said, "yes. I don't know how else to do it."

"Does it really matter?" she asked without looking at me. Her car wasn't locked. She opened the passenger door and reached for the bag.

"Yes, it matters."

"What is this?"

"Chicken and veggies in a pie. Oh. Wait a second." I ran back to my car, reached into the narrow space behind the seat, and pulled out a loaf of Italian bread. I put it on top of the insulated bag that Sabina had placed on her seat.

"It smells good," she said and closed the door. She leaned back against it and looked at me, more puzzled now than annoyed. "Do you *want* something?"

I might have gone into the thing about wanting family. But I had already done that.

"Today, yes," and rushed out the words. "I want your approval. I've decided to go to New York with Phoebe, because I don't think she can handle the buying trip herself."

"Shouldn't Joanne be the one to go with her?"

"She needs Joanne here. It's either me with her in New York, or no one. She's already told me I know nothing about buying, and she's right, but I do know New York, and at least I can make sure she doesn't fall or get lost or make some kind of gross error."

"She isn't that bad," Sabina said with a tentativeness I hadn't heard before.

"I've been here a full week, and she's no better. It's starting to frighten her, but she refuses to see Tom Martin. So while we're in New York, I'm taking her to a doctor."

"Who?"

"I don't know yet. Tom's my next dinner delivery." I grinned. "I'm hoping he'll be pleased enough with the gesture to help us get an appointment with someone who knows about things like this."

"Things like this?" she asked with a grain of distrust. "You aren't still on the mercury thing, are you?"

"First, I'm on the Parkinson's and Alzheimer's thing."

Chewing on a corner of her mouth, she looked off toward the woods and nodded.

"I'm not telling her beforehand about the doctor," I went on. "She'll fight me on it, I know she will. When I do, I want to be able to tell her you agree with me on this."

Sabina took a breath and looked me in the eye. "I'm worried about her, too."

"Then we're in agreement about the doctor?"

"I still think it's psychological."

"If a doctor rules out the others, she may be willing to see your friend the therapist."

Sabina nodded and looked away again. This time, though, she grew alert. I followed her gaze. Aidan Meade was striding down the drive. He came right up to us, bold as brass, put his hands on his hips, looked from Sabina to me and back, and smiled. "Don't let me interrupt."

Don't let him interrupt? I mused. Planting himself right here?

I was thinking of a pithy response, when Sabina said, "You're not interrupting. Annie can't stay. She was just dropping off dinner for my family. She's a pretty good cook."

He gave me a smug look. "We're Betty Crocker now, are we?"

I might have said any *number* of things then, but none of them would have done Sabina any good, and she was my concern. We actually had a meeting of minds here. I cared more about cultivating that than cutting Aidan down to size.

Besides, I had a date with James—well, not a date in the traditional sense, but we were forging a bond. Runners were good at that. But I doubted Aidan knew. So we also had a secret, James and I. That gave me a sense of strength.

I touched Sabina's arm. "It's fully cooked, so reheating at three-fifty for ten minutes should do it." Without another glance at Aidan, I went to my car and drove off.

Chapter 15

AIDAN MEADE didn't like Annie Barnes. He didn't *trust* Annie Barnes. She was up to no good. He had known it from the start. In Middle River for a whole month? That spelled trouble. If he hadn't sensed it from her holier-than-thou attitude when he had seen her in town that first day, he would have known it from the odd things she did. And it wasn't just him. Marshall was concerned, which meant Sandy was concerned, and when Sandy was concerned, he took it out on Aidan. And now Nicole, putting him on the hot seat no more than an hour before. He hadn't been prepared for *that* one, not when everything was going so well.

"Got a minute?" she had intercommed from her desk as she often did when she had material to run through with him, but he knew something was up the minute she entered his office. For one thing, though she usually teased him about it, she didn't even seem to notice that he had his golf club in hand and was chipping balls into a Chip-Mate on the far side of the room. Nor, ominously, did she have a stack of papers in her hand.

Closing the door, she came right up to him and said in an intimate voice, "We may have a problem. I got a call from Hal Healy this morning. He's seen Kaitlin talking with Annie Barnes. He knew I'd want to know. We've both been concerned about Kaitlin, because she's had a real attitude lately. He's worried that Annie is not a good role model for the girls in town." Her eyes held his. "Me, I'm worried about something else."

"What's that?" Aidan asked. Kaitlin's attitude wasn't his problem.

"I'm worried Kaitlin knows."

"About *us*?" he asked in surprise. "How

could she know? We don't do anything outside this office."

"What about Concord? Or Worcester? Or
New York?"

"We were working."

She folded her arms over her chest. "Uh-
huh, working up a sweat without a stitch of
clothes on."

"Come on, Nicki, how could she possibly
know about those times?" he asked as he
set the club aside.

"Who knows—suspicion, rumor, guess-
work, spies?—but if she's angry at me, giving Annie Barnes the scoop would be one
way to hit back. She hates that I hate her father. No matter that she caught him cheating
on *me,* somehow I'm at fault. The problem is,
if word of what we do gets out, Anton will
use it. He's been waiting for something like
this—just waiting for me to slip up so that he
has an excuse to get out of the marriage
without being called a cad by everyone in
town."

Aidan gripped her arms and gave them an
affectionate shake. "You haven't slipped up.
You don't tell friends, do you?"

"*God,* no. I don't tell *anyone.*"

Of course, she didn't. She knew when she

had a good thing going. She was the best-paid assistant in the company, and she was worth every cent. "And you don't know for sure that Kaitlin knows," he reminded her.

"No, and I can't exactly ask. But why else would she be talking to Annie Barnes? What would my daughter have to say to that woman? Kaitlin's been distant lately. I could feel it rolling off her when she was sitting between Anton and me last Sunday in church. According to Hal, that was right after she talked with Annie."

"Get her to a therapist."

"And have her tell what she knows to someone else? Not wise, Aidan. You know how things are in this town. Confidentiality is a pipe dream."

Aidan was starting to tire of the issue. *He* hadn't made the mess of the DuPuises' marriage. "Talk to Kaitlin. Get on her good side. Buy her something. What about a little car?"

"Anton refuses."

"What do you want *me* to do?"

"Shut down Annie Barnes."

Aidan laughed. "Talk about pipe dreams."

Nicole pulled away from him. "This isn't funny." He reached out and pulled her back, but her eyes were flashing. "Aidan, this isn't funny. If Anton divorces me, I'm sunk. Sunk means without money, and that's the whole point of my marriage. I like where I am. I like what I do. I like what *we* do. If Anton finds out, this is done. You think your wife will put up with your playing around? You blew two other marriages because you couldn't keep your pants zipped. Didn't she make you sign a prenup giving her more money if you cheat? Or was that a story you told me, so I'd know to keep my mouth shut? Well, I have. So maybe the leak is on *your* end."

"You really are beautiful when you're angry."

"*Ai*-dan! *Lis*-ten to me!"

But he was feeling aroused. "There's no leak on my end, not yet, but soon. You are so sexy." He kissed her.

"Aidan," she protested against his mouth, but she didn't resist the kiss. So he deepened it. At the same time, he backed her against the door, locked it, and slipped both hands into her blouse.

She always wore a blouse, usually silk, and a bra, usually lacy, and there were times when he played with the feel of those fabrics against her, but he was impatient today. Freeing her breasts in a symmetrical sweep of his hands, he put a thumb to one nipple and his mouth to the other.

"I don't know what to do, Aidan," she said, but it was little more than a breathy murmur. She had her hands in his hair, holding his head close.

"Do *this,* baby," he whispered against her hard flesh. Reaching under her skirt, he lowered her panties, then his fly. In seconds, he was inside her, and she was ready for him that fast. She always was, which was one of the reasons their relationship worked. He wasn't interested in foreplay when the need hit him hard. The way she moved and the sounds she made told him she felt the same.

It was over quickly and here, too, this was good. She was satisfied, he was satisfied. Usually, that was that.

This day, though, she didn't let it go. After righting her clothes, she straightened and said, "The business about your prenup? The only reason I mention it is to remind you that you have a stake in this, too. It's in your best interest to make sure that Annie Barnes doesn't go public with even the slightest hint of what we do. If I go down, I'm not going down alone."

Aidan had been checking to make sure

his pants looked all right. Slowly he raised his eyes. "What does that mean?"

"It means that I refuse to be poor. If my husband divorces me because of what we do here, I'm assuming you'll help me out."

That actually sounded like a threat. "What are you talking about, Nicole? Nothing is going to happen."

"Good," she said, then smiled. "I just needed to say that." And she left.

Aidan stared at the door without moving. The longer he stared, the more annoyed he grew. He didn't like being threatened, especially not by a woman. He was irascible when his phone rang, and grew worse when the voice at the other end informed him that Annie Barnes was cruising around Northwood's grounds. When he set off to confront her sister, he was positively gunning for bear—all the more so when he found Annie right there, and then, as calmly as you please, she talked about delivering dinner, then got in her car and took off.

Annie was a loose cannon. He had no idea what she was planning to do. Sabina was another story. She was on *his* payroll. He couldn't control what she did on her own time, but he'd be damned if she would aid and abet the enemy on Northwood's dime.

"What did she want?" he asked.

"She brought dinner," Sabina replied, so much like her sister—he had never seen it before—right down to the calmness, that it ticked him off more.

"That was an excuse," he said. "She drove all the way up to the plant. Why do you think she did that?"

"I didn't know she had. Maybe she just wanted to see what's changed since she was here last."

"Maybe she's scoping out the place."

"Scoping it out? For what?" Sabina asked, sounding amused.

"Well, you'd have to tell me that," he shot back. "Don't tell me she's just curious about changes. This company means nothing to her. She's looking to cause trouble, Sabina—and I'm not the only one who thinks it. Even Hal Healy is worried."

"Hal? What does Hal have to do with Northwood?"

"Nothing. But he's seen her talking with some of the young girls in town. He's worried she's a bad influence."

"She's a best-selling author," Sabina reasoned. "She's a successful woman. That makes her a *good* influence."

"Not if she fires these girls up. Think of Annie Barnes, and you think of *Peyton Place*. Think of *Peyton Place,* and you think of sex. Let's face it. Your sister lives life in the fast lane. You have an impressionable daughter. Don't you worry about things like this?"

Sabina had the gall to laugh. "Of course, I do, but not because of Annie. I worry about the boys in this town. Look at you. We were in school together, Aidan. I remember the things you did."

Aidan had done nothing more than any other healthy, red-blooded male would do. Yeah, yeah. She would say he did do more because he was a Meade and could get away with it, and she would dredge up the case of Annie and Cooper's Point. But that was NA—not applicable—to the discussion at hand. "I didn't touch your sister."

"I know. You were too busy touching Kiki Corey. And there were plenty of others before and after. Come on, Aidan. Don't play the prude."

First Nicki, now Sabina. He didn't like it

when women bested him with words. "Am I your boss?" he asked.

"What does that have to do with this?"

"I deserve respect. I'm telling you that your sister is looking for trouble. I don't want you laughing in my face."

She didn't reply. Rather, her face grew blank. And that was worse. That was *insolent.*

"Control your sister, Sabina, or there will be repercussions."

Still she didn't reply, just stared at him with that blank look.

"If I feel," he said, "that your relationship with your sister raises a security problem for this company, I'll replace you."

"Replace me?"

"Fire you. Control her, Sabina. I want her the hell out of town."

Sabina watched him walk off. She had known Annie would be a problem. Hadn't she begged her to keep her nose out of Middle River business—hadn't she *pleaded* with her that first morning in Phoebe's house? She had told her what was at stake, and now it wasn't just theory. Thanks to her sister, Sabina might well lose her job. Annie was *the* most stubborn, *the* most bullheaded, ornery, impossible person she knew!

Simmering quietly, Sabina glanced back at the office. She couldn't work now. She was too annoyed.

And why should she work? It was five. Wasn't that her own assistant packing it in?

"See you tomorrow," he called over the roofs of their cars. She raised a hand in acknowledgment.

He left at five every day. She stayed later. Was it pride in her work? A sense of professional responsibility? The desire to please her bosses?

But now Aidan had threatened her job. That galled her. And he wanted her to make Annie leave town. That galled her even more. Annie had as much right to be here as Aidan did. She was nosy, but nosiness wasn't a crime. Sabina would absolutely not tell her to leave.

She might get a little closer to her. Talk with her more. Get a feel for how she spent her days. That would be productive.

Her eyes fell to the dinner that lay on the seat of her car. Aidan was right; Annie had come about more than dinner. She had come about Phoebe, which was actually a good thing. It was actually a *kind* thing. Phoebe needed help. It was time to admit that. If Annie was willing to get the ball rolling, how could Sabina complain?

Chapter 16

I WAS EXITING past those beautiful stone walls, leaving Northwood behind and heading for the Clinic, when I had a thought. It was after five. If Tom Martin had left for the day, my showing up there would create gossip with nothing at all to show for it on my end.

So I pulled over to the side of the road, took my cell phone from my purse, and called just to check. Sure enough, the answering service came on. "*No,*" I replied casually, "*no message, I'll just catch him tomorrow,*" and clicked off. I called Directory Assistance, asked for Tom's home number, and was automatically connected.

I recognized his voice immediately. It had an intrinsic warmth. "Tom? It's Annie Barnes. I cooked you dinner. Can I drop it off?"

His words smiled. "Cooked me dinner? That's *so* nice. When can you get here? We're starved—you know I have a sister, don't you?"

"I do. There's enough for four."

"Then maybe you'll eat with us?"

"I'll sit for a bit, but I promised to be home so that Phoebe isn't alone. Want to tell me where you live?"

He did that with total trust. There was definitely a rapport here.

Smiling, I was returning the cell phone to my purse when the police cruiser came up alongside. There were no lights now, no audience, just Marshall and me. My smile faded. It wasn't that I felt physically threatened being alone with him on a quiet road. Uneasy was more apt. I had passed a few cars, though saw none now. After our last run-in, I was wondering if the man would be better or worse without witnesses.

He didn't get out of his car, just called across the passenger's seat and out the open window, "Got a problem?"

That should be your line, Grace said in a huff.

I agreed, but knew better than to say it aloud. I simply smiled. "No problem, thanks." I started the car.

"We don't like people using cell phones while they're driving," he called.

"I agree. It's dangerous. That's why I pulled over to talk."

"We ticket people for talking while they drive."

Is he kidding? Grace cried, indignant. *These people talk on cells all the time. We didn't have them in my day. Far better, if you ask me.*

I wasn't asking her. She sounded like my grandmother would have sounded had she been alive. They were roughly the same age.

But cell phones were a fact of life—as was the power that went with Marshall Greenwood's badge. I might sass him all I wanted, but he'd only get me again. It was infuriating. *And* if I dwelt on how infuriating it was (as I had done after he stopped me in town), his power only increased.

So, still smiling, I said a pleasant, "I'll remember that."

"I'd advise it."

"Thank you."

He seemed to want to carry on the argument, but didn't know where to go with it. I had deprived him of opposition. It was the wisest thing I could have done.

He frowned, still thinking. Apparently realizing there was nothing more to say, he simply faced forward and drove off. It struck me then that Marshall Greenwood didn't have much experience being a tough guy—and that made me wonder why he was doing it to me.

Actually, I *knew* why he was doing it to me. I had struck a Meade nerve. That alone was reason for me to persevere.

I know. I know. There was the matter of James. But my running with James had nothing to do with the other. Then again, if he and I developed enough of a relationship so that I could milk him for information that would help me with the other, so much the better.

With Marshall Greenwood now out of sight, I started the car and headed for Tom's. He lived in a yellow Victorian not unlike our Victorians on Willow but on a far larger piece of land. I guessed he owned acres, much of them meadowed. At the front of the property was the house, surrounded by grass, the occasional shrub, and two enormous trees. A wooden swing hung from the arm of an oak; a large tire hung from the arm of a maple. A wraparound porch hugged the house, broken only by wide wood steps leading to the front door and to a side door. Not far from the foot of the latter was a picnic table with benches on either side. Potted petunias, a striking violet hue and evenly spaced, hung from the edge of the porch roof all the way around the wrap.

I parked at the edge of the road, lifted my insulated bag, and went up the front walk. I had barely reached the porch when the front door flew open and a young girl appeared. All smiles, she was dark-haired and pretty in an unpolished way. She had Tom's blue eyes and was equally lean, a fact that wasn't hidden by the overall shorts she wore. It was only when I got close that I realized she wasn't as young as I had first thought.

Still, her smile was infectious. "Hey," I said. "I'm Annie."

Though she continued to smile, she seemed to grow shy, hanging now on the post at the top of the stairs. I was wondering whether she talked much at all, when Tom came out the door. He wore shorts, a T-shirt, and sandals, and was handsomely tanned.

"You are a lifesaver," he said, trotting down the stairs to relieve me of the bag. "Mrs. Jenkins took Ruth down to the outlet stores in Conway today, so she didn't have time to cook dinner. I was about to open a can of tuna."

"Then I've *really* saved your life," I remarked. "You never know what's in tuna."

"Oh, I do." He unzipped my insulated bag. "Tuna steaks are iffy. Canned light tuna, eaten in moderation, is fine. Besides, I'm not pregnant, and neither is Ruth." He put his nose down and inhaled. "Chicken pot pie?"

"You're good."

"It smells incredible."

"It may need reheating."

He touched it. "Nope. Ready to eat. Are you sure you won't join us?" When I shook my head, he looked back. "Ruth. Come meet Annie. And see what she's brought."

Ruth came as far as the bottom step, where she sat. If she was hungry, she didn't show it. She didn't spare a glance at the bag, but continued to look at me.

"I'm pleased to meet you, Ruth," I said.

Tom gestured her over. When she gave a quick head shake, he led me to the steps. "If the mountain won't come to Mohammed . . ."

He settled close to Ruth, clearly not forcing the issue of a formal introduction. Gesturing me to sit, he put the bag on his lap. "You said you have to be back for Phoebe. How's she doing?"

I settled on the other end of the step with my back to the post. "Lousy. That's one of the reasons I'm here. I've tried to get her to see you, but she won't. I'm thinking that's because Middle River is . . . well, Middle River, and people will see her going to your office and begin to talk. So Plan B is to see someone in New York. I'm going down with her Saturday to help out with a buying trip. She's depending on me to organize things and make sure she is where she's supposed to be when she's supposed to be there, so I'll be able to get her to a doctor before she knows what I'm up to. Only I don't know who to see, and time is short."

"I know just the person," Tom said, as I had figured he would. "When do you want to see her?"

Her was even better. "Tuesday morning."

"This *coming* Tuesday?" He gave a sputtering laugh. "You only ask for the world."

"I know. I'm sorry. I've been putting it off, hoping she'd get better. But now even

Sabina agrees that something's wrong. And this trip is the perfect opportunity. If you can't—"

"I can," he said. "She a friend. She'll do this for me. For what it's worth, she's in on the cause."

"*Our* cause?" I asked, though unnecessarily. I hadn't talked with Tom since that first day at the clinic, but I sensed instinctively that we were on the same page, same paragraph, same line.

He nodded. "Judith is an expert in alternative therapies. She's helped me treat some of the people here in town with something called chelation."

"Chelation," I repeated, testing the word.

"It's from the Greek word for *claw*. A synthetic amino acid is used as a chelating agent. It enters the body and gloms onto toxic metals that may be present in body tissue, then pulls them right out. The body doesn't like this synthetic stuff. It can't see the metal under the glomming, but knows that the synthetic substance doesn't belong there. So it directs the whole business to the kidneys and expels it through urine."

The scientific explanation made sense. I still wasn't sure about the politics of it. "Do the people you treat with this know you're trying to rid them of a toxic metal?"

"I've explained it, albeit hypothetically— you know, *if* there's a metal, this will remove it. Obviously, I can't point fingers and make accusations about the origin of the metal. And my patients don't ask."

"Don't ask?" I was amazed.

"Don't ask. They're simply glad to be feeling better."

"Then it does work?"

"I've seen improvement."

"Did you ever suggest it to my mother?"

"Yes, but she wanted to give the traditional medicine time to work. Unfortunately, she fell down the stairs before we could know either way."

That saddened me. But again, my concern now was my sister. "I can make Phoebe agree to do this."

"Assuming Judith recommends it," he cautioned. "She'll do a complete workup to rule out other things."

But my thoughts were racing on. "If the cure for mercury poisoning works, isn't that proof of its existence?"

"No. Is it mercury? Or lead? Or another metal entirely? Lead can be detected in a simple blood test, not so mercury and some of the others."

"But if I can find a connection between the people you've treated successfully and something to do with the mill, isn't that proof?"

"That depends on what kind of connection you find."

"Can I talk with your patients?"

"I can't give you names. That would be a violation of confidentiality. I could call them and have them call you if they were interested in talking. But that raises the problem I cited last time we talked."

"Your position."

"Yes," he said, and his voice said he was making no apologies for that. I had to respect him for it.

I tried to bargain. "Okay. If I went down a list of people who've been sick, would you give me a yes or a no?"

"As to whether I've treated them? No. As to whether I feel they'd be willing to talk, I could do that. I wouldn't be speaking as a doctor, just as a resident of Middle River. I'd be guessing."

"Guessing is more than anyone else has been willing to do. I'll take it," I said and looked at Ruth. She hadn't moved other than to lean forward a notch when Tom sat down so that she could see me without obstruction. "Your brother is a good man." I looked at Tom. "Naturally, when I need it most, I don't have my list with me. Can we talk later?"

"Sure. You have my number."

I eyed the insulated bag. "You really will have to heat that up now. And I do have to run. Thanks, Tom. It's so nice to know you're a friend." I stood. "It's been nice meeting you, Ruth."

She didn't reply, just continued to look at me with what struck me as doe eyes.

Tom noticed it, too, and said as he walked me to my car, "She's been skittish around strangers lately, so that's a high compliment. I think she's in awe."

"Why?"

"She wants a sister, and you fit the bill. She likes the way you look."

"She seems very sweet. How old is she?"

"Twenty-eight. She was a late-in-life baby for my mom. A *very* late-in-life baby," he added sadly, then brightened. "But she likes living here. The city was too loud, too busy, too big. She doesn't do well with change, and the city is full of that. I couldn't control it.

Here, I can. My hours are more regular. I go to work, I come home."

"What happens when you attend conferences?" I asked. Some of those reported in the *Middle River Times* were in places he couldn't possibly do in a day.

"Mrs. Jenkins stays over. She's a godsend. It's working out well."

Phoebe professed to love the chicken pot pie, though I sensed she spoke more from an ingrained politeness than her taste buds. She had withered by the time I cleared the table, and was soon on the sofa in the den with the TV on and her eyes closed. I was loading the dishwasher when Sabina called to thank me for the dinner, of which, she said, they had eaten every last bite. She asked how Phoebe was and said that the more she thought of it, taking Phoebe to a doctor while we were in New York was the right thing to do. She asked what I was doing tomorrow.

Mention of tomorrow made me think of James, which created a perverse excitement that was distracting, so that it wasn't until later that I questioned Sabina's sudden interest. At the time, I simply felt good about the call.

I peeked into the den. Phoebe was dozing. Satisfied that I'd have a measure of privacy, I got out my notes and called Tom. The conversation was simple: I tossed out a name—Martha Brown, Ian Bourque, Alice LeClaire, Caleb Keene, John DeVoux—and Tom said yes or no, depending on whether he felt the person or his or her family would be approachable. There were no guarantees; we both knew that. But given the length of my list, it was a start.

Phoebe continued to sleep. So I set up my laptop in the kitchen, connected to the phone jack, and logged on. I skimmed past the spam and found notes from Jocelyn and Amanda. Saving them as a treat for later, I went straight to two others.

The first was from Greg. I opened it eagerly.

HIT THE 13,000-FOOT MARK IN ONE PIECE, BUT IF WEST BUTTRESS IS THE BEST ROUTE, I SHUDDER TO THINK OF THE WORST. BAD WIND AND SNOW HERE. IS THIS AUGUST? ME, I WOULDN'T BE ANYWHERE ELSE, BUT I'M GLAD YOU'RE NOT HERE. YOU'D HATE THE COLD. HOPE YOU'RE FINDING WHAT GRACE SOUGHT. SEND WORD. LOVE YOU.

Smiling, I shot back a note.

*I'm making progress along with a friend
or two, and things are actually better with
my sisters, which is good. The law here is
a problem; I'm beginning to feel harrassed,
which is not good. Am I finding what
Grace sought? What is that? Small-town
caring? Not yet. Acceptance and
respect? Not yet. Family? Maybe. I found
a running pal. You'd die if I told you who.
I'll keep that a secret, one of Middle
River's many. Climb safely. I love you, too.*

I hit "send," then, with bated breath,
opened a note from TrueBlue.

You're getting warm. I'm impressed.
   Yes, I have information on this end.
Here are a couple of dates—March 21,
1989, and August 27, 1993. Go through
your list of people who are sick and see
if any of them were at Northwood during
the week preceding those dates.

What happened on those dates?

First the names.

And if I give you the names and those "witnesses" suddenly start showing up face-down in the river?

O ye of little faith.

Thursday morning I went at it. I pared down my list to include only Tom's yeses, then pared it down again to include only those people whose ailments most resembled the symptoms of mercury poisoning. I did this before Phoebe came down for breakfast. I waited until she had a cup of coffee, hoping that would wake her up some. When it barely did, I went ahead anyway and asked her about TrueBlue's dates.

She couldn't remember. Had I really expected she would? But I did sense it had less to do with what ailed her than with the passage of time. Would *I* have been able to say what I did on a particular day, week, or even *month* more than a decade ago? A big event, yes. A milestone in my life, yes. An everyday occurrence, I doubt it. Would anyone?

Failure on this first test notwithstanding, I set off at ten o'clock with my map and the highest of hopes. In the ensuing three hours, the map remained, but the hopes faded fast.

Of the people I visited, half were either at work (yes, at the mill, apparently well enough to work) or unwilling to talk. The unwillings to talk included two mothers with ill children and a total of six men and women who had been diagnosed with either Parkinson's, dementia, neurological problems, or multiple bouts of pneumonia. Silent. All of them. And I tried hard. Yes, I was tipping my hand. But there was no law against my dropping by to talk with people in town, was there?

That said, Marshall Greenwood was watching. I passed him often enough—and he slowed and stared at me as we passed—to know that he was monitoring my comings and goings. That might well have explained these people's unwillingness to talk, because I certainly made myself innocuous in my introduction.

"I'm Annie Barnes," I would say to each. "My mother was diagnosed with Parkinson's disease last year." Or Alzheimer's. Or pneumonia. I tailored my approach accordingly. "I'm trying to find other people with the same symptoms to see if there may be a mutual cause. I understand that your husband has been sick." Or your wife. Or your son or daughter or mother.

Inevitably, I would get a yes, but that was the extent of the accommodation. "He's fine now," one said and closed the door in my face. Another asked, "How did you get my name? The paper? Well, Sam's a friend, and I'd like to help find your mutual cause, only I don't think there is one, and I'm really busy now." A third, of course, remarked, "I know who you are. I don't think I should be talking to you."

They were more factual than hostile. Several people asked who had sent me or whether I was working for a health organization. But none agreed to talk with me, until I reached the McCreedys. Remember that name? Omie had put them on the slate, and Tom had seconded the nomination. It was one in the afternoon by now, and Marshall was probably on lunch break, which may have explained their inviting me in. Then again, they did indeed have a "string of problems," as Omie claimed, so it was possible that they simply needed to vent.

Tom and Emily McCreedy lived in the same neighborhood as Sabina. Owners of a flower shop and nursery at the far end of Willow, they were in their midforties. Though neither came from families with histories of ill health, Tom had chronic kidney disease that zapped his strength and immune system, while Emily, who was already being treated for bipolar disorder, had just been diagnosed with adult-onset asthma. They had three children ranging in age from fourteen to nineteen. The oldest and the youngest—both girls—were healthy, but the sixteen-year-old—a boy—was autistic. He attended a special school, and though the state paid part of the bill, the McCreedys footed the rest. Given their own medical bills, it was hard.

The fact that I found them at home in the middle of a workday said something about the state of their health. Once I was in their parlor, they described their ailments in depth. They expressed bewilderment that two people who had been in perfect health not fifteen years before could now be so chronically ill. They spoke of their son with genuine anguish. And anger. They were convinced that something in the air had tripped something in their own body chemistries to cause malfunction—and yes, they had run tests on the air and materials in their home and in their shop. All had come out negative, though they didn't particularly believe the results.

When I asked what they thought might have tripped them up, they tossed out a smorgasbord of possibilities. Acid rain, airborne asbestos, bad well water—they were convinced that at some point there had been a problem, and that if the tests didn't show it now, the problem had either been eliminated or was being hidden. In any event, Emily pointed out with visible irritation, the damage was already done.

I asked about the mill.

"What about the mill?" Tom asked blankly.

"It produces toxic waste," I said.

"It's way on the other side of town."

"Do you ever work there?"

"Yes. We do floral arrangements for meeting rooms and conference areas."

"And landscape design," Emily added.

"Could there be something in the air there?" I asked, trying to carry over the theme they had raised earlier themselves. On one hand, it amazed me they hadn't thought of the mill on their own. On the other hand, they might simply have bought into the hype about Northwood's environmental good health.

"The air there is fine," Tom said. "The

Meades aren't sick, and they're there all the time."

"How long have you been doing their floral arrangements?"

"Twenty years. They were one of our first clients, and they've been consistently the best."

That didn't bode well for me. Their first and most consistent client? It was the loyalty issue again. That said, of all the people I had approached today, the McCreedys were my best hope. Every one of their symptoms could feasibly be traced back to mercury poisoning.

So I asked, "Do you have a record of specific dates when you were on the grounds of the mill?"

Tom deferred to his wife, who apparently kept the files but was beginning to look edgy. "We have financial records," she said. "They would show what jobs were done on what days. But I don't see the point of this question."

"What if there had been a toxic spill at Northwood on one of the days you were there?" I asked.

"We'd have known it. The Meades would

have cleaned it up and helped everyone who was exposed."

Yeah, right, I thought. "Do you know that the physical problems you've had are consistent with the symptoms of mercury poisoning?"

"I have kidney disease," Tom said, "not mercury poisoning."

"But where did you get it?" I asked. "And why you? Isn't that what you've been asking me?"

"Bad things happen," he replied, at which point I tried Emily.

"You're being treated for bipolar disorder. As I understand it, mood swings are typical of that. Did you know that they're also a symptom of mercury poisoning? Same with asthma. Same with the kind of neurological damage that can result in the birth of an autistic child."

Emily's expression hardened. "If you think we haven't thought about all these things, you're dead wrong. We've thought about everything. But I was pregnant with Ryan sixteen years ago, I was diagnosed as bipolar seven years ago, and the asthma diagnosis came just last year. So when might I have been exposed to mercury? *All* those times? And no one else in town has been? And none of the doctors are worried? And the mill has covered it all up?"

I explained some of what I had learned. "You didn't have to be exposed multiple times. One major exposure would have been enough. You might have felt sick afterward, like you had the flu, and then gotten better— except that the mercury would have re- mained in your body and settled in your organs, causing you chronic exposure." I looked at my pad of paper, though I knew the dates by heart. "I have evidence of two spills at the mill. One occurred in March of '89, the other in August of '93. Given your son's age, you were probably pregnant with him during the first. If you knew you were at the mill during the week prior to the twenty- first of that month—"

I was interrupted by the doorbell and an almost simultaneous knuckle-rap on the jamb. "You folks all right?" came the gravelly voice of the chief of police. He walked right into the house, entered the parlor with his stiff-backed gait, and eyed me with distaste. "Is she bothering you?"

"No," said Tom. "We're just talking."

"She's been working her way around town all day, bugging people about finding a cause for their illnesses, like we don't know how to take care of our own. Say the word, and I'll get her out."

"She's fine," Tom said.

Marshall looked at me. "Those others this morning weren't happy. They say you were harassing them."

I couldn't let that go. "If they say it, it's because you told them to. I never even *talked* with any of them. They said they didn't want to, so I left."

"Well, I'm sure the McCreedys would appreciate it if you left, too." He gestured with his hand. "Come on. Let's go."

"Am I under arrest?" I asked in disbelief.

"Not yet. Resist, and you will be. We have no use for rabble-rousers around here. I assume you know what that term means?" he asked, seeming pleased with himself.

I might be impulsive at times, but I wasn't a masochist. Quietly, I gathered my things and rose. To Emily and Tom, I said, "If I've bothered you, I'm sorry. If you'd like to talk more, I'm staying with my sister."

"Hopefully not for long," Marshall tossed back to the McCreedys as he escorted me out of the house.

* * *

I was furious. I reminded myself that my anger was what he wanted, but it didn't help. I tried thinking about my real life, tried thinking about my Washington friends and about my book, but still I boiled. What Marshall was doing was grossly unfair. I wanted justice.

I thought of calling Sam and demanding that he print an op-ed piece on police abuse. Only I knew he wouldn't do it. He would remind me that any such piece would be a direct charge against Marshall Greenwood, since he was the only policeman in town. He would remind me that Marshall was backed by the Meades and that I probably didn't want to take them on.

He would be wrong about the last. I certainly did want to take on the Meades—but about mercury, not Marshall. It struck me that I had to keep my focus and pick my fights with care.

The reality of the Marshall business? No one in town was going to want to take him on, either. He was *their* chief of police. They would be stuck with him long after I left. Much as I adored Mrs. Klausson and Omie, they weren't activists. Middle River didn't have many of those. Except TrueBlue. I might mention Marshall to him. But to what avail? TrueBlue was TrueBlue (i.e., anonymous) because he didn't want to publicly buck the tide. He was using me for that. By Middle River's standards, I was expendable.

Discouraged, I drove to the diner. I parked off to the side, under the shade of a big old oak so that the sun wouldn't beat directly on my seats. Parked there, my car was also less conspicuous, which felt like a good thing just then.

The smell of burgers on the grill and the whisper of the fans hit me just inside the door, along with Elton John singing "Candle in the Wind." Its lyrics had always haunted me. This day was no exception.

And wouldn't you know, just when I could have used it, my favorite booth was taken. A group of high school kids sat there, with others of their friends in the booths fore and aft. Summer vacation was nearing an end. This gathering was something of a last hurrah.

There were two free booths right up in front, but I wasn't in the mood to be quite so visible. Walking to an empty one in the back, I passed other townsfolk. They glanced at me, neither friendly nor not. Feeling distinctly alone, I ordered Omie's hash and an ice cream frappe. It wasn't healthful. But I needed comfort.

I know what that's like, Grace remarked. *They used to make fun of me for being overweight, but when you're desperate for comfort, what else can you do?*

A psychiatrist, I replied, would say you were desperate for love.

Was I?

Of course.

Are you?

I never thought so. My life is full. I'm happy.

My food arrived. I was midway through wolfing it down when Kaitlin slipped in opposite me. I had been too engrossed in eating to see her come in, but there were her friends, settling into one of the booths in the front. Elton John had long since given way to Sting and, again, to Gloria Estefan.

Kaitlin leaned forward and spoke very softly. "I keep thinking about what I said Sunday. I feel like such a fool."

"Don't." I finished chewing what was in my mouth and put my fork down. "There was no harm done. Your secret's safe with me. Truly, it wouldn't even matter if it wasn't. No one would believe a word I said. My credibility in this town is zero. I half suspect that if I dared say something negative about anyone here, I'd be jailed."

"I doubt that. You're so lucky. You get to leave soon. I'd give *anything* to be gone from this town."

"But then you wouldn't have Kevin."

She studied her hands for a long minute, before giving a one-shouldered shrug and raising her eyes. "Maybe I'm not meant to have him. I mean, I don't know what I'd do if I didn't have him now, but I'm not sure love lasts. My parents hate each other. Do I want that? Besides, I want to go to college. I want to work in a hotel someday, maybe in New York or London or Paris. I want to be independent."

"Are you talking about money or parents?"

"Both." She sat straighter, with something akin to defiance. "Like, I *know* someone's going to report back to my mom that I'm talking with you, and she'll get all hot and bothered, but why can't I talk with you? What's so bad about it? You're the most interesting person who's come here in ages."

"That," I remarked, "is the nicest thing anyone's said to me all day. If you're saying it just to make sure I don't tell—"

"No. I mean it. You *are* interesting. Look at who you are and what you've done. Were you really a dork when you were growing up here?"

"Totally."

"And ugly?"

"Very—but part of that was my attitude. An ugly attitude is like a zit." I jabbed a fingertip to the side of my chin. When she smiled, I said, "There. Better. You have a great smile."

She blushed. "You can thank Dr. Franks for that." When I didn't follow, she said, "My orthodontist." She darted a look at her friends. "I gotta get back. Maybe we can talk another time?"

"I'd like that," I said and meant it. Don't ask me why—Kaitlin DuPuis was nothing to me—but I felt better for her coming over. Less alone. Which was pathetic. I was thirty-three. You'd think I could rationalize being alone for a time, wouldn't you?

Wrong.

TRUTH #7? It doesn't matter how old you are. Lonely is lonely.

Having acknowledged that, I returned to

my food. I ate more slowly now, listening to a sweet Sarah McLachlan song, and set down my fork while some food was still left on the plate. Sated, I sat back. Thinking of all those other things—my friends, my book, my *life*— did help then. Marshall Greenwood be damned. I was going to be fine.

"Hello, sweetheart," Omie said, her wrinkled face wreathed in a smile, but the smile fell when she looked at my plate. "You haven't finished the hash. Is it not good?"

She was concerned to the point of pallor. "It's delicious, there's just too much." It struck me that the pallor wasn't from concern. There was none of the color I had seen in her face last time I had been in. "Are you feeling okay?"

"A little tired," she said with a small smile. "I'm not as young as I used to be."

"Will you sit with me?"

"Not today. I'm heading home for a nap. Would you like another frappe?"

"Good Lord, no. Maybe some iced tea. But I'll get it at the counter."

Omie held me in my seat with a fragile hand. "I'll have someone bring it. You stay here and relax. You bring class to this place."

"Your place doesn't need any help," I called to her as she walked slowly back around the counter and into the kitchen. Less than a minute later, her grandson emerged with a tall glass of iced tea. Having exhausted my need for calories, I added artificial sweetener.

I glanced at my watch. It was barely three. Four hours left until I met James.

I wasn't in the mood to stop in at the shop, wasn't in the mood to sit home alone, wasn't in the mood to move from this booth, period. Omie's was friendly turf, and I was marginally invisible here in the back. Did I have anywhere better to be? Absolutely not.

So I pulled the latest *People* from my purse and began to read. Billy Joel sang, then the Beatles, then Bonnie Tyler, then Fleetwood Mac. I accepted a refill of tea, added sweetener, and relaxed. Omie's was unique. There were places in Washington where I went when I wanted to take a break from my writing, but none felt like home the way this did.

No matter that I hated home. Home was still home. Another truth? Sad, but true.

Thinking that I was going to have to look harder to find something like this in the District, I closed the magazine and just sat for a while. When I hit the three-hours-to-go point until I met James, I studied the tab, left a tip, and slipped out of the booth.

Invisible no more, I felt the stares of the occupants of every booth I passed, and how to deal with that? How to let these people know that I wasn't an ogre, that I meant them no harm, that I was—yes—one of them at heart? I smiled. I nodded. I even winked at a sweet little girl who couldn't have been more than five.

Kaitlin and her friends continued to talk.

I paid up at the register, then went outside and down the steps to the parking lot, and the convertible was where I had left it, parked under the oak where the sun wouldn't burn my seats.

Only I shouldn't have worried about the seats. What I saw—how not to, with your car sitting all alone there like a sinner on display—was that my tires were flat. Not one tire. All four.

Chapter 17

MY FIRST THOUGHT, given the day it had been, was that I would find Marshall Greenwood on the opposite side of the parking lot, lounging against his cruiser with a stick of grass in his mouth, as casual as you please, claiming not to have seen a thing. And who would contradict him? If he had chosen a moment when everyone was inside, he could have done the dirty deed in private. In the shade of that big old oak, my car couldn't be seen from the street. Nor could it be seen from inside the front half of the diner, which was the only part open for another hour at least.

It was, of course, a writer's imaginative moment. In fact, Marshall was nowhere in sight. And how ironic was that? At the very moment I needed him, he was gone.

Disgusted, I pulled my cell phone from my purse and punched in the number that every schoolchild in Middle River was taught. It hadn't changed. Marshall's voice came right on; there was no dispatcher here. Middle River couldn't afford one, any more than it could afford more than a one-man department. Marshall was it. How frightening was that?

"This is Annie Barnes," I said. "I'm at Omie's. Someone has slashed my tires."

There was a pause, then a bland, "Why are you calling me? I don't change tires."

I sucked in a quick breath. "*Slashed* is the operative word here. Vandalism is a crime."

"Do you see anyone around?"

"No, but—"

"Think I'll find fingerprints? Footprints on raw gravel? Tire marks?"

I was incensed. Forget depriving him of an opponent; I couldn't resist lashing back. "You've thought this out, haven't you? If I didn't know better, I'd think *you* might be the one who did this. Well, that's actually good.

You've inspired me. I need a bad guy for my book."

I struck a chord there, judging from Marshall's sharp reply. "Not me. I am no bad guy. Push that line of thinking and you'll regret it. You've already caused trouble here. Want to cause more? Or don't you care about the welfare of your family?"

I was stung. A threat from a Meade was one thing; a threat from the chief of police was something else. "What does my family have to do with this?"

"They're accepted in this town. Push your current line of thinking, and people here will turn against them, just like they've turned against you. Know what the word *agitator* means? Well, that's you, and it ain't nice. If you ask me, when an agitator shows up here, she needs to be run right out of town."

I had no idea how to address that. Marshall Greenwood hated me. I didn't know why. But there it was.

Shaking with anger, frustration, and—yes—perhaps an element of fear, I returned to the immediate problem. "My car has been vandalized. Will you come here to investigate, or should I call the state police?"

Calm and cool after winning that round,

he said, "No need to call the state police. I'm just finishing something up. I'll be there as soon as I'm done."

It took him twenty minutes—twenty minutes to finish up whatever it was and drive all the way around a single block. I used the time to call Normie at the service station, and to seethe.

I was doing the latter, standing with my arms folded over my chest and my eyes on those pancake-flat tires, when Kaitlin and her friends emerged from Omie's. Three of the five headed for one car. When Kaitlin spotted me and approached, the fourth ran off to join the others.

"What happened?" she asked, eyeing the tires.

"Someone slashed them while I was inside. You didn't by chance see anyone hanging around here when you and your friends arrived, did you?"

"No. And the tires weren't like this when we came. I saw your car right off and pointed it out to my friends. We'd have noticed this."

But they had been inside the diner for more than an hour. That allowed plenty of time for someone to put a knife to good use.

"I called Normie at the station," I said. "He wasn't thrilled at having to come all the way down here." It was an understatement. Friendly Normie hadn't been friendly at all.

"Normie's a total jerk," Kaitlin replied.

Marshall drove up. Kaitlin stayed, and she wasn't alone. By now, other people had left Omie's, seen us, and wandered over. Marshall walked around my car giving due consideration to each of the tires. Then he looked at the people watching.

"Anyone see anything?"

There were head shakes and murmurs, all to the negative.

"I don't suppose you have four spares in your trunk?" Marshall asked me, but looked at his audience and chuckled.

I wasn't answering a stupid question. "Is there a history of this kind of thing in Middle River?"

"You mean, do we have a serial slasher?" he asked with another amused glance at those gathered around. "Nope. You must have annoyed someone. You're good at that. I'd say this is a message."

"A message saying what?" Kaitlin asked. I noticed that she had come a bit closer to me.

Marshall arched a brow. "Saying that Annie Barnes is stepping on toes in this town.
Are you a friend of hers?"

She ignored the last. "Whose toes is she
stepping on?"

"I can name a dozen people," he said and
began shooing away the onlookers. "Okay,
folks, nothing much to see here." To Kaitlin,
he said, "Aren't your parents expecting you
home?"

"No," Kaitlin said, tossing her hair out of
her eyes in a gesture that was nothing if not
defiant.

It wasn't so bad now that the other people
had left. Still, I feared she would make trouble for herself. "You can go on, Kaitlin. I'm
fine."

"Will he do an investigation?" she asked
me.

Marshall answered. "I'll do what's
needed."

When Kaitlin looked about to ask more, I
clasped her arm. "Go on, now. I'll handle
this."

The last of my words were drowned out
when, with a lumbering noise that didn't say
much for the skill of the mechanics at Zwibble's Service Station, the tow truck came

down the street and turned into Omie's lot. Pulling up where we stood, Normie climbed down from the cab.

There was no smile this time, no chitchat as he surveyed the scene. There was no *oooh*ing and *ahhh*ing over my car, no mention of old friends from school. Rather there was a silent, plodding examination, first one tire, then the other, then a walk around the car to the third tire, then the fourth. When he finally returned to us, he was scratching what remained of the hair on the back of his head. He glanced at Marshall before acknowledging me at last, and even then he kept his eyes mostly on the car.

"We got a problem here," he said. "I don't have the kind of tires you need."

"I'm not fussy about brand," I replied.

"Not brand. Size. We don't have any cars like this in Middle River."

"These tires aren't unique to convertibles," I pointed out.

"Most everything here is trucks."

"My parents don't drive trucks," Kaitlin said. "My mom's Sebring's not much different in size from this."

"But I don't have the right tires," Normie insisted.

"When can you *get* the right tires?" I asked.

"I'd have to make some calls."

He stood there.

"Okay," I said. "Can you do that?"

"Yeah, but I'd need your car back at my place, only I can't tow it back with the tires this way. I need a flatbed."

I waited. When no explanation was forthcoming for why he hadn't brought a flatbed in the first place, since I had told him on the phone that all four tires were slashed, I said, "Can you get one?"

"It's back at the shop."

"So you need to drive back and switch trucks. What'll that take, twenty minutes?" I glanced at my watch. I still had time. Granted, unless he could get the tires from another station in the immediate area, my car would be out of commission until tomorrow at the least. But once Phoebe got home, I could use her van. The timing would work. I didn't have to be at the varsity course until seven.

Normie made a face. "Y'know, it's not like this is all I have to do. I was in the middle of a job when you called. I gotta have that one done today." He shot Marshall a glance.

Marshall seemed perfectly content. If I didn't know better, I would say the two were in cahoots about messing me up.

"Tell you what," I said, reaching for my wallet. "Since you don't want to help, I'll call Triple A." I pulled out my card. "I gotta say, between you two guys, I have *great* stuff for my book."

Normie's round face paled. "What stuff?" He shot Marshall a different look now. "What stuff does she have?"

"Nothing," the police chief scoffed. "She's making empty threats, because the fact is that no one in town wants to talk with her. She's a pariah—now there's a good word. Tell you what. Be a sport and get the flatbed. The sooner she gets her car fixed, the sooner she'll leave."

I didn't bother to correct him. "What about who did this?" I asked with a glance at my car.

"Look, miss," Marshall muttered, "I'm doing you a favor here. Leave well enough alone, will you?"

He turned and walked away before I could argue. After a word with Normie, he got in his cruiser and drove off.

Normie climbed back in his truck and rumbled out of the lot.

I checked my watch. It was five. Assuming he came right back, I was okay.

"Can I drive you somewhere?" Kaitlin asked. "I have my Mom's old Jeep. It's falling apart, but it works."

"Thanks. I'm okay for now." I rested against the side of my car and hugged my middle. Talk about feeling like I was on hostile turf. And it was only getting worse. "You ought to go. Hanging around me won't help your image."

"I'm not worried."

"Well, I am." I smiled. "Really. I'm fine."

"Are you sure?"

"Yes. I'm sure."

She started off, then turned and came back. "You told them you were writing a book."

"And I got a reaction, didn't I? Interesting, that reaction."

"But you told me you *weren't* writing one."

"I'm not writing one about Middle River. That's the truth."

Kaitlin relaxed. "In a way it's too bad," she said with a mischievous half smile. "Know what they say about Normie Zwibble?"

I put my hands to my ears. "I don't want to hear this."

She said it anyway, and of course I listened. "He's addicted to playing the lottery, only he can't buy tickets at his father's place or he'd be found out, so he drives over to Weymouth. That's where he'll buy you your tires. It'll give him an excuse to go."

"The lottery, huh?"

"And Chief Greenwood's a junkie."

"I don't believe that."

"It's true. He's hooked on painkillers. My parents talk about it all the time. Like, it's a good thing we don't *need* him for anything, y'know?"

I was wondering about that as she set off again. Four tires slashed. That was violence against my property—violence against *me*. No matter how much I felt the people of Middle River disliked me, I had never been physically threatened before. I felt a trembling in the pit of my stomach as the reality of that sank in.

With each passing minute, I felt more impotent. Needing to *do* something, I called Sam. It was only when the answering machine at the newspaper office came on that I realized it was Thursday. He would be out playing golf or, more aptly given the hour, in the country club bar drinking scotch with his pals, one of whom was the great Sandy Meade.

I shifted from foot to foot. I leaned against one side of my car and studied the diner, then walked around and leaned against the other side and studied the woods. I thought about Sandy Meade and the satisfaction that would come from winning a round with him. It was the only thing that brought me relief as I waited for Normie to come.

When twenty minutes passed and he didn't return, I began wondering whether what Kaitlin had said about him was true. And not only about him. About Marshall. Addicted to painkillers? If he was, and if he truly believed I was writing a book about the town, I could see that he would feel threatened. But to threaten me in turn? If he thought that would shut me up, he had another think coming. I would have no qualms about publicizing his problems if he continued to put me in harm's way.

I wondered if Sam knew about this. I guessed most of the town did.

Ten more minutes passed, and I grew severely antsy. When my watch read five-thirty and there was still no sign of Normie, I turned to Plan B and called Phoebe. She was with a customer, so I waited. I kept thinking Normie would come along, in which case I wouldn't need Phoebe at all. But Normie didn't come. And Phoebe was distracted. "Why are you at Omie's house?" she asked when she finally came on.

"Not her house. The diner. I stopped for something to eat, and my tires were slashed. I'm waiting for Normie Zwibble, but if he doesn't show up, I'll need your help. You close at six. Can you be out of there by six-fifteen?" I knew there were closing chores, but Joanne was there to do them. Six-fifteen would give me enough time.

"Joanne?" Phoebe called away from the phone. "Do we close today at six? Isn't it eight today? Six? Oh, okay." She returned to me. "I don't know if I feel like going out for dinner. I'm exhausted. Should we eat in instead?"

"We agreed on having the chicken pot pie left over from last night," I said, but I was realizing the futility of the reminder. "Listen, Phoebe. I'll call back if I need a ride. Okay?"

I ended the call and tried Zwibble's, but there was no answer. I couldn't believe Normie had forgotten. It didn't make sense. I mean, if what Kaitlin said was true and he thought I was writing a book, wouldn't he be afraid of crossing me, lest I tell his secret to the world?

Of course, if Marshall had told him to make things hard for me, and if he was more afraid of Marshall than of me . . .

I tried Zwibble's again. Still no answer. We were coming up on six o'clock, and I was developing a whale of a headache.

By the way, if you're thinking I was standing in the parking lot alone all this time, you'd be wrong. We were into the dinner hour now. Middle Riverites had been coming and going. They noticed me, noticed my car, and went on inside. When they came out, they noticed me, noticed my car, then got in their own and drove off.

Talk about being stared at. Talk about being the butt of gossip. Talk about being . . . *shunned*. It was as bad as it had ever been.

Now you know, Grace said smugly. ***Do you understand why I drank?***

No, I do not, I challenged. When you drank, you made it worse. When you drank, you couldn't write, and that was your tool. You could have hit back that way.

Like you're doing, Miss High and Mighty? Write a book that exposes all the phonies there, and you'll have your revenge.

But then I lose Middle River and my family completely, I thought and felt stymied.

Omie. I wanted Omie, but she was home taking a nap. Had she been there, she would have waited with me. She was that kind of person. And her children and grandchildren were of like heart—*good* elements of Middle River—except that they were surely up to their ears cooking and serving inside.

Six o'clock passed. I tried to phone Phoebe to get her here, but the answering machine came on at the shop. There was a private number, only I didn't have it. Nor did Directory Assistance. Mom's line at the house just rang and rang and rang, and Sabina's line was busy.

By six-fifteen, I was seriously considering walking home. I didn't like the idea of abandoning the BMW—couldn't begin to imagine what little "mishaps" might occur if I left the car here overnight. But I might have no choice. Still no one answered at Zwibble's, which meant that Normie might well have forgotten all about me, and I couldn't wait forever. I had to be at the varsity course by seven.

By the time six-thirty arrived, my head

was throbbing. I figured I could run home in less than ten minutes, change clothes, and, assuming Phoebe *was* home with the van, get to the varsity course just in time.

That was when Normie finally showed up, and I was beside myself with impatience. It took him ten minutes to get the BMW on the truck, and another ten to get my credit card information, which he insisted he needed prior to ordering my tires.

I wanted to scream. Instead, I gave him what he needed, then quietly asked if he could drop me at home. He refused, claiming his insurance said I couldn't ride in the cab. I offered to ride in the flatbed of the truck, but he vetoed that, too.

By now it was six-fifty. No doubt about it, I was going to be late. I knew James had a cell phone (he had talked on it that day at Omie's), but I didn't have the number. So I got his home number from Directory Assistance and tried him there.

A woman answered.

Stung, I hung up. There was no wife. I was sure of it. A girlfriend? Fine. I was running with James—running—that was all.

That said, I had to *be* there in order to run. He had said seven, and I had agreed. Now I was going back on my word. Meades might do that, but Barneses did not—not even when they had splitting headaches like mine.

"Can I drive you somewhere?" Kaitlin asked through her car window.

I hadn't seen her pull up, though I might have guessed she would come. With my eyes on the road searching for Normie's flatbed truck, I had seen her mother's old Jeep go past several times. Kaitlin was keeping tabs on me.

Grateful that someone was, if only this young woman who might well have a case of misplaced idol worship, I rounded the car and climbed in. "My sister's, as fast as you can drive without getting a ticket," I said, pressing my fingers to my temple to hold my head together.

She got us there in two minutes flat.

"You're a lifesaver." I climbed out. "Thank you." Seeing the van at the side of the house, I ran up the walk, let myself in, and raced upstairs to change clothes. It was seven on the nose when I ran back down and into the kitchen. Phoebe was at the table, eating. She looked up in surprise.

"Annie? What are you doing here?"

To this day, I believe she meant what was I doing in Middle River when she assumed I was in Washington. At the time, I simply said, "Gotta go out for a run. Are you okay with the pie?"

"The pie. Did *you* cook this?"

Call me callous, but I couldn't deal with the mental fuzzies just then. "I'm taking the van." I looked around. "Keys?"

"Uh . . . uh . . ." She rose from the table and, seeming mystified, searched the counter, the drawer by the kitchen phone, then her purse. I was dying, looking everywhere else I could think of, when she said a vague, "In my blazer pocket?"

"Which blazer?" I called as I headed for the closet in the hall.

She didn't answer. I was trying to remember which blazer she had worn to work that morning, when she joined me and said, "I think it was my white linen. It's upstairs."

I ran up the stairs, found the blazer on her bed and, thankfully, the keys in the pocket. Racing back down, I dangled them Phoebe's way and bolted out the door.

I quickly drove down Willow, turned onto School, and flew across town daring, just *daring* Marshall to catch me, though for sure he was at home eating a big meal by way of rewarding himself for hassling me. A handful of cars were in the high school parking lot, close to the school. Dodging them, I cut through to the very rear and turned the corner, but when the parking strip at the edge of the woods appeared, I caught a sharp breath.

No one was there—no person, no car—*nothing*.

Pulling up, I left the van and actually went to the start of the path. Don't ask me why. It was as deserted as the parking strip.

Trembling, I returned to the grass near the van. It was seven-fifteen. I was late. But not *that* late. Wouldn't he have waited? Or just started to run without me? That was what any rational person would do. Wasn't it?

It hit me then, hit me hard. History was repeating itself. I had been stood up.

I should have been furious. But it was one too many turns of the screw on a day filled with them. I was suddenly too tired to process much, and my head was killing me. Sinking down on the grass, I folded my legs, put my elbows on my knees and my face in my hands, and let myself cry. Self-pity? Yes, and I had a right to feel it.

I don't know how long I sat there, only know that the crying felt good. In time, I wiped my eyes with my forearms, raised my head, and took a long, stuttering breath. I released it and continued to sit. My mind wasn't quite revived. I felt numb. That was probably why I didn't hear the approaching vehicle until it rounded that last curve and came into view.

My numbness dissolved. At the sight of that black SUV, I was suddenly rip-roaring mad. I rose to my feet and approached, so that I was no more than three feet from the car when he opened the door and lowered a sneakered foot to the ground.

"You are a major son of a bitch," I cried, rigid with fury. "Do you have any idea what I did to get here—or what my day has been like? I've been threatened and vandalized and stared at and *talked* about, no doubt, because this is Middle River and the average Middle Riverite doesn't have the *balls* to do more than talk, except for whoever slashed my tires, and *that* person is probably on your daddy's payroll like most of the *rest* of the town. I don't know why I thought *you*'d be better, maybe because you're a *runner,* for God's sake, but you're a snake, just like your brother, Aidan—the two of you chips off the old block, because I'm *sure* all this is coming from Sandy, I would put *money* on that. But if you think you can scare me away, you're dead wrong. First Marshall, then Normie, now you—you all are pushing me *this* close," I held up my thumb and forefinger, barely parted, "to writing my damned book and exposing everything possible about this stinking town. Thought you were being cute letting me sit here waiting, like I waited for Aidan back then? Wrong. I have a life. I have connections. I have *power.*" Pressing my palm to my temple, I muttered, "And a massive headache, thanks to you."

"You didn't get my message?" he asked.

I would have screamed if I'd had the strength. "That's the oldest line in the book."

"I called Phoebe and asked her to tell you I'd be late."

"Well, Phoebe *didn't* tell me," I charged. The words were barely out when I realized that Phoebe might well have taken the call and forgotten all about it, but even *that* had a Meade cause. "If she didn't tell me, it's because something's wrong with her mind, no doubt from pollution caused by *your* mill." I waved both hands. "But that's beside the point. I was worried because *I* was late, so I tried to call *you*. A woman answered the phone. Does that give me a hint about why you were late?"

He didn't smile. "I was late because an account I've been trying to land needed another thirty minutes of work. That was the babysitter who answered the phone."

I blew out a puff of air and said to the dimming sky, "Oh . . . Lord. The . . . babysitter." My eyes returned to earth and challenged his. "Do I look like a *total* fool? You don't have a baby."

"I do. Her name's Mia. She's ten months old."

It wasn't his sureness that got to me, as much as the softening of his voice. That gave me pause. "Ten months old, and never a mention in the paper?"

"My father didn't want it publicized. He doesn't buy into single parenthood."

"So you aren't married?" Nothing about that in the paper, either, but I wanted to be sure.

"No."

"And Sandy is *ashamed* of the baby?"

"No. Angry."

"But you wanted to be a parent."

"Yes."

"And you take care of this baby yourself?" I was trying to picture him changing a diaper. He seemed far too . . . big, was the only word that came to mind.

"Other than when I'm at work, I do. She's precious."

He said the last with a half smile, and it did something to me, that smile, moved something inside me that made me forgive him anything that had made him late. It struck me that I had behaved like an imbecile, not the sophisticated woman I wanted to be. And of course he could see that I had been crying. My eyes had to be red.

Bowing my head, I rubbed the back of my neck. "Oh boy." I felt like a fool.

"Did you take something for the headache?"

"No time."

Before I knew it, he had produced pills and a bottle of water. "Here."

I took them without asking a thing. He wouldn't give me anything harmful; my gut told me that. I don't know why *it* was so sure. After all, James was a Meade. But my gut was also reminding me that he was attractive, and I had no cause to doubt that. He was standing beside me now, so much taller and broader than I, though still lean. And strong. Quietly so.

A large, silent, authoritative man like James rocking a baby to sleep? The image was striking enough to tie up my tongue.

I drank more of the water, then lowered the bottle. My eyes met his for a long moment, and for the life of me, I couldn't think of a thing to say.

Then he said, "Want to run?"

I nodded. Running was *definitely* the thing to do.

So he closed up his car, and we stretched. Then we hit the woods. He offered to have me go first, but I motioned him on, and it was wise. For one thing, though the light was starting to fade, he illuminated the path. I don't know whether it was his tank top, which was white, or the reflective material of his running shoes, or the last of that light on his skin or the gray in his hair. But he was easy to follow—quite a *good* runner, actually, I decided, as it seemed I did every time we ran. And why my surprise? He was athletic; one didn't grow up in Middle River and not know that the Meade boys played baseball, basketball, and football. I guess it was just that running was different. If was solitary, for one thing. I would have thought a Meade would want more of an audience. For another, it was grueling. Typically, Meade boys relied on others to get them the ball when they were in a position to score. Running was not a team sport. Your own two legs carried you, or you went nowhere at all.

We went around once. The cutoff to Cooper's Point came and went.

"Head okay?" he called back.

"Better," I said.

So we went around again. It was darker still this time, but I felt safe—certainly safer than I had standing beside my vandalized car back at Omie's. James ran immediately ahead, his feet rhythmically striking the ground, lean legs in a fluid stride, butt tight, back straight, arms pumping easily. He glanced behind every few minutes to make sure I kept up. Had I fallen, he would have been there. I knew that. It was another gut thing.

Did he alter his pace to help me out? Probably. But he built up a sweat anyway. I could see it on his neck and his arms, in the spikes of his hair and on his face each time he looked back, and when we finally reached our starting point again and stopped, he was breathing as hard and deeply as I was.

Gesturing for me to stay, he got two fresh waters from the car. We drank them leaning against trees at the start of the woods. Dusk had fallen; for all practical purposes, we were hidden away. It was incredibly peaceful—peaceful until his eyes met mine one more time, and one more time I felt the pull.

He pushed off from his tree and came toward mine. "Do you know what in the hell this is?" he asked, sounding truly perplexed.

I shook my head. Oh, it was physical attraction of the most basic kind. But it wasn't supposed to happen. Not between James and me.

Between James *Meade* and me? No way.

But I could see it in his eyes and feel it in my body, and when he put a hand on the back of my neck and pulled my mouth to his, there was something live between us that caught fire and burned. The kiss went on and on, turning this way and that, deepening and withdrawing and deepening again, but it wasn't enough, wasn't nearly enough. By the time he drew back, my arms clutched his shoulders and his clutched my waist, and it *wasn't enough*.

His eyes were dark in the encroaching night, his voice hoarse. "Do we want to do this?"

I shook my head, but my hands betrayed me by curving around his shoulders and sliding over sweaty skin to the nape of his neck— and the sweat was as appealing as everything else. Chemistry? Omigod. Chemistry didn't began to explain the need I felt just then.

He kissed me again. His hands didn't have to guide my face this time, so they found my breasts, and what he did with them was incredible, *but it wasn't enough*.

I could barely breathe, could barely speak, but as I drew back enough to grab the hem of my top, I whispered, "Don't want to do it, but will die if we stop."

I imagined that the strangled sound I heard was a laugh, but it was quickly muffled when he pulled his own top over his head. I wanted my breasts against his chest. He wanted them in his mouth. He got his way. With one nipple drawn in and the other rolled between his fingers, my knees went weak. I clutched him now so that I didn't fall.

A final time, in the heat of it, I thought it—James Meade and *me*?—but in the next instant the thought was gone and never returned. Identity had no chance against raw passion, which was what we shared. To this day, I have no idea how we got out of our shorts, what I leaned against as he thrust into me again and again, whether either of us spoke, or how James knew to withdraw in the instant before he climaxed. But it was good. It was incredible. I couldn't speak for James, though the sounds that came from his throat were something of a testament, but for me, it was the longest and most intense orgasm I had ever had in my life. And as if that weren't amazing enough, we just sat there for a while—James on a log, me astride him, both of us buck naked—sat there until the fire had cooled and evening set in.

We dressed then and went to our cars. We didn't touch. He seemed lost in thought, and I didn't know what to say either. When we stopped, our eyes met, and for a minute longer he seemed confused. Then, quietly and with an odd unsureness, he said, "Want to meet her?"

* 　 * 　 *

I followed him to his house. It was a large brick Colonial, as I had expected it would be, only it wasn't on Birch with the others. It was on the south end of town, not far, actually, from Tom's and with the extra land that James wouldn't have found had he bought a house on Birch. There were no lights on either side of his home or even across the street, just those flanking his own front door and warming the windows inside.

The babysitter greeted us. She looked to be about my age, though her face wasn't familiar. James introduced us, paid her, and she left.

I was assuming that the baby was asleep when he led me through the kitchen to an open family room, and there was a very little girl, sitting in the middle of the floor holding the corner of a fleece blanket, while the rest trailed to the side. Her hair was short, thick, and dark, and her footed pajamas were pink. More vivid colors—shapes, numbers, animated animals—came from the large-screen TV to which she was riveted.

We didn't speak, and for the longest time she didn't know we were there. I looked at James. He didn't take his eyes from the child.

Then something tipped her off. She looked back, saw James, and her entire face lit. In an instant, she was crawling our way. Not once did she take her eyes off his face. He was on his haunches when she reached him and caught her up into a hug.

I actually took a step back then. It seemed such a private moment, such an *adoring* moment, that I didn't belong. But then James stood and brought the child closer. She had a small arm around his neck and large eyes on me.

"This is Mia," he said in an exquisitely gentle voice.

Seeing her close up, I caught my breath. She was beautiful, and I told him that. Unable to resist, I took her little hand. "Hello, Mia," I said in a voice that was as gentle as his, but then, there was no other way to greet this child. With creamy skin, tiny rosebud lips, and dark eyes with just the slightest slant to suggest Asian descent, she was all innocence. Her hand was warm, but it didn't quite grasp mine. I was a stranger.

She kept her eyes on me through a tour of the kitchen, where James filled her bedtime bottle, and then it was up to her room, done up in yellows and pinks, where he settled into a rocker and let her drink, and still she looked at me. By the time she was done, she was comfortable enough to give me a smile, but it was a sleepy one. She babbled a little while James changed her diaper, but even the babble was sleepy. By the time he put her in her crib, with her favorite bear in the corner, her favorite square of fleece in her hand, and her favorite mobile playing a lullaby, she was asleep.

He turned down the light and turned up the monitor. Then he took my hand, led me into a master bedroom suite that I couldn't see because he didn't bother with a light there at all, and made love to me again.

He was truly incredible. Totally aside from stamina, which he had in spades for a guy with more gray in his hair than black, he was so strong and at the same time so *gentle* that I was beside myself with need when he finally entered me. He wore a condom this time, which meant that he stayed there through not one, but *two* orgasms—and I'm talking simultaneous ones, which is so rare. But it was like his excitement spurred mine, which then spurred his, which spurred mine more . . .

I'm sorry. I'm not being very articulate,

certainly not poetic here, but you get the idea. Suffice it to say that when we finally separated, I was totally spent. I thought he was, too, until, with little warning, he scooped me up and carried me into the shower.

We needed it. After four miles of running and two bouts of sex, we were pretty ripe. Soap took care of that—lots of soap and lather and scrubbing. Ah-hah, and I know what you're thinking. You're thinking we were going to make love again, there in the shower.

Nope. There was plenty of feeling and touching—he scrubbed me, I scrubbed him—but we didn't go farther than that. Actually, a kind of shyness set in when those oversize bath sheets of his covered each of us up. For my part, I was trying to understand what had happened and why and what I was going to do with it. He must have been thinking something similar because, with his towel hooked around his hips, he straightened his chest, dusted with wisps of newly washed and dried hair, and said with vague amazement, "Annie Barnes? Who'd a thought it?"

I blushed. How flattering was that?

And being waited on for dinner—how flattering was *that*?

After checking in on Mia, he led me to a sleek kitchen with ash cabinets and state-of-the-art appliances, and plied me with a fabulous Pinot Noir while he grilled us a foursome of burgers. I ate my two as quickly as he ate his. I was starved.

By the time we were done, all sorts of little warning bells were dinging in the back of my mind, but I kept them there—muted, way in the back—until I was in the van and on my way home. Then they rang out, and it struck me. If James Meade could have chosen anything that would keep me in the Meade corner and out of mischief, it was this.

After I'd been floating for the last two hours, that realization grounded me.

Then I walked in and found Phoebe in the kitchen waiting, only not in pajamas. She was fully dressed. With more lucidity than I had heard from her in months, she told me that Omie had suffered a massive heart attack earlier that evening and was dead.

Chapter 18

PHOEBE AND I weren't the only ones driving over to the diner that night. The kitchen had closed as soon as word came of Omie's death, so it wasn't as though food was being served, but people streamed in to console the family and also each other. Omie was widely loved in Middle River. She had touched most of our lives.

For Phoebe and me, as well as for Sabina, who arrived soon after us, there was a special mourning. Omie had been a close friend of our mother's. So we were there for Alyssa's sake as well.

Those of us who had come—and then

some—were back the next morning, now bringing the cakes and cookies, casseroles, finger foods, and soups that we had made late into the night. None of us expected Omie's family to work in the kitchen that day. We streamed in and out, heating this, chilling that, putting the other on platters and carrying it out front—and I say "we" meaning Middle River in the largest sense. If ever there was a sense of community, it was here. Coming together in times of need was what small towns did. I could appreciate that. I could even admire it.

So yes, here was another truth. **TRUTH #8: Small towns have their strengths. They can offer comfort and support in ways that a city may not.**

But as long as I was onto this business of truths, there was more. What was it I said earlier—that I looked at my sisters and saw intelligent women whose lives were wasted in a town that discouraged free expression and honest thought? I wasn't entirely right. I had seen Phoebe at work; she provided a vital service for the townsfolk. Likewise Sabina; that Data Center of hers, with its server behind glass and Sabina its mastermind, was impressive. And now here they both were, communicating easily with all of these people, joined in their grief, taking comfort and giving it back. Honestly? I envied it.

That said, for this day, at least, I was part of it all. No one doubted my having loved Omie, or my grief now that she was gone. We worked side by side in the kitchen, feeding each other and all those children, nieces and nephews and grands, that Omie had left behind. Her body rested in the local funeral home, but she was waked here, and her spirit was strong.

Under the spell of that spirit, people seemed to forget that they hated me. Omie's death was a distraction that mellowed them.

It had that effect on me, too. I was able to smile at Marylou Walker, though she had been decidedly chilly last time I was at News 'n Chews. I could nod politely at Hal Healy without staring at that arm around his wife, which seemed more inappropriate than usual. I even acknowledged some of those who had slammed the door in my face the day before.

Kaitlin was a sweetheart. She spent a lot of time helping wherever I was, and my acceptance by her triggered acceptance by her friends. They talked with me, and they smiled. Hal Healy wouldn't have been pleased.

There were moments when new people entered the diner and I wondered if a tire-slasher was among them. There were certainly moments when I wondered if TrueBlue was here. Marshall Greenwood was. I saw him numerous times, but he steered clear of me. I wanted to think that in the presence of Omie's goodness, he was wallowing in guilt.

Most important, though, without the distraction of mourning Omie, I would have been agonizing over James. It was a pretty big thing that had happened out there on the varsity course, and I'm not just talking about erasing the humiliation of my wait for Aidan at Cooper's Point. I'm talking about sex in the woods and again at his house.

I take sex seriously. I didn't have it until I was twenty, and in the thirteen years since, I'd been with three guys, each of whom I stayed with for multiple years. I didn't sleep around, and I certainly didn't sleep with guys I barely knew, and I barely knew James.

Did I make a conscious decision to sleep with the enemy? No.

Was I scheming to ruin him while his tongue was in my mouth? Absolutely not.

Grace would have been disappointed in me. She would have said I had squandered a golden opportunity. But in all honesty, who James was and what he did was the farthest thing from my mind when we were making love. In those moments, he was a guy to whom I was powerfully attracted and who—omigod—was powerfully attracted to me! I felt a tingle just thinking about it even hours after the fact.

So I kept busy. I did talk with people, but I felt most comfortable working in the kitchen. Washing platters was perfect. Same with loading and unloading the dishwasher. It kept me out of the spotlight, with my hands busy and my mind on Omie.

Did James show up? Of course. I think that I sensed it the instant he walked into the diner, and I was in the kitchen at the time, which says something about the sexual antennae we humans have. I had a weird feeling; I glanced at the pass-through, and there he was.

I stayed off to the side so that he wouldn't see me. I mean, what could I say? Even in *spite* of that powerful attraction, I was confused. I would have thought that James's last name alone would have been a depressant. But it wasn't.

He found me in a moment when there were few enough people in the kitchen so that he could say a few words without being heard.

And that's all he said—just a few words— the first of them while he kept his eyes on the lavash I was taking from a large plastic bag and arranging in a dish. "Got your tires straightened out?"

"Uh-huh. Normie managed to locate the right ones over in Weymouth." I looked up in time to see a twitch in the corner of James's mouth that suggested he knew what else was in Weymouth. Then his eyes met mine, and an instant something passed between us.

I could barely breathe, certainly couldn't look away. I was relieved when he finally broke the spell by saying, "I can't run on weekends. No babysitter. Can we do Monday?"

"I'll be in New York Monday with Phoebe. We're due back Tuesday night."

"Wednesday morning, then?"

"Okay."

He nodded, tapped the stainless steel counter, and turned to leave, and I thought that would be it. Then he looked back. "I don't do one-night stands."

Well, what in the devil did *that* mean? That there wouldn't be a repeat of what we had done because it had, in fact, been a one-night stand and he didn't do those? Or that there *would* be a repeat of what we had done because that would make it something *other* than a one-night stand, which was fine?

I couldn't ask, of course, because by the time I wanted to, he was halfway across the room. All I could do was watch him work his way around counters and shelves. His back was straight, salty hair neat, blue shirt neat, gray slacks neat, loafers neat.

To this day, I don't know for sure whether people saw us talk and thus deemed me worthy of trust. But after James left, I did get some questions. People were wondering whether Omie's death was from simple old age, or whether it had been hastened by something external to her. Apparently—I hadn't known this—longevity ran in Omie's family. Her mom had died at ninety-six and her dad (who was much older than his wife, hence his death when Omie was still relatively young) at ninety-eight. Omie had only been eighty-three, which sounded plenty old to me, but not to these others. They wanted to know why she had been sick on and off for the past year, why the bouts of pneumonia, why the massive heart attack at the end.

They were nonchalant about it, asking quietly, almost rhetorically. But they did ask. And they asked *me*. That meant something. It was like they knew I was looking, and that made me their advocate. Like they didn't have anyone else to ask. Like they *trusted* me.

Maybe the last was pushing it a little. Still, by the time I left Omie's at midafternoon, I was feeling a sense of heightened responsibility.

I turned on my computer. My friend Jocelyn had e-mailed to say that she had reached the part in *Peyton Place* where Tomas Makris was stripping Constance MacKenzie with his eyes, and that she thought he was absolutely fabulous. I wrote back telling her that, yes, Tomas was totally sexy but that the *point* of the relationship was Constance coming to accept her own sexuality. How far we women have come in this sense. I could never have done what I did with James if I didn't like sex—but I didn't write Jocelyn this part. My friends knew nothing of James at this point.

Greg e-mailed to say that they had reached fourteen thousand feet but were being slowed by wind and snow, and I have to say I felt an inkling of unease. People died climbing mountains, and Mount McKinley was a biggie. Was being caught in a blizzard at fourteen thousand feet any less dangerous than walking through the war-torn streets of Afghanistan? I wasn't so sure.

My thoughts were brought back to the dangers here at home when, just as I was about to reply to Greg, an e-mail arrived from TrueBlue. *Any luck?* he wrote, and, knowing he was right there at his computer, I was suddenly annoyed.

No luck yet, but you're setting me up to fail unless you give me more information. I can't ask people questions without backing those questions up with facts. I can't win their trust unless I sound halfway intelligent, for God's sake.

What happened at Northwood on the dates you gave me? Was there or was there not a mercury spill?

There were fires on both of those dates. On March 21, 1989, a fire destroyed the Clubhouse, and on August 27, 1993, a fire destroyed the Gazebo.

On those dates, huh? Okay. But the fires themselves aren't news. We all know they happened. Sam gave them front-page coverage. What's the significance here?

Those fires were made to look accidental. The Clubhouse fire was blamed on faulty wiring in the kitchen. The one at the Gazebo was blamed on a torch that hadn't been properly extinguished. But neither fire was accidental. They were set by representatives of Northwood at the order of Sandy Meade. The purpose was to level the structures so that Northwood could make a big deal about rebuilding from scratch. What never made the paper was the fact that prior to the rebuilding, mercury was removed from the soil underneath.

Mercury at the Clubhouse and at the Gazebo, but not at the plant itself?

Remember I told you that Northwood used other methods of disposing of mercury waste? There were huge drums buried under both of those structures.

Under the Clubhouse and the Gazebo? Why? Why not bury them out in the middle of nowhere?

Sandy Meade thought it would be clever—and less suspicious if ever there was a probe—to build something for the good of the town on top of those dumps. In fairness to him, the drums were supposed to hold. There wasn't supposed to be any leak.

But there was?

During the week prior to each of the fires, events were held at each of the places. Following those events, there were outbreaks of something alternately called flu and food poisoning. It was enough to make the Meades test the air. They found evidence of a major spill that had likely seeped into the air vents.

The fires were a cover-up then?

Yes. A cleanup was done in every sense of the word. Internal memos were wiped from the files. There are still records of what events were held during those days leading up to the fire, but there's only a scanty record of who attended each. The outbreak of illness wasn't widespread enough to attract the attention of state authorities. The Meades were satisfied that they had caught the spill early and had totally cleaned it up.

And they thought that absolved them of responsibility? Did they do anything for any of those people?

A few things have been done, but quietly enough so that the people involved don't know why those things were done.

Are there more sites like this?

Yes. The day care center sits on top of one. It was built there under the same premise. For what it's worth, there have been no other leaks. Given the years that have passed since the last leak, Sandy Meade is assuming that those two simply had problems, and that these others will hold.

And they're willing to take the risk, knowing that children are in harm's way?

Apparently. They grow more secure with the passage of time. They see the lack of a leak as proof that there won't ever be one. I see it as a tragedy waiting to happen. That's why I need you.

I was appalled. Even aside from what had happened before, I couldn't imagine what kind of evil mind could do that to children, and that raised a new issue.

If, say, Sandy Meade were to drop dead today, what would happen? Would the sons continue this practice?

That depends which son takes over.

James is in line.

Don't be so sure. By the way, you looked like you were really into it at Omie's.

You were there and didn't present yourself to me? When will we meet?

When we have evidence that's workable. We need names. We need people who were at one of those two sites during the week in question, who have suffered physical problems since, and who are willing to talk.

What's in it for them? Can we offer them help of some sort?

<u>WE</u> can't, but I assume the Meades will as part of a settlement. Isn't that where we're headed?

It was. Of course, before any settlement could be reached, there would have to be a confrontation with the Meades. I didn't look forward to that. Their three to my one would be formidable odds. If TrueBlue came out of the closet, that might make it their three to my two, but I couldn't count on his doing that. By his own declaration, he wanted to continue to work at the mill. That would be impossible if he stuck it to the Meades.

Nor could I count on Tom. Oh, he was being a *huge* help to me. Not only had he gotten us an appointment with his friend in New

York, but he had also given us my mother's files, so that we would be able to show the doctor something of the family history with regard to Phoebe's ailments. But his job, too, would be on the line if he came forward.

The public side of this fight appeared to be mine and mine alone.

Chapter 19

NICOLE DUPUIS was feeling complacent after her talk with Aidan. She had him over a barrel, and he knew it. If her marriage fell apart, he would have to help her out—and it wasn't blackmail, exactly. But the truth was, he had a good thing going in her, totally *aside* from the sex. She held up his end of the business in ways she was startled by herself sometimes. She was Aidan's voice; she made him articulate. Sandy Meade was certainly fooled. He was relying on Aidan more and more to be the front man for the mill. If things continued as they were, Aidan was a shoo-in for chairman of the board after Sandy, and that was in Nicole's best interest, too. With Aidan as chairman, she would be all the more indispensable. Moreover, if he were chairman, he would have unlimited access to funds. If enough came her way, she could kick Anton out on her own.

She would like that, would like it a *lot*. Anton was a lying cheat. They no longer shared a bed, but there were still meals and social events to endure. Nicole hated the constant tension. It was wearing on her.

Precisely because of that, she had put off talking with Kaitlin, because that would invite *more* tension. Kaitlin was touchy enough as it was. Oh, she put on an act. She smiled and nodded and said, *Yes, Mom, I will.* But Nicole could see the anger beneath. Confronting her about what she might have told Annie Barnes could be unpleasant.

Then Omie died, and everyone was at the diner on Friday, and how to miss Kaitlin's connection with Annie? Every time Nicole looked, it seemed, Kaitlin was either talking with Annie or standing close by. Add Hal, pointing it out and expressing concern about what appeared to be a blossoming relationship, even suggesting that Kaitlin had become the liaison between Annie and the other young girls, and Nicole couldn't put it off any longer.

She waited until the funeral was done Saturday morning and Anton had left to play golf. Kaitlin was out by the pool, wearing a bathing suit that was at least a size too small. Her body was smeared with oil. She was stretched out on a lounger, talking on the phone. When she saw Nicole approach, she ended the call, dropped the cell phone onto a towel on the ground, and closed her eyes.

"Who were you talking to?" Nicole asked in a friendly way, one that was actually sincere. She did want to be friends with her daughter. She truly did. It just hadn't happened yet.

"No one," Kaitlin replied.

Nicole smiled. "Couldn't have been no one all that time."

"No one important."

"Kristal?"

"Jen," Kaitlin said. Without opening her eyes, she put a hand down on the far side of the lounger, took a cherry from a bowl, and put it in her mouth. Removing the stem, she dropped it back in the bowl.

"Ah. Jen." Nicole liked Jen. She and her

mother had the kind of relationship Nicole envied. Pulling over a second lounger, she sat on its lower half. "Can we talk?"

Kaitlin was chewing the cherry. "Mm-hm."

"You were with Annie Barnes a lot yesterday. Any special reason?"

She shook her head, spit the pit into her hand, and dropped it into the bowl.

"You've been seeing her a lot."

"No more than anyone else in town."

"They don't talk with her. You do."

Kaitlin slit open an eye and raised her head just enough to look at Nicole. "How do you know? Is someone telling you this?"

"I could see it for myself yesterday at the diner." Kaitlin had looked perfectly comfortable with Annie—not only comfortable, but happy. Nicole hadn't seen her that way in a long time. "What do you two talk about?"

Kaitlin dropped her head back again. "She was telling me how to arrange the chicken fingers on the platter. There's a knack to it."

That was exactly the kind of remark that irked Nicole—innocent on the surface, mocking beneath. "Don't be fresh, Kaitlin."

"I'm serious. That's what we talked about."

"Lots of people in town are concerned

about Annie Barnes," Nicole said in self-defense. "It isn't just me."

"Like who else?"

"Mr. Healy, for one thing. He thinks she's a bad influence on girls your age."

"Mr. Healy would think that. He's *so* full of hot air."

Nicole agreed with her on that, but wasn't about to admit it. For the current purpose, Hal was simply the high school principal. "He's concerned about Annie Barnes."

Kaitlin rose on her elbows, both eyes open. "Concerned about *what*?"

"That she'll worm her way into your lives, then turn around and write about it."

Kaitlin made a face. Sitting all the way up, she reached down for the cherry bowl and settled it in her lap. "She is not writing a book about Middle River." She put another cherry in her mouth, pulled off the stem, and dropped it in the bowl.

"How do you know?"

"She *told* me," she said around the fruit.

"And how did that subject arise?"

She spit the pit into her hand. "I *asked* her, Mom. Half the *town's* been asking. Do you have a problem with that?"

"No problem," Nicole said, because she really *didn't* want to fight. But there was still the whole business about what Kaitlin knew and what she might have told Annie. This whole conversation was about damage control, wasn't it?

So she reversed herself and said with greater force, "Actually, yes, I have a problem. Women like Annie Barnes are devious. They act like they're your friend, then they stab you in the back. She may be using you, Kaitlin. I hope you haven't told her anything you could come to regret."

"Like what?"

"Well, *I* don't know." Nicole wasn't about to mention Aidan. There was always a chance Kaitlin was still in the dark about that.

"I trust her."

"Oh God. You have told her things."

"All I mean," Kaitlin said, "is that she isn't using me. She isn't that kind of person." She took another cherry.

"And you know what kind of person she is?" Nicole asked. "You are seventeen, Kaitlin. You are *not* an authority on who to trust. Do you have any idea what happens when someone like that takes your confidences and spreads them around? Do you know the harm that can be caused? Annie Barnes does. She knows it firsthand. You tell her stories about *any* of us, and once they're out, there's no taking them back. Some things are meant to stay private. And stop eating those *cherries,* for God's sake. They're filled with sugar. That bathing suit is tight enough already."

Nicole knew she had made a mistake the instant the words were out of her mouth. Kaitlin's face grew stony.

"And that," the girl announced coldly, "is why I *like* Annie Barnes. She wouldn't say something like that. She doesn't think I'm fat at all. She was ugly, too, when she was my age. I'd like to grow up to be like her."

Nicole was startled. "You're not ugly."

"No? Then why do we work on my nose, my teeth, my jaw, my hair, and my skin?"

"I have never said you were ugly."

"You don't use that word, but you say it in all sorts of other ways. 'My bathing suit is tight enough,'" she mocked. "'Stop eating those *cherries,* for God's sake.' Mom, I'm not deaf. I hear what you're saying. Well, maybe I'm not *meant* to be as beautiful and thin as you are."

"I *never* said you were ugly."

"Fat is ugly, and you're always telling me

what to wear that will make me look thinner. Now you're even saying I'm dumb."

"I am *not*."

"I am *not* an authority on who to trust," Kaitlin mocked again. Cradling the bowl of cherries, she grabbed her cell phone and stood. "Well, maybe I'm smarter than you think. Maybe I do know who to trust. Annie Barnes understands me more than you do. She *knows* what I'm feeling. She's *been* there. If she says she isn't writing a book, I believe her—and if *you* have a problem with that, it's because you're afraid all *your* secrets will come out. Well, she doesn't *know* your secrets, and it has nothing to do with whether I trust her with them or not. There are people in this town I'd talk about before I talked about my own parents. Do you think I'm proud of what you and Dad do? Why do you think I don't have friends over?"

"You do," Nicole argued, but a hole was opening up inside her.

"Not for dinner. Not for weekends. Not when there's a chance you and Dad will both be here, because it's like anyone can *see* how much you hate each other—and don't try to deny it. I've heard you say things to friends on the phone."

Nicole was appalled. "You had no business listening."

"I was in the *room.* How could I *help* but hear? I'm not three anymore, Mom. You can't just say things and think I don't know what they mean—and besides, I *saw* Dad with that woman. They were in bed, and they were naked in the middle of the day. They were having sex. Do you think I don't know what that is?"

Nicole gave an uneasy laugh. "Well, yes, I was kind of hoping that."

"Because I'm ugly? Because I'm fat? Y'-know, there are some guys who don't care."

Her unease increased. "What are you saying?"

The girl looked angry enough to go on. Then she caught herself, let out a breath, and calmed. "I'm saying that it is *the* most obvious thing how bad your marriage is, and if I had my friends over more, they'd see it in a sec, and it'd be embarrassing. So if I feel that way, why would I ever tell Annie Barnes what goes on here? I am not *totally* stupid. Give me credit for that at least, will you?" Grabbing her towel, she walked off.

*   *   *

Sunday morning, Sabina went to church with her family. She was feeling a particular need for comfort. Part of it had to do with losing Omie, who had symbolized all that was stable and safe. The rest was more complex. It had to do with accepting that Phoebe was sick, with appreciating Annie's help, and with feeling that, deep down inside, something in her own life was about to shift.

She didn't understand what the last was until the service ended and she went around back to visit her parents' grave. The grass had filled in over Alyssa's coffin, the rhodies flanking the stone had grown, and the impatiens that she and Phoebe had put in earlier that summer were a bright orange, pink, and white—all attesting to the strength of the growing season in Middle River this year.

The setting was lush, and so, in its way, was her life. Ron and Timmy had stayed out front, but Lisa was with her, holding her hand in a way that Sabina knew was precious, given how grown-up the girl was becoming. Sabina was truly blessed with her children, certainly with her husband, and, yes, with her sisters. Her parents were pleased that she understood it; she could feel that here in the warmth of the sun.

After a few minutes, they headed back. Just beyond the graveyard gate, they bumped into Aidan Meade and Marshall Greenwood. When Aidan spotted her, he broke into a crooked smile. "We were just talking about your sister," he said.

Sabina gave Lisa's hand a squeeze. "Go on out with the others. Tell your dad I'll be right there." When Lisa ran ahead, Sabina turned to Aidan. "Which sister?"

"The troubled one."

Sabina gave a curious smile. "Which one is that?"

"Annie," Marshall said in his sandpapery voice. "I had quite a time with her on Thursday. She's riling up people all around town."

"Huh," Sabina mused. "No one seemed upset with her Friday at Omie's."

"They were being polite," Aidan advised. "After all, it was a wake. But she annoyed someone enough to get her tires slashed the day before. I'd call that a message."

That was when Sabina felt the shift. It was subtle, but distinct. "Kind of like the one you gave me on Wednesday, the one where you threatened to fire me if I didn't get Annie to leave town?" She looked at Marshall. "If I didn't know better, I'd say you guys were behind those slashed tires."

"Uh-huh," said Aidan. "And have you shared that theory with anyone?"

"Not yet," Sabina replied with a smile. "It just came to me now, seeing you two out here talking, by your own admission, about Annie."

Marshall tugged on his belt. "So she's gone off to New York? How long's she staying?"

"She's with Phoebe, so it's *they*. *They* will be back on Tuesday." Sabina raised her brows. "What's next? Think someone'll take a potshot at the house? String up roadkill on the porch? Burglarize the store?" Looking from one to the other and feeling a distaste for both, she identified that little shift inside. It had to do with loyalty. "Here's a promise. If something else happens, first thing *I'm* doing is calling the state police."

Aidan drew himself up. "That sounds like a warning. I don't like warnings, Sabina."

"Yeah, I know," Sabina commiserated. "They really make you feel small and powerless, don't they?" She smiled again. "You, though, are big and powerful. All it takes is a word from you, Aidan. Let Marshall here know that Annie is to be protected, and she will be protected."

"I want her gone," Aidan stated flatly.

"Well, she will be that, too. She's only here for the month, and nearly half of that's passed. Two more weeks, Aidan. That's all. Can't we make this work?"

Chapter 20

PHOEBE AND I left for New York immediately after the funeral, and the timing couldn't have been better. For the tiniest while I had actually felt *comfortable* in Middle River. But feeling comfortable was a luxury I couldn't afford. My life was elsewhere. I needed a reminder that there was, indeed, life beyond.

That reminder came as soon as we passed through airport security in Manchester. Had I truly felt unsafe because someone had slashed my tires in Middle River? The thought of bona fide terrorism put that in perspective and, emotionally, it was uphill from there.

New York was my kind of town. When I had decided to join Phoebe, I had upgraded our reservations, not because I didn't want to stay in the more modest hotel she had initially booked, but because I had promised her a good time, and location was everything. Our room overlooked Central Park and couldn't have been more lovely. We had barely unpacked when a bottle of wine and a bowl of fruit arrived, and Phoebe felt pampered. The feeling remained when we shopped. We hit Fifth Avenue and did more window-shopping than anything, though I did buy Phoebe a beautiful scarf at Bergdorf's and, at Takashimaya, bath salts and body lotion.

We also stopped at the Godiva boutique at Rockefeller Plaza. You already know I love chocolate. What you don't know is that I have a thing—a *real* thing—for Godiva truffles. And not just *any* Godiva truffles. I would beg for a roasted almond truffle and steal for a smooth coconut, but for a hazelnut praline, I could be tempted to kill.

I'm kidding, of course. But I was happier

once my shoulder bag contained that little box of truffles.

We walked over to Madison and shopped more. I kept an elbow linked with Phoebe's to steady her, but it brought us closer even beyond the physical. We talked as we walked—about Mom and Daddy, about Sabina and rivalry and happy times, too. Then Phoebe flagged. We returned to our hotel for high tea in the lobby, and, reinvigorated, visited the Central Park Zoo. Phoebe needed a nap after that, but she loved the restaurant where we ate dinner. I was gratified. It had always been a favorite of Greg's and mine.

We spent Sunday and Monday at the buyer's show, held at the piers on the Hudson, and I gained a new respect for what Phoebe did. This was hard work. I have never seen so much merchandise of so many different kinds, with so many people so skilled in the sell. How much of what to buy and in what colors? Amazingly, Phoebe was able to sort through it all.

I know what you're thinking. You're thinking that her problem wasn't physical at all but purely mental, that it was depression, and that removing her from Middle River had lifted her spirits. I had been hoping that, too. But it wasn't the case. The physical problems remained—the odd gait, bad balance, fine tremor. She still easily lost her train of thought and had trouble finding words. What she did with regard to work, though, was to compensate. She had brought along notes and reminders, and not only checked them often herself but had prepped me on them, so that I could assist her as smoothly as possible.

Using the past as a guide, she moved methodically down a list of known vendors. And she was remarkable. She might have faltered at first, but she always ended up asking the right questions of each, expressing reservations when the pitched merchandise was inappropriate, approaching new vendors when their goods caught her eye. She drew on every bit of her inner wherewithal to stay focused. But it exhausted her. We had room service deliver our dinner both nights.

Then came Tuesday morning. Phoebe had expected that we would sleep late and leisurely return to the airport. She was not at all happy when I told her what I had done. I do believe that if she had been rested, she would have flat out refused. She protested, but feebly. She said she didn't need a doctor. She said that I had no business making an appointment behind her back, and that I was causing a rift between Sabina and her. She argued that Alyssa had not had mercury poisoning and therefore why in the world should she?

But she didn't have the strength to fight for long. Any last protest died when Sabina called on the phone in support of my plan.

Judith Barlow was a surprise. Anticipating a specialist in alternative medicine, I had expected someone close in age to Tom, but she was easily approaching sixty. I had expected someone who looked eccentric, but she was elegant in a simple and conservative way. Likewise her office, which, contrary to my expectation, was in a prime (read "high rent") area, and was large, professional, and polished. The woman clearly had a successful practice.

She smiled when I was drawn to the diplomas on the wall. "People come here half expecting that I got my degree from Sears Roebuck."

"Harvard?" I was impressed.

"Undergraduate and medical school, with

my training at Mass General. The fact is that I practiced traditional medicine for twenty years before I moved into the other."

"Why the move?"

"Too many people came in with problems that I couldn't touch with traditional treatment. When you reach the point of having tried everything and getting nowhere, you have to look farther."

"We're not quite at that stage," I cautioned with a glance at my sister. She was sitting in a chair, looking frightened. "Phoebe hasn't seen a doctor, but her symptoms are identical to those we saw in our mother. I was hoping we could skip treatment that didn't work on Mom."

Judith sat down at her desk, put on wire-rimmed reading glasses, and studied the file I had brought. When she was done, she opened a clean file on Phoebe and proceeded to ask more questions than I had ever heard a doctor ask at a single sitting. More important, she asked the *right* questions. I'm still not sure whether Phoebe had decided to cooperate or was simply worn down by what ailed her so that she had no defense. But I was stunned by some of her answers. I knew about the obvious symptoms. But joint pain? Itchy skin? Night headaches? These were all news.

There was a physical exam, followed by more tests than I thought a single doctor's office was capable of performing. But Tom had known what he was doing in recommending Judith. She had a remarkable facility, with nearly every diagnostic tool at her fingertips, and those that she didn't have were in nearby office suites. Moreover, she was able to get quick readings on tests, so that she could decide which ones to do next.

We missed our plane. And Phoebe grew increasingly cranky. Taking one test after the other was grueling and, of course, there were waits between tests. But Judith was methodical. At every turn, she explained what she was testing for and why. As Tom had told me she would, she ruled out every conventional ailment before she ever mentioned mercury.

That came in the final sit-down. "If these symptoms were new," she told Phoebe, "say, within the last three or four months, I'd test a hair sample for the presence of mercury. Given how long you've had symptoms, though, mercury from an initial exposure would have long since grown out. Absent our ability to actually isolate it through other tests, I can only make an educated guess. Mercury is the leading suspect. Things like grogginess and itchy skin aren't indicative of Parkinson's or Alzheimer's. Nor is asthma, which is what that lingering cold of yours appears to be. So we have a choice. We can treat the symptoms. Or we can go for a cure."

I looked at Phoebe, but she didn't seem to know what to say. So I asked, "What does a cure entail?"

Judith explained the general premise of chelation therapy to Phoebe, much as Tom had to me. But she went further, patiently explaining the details. "The specific protocol for this treatment is evolving as the results of new studies come in. Normally, I would recommend the oral route, which would entail a varying schedule of pills taken every four hours on alternate weeks, then, depending on your progress, a switch to pills taken three to four days every week or two. We do it slowly over a period of months, and we give you breaks. It's gentler. But it does take a while before you feel better."

"What's a while?" I asked.

"Two to six months. On the other hand,

given your proximity to Tom, he could administer a newer protocol that involves an eight-hour intravenous infusion. Some believe that this is the only way to remove mercury from the brain. The infusion hits your body hard. Following it, you have a bad couple of days during which you can't do much of anything. But then suddenly you *do* feel better. We repeat the infusion every few months, until the mercury is gone."

"How can you tell when it's gone?"

"When mercury is pulled from the organs, as happens in chelation therapy, it is expelled in the urine."

"I thought mercury wasn't expelled that way."

"On its own, it isn't. Attached to a chelation agent, it is. In those few days following the infusion, we closely monitor the urine for its mercury content."

"How long until it's gone?"

"Not knowing exactly how much is there, I can't tell for sure. In general, though, it takes one to four years until you're free of the metal."

"*That* long?" Phoebe cried feebly.

Judith smiled. "It isn't so long when you stop and think about how long you've been experiencing these symptoms." She patted Phoebe's file, which had mushroomed in a day. "According to what you've told me, you've been symptomatic at some level for several years." She shot me a glance. "Now the thing to do is to find out when and where you were exposed."

I called Sabina from the cab on the way to the airport. She had been trying me every few hours and was waiting for word. To her credit, she listened. More than that, she *heard*.

"Not Parkinson's?" she asked.

"The doctor doubts it."

"Does that mean Mom didn't have Parkinson's either?"

"It's hard to tell. Mom was older than Phoebe when she got sick, and in fairness to Tom, the symptoms could have fit Parkinson's. But Phoebe has other symptoms." As I spoke, my poor sister was tucked into a far corner of the tattered leather seat, staring out the window in something of a fog. I'm not even sure she heard what I said. I asked Sabina, "Did you know that she gets intense skin rashes on her stomach?"

"No."

"Or that she has pain in her joints?"

"She thought that was the flu."

"She's had those pains on and off for several years."

"Several years, and she didn't *say* something?"

I had been frustrated, too. But I could see Phoebe's point. "She thought it was age. Early arthritis. Then Mom got really bad and Phoebe was frightened. Mom died. It's scary when you have the same symptoms. But there's hope for Phoebe." I explained what chelation was about.

"Attaches to metal?" Sabina echoed. "Couldn't that metal be lead?"

"The blood workup says no. No lead there. Mercury won't show up in the blood, but that's what the doctor thinks it is."

"The doctor, or you?" Sabina asked with the old distrust.

"I was quiet, Sabina. I didn't say the *m* word once. She came up with mercury all on her own." When Sabina was silent, I said, "I know you don't want to hear this. It complicates things, because if mercury is the problem, there's only one possible source." I waited for her to argue, and still she was silent. "Do you know what I mean? Sabina?"

I checked the bars on my cell. There was barely one.

We weren't able to reconnect until Phoebe and I reached the airport, and then I couldn't talk. There was too much to do, what with supporting Phoebe and our luggage. We managed to get seats on the last flight to Manchester, but time was short and the security lines long. I had visions of our missing this plane, too, and sweated it out. We actually made it to the gate with a handful of minutes to spare before boarding began. Moving off to the side, I called Sabina.

"Hey," I said. "Can you talk?"

"Yeah. I'm trying to take in what you said. How's Phoebe handling it?"

"Hard to tell, she's so tired. She says she's relieved. Poor thing must have been terrified of getting worse and worse."

"Mercury? Are you *sure*?"

"Yes. And I have to tell you more, Sabina." It was time. I needed an ally. Sabina's mind seemed open. I spoke quickly, wanting to get it all out before we boarded the plane. "Remember the fires that totaled the Clubhouse and the Gazebo? They were set." Without mentioning TrueBlue, I told her what he had said.

Sabina was fast to respond. "I find that very hard to believe. There was a clandestine clean-up, and *none* of the people involved told *anyone* about it?"

"What if they weren't told what had spilled? Or what if they had been sworn to secrecy? The Meades give bonuses for that kind of thing. *Big* bonuses."

"Wouldn't they have gotten sick themselves?"

"Not if they were wearing protective gear."

"But you're talking arson—and *fraud*. That's big-time criminal behavior. It's a lot to swallow. A lot. Who's your source?"

"I can't say yet, but think about it, Sabina. If this plot was in a book, it would make total sense, wouldn't it?"

"But do you know how many events take place at the Clubhouse?" she continued to argue. "If there was a mercury spill beneath it, wouldn't *all* those people be sick?"

"Not if they weren't there on one of the few days between the spill and the fire. What if we found that Mom was? What if Phoebe was? She can't remember. At the time of the Clubhouse fire, she was already working with Mom. We need to try to reconstruct her schedule. Same with other people who might have been there. I talked with the McCreedys."

"Annie—"

"Don't worry. They deny any connection. But look at the physical problems they've had. They're florists. They do work for the mill. Isn't it feasible that they were at the Clubhouse delivering flowers for an event that Mom attended?"

"I don't think you should be involving other people yet. This is *so* off the wall."

I didn't snap at her, just stayed calm and cool. "Look at it this way. If Phoebe does have chronic mercury poisoning and can be helped with Dr. Barlow's protocol, how many other people in Middle River could use the same?" I had been listening to airport voices with half an ear, but suddenly heard the right words. "They're calling our flight, Sabina. I have to get Phoebe. We won't be arriving in Middle River until late, and I want to go running early tomorrow. I'm feeling emotionally stuffed up, if you know what I mean. Can we talk later in the morning?"

Emotionally stuffed up? Well, that was one way to put it. Another was curious. I wanted to see James.

Wanted to? Was *dying* to. What can I say—he was an incredible lover. I would have thought that being a Meade, he would have made love in an egotistical, *Me, Tarzan!* sort of way. And he was strong. And forceful. And definitely took the lead—but he also took care to make sure he satisfied me.

Granted, it wasn't a hard thing to do. One look at James, and my insides quickened.

About now, you're probably thinking that I am a total hypocrite. After all, since I'd last seen James, I had learned that he and his family had been involved in the dirtiest, most immoral of activities. I mean, knowing that people were exposed to mercury and not telling them what it was that they had? That was as low as it got.

I could rationalize and say that something didn't fit about all this, and it didn't. James didn't seem all bad. Hadn't he tried to separate himself from his father and brother that first day we talked at Omie's? Besides, could a man who had adopted a baby he clearly adored actually raise her in the shadow of poison? Granted, he didn't bring her to the day care center at the mill. He had a nanny. Still, I had to believe he cared for his child.

So that was one reason I wanted to see

him in spite of all that I had learned. But there was his parting shot at Omie's last Friday. *I don't do one-night stands.* I didn't know what he had meant by that, but I also didn't know what I *wanted* him to mean. He was my summer fling. That's all. We had no future, James and I. James was a Meade. End of story.

That said, as summer flings went, he was hot and heavy—and that was *not* my imagination. So maybe I was curious to see if I could make him hot and heavy again. Was my ego involved? Of course, it was! Remember, another Meade had made a fool of me once. Egos don't forget things like that.

Despite getting home after midnight, I was up at six Wednesday morning. By six-thirty, I was sitting in the yard with Mom's flowers, the willow, and my second cup of coffee. I called Tom at seven to tell him to expect a call from Judith. At seven-thirty, I checked on Phoebe, who was sleeping still. I left her a note, propped it on the kitchen table, and set off for the varsity course.

The high school parking lot had a sprinkling of cars now, more than last week, fewer than next week. By then, if we continued to run at eight, we would be seen. Of course,

James had no choice but to run at that hour.
I understood that now. He couldn't leave the
house until Mia's nanny arrived.

Skirting the cars in the lot, I drove to the
back and rounded the corner. James's SUV
was there. The man himself was stretching
on the grass.

My pulse gave a hitch, then raced on
ahead. I parked, climbed from the car. He
shot me a small smile, which I returned.
Then I joined him stretching.

You are such a coward, Grace declared.
*He is the man to talk with. Are you going
to stand there and stretch, when he has
answers you need? 'My pulse gave a
hitch, then raced on ahead.' Oh please!*

Whoa. Weren't you the first one to tell me
how sexy he is?

*I never slept with the enemy. For that
matter, I never slept with men I didn't
love. But all that is beside the point. The
point is that the man stretching right
over there is someone to challenge. His
family is bad. You need to talk to him
about this.*

But we don't talk, James and I. We run.
And we make love. We don't bring in the rest
of the world. Hey, he could be asking me

about my trip right now, but he isn't. I'm okay with that. It works on my end.

That's why I say you're a coward. Life is about confrontation. I lived in a time when women didn't dare speak their piece, and I did speak my piece. That took guts.

And look where it got you. You drowned in a bottle and died. What do you want from me?

Confrontation. You live in a time when you can speak out against wrongs, and you do have the goods on the guy. So speak. You owe it to me.

To you? I was startled.

Yes. I coddled you when you needed it. I gave you your identity, for God's sake.

Hah. It's been a yoke.

It's been a boon. So now I'm asking for your help. I need the Meades destroyed.

Why?

Because I can't do it myself. Because I died too young. Because I had too many other problems, and those problems mucked things up. You don't have those problems. Don't you see, Annie. If you do this—if you go home to your little New England town, turn over those rocks, and expose the dirt—it'll validate what I did myself. I was vilified. That wasn't fair. I was living in the wrong place at the wrong time. I need you to set things straight.

"Ready to run?" James asked.

The sound of his voice silenced Grace. I knew she'd be back. She had raised a valid point. We needed to discuss it, she and I. But not now.

Pushing myself to my feet, I nodded. We headed for the woods and set off on the path—and it was a relief from the start. I did need the airing-out. It was only as I ran that I realized how tense I had been those days with Phoebe. Running now, the tension began to melt from my limbs. I focused on the warm air and the trees, on the pine needles and the sun that pierced them. I focused on being back here doing something I liked—and when we passed the cutoff to Cooper's Point, I focused on how far I'd come. Feeling strong, I lengthened my stride.

That was when I hit James—just ran smack into him—because he stopped short. Turning quickly, he caught my arm and steadied me. We were both breathing hard.

Our eyes met, and it was there again, every bit of what we had shared the Thursday before and then some, because this time we knew where it could go. He dipped his head and caught my mouth, and it was almost comical, trying to kiss that way with the two of us breathless from running, but amazingly it worked. He deepened the kiss—or did I?—because a simple kiss wasn't enough. Nor was not touching. My back found a tree that fit, and he pressed me there. His hands moved from my thighs over my middle to my breasts—both hands, both breasts—while I worked my own against his groin.

I heard a cry. I might have thought it had come from James, if it hadn't been so high. And it wasn't from me. At least, I didn't think it was.

Apparently, neither did James. He stopped what he was doing and lifted his head. "There it is again," he said in a raggedy way. "I thought that was you."

"Again?"

"I heard it before. That's why I stopped."

"You thought I—"

"Wanted me." He looked off into the woods.

I might have been embarrassed—like he had stopped only because he thought *I* needed *him,* like I was a charity case—if I hadn't felt his erection against me. It had been there all along. The need was mutual.

But that cry came again, and I, too, looked toward deeper woods.

"Trouble?" I whispered, quickly forgetting my own need.

"I don't know," he whispered back, apparently past his own need as well. Taking my hand, he led me off the path and into the woods. We worked our way between trees, over pine needles and moss, around rocks.

Still holding my hand, James stopped and waited. The cry came again, definitely human and female. Our eyes met.

"Distress?" I whispered.

"Doesn't sound it," he whispered back, "but can we be sure?"

We couldn't. On the chance that there was indeed trouble, James put me behind him. I clutched the back of his shirt and measured my steps to his and we moved on.

We didn't have to go much farther. Just when another cry came, we rounded a tall patch of ferns to an opening in the woods— and, looking around his arm, I caught my breath. A woman was there, bound to a tree.

No, I realized. Not bound. Her back was to the tree and her arms curved around it behind her, but there were no ropes. Her eyes were closed, head turned to the side, long dark hair draped over her shoulder. No, there were no bonds. Ecstasy held her there, aided by the dark head and pale male body that covered her lower half.

"*Omigod,*" I whispered and would have backed up as silently as we had come—we were intruding on *the* most intimate of activities—if James hadn't been so firmly planted where he stood. I gave a tiny tug at his elbow, but he wasn't moving. Even more startling, he put his hands on his hips.

The wait wasn't long. I don't know whether something tipped off the woman (whom I didn't recognize) or whether she had simply reached a break in her ecstasy-driven bliss, but her head rolled to the front and her eyes opened. She didn't cry out then. She screamed. Her arms crossed her torso to cover her breasts. Her lover looked up, then around.

I did recognize him. Instantly. It was Hal Healy—Hal, who was so in love with his voluptuous wife that he couldn't keep his eyes or hands off her; Hal, who had accused *me* of being a negative influence on the impressionable young girls in town; Hal, the illustrious principal of good old Middle River High. We had stumbled across another of Middle River's dirty little secrets.

"You self-righteous prick," James said.

Hal seemed to sink lower toward the ground, trying to hide his buttocks from our view, though he couldn't turn without exposing himself more. He couldn't run, couldn't hide. All he could do was to go red in the face, which he did. It would have been humorous, if it hadn't been so pathetic.

"It's not what you think," he managed in a shaky voice. His eyes were on James, not me. James was clearly the threat.

"No?" James asked. "Then what is it?" To me, James said toward his shoulder, "I assume you know Dr. Healy. The lovely woman with him is Miss Eloise Delay, the guidance counselor he hired last year. She's fresh out of college, which makes her substantially younger than our prinicpal. She gives advice to troubled students."

Miss Eloise Delay was frozen against the tree, covering herself as best she could, her eyes large and filled with horror. She looked like she would have welcomed death, if that were her only means of escape.

"Don't take this out on her," Hal begged.

"Was she unwilling? It didn't look it. I'd heard rumors, Hal. People are wondering why you're all over your wife whenever anyone is looking, at the same time *she's* complaining that you're at school night after night. The rumors hinted that it was someone at school, but could never quite home in on who. So it's you and Eloise, working late evenings? That's bad enough. But here, in broad daylight? What if I'd been one of your students? What if I'd been a whole group of your students?"

Lacking an answer, Hal begged, "Back off so we can get dressed, at least."

"When I'm done," James said with every bit of the authority I had always attributed to him. He didn't have to raise his voice to imbue it with thunder. "What are we going to do with you?"

"It will never happen again," Hal said.

"How many times have students said that, just prior to being suspended or, worse, expelled? How long's it been going on, Miss Delay? Was the spark there when you interviewed for the job?"

Eloise didn't speak. I suspected she couldn't have found her voice if she'd tried.

"Hal?" James prodded. "How long? I remember when you hired her, there was some talk about her being inexperienced. You were her champion before the school board. What was it you said? Stellar academic record? Brilliant recommendations? Personality plus? I guess the *plus* takes on new meaning now."

I touched James's arm. "Let's go," I whispered.

But James wasn't done. "What about your wife, Hal? She's lived in this town all her life. Do you have any idea what happens when the rumors go to work? Think it's fun for her? Okay, so you resign and take a job somewhere else, but what about her? This is her home. Are you going to uproot her and take her along? Is she your cover? Is that it? Or will you divorce her and scoot off with Miss Delay?"

"James," I repeated. I detested Hal Healy, but I was beginning to squirm.

My plea registered. James made a scornful sound in his throat, turned and strode off. I followed gladly.

"What are you going to do?" Hal cried.

Calling back over his shoulder now, James thundered, and I have to say it was awesome: "Let you *wonder* what I'm going to do."

Back on the path, he continued to stride, taking the shortest route to the parking lot.

"James?" I called from several paces behind.

He held up a hand and went on, stopping only when we reached the grass on the far side of the trees. Bending over, he put his hands on his knees and hung his head. I came alongside, but just stood there and waited until he had regained control. Then he straightened.

His eyes met mine. "I'm not sorry for one word of that. He deserved it. He's a pompous ass. But what about us? Are we any different?"

I knew exactly what he was thinking. Lord knew I had my faults, but I refused to group myself with the Hal Healys of the world. "Yes, we're different. For one thing, when we did it here, it was dark. For another, neither of us is married. For a third, *we* don't preach abstinence and restraint. He does—all the time, from what I hear. My niece was telling me that he's even cracked down on the showing of bare skin in school, and I agree completely, except that he has some gall to be calling for that, when he's out here without a stitch of clothing on *and* with someone who isn't his wife."

James's eyes were dark. "Are you Greg Steele's lover?"

"No."

"You live together, but you don't do it?"

"No."

"Did you ever?"

"No. We're very good friends. The cost of condos in the District has gone through the roof, but because we pooled our resources, we could afford something good. Neither one of us wanted to live alone. We'd rather live with each other than with anyone else. We see other people; if one of those relationships ever gets serious, we would sell the condo, but for now, this works. We have separate bedrooms on separate floors. We've never been sexually involved."

I'm not sure if my answer pleased him. His eyes remained dark, not so much doubting as haunted. "So what we did hasn't broken any moral code. And yes, we did it before in the dark. But just now, would we have stopped what we were doing if we hadn't heard her cry?"

He had me there. I couldn't answer. The truth was that when I was with James like that, the rest of the world didn't exist.

He grunted. "Yeah. I know. So what are we going to do about it?"

"Do we have to do anything?" I asked with a half smile. "Can't we just enjoy it while it lasts?"

He studied me for another minute, then pushed a hand through his rumpled hair. Leaving that hand on the back of his neck, he regarded me with what seemed like amazement and returned my half smile. "I keep saying, is this really Annie Barnes? The old Annie Barnes, who was such a royal pain in the Meade butt way back when? I don't understand why I'm attracted to you."

"Thanks."

"You know what I mean. Don't you feel the same way?"

Suddenly I was very serious. I wouldn't have picked this time or place. I actually would have let it go awhile, because it was so nice just . . . feeling James Meade against my body . . . and surely this kind of discussion would ruin that. But here was another truth, but I've lost track of the count, so what is it, maybe **TRUTH #18? Time and place**

have a mind of their own. We can decide one thing. Then another thing happens, and our decision is moot. We can't go back, only ahead.

"You bet I do," I said, and what came out of my mouth then had nothing to do with Grace. She had been a victim of her own time and place, but this was mine. "You're the bad guy. I should *not* be attracted to you, especially after what I learned in New York. My sister went through a whole barrage of medical tests there, the upshot of which is that she has been diagnosed with mercury poisoning. The only source of mercury here is your mill."

"I was under the impression that mercury poisoning could come from dental amalgam."

"It can. But she doesn't have silver fillings. Besides, there have been spills."

He stared at me. Doubting? Daring?

"Phoebe isn't the only one who is sick," I said.

"I know."

I pounced on that. "You know, and you do nothing about it?"

"That isn't exactly correct."

"Enlighten me, please."

He looked off toward the field house,

which was the only part of the high school that was visible from where we stood. "Not here. Not now."

"But you raised it. If not here and not now, *when*?"

"Tonight."

"I have to take care of Phoebe. She is really bad, James, and it'll get worse before it's better. She goes in tomorrow for the first of the treatments. Tom's doing it—and, so help me, if you people cause trouble for him because of that, I *will* write my book."

"No trouble. When will Phoebe be in bed tonight?"

"Nine."

"Come then."

Chapter 21

SABINA MADE breakfast for Ron and the kids, but she was distracted. As soon as Lisa and Timmy had left the room, Ron sat back in his chair. "Calling Sabina," he teased. "Come in, Sabina."

Her eyes flew to his. She smiled sadly. Ron did know her well.

Helping herself to another cup of coffee, she sat kitty-corner to him. "I think . . . we have a problem." She passed on what Annie had told her the night before.

"The first problem is insomnia," Ron said. "You barely slept. Why didn't you tell me this last night?"

"Because I didn't want to believe it—and then, when it kept on nagging at me, I wasn't sure what to *do* with it. There's so much at risk. On one hand, there's Phoebe and all those others who are sick. On the other, there's the mill. Aidan is watching me. The slightest provocation, and he'll fire me." She frowned. "And then there's Annie."

"Sweet, successful, troublemaker Annie," Ron said.

"Troublemaker Annie is right."

"Our very own Grace, stirring things up."

"Yes, and I keep telling myself that, only this time, what she says makes sense. I've been going over it in my mind all night, looking for every reason to say she's simply out for revenge. Except that she's right—as plots go, this one is plausible. Not only that, but it explains so much." She raised her mug and let the scent of the coffee soothe her. In time, she sipped it.

"Did you decide what to do?"

"Actively decide? No. But it's like I'm being swept along with something that I can't drop. This isn't just Annie telling the world that Daddy did most of my sixth-grade science project, or that Phoebe sucked her thumb until she was ten. It isn't just Annie accusing

Aidan Meade of being a liar or Sandy Meade of manipulating Sam Winchell. This is poison. It's innocent people being robbed of their health. It's Mom—and maybe Omie and a slew of others—*dying* because of something that could have been prevented." She paused, struck again by the weight of it all. "I mean, it really is bad. Can I turn my back and walk away without saying a word?"

Ron chewed on his cheek.

"Yes, I may lose my job," she went on. "You, too. Would you hate me if it happened?"

"I'd be worried about how we'd survive."

"But suppose we figured that out, would you hate me for what I'd done?"

"I might wonder what it accomplished. Look, Sabina, even if Annie's right, can she really take on the Meades and win?"

Sabina had spent the wee morning hours considering that question. She kept trying to find alternative answers, to give herself *choices,* but only one emerged. Could the Meades actually be beaten?

"This time, yes," she said, "because if what Annie says is right, this is bigger than Middle River. This isn't about who's sleeping with who. It isn't about who has power and who doesn't. It's about who's sick and dying, and whether it can go on." She spoke plead-ingly, desperate that her husband under-stand, because, of all the other arguments, this one tormented her the most. "Our kids were in that day care center, Ron. They were there, each of them, from the time they were three months old until they went to school. What if something had happened during those years? Or before they were born? You're up there by the shipping docks all day, but I'm going to different buildings all the time. What if I had happened to walk through an area where there was a spill while I was pregnant? Getting mercury out of Phoebe's body is one thing, but when there's fetal damage, it's permanent."

Ron countered, "What if I fell off the ship-ping dock, like Johnny Kraemon did last week? He broke his back. He'll never walk again. Bad things happen."

"Accidents do. But what if we have a chance to prevent them? Don't you see, there's my sister Annie, who has a life some-where else, and she's committed to this cause—and here I am, with a life right here, and I'm on the fence? Who stands to lose more? How will I feel if she actually *wins* this?"

Ron was silent for a moment, his expression grim. Then, with a chill she rarely heard, he said, "Is it still a race between you two?"

Sabina was stunned. "Low blow, Ron."

"Answer the question. Is the motivation Annie, or is it the cause?"

Taking cover in indignation, Sabina rose from the table, noisily put her mug in the sink, reached for her briefcase, and, without another word to her husband, set off for work.

Middle River was its bucolic self, but she saw little of the sun, little of the trees, flowers, or homes. What she saw were the Waxman twins, age five, riding bicycles in their driveway—and the Hestafield boys, ages eight and eleven, throwing a ball into a pitch-back on their lawn—and the Webster children, age one and three, being strapped into car seats by their mom. Middle River was full of children, all innocent of worry and dependent on the goodness of the town. She wondered if Ron realized that when he questioned her motives—wondered if Annie realized it when she pricked Sabina's conscience—wondered if Aidan realized it when he issued ultimatums.

She arrived at the Data Center feeling entirely unsettled, and that, before she discovered a major electrical problem. Yes, there was a backup generator, but it was nowhere near as powerful as the regular current, which meant that anything she tried to do would be painfully slow.

She called Maintenance, then put several backup CDs in her briefcase. Electrical problems happened. She could always work in one of the other buildings.

But not now. She was *so* not in the mood. Once the electrician arrived, she went outside, crossed through the parking lot, and sank into an Adirondack chair.

Northwood Mill was waking up. She saw Cindy and Edward DePaw pass in their pickup, saw Melissa Morton, Chuck Young, and Wendy Smith in their vehicles, all on this road as opposed to the plant road, because strapped safely into the backseat was a child—destination, the day care center.

Antsy, Sabina pushed herself up and headed across the grass, over a knoll and through a cluster of trees to the marketing department. Part of the Administrative Campus, it was housed in a redbrick Cape comparable with Sabina's Data Center. Selena Post worked here. She and Sabina had become co-conspirators in school when aging teachers confused their names. They still had coffee together from time to time.

Selena's office was warm after a night without air, and filled to the brim with stacks of promotional materials. "Major mailing this week," she told Sabina by way of explanation for the mess.

Sabina slid into the chair by her desk and said in a very low voice, "Mercury. Know anything about it?"

"Oh yeah. We're clean. That's always been part of our marketing claim."

"Always?"

"Well, for the last few years. It was a while before people realized the harm mercury could do. Every new piece of equipment we buy for the plant is either toxin-free, or comes with the mechanism to nullify whatever toxicity is produced."

"What about cleaning up after the mercury produced years ago?"

"Done back then. Northwood is cited by the state for its environmental considerations. We rank among the top in all of New England for the cleanliness of our plant. That's one of the reasons we can afford to provide such extensive health care coverage for our workers and their families."

Sabina knew there was a chance Selena was right—which left Alyssa's death, Phoebe's illness, and the ailments of so many in town unexplained. But there was also a chance that Selena had fully bought into the hype, or that she was truly ignorant of the problem. In any case, Sabina knew a hard sell when she heard it. Her gut told her that further discussion would be fruitless.

Closing the conversation with Selena, she proceeded on, this time walking along the side of the road, one building back to human resources. Her friend Janice worked there. Fortuitously, she was just climbing from her car when Sabina approached.

"Hey," Janice said with a smile, but the smile faded when Sabina drew closer. "What's wrong?"

"I have to ask you something, Jan," Sabina said in a voice that wouldn't carry past where they stood. "Do you ever hear anything about mercury here at the mill?"

"Not lately. Years ago, maybe. But it's been cleaned up."

"How do you know?"

"I have it from the horse's mouth."

The horse's mouth meant one of the Meades. Sabina tried a different angle. "You handle requests for disability benefits, right?"

"Yes."

"Permanent disability?"

"Some."

"Have there been any suspicious ones? Any trends?"

"Like . . . ?"

"Parkinson's. Alzheimer's. Chronic fatigue. Depression."

"A few. That's where the Meades shine. They come in and help. Not only is the basic benefit package solid, but they add things when the severity of the problem merits it."

Sabina hadn't heard about this before. "What do you mean, 'add things'?"

"Money. It's done in a quiet way, so that the recipient doesn't feel as though people are looking and saying they're pure charity cases. The way it's done, they maintain their dignity. But they are taken care of."

Sabina just bet that they were. It was called "hush money." "Do they sign papers promising not to tell?"

"No," Janice chided. "I mean, I have no

idea what's said face-to-face in the privacy of Sandy Meade's office. He handles it himself. He says it's one of the best parts of his job, putting people's minds at ease." She hesitated. "Why do you ask?"

Because, Sabina wanted to say, *you have a little girl in the day care center as we speak, and she may be in danger.* But she didn't say it, because she didn't have proof, and it would be wrong to create unnecessary fear. Just because Annie had presented a convincing scenario didn't mean it was so. *Troublemaker Annie,* Ron had called her. *Our very own Grace, stirring things up.*

Sabina couldn't count the number of times she had called her sister these things herself and felt perfectly justified doing it. But this time it didn't feel right. She didn't know why—Annie hadn't revealed her source—but Sabina believed her this time.

She couldn't explain all this to Janice, though. So she simply shook her head, said a teasing, "No reason. Just wondering what to expect in *my* doddering old age. Catch you later," as Janice turned back to take her briefcase from the car.

Sabina headed back to the Data Center, but her thoughts had stalled. No, this was *not* a race between Annie and her. For the first time, truly it wasn't. The dynamics between them had changed. This playing field was new.

Is the motivation Annie, or is it the cause? Ron had asked, and it was actually both, though in a surprising way. Sabina believed in the cause, but she also, for the first time, believed in her sister. If Annie had a source, it was a good one. Sabina wanted to be *with* her on this.

Call it a reaction to losing Alyssa. Call it a reaction to Phoebe's symptoms. Call it the culmination of years of disdain for Aidan Meade. But there it was.

Several hundred yards shy of the Data Center, she abruptly shifted course. She headed north toward the picturesque little threesome just shy of the plant itself, and picturesque it was. There had always been something special about this area. It was typically the first to be cleaned, polished, landscaped. Sabina had always attributed that to the fact that these three—Clubhouse, Gazebo, Children's Center—were used by Middle River. Suddenly there was another possibility. It was possible that the Meades kept the surface here utterly pristine quite literally to compensate for bad stuff that lay beneath.

What was it that Grace Metalious had said so many years ago—that to turn over a rock in any small New England town was to find bad stuff beneath? Middle Riverites had heard that quote more times than they could count. Each time, they nodded, chuckled, and went on with their lives.

Sabina couldn't do that this time. There was merit to the quote, and if Annie was the one to turn over that first stone right here, Sabina was proud of her.

The Children's Center was still dressed in the yellows and greens of summer—decorative flags out front, painted sign, protective awnings. She held the door open for a pair of mothers delivering their children, then followed them inside. It was a minute before she located Antoinette DeMille. Toni was the center's director and one of Sabina's most respected friends at the mill. Sabina found her in the infant room, holding a crying child that couldn't have been more than three months old.

"Separation problems?" Sabina asked.

"Nope," Toni said dryly. "Colic. Should be easing up any day, but until then, it's tough.

Breaks your heart to feel these little legs scrunch up against belly pain."

Sabina nodded. She waited until one of the other teachers was free and took the child from Toni. Then she followed Toni out of the room.

Toni glanced at her as they walked. "You look like you need to talk."

"Yes. But in private."

Toni gestured her into her office and closed the door once they were inside.

Sabina didn't mince words. "Not so long ago, Northwood used mercury as part of the process of bleaching the wood. That produced toxic waste, which was disposed of, and the newest processes don't use mercury at all. Still, it's in the river. Know how people aren't supposed to eat fish?"

"They do anyway."

"Yes, because the warning is always half-hearted. But I've been told there's another problem, that some of the mercury waste was buried in large drums here on the campus, and that several of the drums have leaked." She tied the fires and subsequent rebuilding of the Clubhouse and the Gazebo to this. "The Children's Center was built on a burial site, too."

Toni's jaw dropped, but she quickly recovered. "Oh, I don't think so. The Meades wouldn't do that. There are children here."

"What better place to hide illegal disposal of toxic waste?"

"And put children in danger?" She shook her head. "I don't think so. The Meades wouldn't do that."

"Do *they* have any children in the center?" Sabina asked, then answered, "No. So maybe they know something we don't?"

"But the center's been here for more than twenty-five years, and there's been no unusual illness."

"Because the drums are holding. But what if there's a leak, like the leaks at the Clubhouse and the Gazebo? There would be a fire and a cleanup, but what harm would be caused in the meantime?"

"Those fires, Sabina. There were logical causes for both. Who told you they were set?"

Sabina hesitated.

Toni guessed. "This is coming from your sister, isn't it? She's been trying to nail Middle River for years."

"Actually, not," Sabina said. "She's been gone a long time. She could have skewered us in a book, but she hasn't, and she doesn't plan to. This is for our mother and all those other people who've been sick."

"Do you know they were here at the time of the leaks?"

"No."

"There you go. I know you love your sister, but this time she's out in left field. If you'd been talking about a mercury spill at the plant that was covered up, that's one thing. People at the plant know the risks. But to say that the Meades would put innocents at risk is something else. I mean, the Clubhouse and the Gazebo are one thing. They're used only occasionally. But there's no way the Meades would build a day care center over a waste dump, not something that's used every day and by *children*. If they realized there was the potential for harm here, they'd have razed the place early on, cleaned up, and rebuilt. We're safe. Truly we are."

Toni DeMille loved children. She had raised six of her own and had earned a degree in early childhood education prior to being named director of the Children's Center at Northwood Mill. She prided herself on being one step ahead of the Children's Center's insurers, and was putting tots in for naps on their back even before it became the norm. She didn't believe in taking chances when it came to the health of her kids.

For that reason, she found herself thinking more and more about what Sabina had said as the morning passed. As the afternoon progressed and the children of part-time moms started to leave, she was taking good, long looks at those children and thinking about it even more. She debated calling her husband, who worked in the plant. Or her neighbor, who worked in sales. Or her cousin, who worked with Sabina's husband in shipping.

But that was how rumors started, and the Children's Center couldn't function in a maelstrom of rumors. So she went to Aidan Meade and told him what she had heard. She didn't mention Sabina's name until Aidan did, and then she couldn't lie. When he made mention of Annie Barnes, they shared a chuckle. Annie did have that element of Grace in her. But Middle River was no Peyton Place. They agreed on that.

On the matter of mercury waste being buried under the Children's Center, Aidan put her worries to rest. He denied there was anything buried anywhere. He said that the Children's Center was the pride and joy of the mill, largely thanks to Toni's vision and care, and that his family would never, ever do anything to jeopardize that.

Aidan Meade wasn't upset. Quite the opposite, he was very pleased. He was a master at trumping up charges, but when real ones presented themselves, it saved him the work. When *real* ones presented themselves, he didn't have to explain himself to anyone.

He stood at his office window until he saw Toni DeMille cross the road and head over the grass toward the Children's Center. Then he picked up the phone and called the mill's security chief. He didn't check with Sandy; Sandy would only ask why he had waited so long to deal with the problem. Nor did he need Nicole punching out the extension for him. He knew what to do and wanted the joy of doing every part of it himself.

Within five minutes, the chief and two of his trusted guards reached Aidan's office. Setting off on foot down the road, Aidan explained what they were going to do. When they reached the Data Center, he led the way inside and went straight to Sabina's desk.

The two others in the office had looked up, but she was so engrossed in something on her computer screen that she didn't appear even to know he was there. That clinched it. He felt no mercy at all.

"Sabina," he said sharply.

She looked up in surprise. Her eyes strayed to his companions, then returned to his. She frowned.

Feeling immense satisfaction, Aidan said, "You're fired. You have five minutes to take whatever personal belongings you have from your desk. Joe and his friends here will watch to make sure that's all you take. Then they'll escort you to your car and follow you out."

Sabina drew back her chin. "Fired?"

"Done. Gone. Outta here."

"*Why*?" she asked with what actually sounded like indignation.

"You're spreading rumors about Northwood that have nothing to do with reality. Mercury? Excuse me. There's no mercury here. Check with the New Hampshire DES. They'll vouch for that."

"I'm not spreading rumors. I'm asking questions."

"Same thing," he said and shot warning looks at her coworkers. If they wanted to keep their jobs, this was a lesson for them. "We expect loyalty. If you have a problem, you take it to the top. You don't go skulking around asking little people here and there. Honestly, Sabina, I thought you were smarter than that. You aren't good for this organization." He shot a thumb toward the door and couldn't resist adding a firm, "*Now.*"

She stared at him for a minute, and he could see her mind working. She was trying to decide how she could keep her job, whether apologizing would work, whether she ought to actually *grovel,* like a Barnes woman knew how to do that. Aidan didn't envy Ron. He had his hands full with this one. She knew computers. Aidan had to give her that. But she was no match for a Meade.

Averting her eyes, she opened a drawer, removed her briefcase and purse, then stood. She had the gall to look amused. "Who'll take care of your computers?" she asked.

"We have two others right here who can do what you do."

"Is that so?" she asked with a crooked grin.

"What's in the briefcase?" asked the security chief.

She opened the briefcase wide. Aidan watched while the chief thumbed through a few papers. From where he stood, they looked pretty worthless. He guessed that anything of value was there on her desk. Of course, if he and the guards hadn't been there—say, if he had been stupid enough to fire her in his office and let her return to her office alone—she would have been filling that briefcase with everything she could that would help her double-cross Northwood. The Barneses were vindictive. This all went back to Annie and Cooper's Point.

"She's clean," announced the chief.

Looking at Sabina, Aidan put his hands on his hips and hitched his head toward the door.

She actually smiled. "Thanks, Aidan. You've been a help."

"A help in what?" he asked. Helping her was the *last* thing he wanted to do.

"Sorry. I gotta go." She started off.

"A help in *what*?" he repeated, turning as she went.

She stopped at the door. "Since you are no longer my employer, I don't need to answer that." She raised a hand to the two workers at their desks, who had been watching every minute of the show. Then she winked at the guards, gestured them along, and, leaving Aidan totally up in the air, was gone.

Sabina drove straight to Phoebe's. She found Annie in the kitchen making balsamic chicken salad for dinner.

"Can't stay long," she said at the end of the same breath in which she had told Annie she was fired. "Ron'll hear through the grapevine, and I want to be home when he gets there. But Aidan was a huge help—so arrogant and heavy-handed and condescending and *defensive* that if I still doubted your theory, I can't anymore. He's despicable. And he's hiding something. So I'm with you, Annie. Tell me what to do."

Chapter 22

I WAS STUNNED—oh, not that Sabina had been fired. Aidan would do that, if for no other reason than to flex his muscle. What stunned me was that she hadn't blamed me. After the way she had warned me off when I first came to town, this was a turnaround. She wasn't even angry—at least, not at me. At Aidan, yes. She called him a first-class asshole, a word I'm reluctant to repeat, but which gives you a sense of where my sister was at. I don't think she had thought out the long-range implications of being unemployed. Right now, she was actually happy about it. I never would have imagined that.

Nor would I have imagined that she would have taken up my cause. Of course, there was always the possibility that she was doing it to validate having been fired, or that she was simply filling a void. But that she had come to my side after a lifetime of dissension? That made me feel good. Made me feel *really* good. Like she was validating what I was doing. Like I wasn't bucking the tide in this. Like I wasn't *alone*.

I was in a good mood making dinner for Phoebe and myself that evening. My chicken salad was another of Berri's recipes, given to me spiritedly that afternoon. Berri was in love with her philanthropic lawyer, with whom she had spent the entire weekend. I suspected she would have even given me her great-grandmother's top-secret waffle recipe, she was feeling that magnanimous.

The chicken salad was enough. Made with poached chicken, chopped peanuts, and sliced green grapes, drizzled with a uniquely sweet dressing that made all the difference, and served with mini corn muffins, it was delicious. Even Phoebe liked it, and she did eat, hungry after spending the day at the store. Tom was doing her first IV infusion the next morning. Knowing she might miss a few days of work, she had been trying to get things in order.

We talked about that possibility over dinner. I had promised her that I would fill in, and now Sabina planned to join me. Though Phoebe was tired this late in the day, which meant that her mind was more cloudy than clear, she was able to answer most of my questions.

I took notes. After dinner, I cleaned up the kitchen. I checked my e-mail. I looked at the clock.

By nine, Phoebe was upstairs. Sitting on the edge of her bed, I said, "I'm going out for a little while. I'll just drive around, do some thinking. Will you be okay?"

Dryly, in one of those clear moments, she said, "I'd better be. You won't be staying here forever."

I felt a twinge. Guilt? Nostalgia? Regret?

Whatever, she was right. She would be on her own soon enough. "But you're going to be feeling *better,*" I said in an urgent whisper. When she simply looked concerned, I asked, "Are you nervous about tomorrow?"

"Yes. She said it would get worse before it got better."

"That's why I'm here. That's why Sabina's

here. That's why *Tom's* here. You'll be better in no time, Phoebe. I *promise*." Giving her arm an encouraging squeeze, I left the room.

You may be thinking that I had some gall being so positive about something that was really quite iffy. But I didn't think it was iffy, and Phoebe needed my confidence.

Then again, that's probably not at all what you're thinking. You're no doubt wondering why I didn't tell her I was going to James's house. Why I *lied*.

Well, I didn't really lie. I did plan to drive around and do some thinking. I just chose not to tell her that I was making a stop. And I had good reason for that. James was a Meade, and the Meades were the enemy, not only in terms of abuse of power, but in terms of mercury leaks and, now, Sabina's firing—which James *had* to have known about. He could have stopped it in a second. He was higher on the ladder than Aidan.

At least, that was the perception. For the first time, though, as I crossed town to his house, I wondered about that. What was it TrueBlue had said with regard to which of the Meade sons would take over after Sandy? I had said James was in line, but he had written, *Don't be so sure.* TrueBlue worked at Northwood. Maybe he knew something I didn't. Maybe he knew something most of the *town* didn't know.

Maybe James wasn't the crown prince after all. Hadn't he cautioned me not to clump him with his father and brother? He did seem different—for starters, in temperament. I needed to ask him about that. All in all, we had *lots* to discuss.

Naturally, I forgot about everything else the instant I went up the walk and was greeted by James holding Mia. The child was absolutely precious in her pajamas, a slim little girl with diaper padding, her small legs straddling James's torso, one arm around his neck. Her face was already familiar to me—creamy skin, dark eyes, rosebud mouth. She had a tiny barrette in her dark hair. I leaned closer to see it.

"Mia," I read, then said to the child, "That is the prettiest barrette I've ever seen." She smiled. "On the prettiest little *girl* I've ever seen." To James, who was smiling, too, I said, "She really is the sweetest thing. Can I hold her?"

When he started to shift her, I held out my arms, but Mia's lower lip went out and began to tremble. Wary eyes stayed on me, but she leaned closer to James.

"Another time," he said. "Come on in."

We went to Mia's room and played with her for a few minutes. By the time he put her on the dressing table and changed her diaper, she was warming to me, playing with the furry puppet I held, even laughing aloud when I made it tickle her neck.

There is nothing like a child's laugh. I had discovered that when Sabina's kids were young, and was reminded of it now. A child's laugh is infectious. It is uplifting. It is innocent and pure and filled with hope and light. It is about prioritizing life, putting the good things first, placing exquisite value on moments of harmony.

I say all this to explain why, after Mia was settled in and the monitor turned on, James and I went right down to the leathery den on the first floor to talk and made love instead, which actually answered one of the questions I had. What he had meant about not doing one-night stands was that this wouldn't *be* a one-night stand. And it wasn't me who started it. He was the first to make a move—a hand on my neck as we left the baby's room, fingers linking with mine as we went down the stairs, an arm circling my back as he guided me to the den. He couldn't keep his hands off me, and after his hands, it was his mouth. That mouth touched *everything*.

Did I object?

Of course not. I loved what he did.

I didn't hear Grace once, but then, she had no need to speak. She approved of this, probably wished *she* could have done what we did with the impunity we had. A free spirit in the buttoned-up fifties, she had taken flack for the role sex had played both in her own life and in the lives of her characters. To this day, I'm convinced that much of the negative reaction to *Peyton Place* had less to do with sex, per se, than with the sexuality of its women. Those women threatened people. Take Betty Anderson. After dating *her* for months, Rodney Harrington had turned around and taken Allison MacKenzie to the school dance. Furious, Betty went to the dance with John Pillsbury. Halfway through the evening, though, she lured Rodney to John's car for some heavy petting, got him thoroughly aroused—then kicked him out of the car onto the ground and stormed off.

Call her the worst kind of tease, but she

had the guts to make a statement. It was a statement of strength, and it was threatening both to men, who feared being on the short end of a stick like that, and to those women who didn't dare do it themselves.

I dared. But then, I lived in a different time from Grace. Part of the transformation I experienced in college had to do with discovering my sexuality. With the right partner, I thoroughly enjoyed sex.

There in the den of his house, James was the right partner. End of discussion.

Actually, *beginning* of discussion. Once we were satisfied (make that, physically exhausted), we had to talk, and I did start that part of it, but he didn't object. We sat in the dark on the Berber rug in the den. Still naked, we faced each other, but we no longer touched. I could make out his general features, but details were lost, and that helped. Between the scents of leather and sex, and the dark, I was emboldened.

"Sabina was fired," I began.

"So Aidan said. I'm sorry he did that."

"You disagree, then?"

"Yes, I disagree."

"But you didn't stop him from doing it."

"I couldn't. I don't control Aidan."

"Who does? Sandy?"

"When he wants to. He's getting older. He likes it when Aidan makes the tough decisions."

"Like that was the *right* decision to make?" I asked in alarm.

"Tough doesn't mean right. I've already said I disagree. But Sandy doesn't. He doesn't like dissension within the company."

"And Aidan's his hatchet man."

"You could say."

"What are you?" I asked.

"What do you mean?"

"What role do you play for Sandy?"

"I do product development, and not for Sandy. It's my own thing."

"Isn't Sandy involved?"

"No."

"Or Aidan?"

"No. I work independently."

"I was starting to get that impression. Don't you three get along?"

"We've had issues."

"Like mercury?" I asked. After all, that was the crucial issue for me.

James was silent for the longest time, during which I would have liked to have seen the details of his face. I could tell that he was frowning. At least, I thought he was.

Finally he said, "Like mercury. I'm sorry about Phoebe. Are you sure that's what's wrong?"

"We won't know conclusively until she has the treatment. James, I meant what I said about Tom. I don't want him being made a scapegoat. He's an innocent party in this. He just happens to be the one who can administer the treatment."

"Tom won't be punished," he said with the quiet authority that made me believe.

"And if he treats other people after Phoebe?"

"I'm all for it."

"What if word spreads that the mill has a mercury problem?"

"The mill doesn't. Not now."

"But it did, and it could again. I know about the toxic waste buried under the clubhouse and the gazebo."

"There's no toxic waste there."

He was playing with words. Annoyed, I reached for my blouse. "Not anymore," I conceded, "but there was. And what about under the Children's Center?"

"There's no problem with the Children's Center."

I put my arms in the sleeves. "Did you say that about the clubhouse and the gazebo, too?"

"No. But I'm saying it about the Children's Center. It's being monitored. If there's a problem, we'll know before anyone is harmed."

"What about all the people who have *already* been harmed?" I asked, buttoning up.

He sighed. "Annie, I'm doing what I can."

"What does that mean?"

"It means I'm doing what I can."

"What does that *mean*?"

"It *means,* I can't *say* any more right now."

I groped around in the dark, found my shorts, and pulled them on.

"You don't have to leave," he said quietly.

Standing, I stepped into my sandals. "I do," I said. I pushed my hands into my hair and held it off my face. "Talking about this tears me apart. If you can't *say* any more, it's because you don't trust me—"

"I trust you."

"—and if you don't trust me, that demeans what we just did. So that leaves me thinking about why I'm here—I mean, not only here,

in your house, but here in Middle River. I
came to solve a mystery, and I have. It's
starting to look like my mother died,
James—*died*—because she was exposed
to mercury from your mill. Phoebe has been
suffering from the same thing. God knows
how many others are, too. I don't know how
you can let that happen. I don't know how I
can *be* with you, knowing you let it happen.
How do you live with yourself? How do you
sleep at night?" I pointed toward the stairs.
"And how do you justify letting that child stay
with a nanny here in the safety of your home,
while the rest of the children in town risk being poisoned every time they go to day
care?"

A shrink would say that I deliberately provoked James—deliberately painted him as
the bad guy—because I needed to put distance between us, and I suppose it was true. I
was feeling too drawn to him. I was liking him
too much. And how not to? He had never
been anything but respectful to me, had never
been anything but solicitous and caring,
whether we were running or making love.
Moreover, he was an exquisitely gentle father.
My heart melted when I saw him with Mia.

Fine. I could blame the Northwood problem on Sandy and Aidan. But James was there. He knew what was happening. And no, I *didn't* think he was doing enough.

I told myself that at least a dozen times on the drive from his house to Phoebe's, and then I turned on my computer and found two e-mails that helped me focus again. The first was from TrueBlue.

Do you have people to connect with those dates yet? We can't do anything until you do. Even if they don't agree to talk, if you find enough sick people who were at the sites in question on or around the dates in question, the circumstantial evidence may suffice. Finding those people was your job. Are you going to do it or not?

He sounded impatient. He sounded *annoyed*. But he was right. I was slow.

I'm working on it. I have reinforcements coming in to help tomorrow. And what about your job? I need copies of internal memos showing that the Meades knew about the mercury. And then there's the

*issue of the showdown. Do you plan to
show up when I confront the Meades?*

I sent it off with a determined click and
opened the second e-mail. It was from Greg.

WE SUMMITED THIS AFTERNOON,
AFTER THE MOST UNBELIEVABLE
ASCENT. WHAT AN EXPERIENCE!
WAIT'LL YOU SEE MY PICTURES.
THEY MAY BE THE LAST I TAKE FOR A
WHILE, THOUGH, BECAUSE SNOW'S
PICKED UP. GOTTA CONCENTRATE
ON THE DESCENT. BESIDES, PHOTOS
TAKEN UNDER WHITEOUT CONDI-
TIONS DON'T SHOW A HELL OF A
LOT. WOW. THIS IS AMAZING. CAN'T
WAIT TO FILL YOU IN.

Knowing Greg as well as I did, I felt his triumph. And I couldn't wait for him to fill me in,
either. I missed seeing him, missed bouncing ideas off him, missed our dinners together and evenings spent with friends at
local bars. I had a *great* Washington life.
Soon enough I would be back.

In the meantime, he was an inspiration for
me to plod on.

Chapter 23

I RAN AT six the next morning, in part because I knew that I had to deliver Phoebe to
Tom at eight and that this would allow me to
spend time with her before we left, and in
part because I knew James couldn't run that
early. I was annoyed at him. I was annoyed
at men, period, because TrueBlue hadn't answered me yet. He wanted results from *me*
but wasn't willing to produce any himself. Did
he distrust me, too?

The old feeling of being alone returned. It
was me against the world.

Good, said Grace. ***Maybe now you'll do
what needs to be done.***

Forget it, I argued. I am not doing your bidding. I am not writing a book. I am not destroying the Meades just to validate you.

Then you're a fool. James Meade didn't prove to be so perfect, did he? What did I tell you? They're all two-faced rats. They take what they want and then, by Jesus, they're gone.

Well, I'm not gone. I'm still here. The Meades will answer for what they've done. But a book is a wasted effort. Times have changed, Grace. Tell-all books are a dime a dozen. No. I want direct confrontation.

Oh, honey. Come on. Do you truly think that will work? You need the masses behind you. You have an audience. Use it.

I didn't need the masses behind me, I realized. I needed the *law* behind me.

Inspired on that score, the first thing I did when I reached Miss Lissy's Closet was to go up to the office and phone Greg's lawyer friend Neil in Washington.

I know. I had an agreement of sorts with TrueBlue not to take the lawyer route, but there were different lawyer routes to take. TrueBlue decried the splashy, very public class-action one that might destroy the town in the process. What I had in mind was more subtle. If I could gather enough evidence to show the Meades that they stood to lose big if there ever *was* a public legal case, I might get them to deal. Call it blackmail if you will. I call it gentle persuasion.

As it happened, my phone call was unproductive. Neil was on trial, and I knew how those trials went. He could be tied up in court for weeks. The fact that this trial was ongoing in August, when most law-enforcement personnel—not the least of whom being judges and their clerks—wanted to be on vacation, spoke of the demands of the case. Neil would definitely return my call. I just didn't know when.

I might have been discouraged if Sabina hadn't arrived at the store just then looking utterly defiant. "Ron is furious with me," she announced, plunking her shoulder bag on the office desk. "He says I was irresponsible talking to *anyone* at the mill about something that may not be true. He says I put my own needs before those of his and the kids, and that *infuriates* me. In the first place, if Aidan is going to fire *him* because of me, why would Ron want to work there, period? Isn't my speaking up a matter of principle? In the second place, why do *I* have to be the major breadwinner in the house? What about *his* responsibility? If he's worried about money, let him get a second job. Me, I'm worried about the health of my children—because something's up, Annie. You hit pay dirt this time. Aidan wouldn't have fired me if I hadn't touched a nerve."

"We need proof," I warned.

She grinned. With a conspiratorial gleam in her eye, she reached into her bag and pulled out a handful of CDs. "We had an electrical problem at the office yesterday morning. I was thinking I could work in one of the other buildings until it was fixed, so I stuck these in my briefcase. I didn't use them after all, but I never removed them. When Aidan and his goons arrived to escort me out of my office, the security chief searched my briefcase, but apparently he isn't familiar with CDs in skinny little jewel cases. I had them in a side pocket. He never even saw them." She waved the CDs in delight. "These are backups of some of my most important information, including access codes. I can get into any account I want, Annie. Anything that is in the mill's computer system is ours."

"Even e-mail?" I asked with growing delight, because my gut told me that's where valuable information might be.

Her grin widened. "Piece a cake."

The ease of it gave me pause. "That includes personal e-mail?"

"*Everything*. If it's in the company system, it's the company's property."

"If the government went to the company, could they see everything, too?"

Sabina nodded. "It's scary."

"But legal."

"Based on recent legislation, yes."

"What about what we're doing? Is this legal?"

Sabina didn't seem concerned. "Probably not, but I could always say I was looking for material of my own in there. Besides, they won't find out. Aidan is too arrogant to think I'd dare hack into the system, and Sandy isn't savvy enough to realize it's possible. The only one of the Meades who knows anything is James, and he sticks to his own end of the business."

I jumped at that opening. "So James is separate? Who takes over when Sandy dies?"

"Aidan."

"Not James?" I asked, wanting to be sure, because though TrueBlue had implied there was doubt, this went against conventional wisdom. "James is the older."

"James might succeed if there's a power struggle. The mill is what it is today because of him. He's been behind everything new in the last ten years, and without new direction, the mill would have lost ground. So if Aidan takes the helm, will James leave? And if he does, what happens to the mill? They're troublesome questions. And I don't know the answers."

"Do you think Sandy does?"

"I think Sandy tells himself what he wants to hear."

"Even if it's not for the good of the mill?"

Sabina shrugged. "He likes Aidan better than he likes James. They're two peas in a pod."

"Was it always that way?"

"Sandy preferring Aidan? No. It's developed in the last couple of years."

"Why does the town still talk of James as the heir?"

"Wishful thinking. Most everyone hates Aidan."

Joanne arrived to prepare to open the store, seeming to be totally on top of everything

Phoebe had told me to do. While she rechecked the closing charge tallies from the day before and packaged several customer sends, I put on a pot of coffee, vacuumed, and refolded sweaters. I finished just as the store opened, at which point UPS delivered four huge boxes of goods. While Joanne waited on customers, I inventoried the boxes and found an immediate problem with a dye-lot discrepancy, tops mismatched with bottoms. So Joanne and I switched; she got on the phone with the supplier while I waited on customers.

It was enlightening. For one thing, I was greeted warmly, as though I belonged there. For another, word of Phoebe's treatment had spread. Indeed, the doorbell *ting-a-linged* often enough to suggest that people were stopping by as much to get news as to shop. Apparently, Phoebe hadn't been as adept at hiding her symptoms as we thought. The townsfolk knew she wasn't well. They were concerned.

The circle of customers around me grew until, at one point, there were six women asking questions about Phoebe, and I tried to be vague. But these women weren't dumb. They could tell an answer from an evasion. When yet one more of them asked what Phoebe was being treated for, I gave in. Quite honestly, I didn't see why I shouldn't. They wanted to know. And I *wanted* them to know.

"We suspect she has mercury poisoning," I said.

There was a collective gasp, then a flurry of questions of the when, where, and how type.

"We won't know anything for sure until the treatment is done," I said. "It's all very tentative."

There were a few more questions, but they generally tapered off to expressions of support. I shouldn't have been surprised when every one of those women left Miss Lissy's Closet and began to spread the word.

After an hour or so, Sabina took over helping Joanne, while I drove to the clinic to sit with Phoebe. She was in a large room with two patients with chemotherapy drips, and there were more people checking on the three of them than I would have imagined, including Tom. He and I talked for a few minutes—talking with him was so easy that yet again I thought of Greg—before I returned to the store.

The first person to greet me there was Kaitlin. She had heard that Phoebe was at the Clinic and that Sabina and I were at the store, and she wanted to help. Joanne put her to work unloading additional boxes of goods that had arrived. I sent Sabina back to the office, and so it went. We took turns helping Joanne work the store and running to the clinic to be with Phoebe. But Sabina was as focused as I was when it came to what had now become our shared mission. Operating on Phoebe's computer with code from her CDs, she infiltrated Northwood's system and went looking for dirt.

Me? I had to find people who would talk. To some extent, my past efforts had been blind. If I hoped to find something soon, I had to narrow the list of those I approached. That meant finding people who had in fact been at either the Clubhouse or the Gazebo in the days immediately prior to the fire.

I began with a visit to Sam, catching him as I had two weeks before, just as he was setting off to play golf. "You're ruffling feathers," he said around his cigar as he gathered his things, "but I love your spunk. Whaddya need?"

"I want to know who held events at either the Clubhouse or the Gazebo immediately preceding the fires at each. That's the kind of thing you print in the *Times.* When you report on town events, you tell where they took place. I want to go into the archives again."

"You have the dates?"

"I do."

Sam went to his office door and bellowed, "Angus!" I had thought the place was deserted, what with the *Times* having come out that morning, but a fair-haired, bespectacled boy showed up at the door. Sam pointed at me with his cigar. "She needs information. Research the archives for her, like a good kid. She'll give you the dates and the place."

Pleased, I not only gave him that information but also my cell phone number. While he went to work, I visited Marsha Klausson at The Bookshop. As always, the scent of honeysuckle welcomed me, but if that hadn't done it, her personal warmth would have. "Did I tell you I had to reorder all three of your books?" she asked the instant she reached my side. "We've had a rush on them since Omie died. Must have been seeing you there at the diner that made people curious."

"Is curious good?" I asked cautiously.

Mrs. Klausson nodded. "They're intrigued. You were a big name before, but they've always associated you with Grace. I think they're starting to see you as different, and it's about time. You're *very* different from Grace."

I was starting to think so myself. When I was a kid, I never disagreed with Grace. Now I did.

But I wanted to hear Mrs. Klausson's take on it. So I asked, "In what ways?"

"Grace had an edge. It showed in her writing. People were either really good or really bad. You've always been more nuanced. You see the middle tones in people. You're more constructive."

"Constructive?"

"Perhaps practical is a better word. Yes, you rebelled. But you were always more into finding solutions to things. Your books do that. Your characters grow."

"So, here's a question," I said, because there was definitely a solution that I needed right now. "You helped organize the Middle River Women in Business, didn't you?"

"I certainly did. There were six of us who got it going—Omie, Elaine Staub from the home goods store, the realtors Jane and Sara Wright, and, of course, your mom."

"Elaine died," I realized the instant she mentioned the name.

"Yes. Several years back. She had a rough time at the end. If it wasn't the flu, it was colds, even pneumonia or *shingles*. That's painful, you know."

So there was another person whose death might be related to a long-ago exposure to mercury. "And the Wrights moved away," I said, pulling that vague thought from my past reading of the *Middle River Times.*

"Yes. They weren't very old, but Jane suffered from rheumatoid arthritis. When it got bad enough so that she had to struggle to work, they moved to Arizona for the warmth and the sun. Jane died a year or two ago—nothing to do with the arthritis, it was her heart. Sara has stayed out there."

"And she's well?"

"Far's I know. We exchange cards each Christmas, but she's made a whole new life for herself. I believe she found a well-to-do widower. She sounds happy."

"So of you six women, four have died and two are healthy."

"Yes." The bookseller frowned. "It is frightening, the fragility of life, isn't it? Jane wasn't old. Your mother wasn't old. Nor, come to think of it, was Elaine."

Omie had been old, I reminded myself, but not as old as her parents had been. I was feeling like I was getting somewhere. "Do you remember when you started meeting at the Clubhouse?"

"It was before the fire."

"*Right* before the fire?"

"Oh yes. I remember thinking how lucky we were that we weren't there at the very *time*. The fire started in the kitchen while the meal for a meeting was being prepared. It wasn't for *our* meeting, but it might have been. We did meet that week." She frowned again. "Twice? Why am I remembering that? Was the fire on a Monday?"

"Tuesday."

Her eyes brightened. "Tuesday. I remember now. We were going to meet for breakfast *that* day—Tuesday—but called it off because a stomach bug was making the rounds. The others—Alyssa, Elaine, Omie, and Jane—all had it. Sara and I had been sick the week before, so we missed the lunch meeting on *Friday. That's* how it was."

"If that's how it was," I said, repeating the story to Sabina at noon, "there's a distinct possibility that one of those buried drums sprang a leak that Friday. Mrs. Klausson and Sara missed the meeting because, coincidentally, they had the flu. When the others got sick, it was assumed they had the flu, too—but remember, acute mercury poisoning produces flulike symptoms. Maybe others had similar reactions—you know, the cook, the waitstaff. When the powers that be at Northwood saw this, they knew they had a problem, and they knew just what it was. So they burned the place down. That gave them an excuse to clean up underground and rebuild with no one the wiser."

"And Mom and the others who were sick?"

"The acute symptoms let up, the mercury drifted to different organs and lay dormant for years, then erupted as chronic mercury poisoning in the form of immune deficiency disease, rheumatoid arthritis, and, in Mom's case, Parkinsonian symptoms. All four of the women who were at the Clubhouse that day are dead. That says something."

"But not about Phoebe. Do we know if she was at that Friday meeting?"

"I asked Mrs. Klausson. She thought and thought, but she wasn't there herself and had no recollection of hearing that Phoebe was. Phoebe wouldn't have worked all that long for Mom. She certainly wasn't a seasoned businesswoman at the time. So why would she have been there?"

"I don't know. So we still don't know where she was exposed."

I glanced at the computer. "Have you had any luck there?"

"I've found lots of chatter, but nothing usable. I was hoping to find e-mail from Aidan either to his father or to James mentioning mercury, but there's nothing yet."

"You said Sandy isn't a computer person. Does he e-mail?"

Sabina smiled dryly. "He e-mails like Aidan does—tells his secretary what he wants sent and she does it."

"Maybe the secretaries would talk with us?"

Sabina was shaking her head even before I finished. "Dire loyalty," she said as my cell phone rang. "They like their jobs."

I fished the phone from my purse. It was Angus calling from the newspaper office. Pulling up a pad of paper, I wrote as he spoke. According to the *Times* archives, four events had been held at the Clubhouse and the Gazebo in the days immediately preceding their fires. Thanking him profusely, I closed my phone and put the list before Sabina. At the same time, I pulled out my original list of ill people and set it alongside the other.

"Can you match any up?" I asked.

Sabina didn't have to look long. She placed a finger at a spot on each list. "Here. Susannah Alban has had one miscarriage after another. And here. The Alban-Duncan wedding was held at the Gazebo two days before the fire." Stricken, her eyes met mine. "I know about Susannah's miscarriages, because the first of them coincided with one of Phoebe's miscarriages."

"Was Phoebe at the wedding?" I asked, hoping, hoping.

But Sabina shook her head. "Susannah's more my friend than hers. Ron and I would have been there, except that the wedding was really small, and afterward we were grateful. The guests all had food poisoning. It was a hot day in August. The food was sitting in the sun. It made perfect sense."

It certainly did, just like the WIBs having the flu, which was going around town at the time. "Any others?" I asked with growing excitement. TrueBlue was going to love this.

Sabina refocused on the lists. A minute later, she pointed again, one finger to each list. "Sammy Dahill. Rotary Club meeting. Sammy was chairman of the club a while back. I only know it because he lives on our street. He's had kidney problems."

I knew about those kidney problems. They had been reported in the "Health Beat" column of the *Times*. "Omie said that runs in his family. According to the paper, three generations of Dahills have had kidney problems."

"That's true," Sabina acknowledged, "but what the newspaper doesn't say is that Sammy was adopted."

"Adopted? Why didn't Omie know that?"

"A senior moment? But it makes kidney problems mighty coincidental, wouldn't you say?"

I struggled to contain myself. "More importantly, what would Sammy say?" I gathered my things. "I'm going to go see him. And Susannah. If one or both are willing to talk, we're in business."

Neither would. Susannah, for one, had her hands full caring for three children under the age of five, all adopted after so many miscarriages. She was convinced that the problem at the wedding had been nothing more than food poisoning. After all, she said, she and her husband both worked at the mill at the time, the Meades had paid for the wedding, and those in attendance really couldn't complain about a few stomach cramps. Moreover, the Meades had been extraordinarily generous when the children arrived, even transferring Susannah's husband to the Sales Department and letting him work out of the house when a sleeping disorder (yet *another* possible by-product of you-know-what) made work at the mill a risk.

As for Sammy Dahill, he owned the printing company that printed not only stationery but annual reports and marketing material for the mill. He claimed not to remember whether there had been a general illness following a meeting of the Rotary Club in March of '89. Too many years had passed, he said.

Personally, I didn't believe him. Memory was a convenient excuse.

Starving by now, I picked up a bag of chocolate pennies and, eating as I went, headed for the clinic. But I'll bet you're wondering where Marshall Greenwood had gone. I certainly was. He had been harassing me right up until Omie died. It didn't make sense that he would suddenly stop.

So I asked him. No, he wasn't following me. There was no sign of him at or around Susanna Alban's or Sammy Dahill's, but after I left News 'n Chews, I spotted him parked in front of the Sheep Pen. He actually appeared to be minding his own business. Swinging around the corner, I pulled over behind the cruiser and swallowed down a chocolatey mouthful as I parked and walked up.

"Hey," I said, smiling at him through the open window. I could see on the passenger's seat the crumpled Sheep Pen bag that suggested he had just eaten some kind of bar fare—likely chicken fingers or nachos; still, I held out the bag of pennies. "Want one?"

Marshall eyed me with distrust. "What's in 'em?"

"Pure chocolate."

"You didn't add anything?" he asked in his usual rasp.

I couldn't decide if he was serious. But I was remembering what Mrs. Klausson had said that morning about my being more nuanced in my judgments of people. So I smiled again. "I should have, after the trouble you gave me last week. But I really am not an ogre." To show him there was nothing but pennies in the bag, I ate one myself.

This time when I offered, he helped himself to a few. He didn't thank me, but I didn't need thanks. "Are we friends now?" I asked.

"That depends," he said.

I leaned closer, spoke more quietly. "On whether I'm going to tell the world what I know about you or Normie or Hal Healy?"

Trite as it sounds, Marshall did have a deer-in-the-headlights look just then. "What's Hal Healy done?"

"I'm not telling, because that's my point. I don't *do* that, Marshall. We all have our personal little secrets—our personal little *problems*. I don't care about those. It's not why I'm here. I am not writing a book, and even if I were, I wouldn't write one about you. So if that's why you were after me, please don't be. You are perfectly safe."

He thought about that for a minute, then said a grudging, "You're safe, too, even without that assurance. You have friends in high places."

I straightened. "Who?"

He shot me a crooked smile. "You tell me about Hal, I'll tell you about your friend."

It was a test. Did I betray people, or did I not?

I did not. And it was far more important that Marshall know this, than that I know who had spoken on my behalf. Besides, it wasn't that great a mystery. I knew it wasn't Sandy Meade, and it sure wasn't Aidan. There was only one Meade left who had enough clout to get Marshall off my back.

James. I had been trying not to think about him.

Not wanting to now, I simply said, "Checkmate." I held out the bag. "A few more for the road?"

Marshall took a handful, which left precious few for me, but my hunger had been eased, my craving satisfied for now. Returning to my car, I drove on to the clinic. I hadn't been with Phoebe for more than five minutes when Tom appeared at the door and hitched his head. I joined him in the hall.

"How's she doing?"

"Fine," he said. "Her vital signs are good. She's starting to feel zapped and disoriented, which is the part that gets worse before it gets better. The IV drip is just about done, but I'd like to keep her overnight. I want to monitor everything that comes out, and she's apt to be unable to do it herself. Are you okay with that?"

I was fine with it. I was actually *great* with it. If Phoebe stayed over at the clinic, I could work longer with Sabina, and the longer I worked with Sabina, the more evidence I might amass and the less I would think of James. "Totally. Has any mercury come out yet?"

"It takes a little while for that to start. Late tonight, early tomorrow—those are the critical times."

"What'll you do if the tests are positive?"

He looked me in the eye. "Face a moral crisis. There are lots of people here who are sick. Do I treat them for mercury poisoning, or do I not? I need evidence that they were exposed."

"You didn't have evidence that Phoebe was exposed."

"I know. That's the problem. Can you find it?"

* * *

Determined to do just that, I returned to the store, but the place was humming. Sabina was helping Joanne, and they needed me, too. It wasn't until we had locked the door, closed the register, and tallied up cash, checks, and credit slips, that Sabina and I had time to think, and then we were hungry. Sabina had already left a message at home that she would be staying here at the store with me awhile. Needing air, she volunteered to pick up dinner at Omie's.

While she was gone, I went up to the office and logged on to check my e-mail. There was nothing from Greg, though I pictured him sliding down Mount McKinley in the snow. Berri sent a note saying that she and John were still hot and heavy. There was a note from my editor saying that my revisions were fine and she was clearing out of the city until after Labor Day, so I wouldn't hear from her again until then, which suited me fine. I was feeling distanced from work. Not so from my friends. I felt a warmth inside reading a note from my friend Jocelyn, who was nearing the end of *Peyton Place* and wondered how much was autobiographical. She was trying to picture me there, she said. I clicked on "reply."

Grace wouldn't say that <u>Peyton Place</u> was the most autobiographical of her novels. She would say <u>No Adam in Eden</u> was that. But there are similarities between Grace and Allison MacKenzie. Both grew up without fathers in the house. Allison was something of a social outcast, as Grace was. Allison was slow to reach puberty, as Grace was. Grace was actually called "slats" when she was a teenager, because she was flat as a board. By the time she was sixteen, she had made up for it, but being called names may have enlarged the chip on her shoulder.

As for trying to picture me in the town of Peyton Place, don't. Middle River was like Peyton Place when I was a child, but returning here now, I see how different it is. For one thing, there's fifty years of modernization. Middle River hasn't stood still. Everyone has cell phones, Omie's plays popular music, businesses are computerized, FedEx comes twice a day. Even the Sheep Pen has gone modern with beer from microbreweries.

Funny, but I don't feel isolated here

the way I thought I would. Isn't e-mail a
fabulous thing?

I clicked "send," then pulled up TrueBlue's
last note—the impatient one—and reread it.
Rather than impatience now, I sensed urgency. So I clicked "reply."

The good news is that I've located two
people who were at the Clubhouse and
the Gazebo prior to those fires, and who
have had chronic problems in the years
since. The bad news is that neither of them
is willing to talk. I put the bug in their ear
and am hoping they'll come around. In the
meantime, I'm working on finding others.
    You've probably heard that my sister
is being treated at the clinic. If it proves
that she did have mercury poisoning—
and if the treatment works—her mind
may clear up so that she'll remember
more of the when and how. But that
could take a while.
What are you getting on your end?

I sent the note, and quickly got one in return.

Lots of flack. The Meades are feeling
threatened. Aidan is using your sister's
defection as a sign of looming trouble. A
board meeting's been called. You need to
work fast.

I was reading his note when Sabina returned, and I might have hidden it from her. I
had enough time. But she was my ally now.

"He's my source," I explained. "He has
given me information on the mill's use of
mercury. He was the one who told me that a
toxic cleanup was done after the fires."

She finished reading. "Who is he?"

"He won't say. He works at the mill, has for
years and hopes to for a lot longer, but his
user identity is all I have. TrueBlue. Any
ideas?"

Sabina looked at me. I could have sworn I
saw a flicker of smugness. "Actually, yes. I'd
guess it was James."

"James." My heart tripped. "James
Meade? Why do you say that?"

There was more than a flicker of smugness now. She seemed downright lordly.
"I've spent much of today reading e-mail
that's come into the system. James is currently going back and forth with a friend who

lives in Des Moines, a lawyer, from the gist of the conversation. I gather they're old college buddies. The guy calls him Blue. Sounds like a college nickname."

Hadn't TrueBlue said that his college roommate was a lawyer?

"That's bizarre," I argued. I did *not* want TrueBlue to be James. James and I had been too intimate for him not to have told me something as important as that. Besides, Sabina's tone grated. I didn't want her to be right. It was like we were kids again, and she was one-upping me. "Are you sure James's friend was calling *him* Blue? Maybe he was referring to someone else."

Sabina had an answer for that, too. "He did it multiple times—as in *Hey, Blue, good to hear from you,* and *So, Blue, how's Mia?* and *Good thought, Blue. Don't e-mail. I'll call.* James had legal questions that he wanted to keep confidential. Is it a coincidence that he's asking a pal legal questions at a time when Aidan and Sandy are getting nervous about mercury?"

No, it wasn't a coincidence. So was James asking legal questions on behalf of Aidan and Sandy? Or was he asking on behalf of TrueBlue?

"And he has information on a board meeting," Sabina said. "The rank and file wouldn't know that."

"Why would his college friends call him Blue?" I asked, resistant still.

Sabina put the bag of food on the desk. "Because, Annie, he wears blue all the time. Blue shirts, blue jerseys, blue sweatshirts and jackets. Ask me, and I say that's boring as hell, but hey," she opened the bag, "each to his own." She removed take-out containers and handed me one. "Omie's hash, in memory of Omie. She always told my kids that her hash was brain food. Maybe it'll help us. If a board meeting is coming up soon, we need something more than we have."

We sure did. Phoebe was our best bet, but she wasn't thinking straight, and now I had to think about James, too. There were oh-so-many reasons why his being TrueBlue made sense, and the more I thought about it, the more sense it made. But it raised the major question about whose side he was on, and how could I possibly know? James didn't trust me enough to share more than a very general *I'm doing what I can.*

Well, from what I could see, that wasn't enough. Bottom line? I didn't trust him either.

* * *

As soon as Sabina and I had eaten, we phoned Susannah Alban, Sammy Dahill, and Emily McCreedy in the hope that maybe if Sabina voiced the appeal, one of them would talk. No luck. We did get the full Rotary Club roster. It was in the town hall files, and since the Meades so generously allowed the town to use Northwood's server, Sabina was able to pull it up. We matched three other names from that list to mine, but two were dead, and the third wouldn't talk.

Moreover, the fourth event held prior to one of the fires was a photo shoot at the Gazebo for a clothing catalog. Since the *Times* hadn't listed names (I asked Angus that on the phone), and since most of the participants were from out of state, Sabina was going to have to dig into the Northwood system for details.

For now, we scrutinized more of the mill's e-mail. It seemed that this very day Aidan had initiated a dialogue with an agent of the state Department of Environmental Services suggesting that Northwood sponsor another of the public information days that the department periodically held. Northwood had done this before. The mill was a best friend of the DES, enough to make one wonder. And then, there was the timing. It was fishy. But the correspondence was clean.

Frustrated, we left the shop at ten. We stopped briefly at the clinic to find Phoebe in a private room now, sound asleep. It was past ten-thirty when I finally got back to the house. I had barely set my things on the kitchen table when the phone rang.

Caller ID told me it was James, and I was suddenly furious, hurt, and confused, all at the same time. I debated not answering. I simply didn't know what to say to him.

But I couldn't just let it ring. So I picked it up. "Yes, James."

There was a pause, then a quiet, "How's Phoebe?"

"She's sleeping at the clinic."

"How's she feeling?"

"Lousy, I'm told."

"I'm sorry."

"That she's sick? That half the *town* is sick? That *you* can't trust me enough to confide in me?" I could *feel* Grace telling me to confront him about being TrueBlue, but I didn't want to. I wasn't ready.

"We need to talk," he said in that same quiet voice.

I rolled my eyes. "Have I heard that before?"

"We need to talk about us."

"Us?" I cried, realizing in that instant that I wouldn't be asking about TrueBlue at all, because this phone call had nothing to *do* with TrueBlue. This was between James Meade and me. "What us? There is no us."

"I think there is. That's why this is so hard."

"What's so hard?" I shot back. "Lying?"

This time he was silent so long that I half expected he had hung up. I was surprised when he said, more quietly than ever, "I want to see you."

Fingers closed around my heart. Eyes tearing, I said, "Well, you have a child, so you can't leave your house, and I have obligations to my family and dozens of other people in this town, so I need to be up early tomorrow to work, and I'm not sure there's any *point* in our seeing each other, because I'm going back to Washington in two weeks, so even if there *was* an us here, it's done. It's been fun, James, y'know?"

As I waited for him to reply, I cursed myself for sounding like a petulant twelve-year-old.

When he remained silent, I rubbed my chest. It *hurt.*

When still there was nothing one minute, two minutes, three minutes later, I realized that this time he truly had hung up.

Chapter 24

NATURALLY, I didn't sleep well. I ran early to work out the kinks, but James spoiled that, too. I kept looking for him. Yes, I knew he couldn't leave Mia until the nanny arrived, but I kept thinking that if it mattered enough to him to see me, he would have found a way. Hell, he knew Phoebe was at the clinic and that I was alone in the house. James was a resourceful guy. He could have found a way to come at two in the morning, if he chose.

He didn't, which was probably just as good. My mission was never more narrowed than now. With no one else in town willing to talk, Phoebe was our only hope. Assuming the tests proved mercury in her body, I had to link her to an incident of actual exposure.

After showering, I dressed quickly and went to the clinic. Phoebe was groggy, dozing on and off throughout the *Today* show as it came from the monitor high on the wall. She was still catheterized, but she seemed comfortable. Comfortable? I half wondered if she was on morphine, she was that spacey.

"Hey, Phoebe," I said.

She was slow to turn her eyes to mine, then a minute in recognizing me. "Hello," she replied in a distant voice.

"Feeling okay?"

"What's it doing out?"

"It's warm and humid. They're predicting thunderstorms for later."

"Oh dear," she murmured and closed her eyes. "No golf."

"You don't play golf."

"Michael does."

"Oh, sweetie." Michael was her former husband. He lived three towns over, had remarried since their divorce, and now had two kids. "Do you ever see him?"

Phoebe didn't open her eyes. "Who?"

I let it go. But I clutched her hand. "I have

to ask you something. I know you're feeling kind of out of it, but remember I asked if you were with Mom at the Clubhouse right before the fire?"

Phoebe opened her eyes wide. "Mom's dead. What are you talking about?"

"The Clubhouse fire. A couple of days before it, Mom was there with the WIBs. Were you there?"

Phoebe closed her eyes again. "Tell them I can't move," she said, back to murmuring. "I'm sorry. If there's a fire here, it'll have to go on without me."

"The *Clubhouse* fire."

"What Clubhouse fire?"

I didn't say anything else. As the minutes passed, I'm not sure Phoebe even realized I was still there. In time, I left and went on to the store.

Tom called shortly after noon to say that the tests were positive, that there definitely was mercury leaving Phoebe's body. I found the news almost anticlimactic; I'd been that convinced even before all this. That said, I was definitely relieved. We would never know for sure whether Mom had mercury poisoning, too, because we weren't about to exhume her body for an autopsy, but in my heart I knew. Though the satisfaction was bittersweet, it was satisfaction nonetheless. Wasn't this why I had returned to Middle River? Granted, my mission had changed. But I did feel a certain vindication on Mom's behalf.

I e-mailed Greg with the news. I knew he was somewhere in his descent and wasn't sure whether he could send or receive, but I wanted the message waiting for him.

I did not e-mail TrueBlue. The idea that he was James still had me reeling. I kept remembering remarks in his earlier e-mails—had even reread a few that morning. The clues were there. He slept alone; he had to be up early; he liked the anonymity of being TrueBlue so that he could be free of who he was and what he did. Even the flirting felt familiar, though with James in the flesh, it had been—well—in the flesh.

Besides, James had hung up on me. And now TrueBlue wasn't e-mailing. Did either of them have to know that the mercury theory was a go? Not from me, they didn't. At least, not right now. After all, the theory would only be good if I could show that Phoebe—or *someone*—had been exposed to mercury at the mill.

We spent the day at it, Sabina and I, but it seemed that no matter how many Rotarians we called, no one was willing to talk. Likewise, no matter how many leads we followed trying to track down Sara Wright (now married and presumably using another name) in Arizona, on the chance that she might know whether Phoebe had been at the WIBs meeting that March day, we had no luck.

And through it all, we had to help in the store. It seemed that we were no sooner into something in the office, when Joanne gave a yell—and this, even with part-timers at work and Kaitlin and Jen in the back room. August was one of the busiest months of the year, Joanne kept saying by way of apology, and I'm sure that between back-to-school wear and general fall merchandise, it was true. What became apparent as the hours passed, though, was that, once again, many of the customers came by in a show of concern for Phoebe. They wanted to talk as much as to buy, and for that, they wanted Sabina and me.

I was surprised by the last part. Totally aside from Omie's wake, where emotion, compassion, and grief snuffed out other realities, I would have thought that Middle Riverites had such a deeply ingrained image of me as a wretch of a writer that they would never have been able to warm to me as a person, much less *trust* me in the retail setting. Enter **TRUTH #9: All people have the capacity to change their minds, even people who live in small, insular, parochial towns.**

In fact, with regard to that issue of trust, I swear they came to me more. It was as if they wanted my advice on what they should buy, whether one size or color or style was better than another, precisely because I was now from the big city.

Perhaps it was my own need to think that. But the end result was that I was every bit as busy on the floor as Sabina.

So she and I talked with the people who came in. We sold lots of clothes. We ran up to the office to catch a minute here or there at the computer, and, through thickening clouds, dashed over to see Phoebe. Tom wanted to keep her at the clinic one more night, more for his own education than anything else. He figured that if he might be treating others this way, the data the nursing staff collected with regard to vitals would be a help. And again, neither Sabina nor I minded.

She headed home at seven for dinner with the kids, but I stayed on. I had a bag of chocolate pennies that (surprise, surprise!) Marylou Walker had brought over herself, and I still had the second half of my lunch sandwich. I wasn't in the mood to go out, because distant thunder was rumbling and the air was as thick as it could possibly be without rain.

The thunder neared, the rain began, and Sabina returned—at least, I thought it was her. Shortly before nine, I heard the downstairs doorknob jiggle, then the front door open.

"Up here!" I called. I was following yet another lead in trying to locate Sara Wright, and didn't think anything of it when Sabina didn't answer. At the sound of footsteps on the stairs, though, I was frightened. This wasn't Sabina. These footsteps were heavier, definitely male.

I stood and looked around frantically for something—letter opener, scissors, *any-*

thing—with which to protect myself, which was, of course, what my Washington self would have done. That this was Middle River, hence the open front door, was a minute in registering. By the time it had, James was on the office threshold, shrugging off a hooded rain parka. Underneath, he wore sneakers, shorts, and—yes—a dark blue T-shirt. His limbs were long, lean, bare, and hair-spattered, but the most notable aspect of his appearance was Mia. Wearing a thin pink sleeper, she was strapped to his chest, facing out, arms and legs dangling free. She was wide awake, looking straight at me, and I must have been starting to look familiar to her—she certainly was looking familiar to me—because she smiled. I saw two teeth on the bottom and two on the top, plus eyes that had narrowed with the smile but were beautiful, indeed. That smile plucked at my heart.

"What a coward you are," I scolded, speaking to James but looking at Mia. "Hiding behind a *child*?"

He might have said, *I told you I was a coward at the start,* which would have acknowledged that *he* knew *I* knew that he was TrueBlue. So maybe he didn't know I knew?

Well, I wasn't clueing him in just yet. I wanted to see just how long he would drag out the ruse.

"I had no choice," he said, hooking the hood of his jacket on the coat tree by the door. "The nanny's gone, and I needed to see you."

"Shouldn't the baby be asleep?"

"Yes, but this is urgent. She'll probably doze off right here. We need to talk, Annie."

I sat back down at the desk, swiveled to face him, and folded my arms over my chest. Thunder rolled, though not loudly. When it was gone, I said, "I'm listening."

Not only hadn't Mia been alarmed by the thunder, but she didn't look at all ready to doze off. She was kicking one leg and craning her head back, looking up at James now. He took both of her small hands in his, but his eyes were on me.

"I didn't expect this," he began and stopped. Those dark eyes—dark brown eyes, actually, and quite deep—suggested confusion.

"This what?"

"You. This attraction. I'm not supposed to be attracted to Annie Barnes. She's trouble."

I grinned.

"See?" he said, pointing at me. The spontaneity of the moment made him look younger. "There. That grin. It *does* something to me, and I don't know why. But I keep thinking about things like that and telling myself that there's no way anything between us could work. For one thing, we live in different places, which I *swore* I would never do again. And for another, you'd like nothing more than to bring down my family business."

"That's not true," I argued. "If I wanted to bring down the business, I'd have gone to the press long before now. It's a hot issue, so there'd be no trouble getting coverage. There'd already be producers from *60 Minutes* up here and investigative journalists from the big papers. What I *want*," I said with emphasis, "is for Northwood to finally take responsibility for two very serious mercury spills, compensate the people who've been hurt, and make sure it doesn't happen again."

He didn't deny that there had been two spills. Nor did he ask how I knew. He was distracted, because Mia had started to squirm. Perhaps it was the suddenly heightened slap of rain on the roof, perhaps it was simple restlessness, but she clearly wanted out of the carrier.

Holding the child, he unhooked the straps. "Do you think I haven't done things? Do you think I haven't compensated the people I knew who were hurt?" He set the child on the floor and straightened. "Why do you think those people won't talk with you? They won't talk because I've set up funds to help them with the things they need."

So here was another thing. TrueBlue knew I was having trouble getting people to talk. James and I hadn't discussed it.

But suddenly the distinction didn't matter. "You, personally, set up funds?" I asked. Mia was crawling toward the far end of the room.

"Me, in the name of the mill," he said, watching the child. "We call it a civic disability package, but I know what it is. So do my father and my brother."

"Then you know that those spills have caused illness?"

"We've known since it happened back at the Clubhouse."

"And some of that illness has caused death?"

He nodded slowly, grimly.

"Did you set the fire?" I asked.

"No. If I'd had my way, I'd have been blunt and just torn down the building and cleaned up the spill. But it was a fight. My father wasn't going to do a damn thing at all, until I badgered him enough."

"He'd have *ignored* it?" I asked, appalled.

I had barely finished speaking when a loud clap of thunder came. Seeming more puzzled than frightened, Mia looked back at James. Picking up on his lack of concern over the noise, she resumed crawling.

"He'd have prayed it would go away, which, of course, means it would seep deeper into the ground and move toward the river with each rainfall until the river was more polluted than ever." He sputtered scornfully. "Talk about burying your head in the sand. So I threatened him. I talked about lawsuits and media attention. I told him we had to do something. But I was as surprised by the fire as the rest of the town." Stretching forward, he scooped Mia up and returned her to his spot. The instant he set her down, she took off again, this time toward the bookshelves beside my desk.

"Why didn't you speak up then?" I asked, because what James was telling me implicated him in deception, and I didn't want that. "Couldn't you have told *someone* the truth?"

"Oh, I tried," he said dryly. In a gesture of frustration, he scrubbed the top of his head, ruffling salt-and-pepper hair that was already ruffled—and in this, too, he seemed younger than his usual, composed, authoritative self. His eyes were on the child. "And every time I did, my father would cover my tracks so that it was like I hadn't said a goddamned word." His voice rose. "At one point he told his golf buddies that I was full of radical ideas I got at business school. How's that for a smear? He knew they would spread the word, but in the kind of subtle way that has more weight for its semblance of confidentiality. Sandy is a genius at creating spin and making it stick. It's a natural marketing skill. And yes, it's the power thing that you hate."

I wanted to believe him. He sounded genuinely upset—guilt-ridden, angry, and dismayed. "What about the Children's Center?" I asked.

He showed no surprise that I knew about that, either. Mia had begun taking books from the shelves, a soft *thu-thunk* of each, audible only because the rain had eased again. He went after her and began putting back the books. "Yeah, well, that's an ongoing argument," he said. "Sandy claims that the other leaks were a fluke, and that since there hasn't been a problem under the Children's Center so far, the drums there are solid and that to do something will only draw attention to the presence of those drums, which are illegal, by the way." He sat back on his heels, looking up at me from there, "And before you lambaste me for that, please remember that those drums were buried before I became part of the company. I was away in college at the time. Sandy had no reason to ask my opinion. I sure as hell wouldn't have built civic buildings over a toxic waste dump. Is that *idiocy*?" he cried.

"The Children's Center is a tragedy waiting to happen."

"Yes," he shot back, "and I have nightmares about that. But you have to understand, my relationship with my father is tenuous. That's why I keep to myself and run my own part of the business. He and I don't see eye to eye on things, and the more I push, the more he marginalizes me. So I have a choice. I can let myself be marginalized so much that I'm no good to the company. Or I can pick and choose my fights—which is what I do. The mill is the lifeblood of Middle River. If it folds, the town is in dire straits."

Hadn't TrueBlue said something similar?

Thunder struck. Mia shot a frightened look at James. His hand on her head soothed her.

"But what about the Children's Center?" I pushed. "Would you want Mia there?"

"Why do you think she *isn't*?" he asked with a tortured look. I had touched a raw nerve. His voice was impassioned as the thunder rolled off. "Do you think I don't want her with other children? She needs to learn how to play. She needs to learn how to *share*. She needs to have friends, and play-dates with children with mothers, because, Lord knows, I'm no mother, but I'm doing the best I can, damn it. Okay, she's only ten months old now, so keeping her home with the nanny is okay, but in another few months I'll have to do something. My father would be perfectly happy to keep her hidden away. I refuse to do that."

"Hidden away because she's adopted?"

"Hidden away because she's *not*

adopted!" he said so sharply that Mia began to cry. Quieting instantly, he drew her up into a hug. "Shhh, I'm sorry, baby," he murmured against that short, shiny black hair. "I'm not angry at you, never at you."

Mia cried for another few seconds before responding to his calm.

"There," he said by her ear. "Want to take the books out again?" With an economy of movement, he put every last book back on the shelf and sat her down. She went right to it with a resumption of the same soft *thu-thunk, thu-thunk* as each book hit the floor first with the bottom of its spine, then with its side.

Me, I was stunned. Arms no longer crossed, I had my hands on the edge of my seat. "She isn't adopted? Does that mean what I think?"

There was no thunder to punctuate his revelation. It was striking enough on its own.

James bowed his head toward Mia, then raised it and faced me. "Up until last year, I was on the road often. When you develop new products for a company like ours, there's a lot to consider, from the machines that produce it, to the raw materials those machines need, to the markets that will purchase the product when it's complete. I've been doing this kind of thing for ten years. For six of those years, I had a relationship with an advertising executive in New Jersey. It started off as professional—she created an ad campaign for us—but then it became personal. She wasn't supposed to get pregnant, but it happened. She wanted to abort; I wanted the baby. I finally made her an offer she couldn't refuse."

"You bought her off?"

"Don't judge me," he warned.

"I'm not," I said quickly. "I'm just amazed. Was she after money all along?" It was a definite possibility. Anyone dating a Meade of Northwood Mill knew she was dating money.

His eyes held mine. "No. She wasn't after anything. When push came to shove, she didn't *want* anything—not me, not my baby. She had a town house she liked, and a job she liked, and there was no way in hell she was giving up either, much *less* moving to a small town where she would stick out like a sore thumb because she was Asian American. Well, she was right about that part. I argued that it wasn't so, because she was such a dynamo of a woman that she would have made friends here, and the color of her skin wouldn't matter. But Middle River isn't ready for someone like her, not so long as Sandy Meade is at the helm, because he sets the tone for the town. He's not exactly a bigot; it's just that he's threatened by anything different. He went berserk when I told him about April—that's Mia's mom—so I knew how he would react to the baby. I picked her up at the hospital when she was two days old, and my father dictated the spin."

"That she was adopted? How could he do that? And how could you let him? You're her biological father."

The *thu-thunk* of Mia's mischief had kept up, but I'm not sure James even heard. He looked disgusted. "Well, I can be bought, too. He told Middle River the story he could swallow, and told me that if the truth came out, he would disown me."

"*Why?*"

"Because Mia's different, and Middle River doesn't do different. Adopting her is acceptable because it makes me a remarkable person who is giving a poor orphan a good home. Siring her myself is something else."

"I think you're wrong," I said. "I think Middle River would love that you insisted Mia be born. Doesn't Sandy see that?"

"No. And if you're going to accuse me of going for the money like April did, all I can say is that Mia is my flesh and blood, and I want her to have every advantage money can buy. It's bad enough that her mother wants no part of her. Amazing how history repeats itself, isn't it?"

"Your mother died."

He was looking at me, slowly shaking his head. It was as though Pandora's box had been opened, and he just couldn't close it back up. "That's what Sandy would have Middle River believe. Aidan even believes it, but the truth is that my mother is alive and well and living in Michigan with her second husband and the three daughters she had with him."

My jaw dropped. The dead mother had been part of Middle River lore. With a concerted effort, I closed my mouth. Another round of thunder grumbled past. "Do you ever see her?"

"I go there once in a while."

"And you never told Aidan?"

"Oh, I told him, but he chose to carry on

the myth. That means he doesn't have to think of visiting or sending a note. It means less responsibility for him. And hey, I hear what he says. He doesn't owe her anything. She wasn't here when he was little, and he was younger than I when she left. It was lousy growing up without a mom. But Sandy's even more at fault. He was miserable to her while she was here. I want him to pay for that. One way is through my daughter. I want Mia to have a piece of his estate. There'd be poetic justice in that on several scores. So I'll keep my mouth shut until the inheritance is safe." Seeming purged, he turned his attention to the child. She had crawled farther along the floor to the next bookshelf and, with more rhythmic *thuthunks,* was pulling those books off, too. "Oh, little girl," he said, making a face, "look at this *mess*." He began to pick up the books.

Beak of a nose? Had I really thought that once? Yes, his nose was straight and more pointed than some, but when he made a face for his daughter, there was nothing beaky about it.

"Leave them," I said. "She can play. There's no harm." I was trying to ingest what he had said. "Who else knows she's biologically yours?"

He kept his eyes on the child. "No one."

"And you told me? Annie Barnes? The town loudmouth?"

His eyes met mine. "I trust you."

The simplicity of it—the honesty I heard and its implication—made me cry out, "Then why *TrueBlue*?"

He didn't blink. "When I talked with you that first day at the diner, I thought maybe you could help. But it was awkward, given who I was. You didn't trust me. And I didn't know if I could trust you. I sensed that TrueBlue's being anonymous would appeal to your novelist's imagination." He gave a crooked grin. "Besides, you're from Washington, aren't you? That's Deep Throat country."

"Uh-huh, and Woodward and Bernstein got a book out of it, but I'm not writing a book. I've said that from the start, and still you hid behind TrueBlue."

"Hid?" He considered that, then lightened up. "I actually thought it was kind of fun. And it was a relief. I said that—or TrueBlue did. Because I wasn't me, I could ask you questions without your asking *me* questions that I wasn't comfortable answering."

I was feeling vaguely used. "So I'm asking them now. Why do you need me to find people who were exposed at the time of those two spills? You know who they are. Isn't that what your 'civic disability package' is about?"

"There are people we don't know about. I want an inclusive list."

"Ask the guard at the gate."

"I would if I could, but sixteen years ago, he was a token presence. He wasn't there all the time, and when he was, he didn't keep a paper trail of who went where. He does now, because security is an issue everywhere, but he didn't do it then. So we don't know for sure who was at those sites at the time of each spill. I can't exactly put an ad in the paper. And I can't go around town asking people, because that would create a rumor that could grow out of control. Beside, we've never had proof. We never wanted it—and I'm using the royal *we,* meaning the voice of the company. When we created those disability packages, we suspected illnesses related to mercury poisoning, but we give generous disability packages to our own employees anyway, so this was simply perceived to be an extension of that. We never made connections to mercury, certainly not publicly. So we never got that proof. Did Tom find it in Phoebe?"

I was saved from answering by Mia, who chose that moment to make a pitiful sound. It wasn't from the storm, because the thunder was more distant now and the rain a bearably light patter on the roof. She was sitting in a sea of books, rubbing one tiny fist back and forth over her nose. Her other hand was pulling her hair. She was tired.

James caught her up and cradled her on his shoulder.

Me, I wasn't quite there yet. I was still looking at that sea of books. Well, yes, they were books, but not like novels or reference books or even sample books. For the first time, I realized that they were my mother's notebooks, apparently displaced during Phoebe's recent reorganization frenzy. Hidden down by the floor, they hadn't made it either to my radar screen or to Sabina's. Now it struck me that they might tell us whether Phoebe had been with the WIBs that fateful day.

I didn't say anything to James.

And don't ask me why, because I don't

know. Yes, he had said he trusted me, and given what he had confided in me earlier, I believed him.

The problem was me. Maybe I didn't know what to do with that trust?

Mia was quiet, her eyes closed, cheek on James's shoulder. Putting my elbows on my knees, I leaned forward and whispered, "She's very sweet."

James stroked the child's head with a large, strong, masculine hand that reminded me of things we had done in the dark. Not knowing what to do with that thought, either, I simply met his eyes and said in a voice soft enough not to wake the baby, "Why do you need proof now?"

"Proof is the only surefire lever I have."

"For what?"

"A bloodless coup. There's a board meeting Monday at four. Can you come?"

He didn't kiss me. I knew he wanted to. I could see it in his eyes, in the way they looked at my mouth. I could also hear it in his breathing, which was less steady than it had been. Mia was back in her carrier, asleep now and totally limp, hanging over his front so that if he was aroused, I couldn't see.

But he had to leave. I wanted him to leave. Something was happening too fast. I needed time to focus.

I got it, though in an unexpected way. James was no sooner out the door when my cell phone rang. It was Greg's friend Neil, calling not from Washington but from Anchorage, where he had flown after being granted an emergency recess in his trial. It seems Greg had fallen during the descent of Mount McKinley and smashed his leg. He had already had surgery. Now Neil wanted to help him return home, and fast.

My first sense was sheer envy that Greg had someone who cared enough about him to drop everything he was doing to fly to his aid.

My second was that I wanted to be there when Greg got home. I could help make him comfortable in the condo, could stock the fridge, do some cooking, air the place out after it had been closed up for two and a half weeks.

I told Neil I would be at the condo by noon the next day.

Then I called Tom. "Too late?" I asked.

"Nah," he said easily. "I'm reading."

"I was wondering what your thoughts are about Phoebe coming home. There's an emergency in D.C. I'm flying out early tomorrow. If she's coming home, I'll have Sabina take care of her."

"Have Sabina take care of her," Tom said. "She's starting to feel better. Now that we've identified the problem as mercury, there's no need for constant monitoring. The occasional test next week is fine. Is the emergency something to do with this?"

"No. It's my roommate, Greg. Broken leg. His partner flew out to be with him in Anchorage. I think they're already flying back."

"Huh," Tom said in a tentative way. "When were you thinking of leaving?"

"Here? No later than seven."

"And coming back?"

"Probably Sunday night."

There was a pause, then, "Want company?"

"Sure. I'd love it. But can you do that?"

"I don't know," he said, but the tempo of his voice had picked up. "I have a friend in the EPA who could answer some questions I have. I'm long overdue for a visit anyway. Let me see if Mrs. Jenkins can come. I'll call you right back."

He called five minutes later to say that he would be with me. I called the airline and booked tickets, then called Sabina to let her know.

I didn't tell her about the notebooks, didn't want her getting her hopes up. Honestly? I didn't want her asking to see them. As the last-born child, I had always had to share. These journals were a link to our mother. For now, I wanted them all to myself.

Chapter 25

KAITLIN WAS UP early Saturday morning,
showering and putting on clean jeans and a
new jersey and clogs from the store. She
took care with her hair and applied only as
much makeup as her mother liked—but not
to please Nicole. She didn't care what her
mother thought. She wasn't dressing for
Nicole. They had barely talked in a week.

Leaning closer to the bathroom mirror,
she studied a little raised dot on her chin. It
was a zit in the works, no doubt about it. Taking a tube of acne medicine from the medicine chest, she dabbed on just enough to
zap the zit without spoiling her makeup. It

used to be that her whole *face* was covered with zits. She loved her dermatologist for helping with that. Actually, she loved him anyway. He was always warm and smiling. He looked at *her,* not at Nicole. He acted as though *she* was the important one there, not Nicole. And in follow-up visits, he would tell her how *fabulous* she looked.

She didn't necessarily believe him. He told that to all the kids he treated. But it was nice to hear it, anyway.

"You're up early."

Kaitlin jumped. The slight movement brought her mother into the mirror's frame. Nicole was at the bathroom door in her silk robe, shoulder to the jamb, arms crossed. Kaitlin leaned in again to block the view. She pushed her tongue against the spot with the zit. Definitely a zit.

"Where are you going?" Nicole asked.

"The store."

"Again? What's with this? You've been there every day this week. Wouldn't you rather be out by the pool than stuck in the back room of Miss Lissy's Closet with cartons of clothes? Everyone knows that's where you are, Kaitlin. They're not putting you out front, are they." It wasn't a question.

"Today they are," Kaitlin said with pride. "Joanne's been showing me how-to stuff all week. Annie can't be there today, because she's going to Washington, and Sabina will be with Phoebe, so I'll be on the floor." That was the expression Joanne had used when she called last night. On the floor. It meant actually selling clothes to customers. Kaitlin still couldn't believe Joanne had asked. Salespeople had to look good in the stores clothes. Wasn't that a major selling tactic?

"Joanne, Annie, Sabina, Phoebe—on a first-name basis with *all* of them?"

"Yes," Kaitlin said. Finished in the bathroom, she headed for her bedroom. That meant slipping past her mother. Fortunately, Nicole stepped aside. Unfortunately, she didn't go away.

Following Kaitlin to her room, she said, "I'd be careful of aligning myself with those people, if I were you."

Kaitlin was in the process of reaching to neaten the bed when she stopped and straightened. "Those people?"

"The Barnes women. They're not in the mainstream in this town. I'm sure you know Sabina was fired."

Kaitlin stared at her mother. "By your

boss, because he's *so* afraid of what she knows. I'm sure *you* know Phoebe has mercury poisoning. What does your boss have to say about *that*?"

Nicole frowned. "What are you talking about?"

"Mercury poisoning. She's being treated for it, and it's absolutely pouring out of her body."

"I don't believe you."

"Call Dr. Martin. He's the one who's treating her. And there's only one place mercury could come from in Middle River, and that's the mill. If I were you, I'd be worried about working there. But your boss hasn't told you that, has he?"

Nicole looked momentarily flummoxed. It was all Kaitlin could do not to smile. And, of course, her mother recovered. Nicole *always* recovered. "See?" she charged. "That's the foolishness you're getting from those Barnes women. Hal Healy is right. They're a bad influence on this town."

Kaitlin did smile this time. She couldn't help it. "Hal Healy is history."

"What are you *talking* about?"

"He submitted his resignation yesterday."

"How do *you* know that?"

"Because I'm at the store, where Middle River comes and goes, while you're at the mill hearing only what the Meades want you to hear. Mrs. Embry came by the store yesterday to ask about Phoebe. She's the vice chairman of the school committee, second only to *your* boss's father, so either Sandy is keeping Aidan in the dark, or Aidan knew about it and chose not to tell you. Mr. Healy resigned."

"Why did he do that?"

"It wasn't only Mr. Healy. It was Miss Delay, too. Isn't that a coincidence?"

"Kaitlin."

"Oh, come on, Mom. Everyone knows they're fooling around. Like he's in her office *all* the time talking about students, only no one believes there are *that* many students with *that* many problems, and if it's only school business, why do they lock the door?"

Nicole was quiet. She looked confused. It was such a rare thing that Kaitlin almost took pity on her—almost, but not quite. The moment was too good to waste. "So, if I were you, I wouldn't worry about the Barnes women being a bad influence. I'd worry about people like Mr. Healy and Miss Delay, and I'd worry about your boss, because if it's true that the mill is poisoning people, he'll get his. You may get yours, too. Want me to tell you the symptoms of mercury poisoning?" She raised her voice to follow Nicole, who had turned and was walking down the hall. "Because I know about those, too. It isn't Alzheimer's, and it isn't Parkinson's. It's sometimes both, and it isn't *pretty,* Mom!"

Since Anton left early for golf or whatever it truly was that he did, Nicole liked her Saturday mornings. Weather permitting, she spent them out by the pool with coffee, strawberries, and the catalogs that had arrived in the mail that week. Last night's storm had left thick clouds and dense humidity, so the pool would have been a bad idea today, but the sunroom at the back of the house would have been fine. It was air conditioned.

But she didn't go there either, instead just sat at the kitchen table. She had the coffee, but no strawberries and no magazines. She wasn't in the mood to eat or to read. She was bothered by what Kaitlin had said— bothered that Kaitlin knew more than she did, but not simply because as the mother, she should be the one to know things first.

She was bothered because Kaitlin was right. *You're at the mill hearing only what the Meades want you to hear.* It was insulting. It was *humiliating.*

Aidan knew things he wasn't telling her, and it had to be deliberate—a concerted effort to hide certain things—a calculated absence of the word *mercury* in any e-mail she might see—a willful withholding of the news about Hal. There had been phone calls yesterday. She knew there was a board meeting on Monday, but was it anything out of the ordinary?

Aidan had said it was not. He had specifically said that, when she asked. Apparently, he had lied. Apparently, he hadn't thought she was important enough to know.

So where did that leave her? Was she or was she not his executive assistant? Was she or was she not his right-hand man? Was she or was she not the one who had to know *everything,* if she had *any* hope of representing his interests in the best possible way?

Picking up the phone, she dialed his cell number. He didn't answer, but she knew how that went. He didn't answer his cell when he wasn't in the mood to talk.

But she was. She had too much invested

in their relationship to let this go. He owed her an explanation or two.

So she called on his home line. Beverly answered, which, had Nicole been a lesser woman, might have been reason to hang up. But she was used to talking with Aidan's wife. The woman often called the office.

There was a bedlam of children's voices in the background. Nicole spoke loudly and with authority. "Hi, Bev, it's Nicole. I'm sorry to be bothering you at home, but I've run into a problem with one of the spreadsheets I'm preparing for Monday's board meeting. Is Aidan around?"

"Are you sure this can't wait?" she asked, sounding annoyed.

"Yes," Nicole replied firmly.

"*Aidan! It's your secretary!*" The phone hit the table with a clunk. The cacophony of background noise went on and on.

Nicole simmered. Secretary? She was a lot more than Aidan's secretary. This woman had no *idea* how much more than Aidan's secretary she was.

She waited for what seemed an eternity through yet more bratty child sounds, and she knew exactly why it was taking Bev Meade so long to get Aidan. He was hiding somewhere in the house where he wouldn't have to hear the kids' noise.

"Yuh," he finally barked into the phone.

"I need to see you."

"What spreadsheet?"

"Mine. Everyone here is gone for the day. Can you come over?"

"And this can't wait the hell until Monday morning?"

"Absolutely not," Nicole said and slammed down the phone. Then she sat back, fuming, thinking every vile word she could about Aidan Meade until he pulled into her driveway and stormed out of his car—and still she sat. Unshaven and uncombed, he let himself in the kitchen door and slammed it shut.

"What spreadsheet?" he asked, looking at the half of her silk robe that showed.

She ignored the look. "Why didn't you tell me about Hal Healy resigning?"

His eyes rose to hers. He seemed taken aback. "Why should I? What's he got to do with the mill?"

"My daughter knew he'd resigned before I did. But more important, what's the business about mercury?"

"*What* business?"

"The *business* about Phoebe Barnes having mercury poisoning."

Aidan scowled. "She doesn't have mercury poisoning."

More lies? Nicole fumed. "Yes, she does. She's being treated for it at the clinic. I thought the mill didn't use mercury anymore."

"It doesn't. Phoebe Barnes is a bizarre person. She hasn't been right for months."

"Exactly, and now they know why. So there I am sitting up in that office at the mill five days a week, forty-nine weeks a year."

"There's no problem with the office."

"Then the plant. I'm up *there* three, four, five times a week doing your bidding. Do I stand to become a . . . a . . . bizarre person, too?"

Aidan made a face. "Geez, Nick. I can't believe you got me over here for this. You know who started the trouble, and you know why. Annie Barnes was waiting for an excuse. Sabina being fired was it."

Nicole might have bought into that excuse before. But to do it now went against common sense—not to mention Tom Martin, Sabina Mattain, and everyone else walking in and out of Miss Lissy's Closet. To do it now said that her daughter, Kaitlin, was as much of a troublemaker as Annie, and somehow, suddenly, Nicole didn't believe that. Suddenly, Kaitlin seemed perfectly reasonable.

"And if you got this from your daughter," Aidan was going on, "it's because she is furious at you because *you* have failed to hide your lousy marriage from her. By the way, did you ever find out if she knows about us?"

By the *way*? Like it was an inconsequential aside? Even *that* annoyed Nicole. "She doesn't know."

"Are you sure?"

She rose, went to the sink, and put her mug down none too gently. She turned, her back to the counter. "For God's sake, Aidan, I couldn't exactly ask directly, or she would have known something she might have only guessed up until now. *Are you sure*? Pu-leeze."

"She knows."

"She does not. She talked about Hal Healy and Eloise Delay and would have talked about us in the same breath, if she'd known."

"Good," Aidan said. "Wouldn't want the girl ruining a good thing." His eyes fell to her robe. "Where is she now?"

"Out. Define *a good thing*."

He took a step closer. "And Anton?"

"Out. Define *us,* Aidan."

Smirking, he came up close and gripped the tie of her robe. "What's under this?"

"Absolutely nothing," Nicole said, as much in defiance as anything else. She wasn't turned on. She didn't know whether it was the fact that this was Anton's house, or the fact that Kaitlin had been eating breakfast in this very room not that long before, or the fact that Aidan had just come from his wife and kids, and Nicole had never set out to cheat them. Perhaps it was just the fact that here in her own home, where there was no pretense of work, where Aidan didn't have the aura of the would-be heir to gild him, he wasn't as attractive to her. "I need you to tell me this, Aidan."

"Tell you what?" he said, distracted as he untied the robe.

"What *we* are. You just called us a good thing. What does that mean?"

He opened the robe. "Quick. Easy. Sex." His eyes were on her along with his hands, and she let him do it. She let him fondle her breasts, bury his mouth against her neck, put a hand between her legs. She even moved against that hand, moved against his crotch, until he began to breathe in that half-gone way he had, "Ooooooh, baby. Ooooooh, that's good."

"Is it, Aidan?" she whispered breathlessly, pressing her hips forward.

"Oh yeah, it's good. It's always good."

"Is it good and hard?"

He grabbed her hand and showed her just how hard it was, and that made it easy for her to squeeze, to squeeze tightly and twist, then shove him away.

"Bitch!" he cried hoarsely, doubling over.

She pulled her robe together and straightened. "A good *thing*, Aidan? Is that what I am, a *thing*?"

"Bitch," he repeated, adding several other pithy epithets under his breath. He raised threatening eyes, but before he could say a word, she spoke.

"Don't even think it, Aidan," she warned. She was the brains behind their duo. It was about time he realized that. "You fire me, and I'll say it's in retaliation for my rebuffing your sexual advances. I'll say that what you just did was attempted rape."

"It wasn't any freaking rape," he snapped, still trying to catch his breath, now with both hands on his knees.

"Okay, then," she said calmly. "Let's call it

consensual sex that's been going on for three years, and how will your wife take to that, Aidan? How will Daddy take to it?"

"You can't prove a thing."

"But I can. I have hotel receipts, plus the names of clerks everywhere we went. We were a handsome couple. They'll remember us."

"You go public about us, and Anton will divorce you."

"Well, maybe that wouldn't be such a bad thing," Nicole said. She was thinking about Kaitlin, who was no longer fooled. And she was thinking about herself. Aidan was right. What they had was quick, easy sex. Having Aidan gave Nicole a good excuse to stay with Anton. So she was beholden to two men, neither of whom she personally cared for. Didn't she deserve more? Wasn't she *worth* more? There had to be someone who would see that, someone who would worship the ground she walked on.

"This relationship is taking its toll," she said with both men in mind. "Maybe it's time we call it quits."

Chapter 26

I WAS UP late Friday night reading Mom's journals, then up again at dawn to pack, which meant that two nights in a row I hadn't gotten much sleep. That might have explained my heightened emotionalism, though I suspect that between James's declarations and those of my Mom, there was good enough reason for it.

I was ready when Tom came by to pick me up for the drive to the airport in Manchester, but I wasn't in the mood to talk. He seemed to know that without my having to say it, which in turn made me wonder why I couldn't fall in love with *this* man. He was good, decent, and compassionate. He was attractive and intelligent. He was a success at what he did, and he wasn't enamored of *my* success. Admiring, yes. But at no point did I feel that he was playing up to me because I had a big name and a solid stock portfolio.

Does that sound arrogant on my part? Well, believe me, it's happened. In the last year especially, after I hit it big with my books, I suddenly became visible to men who hadn't seen me before. I met them at signings, charity events, even Greg's functions. Most were subtle, but others were frank. *I think,* one said, *that it has to be every man's dream to be hooked up with a woman who can keep him in style.* Can you believe he said that? Naturally, I told him that I wouldn't be caught *dead* with a man who wanted to be "kept," and he quickly slunk off, but I'm sure he wasn't alone. I was worth more money this year than last. There were some men to whom that mattered.

It didn't matter to Tom. He fit into my life as a friend in ways that I knew would outlive my return to Washington in the fall. He was going to be a permanent friend. But nothing more.

Had I not been preoccupied, I might have spent the entire trip wondering why that was so, and I might have come up with a handful of possible reasons. But I was thinking about my mother's journals, feeling close to her in ways I hadn't even felt at the cemetery, close in ways I hadn't felt in years.

———

I know I'm supposed to be reading about market trends, she had put down in her neat script, but the bulletins sitting here on my desk will have to wait. I just finished reading Annie's book for the second time, and it was even better than the first. She's such a talented writer. Maybe if I had been as talented as she is, I would have succeeded as a writer. But no. It takes more than talent. It takes drive. Annie has that.

I always knew that about Annie, but for the longest time I didn't know where the drive would take her. She could have gone in so many different directions, and a lot of them would have been self-defeating. But she took one of the good roads, which put her in a good place, and she did it no thanks

to me. I didn't give her the attention I should have. I suspect she took to writing when she was a teenager in part to get my attention. Well, she has. I'm more proud of her than I can say.

———

I mean, what do you do when you read passages like that? You sit there in a public place, beside a person who is attuned to your moods, and you bite the inside of your lip to create pain as a diversion, because otherwise you'll break down and cry. But you can't stop reading, can't stop feeling, can't stop wishing your mother were still alive so that you could talk about these things.

So you read on, but you have little nuggets of new knowledge tucked in a corner of your brain, and it suddenly occurs to you that if the plane takes off with you and your journals, and something unspeakable happens, everything will be lost. So you wait until the last minute, until you can see the airline personnel preparing to board passengers with children or those who need extra help. Then you buy privacy by moving off to the side, and you take out an insurance policy by calling Sabina's cell phone.

"How's Phoebe?" I asked.

"Feeling okay. I have her home."

"At your house?"

"No. At Mom's. The nurses agreed that she'd be better in the place that's most familiar. Besides, I'm still annoyed at Ron."

"Still not talking?"

"Nope."

"Oh, Sabina."

"It's okay. I believe in this cause, Annie. I look at Phoebe, and I see just this little bit of improvement, and I know it's only the beginning. I think we're doing the right thing."

"I know we are—know it for fact. I found our proof."

I heard a gasp on the other end, then a breathless, "What?"

"That March Friday when the Middle River Women in Business met at the Clubhouse? Phoebe was there. Mom, the would-be writer, was the unofficial scribe of the group, but she had inadvertently picked up the wrong folder on her way out of the store to the meeting. She called Phoebe, who brought over the WIB folder, and it only made sense that she should have something to eat while she was there, so she ended up staying through the rest of the meeting."

I could hear tempered excitement in Sabina. "How do you know this?"

"Remember Mom's journals? The ones she kept on the shelves at the store?"

"Those were Gram's journals."

"Nope. Mom's."

"Are you sure? I haven't seen them lately."

"I know. Phoebe moved them. I hadn't seen them either and completely forgot they were there until last night."

"And it says all this inside? In Mom's handwriting?"

"Mom's handwriting, with the entry clearly dated." The general boarding was starting. I could see Tom rise and glance my way. "I have to run, Sabina. Is this good news?"

"It's great news!"

"So if the plane crashes, you'll know what to ask Phoebe when she comes to—"

"The plane won't crash."

"—and if she still can't remember, have her hypnotized."

"The plane won't *crash*!"

"And the best?" I said as I hoisted my backpack to my shoulder.

"There's more?"

"The Northwood board is meeting Monday at four. I've been invited. Is this cool?"

We had climbed through broken clouds, leveled off at thirty-three thousand feet, and left New Hampshire airspace, heading south, when I heard from Grace. I was startled. So much had been going on that she and I hadn't talked in a while. I truly thought I had left her behind.

Why are we here? she asked, sounding disgruntled. *I nearly died once on a flight from Dallas to Atlanta.*

That was in 1961. Aircraft were entirely different then.

Why couldn't we drive? I would have much preferred that. This is not where I want to be.

We couldn't drive because it takes too long. Besides, if not here, where?

New York. I always loved the Plaza. I want to go back there. Pay your money, and they love you. They don't care how frumpy you look. I could be myself there. Or Paris. I love Paris, too. Or Beverly Hills. I could live in any one of those places.

In a hotel, I said to make sure I understood what she was saying.

Why not? Hotels are the best. You don't have to cook or clean or make your bed, and no one faults you for it.

That's all fine and good, I reasoned. But for how long? You can't live permanently in a hotel. Hotels are cold.

And towns like Middle River aren't? Oh boy, do you forget fast. I'll take a cold hotel any day over that. Get tired of one hotel, you move on to the next. It's a gypsy's life, but I like it.

You *hated* it, I argued. You were miserable. You were alone, even when there were people around, so you filled the void with booze. Y'know, Grace, maybe your problem *was* that gypsy's life. Maybe you would have been better off staying in one place and putting down roots.

Weeds have roots. I wanted better.

That's right. The grass was always greener somewhere else. You were like your mother that way, always itching for something more. Well, maybe you didn't know what you had. Maybe if you'd let people get to know you, they'd have accepted you. Maybe you could have made friends. Yes, yes, I know. You had friends, but really only ones who reinforced your separateness. You expected friends to be loyal to a fault, but that doesn't happen right away. It takes time. Maybe if you'd made the effort, you'd have built the trust. You'd have been part of a community. You wouldn't have been such an outsider.

"Annie?" My arm was jiggled "Annie?"

Eyes wide, I looked quickly at Tom. He was seated beside me, regarding me with concern.

"Are you all right?" he asked. I nodded, but he said, "You were looking fierce. And you were moving your lips."

"Was I?" I asked, mortified. "Oh dear. I'm sorry. It was an imaginary conversation. Argument, actually. Writers do this kind of thing a lot." I waved a hand. "But done. Over. Forget it happened. Please?"

"Can I help with the argument?"

I smiled and shook my head. "Not this one."

"Then the other," he said, and I knew what he meant.

"Time to put our money where our mouth is?" I asked.

He nodded. "That's why I'm here."

* * *

It was why we were both here, though that particular meeting would take place on Sunday morning. Saturday was for our personal agendas. When we landed at Reagan National, we got in the same cab, but after dropping me at the condo, Tom continued on to meet his own friends.

Was I thrilled to be in the condo? Not a hundred percent.

But you guessed that, didn't you? You knew that the place would look different to me, that its familiarity would represent only half of what was fast becoming the rooted part of my life. You knew I wouldn't be able to walk in that door and not think of James or Phoebe or Sabina or the backyard willows that overlooked the river.

That said, it was *so,* so good to see Greg when he finally hobbled into the condo with Neil. But if I had expected that he would be bedridden, I was mistaken. Recent surgery notwithstanding, Greg had adapted to the use of crutches with the same ease with which he did most physical things. Yes, there was discomfort. He had been told to expect some swelling and had been advised to keep the foot elevated, but he tired of that quickly. Truth be told, the downer of breaking his leg couldn't begin to compete with the upper of having summited, and that was what he talked about for much of the afternoon. He had more film cards than I could count, and he showed me every shot he took. He claimed he had deleted dozens of bad shots along the way. Had he not done that, I shuddered to think how long we'd have sat there on the sofa, tucked together—me, Greg with the elevated leg, and Neil—looking into the monitor of Greg's camera while he went from image to image with a corresponding narration of the trip—shuddered to think of not finishing in time for me to tell him all of my news, shuddered to think of not finishing in time to allow for leisure—because I wouldn't have missed Saturday night for the world. All Greg had to say was that he wanted to go out, and I was the concierge planning it all.

We started with wine and cheese on the steps of the Lincoln Memorial, because this had always been a fun adventure of ours. We had the cheese in a baggie and the wine in a thermos—I know, I know, not the way to treat good wine, but (a) this wasn't good wine and

(b) the wine wasn't the point. The point was feeling that we were elevated physically and spiritually, just that little bit above the city, and from there we went to our favorite restaurant, a burger place in Georgetown where we were known enough so that the waitstaff catered to Greg's casted leg.

Had it not been for that leg, we might have gone to a succession of bars after dinner. Instead, we settled for an old standby in Dupont Circle, where we could drink beer, watch the Orioles battle the Yankees on the plasma TV on the wall, and reconnect with people we knew. Like Neil, these were originally Greg's friends, but I had long since been accepted as one of the guys, which meant laughter, good talk, and total ease. It was a dark-wooded and discreet place; there might be handholding or a hug here and there, but an outsider would be hard-pressed to know who was with whom. And I wasn't the only woman present, because women in the know who wanted a carefree night understood that even beyond the friendship, the laughter, and the beer, the neat thing about being in a gay bar was that there was no risk of being hit on.

That said, it was not an entirely shock-free night for me—and please don't jump to conclusions here. You know my sexual orientation. You know it quite intimately by now. What you don't know—because I didn't know, didn't guess, should have suspected but was too preoccupied with my own world—is the sexual orientation of Tom.

Yes, Tom. Tom Martin from Middle River. To this day I don't quite know why I looked up at that particular moment. But my eyes penetrated the dim light in the bar and spotted Tom, standing with a group on the other side of the room. He was every bit as discreet as Greg, his arm linked with that of a friend in a way that might have been casual had we been in, say, Paris, where physicality between men was more common. But not here in D.C. This linked arm had meaning, but it was understated, and I'm sure it was done that way for the very same reason that Greg took such care with his public image, including living with me. To be outed as being gay might negatively impact the career of a man like Greg, who was seen as something of a sex symbol to women in the heart of America. Likewise, to be outed as being gay might negatively impact the career of a man like Tom, whose dedication to his retarded sister and to his patients had endeared him to the heart of Middle River.

Tom looked around. When I asked him about it later, he admitted to feeling a prickle at the back of his neck when his eyes met mine. Even through the dark I saw a moment's panic. That was what compelled me to worm my way through the crowd to where he stood. By the time I got there, he had separated himself from the man he was with and was awaiting my judgment. I imagine that in those brief minutes he saw his career roll to the edge of a precipice and hang by a thread. Middle River could accept a bachelor doctor. A gay one? Not now. Not yet.

If he feared me, then he had underestimated me. But then, he was so stunned to see me that he didn't see me with Greg. Had he done that, he'd have understood.

I wrapped an arm around his neck and held on tightly. "This explains so much—how comfortable I felt with you from the start, how much you reminded me of Greg, why there was a lack of sexual chemistry between you and me."

"Are you disappointed?" he asked.

I drew back. "Yes, but not in you. I'm disappointed that Middle River hasn't the goodness to accept people who are different."

"But they are good people, Annie. If I didn't believe that, I wouldn't be there."

Which was one of the things I was coming to love about Tom. He had a big heart, certainly bigger than mine. I was quick to judge; if I could learn tolerance from him, he would prove a valuable friend.

I was up early the next morning, first running through familiar streets in the familiar heat, then going on from there to all the other familiar things that I loved—to brunch with Jocelyn and Amanda, to iced coffee with Berri and John (whom I liked a lot, by the way), to iced tea with another friend. I interspersed all this with errands for Greg and with stops at favorite places, like the Air and Space Museum at the Smithsonian, the Tidal Basin, the shops at Adams-Morgan.

I didn't stay long at any one place or with any one person. There was too little time. But I felt a need to see and hear and feel all that had come to mean so much to me in the last fifteen years. I was crowding it all in, waging a campaign of remembrance, because I was starting to fear that I had forgotten that these things mattered. When I was in Middle River, the town was all-consuming. Here, now, I needed to remind myself of all that I loved that Middle River couldn't provide.

We met at the condo at two in the afternoon. There were Greg and I, and there was Tom, who, after our unexpected meeting, had spent a while with us at the bar the night before. Neil arrived at two, taking time from preparation for his return to court the next day to give us legal advice on the victims' behalf. Nancy Baker, who was a pharmacologist at the EPA and a lawyer herself, was also there. She and Tom had known each other for years. They had been discussing the mercury issue since Tom first suspected a problem at Northwood. Her role was to advise us, strictly as a friend, on where the government would stand, should word of this get out.

For two hours, we went back and forth. I learned what I needed to know so that I could be better informed at the board meeting the next day. I won't bore you with the details, but when Tom and I were finally back at Reagan National for our return flight, I turned to a fresh page in my mother's journal and made my list. The prospects were quite bleak.

First, Northwood could not be prosecuted for violating regulations for the abuse of toxic waste, because the statute of limitations had expired.

Second, according to the files Nancy had checked, Northwood was in complete compliance with current regulations, which ruled out new criminal charges.

Third, if illegal dumping could be proven to have occurred, civil charges could be brought in the form of a personal injury suit by either individuals or a group, but this would require proof of mercury poisoning in every one of the plaintiffs.

Fourth, the waters were muddied by fish containing high levels of mercury. For years, the state had posted warnings to residents of Middle River not to eat what they caught. For years, th'other side had ignored the warnings, choosing possibly dangerous food over no food at all.

Fifth and finally, the only chance of criminal charges being brought against Northwood was if it could be proven that illegal dumping had resulted in one or more deaths. That would constitute murder. There was no statute of limitations for murder.

"It'll never happen," Tom advised from my shoulder, where he had been reading my list as I wrote it. "Making the connection between illegal dumping and any one particular death will be impossible to prove. We don't know for sure that your mother had mercury poisoning. We can surmise it, based on Phoebe's case. But even if an autopsy were to prove that there was mercury in Alyssa's body at the time of death, she didn't die of it. She died of asphyxiation after a fall down the stairs."

"That's a technicality," I argued.

Our flight was called.

"She was sixty-five," Tom countered, tucking a magazine in the outside pocket of his leather duffel. "She might have fallen on her own. Omie was eighty-three. Older people often get pneumonia. In time their bodies wear out and their hearts stop. Was this because of mercury? It's circumstantial evidence, Annie."

"But what if we were to gather *all* the circumstantial evidence, and it points to a connection? What then?"

"Then we have headlines and lawyers and media, all the things you just told me you didn't want."

"I know," I said as I reached for my bag. "And I don't want those things. But *they* don't know that, and it's a good threat, don't you think? Threats are what it's about. It's about giving the impression that you have a fool-proof case. Northwood has to understand what they stand to lose if they choose to fight."

I was, of course, referring to Northwood as though it included the triumvirate of Sandy, Aidan, and James, when in fact James planned a coup. A *bloodless* coup, he had said. I wondered what he had in mind, whether he could pull it off, and what would happen if he didn't. I guessed he would have to leave town. He would get another job. He could get a *better* job. He might even move to Washington. That would be promising.

Promising for me. Not so for Middle River. The town needed him. For that reason, I was praying that his bloodless coup would succeed. In that spirit, I knew that any evidence I could bring to the board meeting would be a help.

"We have Phoebe," I said, mustering hope

as we stood and shouldered our things. "There have to be others like her." When I started forward, Tom held me back with a light hand on my arm.

"I haven't thanked you," he said quietly.

"Thanked me for what?" I asked, though I knew.

"Being exactly the same today as you were yesterday."

I was touched by the way he said it, and felt the need to make a point. "Are you a different person today from who you were yesterday? No."

He smiled sadly. "No. It's the perception that changes. In a town like Middle River, that's the name of the game. Y'know?"

Chapter 27

OUR FLIGHT was delayed. We sat on the runway for forty minutes while mechanics tried to fix a problem. When they failed, we returned to the jetport, deplaned, changed gates, replaned, and finally took off ninety minutes late. By the time we had landed in Manchester, found Tom's pickup in the parking garage, and made the drive to Middle River, it was nearly as late as it had been when I had first returned to Middle River for my vacation on a day that seemed like an eon ago.

I would be lying if I said I wasn't pleased to be back. Of course I was pleased. Returning tonight was different from that other night. There was excitement this time, anticipation.

It was stagnantly warm; still, we rolled down the windows of the pickup as soon as we crossed the town line. We had barely passed Zwibble's when I smelled something odd.

"What's that?" Tom asked, clearly smelling it, too.

"Not gasoline," I said. "It almost smells like the kind of wood fire you'd burn in the middle of winter—almost, but not quite." I studied the buildings on Oak as we passed, but they were their usual inert nighttime selves, lit only to show their wares.

Something was definitely burning. A fine whisper of smoke had begun to collect in the air. As we drove on, the whisper grew louder and the smell more acrid.

We were nearly at Cedar, where Tom would turn to drop me at Phoebe's, when I saw the glow above the treetops ahead, up on the north end of town.

"The mill?" I asked in dismay.

Without a word, Tom drove straight ahead rather than turning. We were passing homes now, too many of which were lit this time of night. Coming up behind the taillights of another car, we drove around it and on. Its headlights grew smaller behind us, but they didn't disappear. Someone else was headed for the mill.

"It's brick," Tom said. "How can it burn?"

"Maybe the insides?"

"There's a sprinkler system."

"Maybe the woods around it?" I didn't know what else to suggest. The Gazebo was the only structure made purely of wood, but that had already been torched and rebuilt.

Torched. It was a harsh word, though it did describe what had been done both to the Gazebo and the Clubhouse. There was only one suspicious building left to burn.

"Omigod," I said as we turned in at the stone wall that marked the entrance to the mill. The smoke was heavier here, the smell sharper. "The Children's Center," I said. "Go there."

We didn't get half that far. Too soon we found ourselves behind a line of cars, with a mill guard flagging us down. We saw flames ahead—heard their crackling, the pounding of water from hoses, the bark of a bullhorn. Parking behind the other cars, we continued on foot.

The crackling actually came from the

trees that surrounded the Children's Center. There was nothing left to burn of the center itself. It was little more than a brick shell.

We passed clusters of people. None seemed hurt. Still Tom ran on.

I caught his arm. "A fire here can mean only one thing. There's a mercury leak. Don't go on. It's not safe."

"I have to," he said. "I need to make sure no one's hurt. You stay here. I'll be back." He held up a finger in promise, then disappeared into the glare of spotlights and fire.

I had just lost sight of him when someone came up on my right. It was one of Omie's grandnieces. "The building was completely engulfed when the trucks arrived," she said. "It spread to the wood equipment in the playground, went along the whole row of them and up the kids' tower into the trees. That's what we're seeing now. The trees."

"How did it start?"

"No one knows," she said. Several others had joined us. I recognized more of Omie's relatives, Marylou Walker's son, a Harriman or two.

"Is anyone hurt?" I asked.

"No," said the Harriman. "I was up at the front a couple minutes ago. All's left is to contain the blaze. They've got it, I think. Big flames, short life, but more excitement than we've had here in a while."

I was thinking that they should only know what's coming, when I saw James emerge from the smoke. Remaining separate and off to the side, he put his hands on his hips and turned to watch.

I slipped away from the others. He didn't see me until I was close to his side, and then he was startled, but only briefly. His eyes returned to the fire.

I turned to watch with him, moving close enough so that we were arm to elbow. "Is there a leak?" I asked softly.

His voice was tight. "My monitors said no. They're checked twice a day."

"Your monitors. Ah-ha." I should have guessed. Hadn't he told me he would know if the Children's Center drums sprang a leak? Monitors would tell him that. Still, he sounded angry. "Are kids getting sick?" I asked.

"No."

"Did you set the fire?"

"No."

"Your father?"

"Not personally, but I'm sure he ordered it done."

Angry? James wasn't angry. He was *furious*.

"But isn't this a victory for you?" I reasoned. "Isn't it an admission from him that the drums underneath need to be removed?"

"Oh yeah, only no one will ever know it," he said through clenched jaws. "It's another coverup in too long a string. He'll get those drums outta there so that no one'll ever be able to claim his child got sick from something leaking into the water supply in the playground bubbler. He'll tell the board there are no toxic sites at the mill—and it'll be a lie, because there *are* other ones, only he ran out of pretty little diversionary buildings to put over them, so they're away from where people go, which is good, at least—and the board will go home and sleep well with the reassurance that Sandy Meade is on top of things. He knew they were starting to talk. Word about your sister spread faster than this friggin' fire. He was starting to get phone calls."

"Was he the one who called tomorrow's meeting?"

For a minute James's mouth was a rigid line. Then he took a breath. In the orange sheen of the fire—diminishing now—I saw the brief flare of his nostrils. "Nope. It was me. Only he reframed *that* fact by calling each of the board members and setting forth his agenda. The stakes are rising. This is an anti-me move. But it's typical of my father— trying to one-up anyone who crosses him. Well, fuck it, I'm not going away. I'm tired of coverups. I'm tired of working at a place that is dishonest enough to put its employees at risk. There's so much that's good in this mill, but it all gets diminished by the stain of graft."

"Graft."

James shot me a look. "The guys who do the cleanup? Ever wonder why they don't tell what they've done? The answer'll make your blood boil."

"Money for silence?"

"Big-time."

"Do they know they're handling mercury?"

"No. They're told it's 'production waste' that may have turned toxic over time. They're protected—masks, coveralls, you name it. And the work is done after hours, so if you're wondering why the rest of Middle River doesn't guess what's wrong, don't."

He went silent. I might have said he had run out of steam, if I didn't continue to feel his anger. It came off him in a rigid pulse. That arm cocked on his hip? It was glued there, and it was iron-hard. I knew that, because I touched it. It was my pathetic attempt to give comfort.

Very quietly—timidly, actually, because I didn't know him well enough to know how he reacted in situations like this—I asked, "Are we set for tomorrow?"

"I am," he said tersely. "Are you?"

"Yes."

"Good."

That was all he said before he vanished into the night.

Phoebe was upstairs in bed, but Sabina was in the parlor, asleep on the sofa in weirdly similar fashion to the way I had found Phoebe the very first night I had come. The difference came when Sabina awoke. She sat up, instantly alert.

"There was a fire at the mill," she said. "I'd have gone over, but I couldn't leave Phoebe."

"Tom and I just came from there," I said and told her what we had seen. We moved into the kitchen for tea, and it didn't seem to matter that it was one in the morning. I wanted to know the latest on Phoebe; Sabina wanted to know about my afternoon meeting. I wanted to know if Sabina had talked with Ron; Sabina wanted to know how it had gone with Greg.

We went through several cups of herbal tea, went on talking longer than either of us would have imagined, but Sabina didn't seem to want to go to bed any more than I did. We had never done this before, she and I. I mean, *never*. For the first time, we seemed to be more alike than different. For the first time, we were friends.

That was why, when a telltale ring came from my purse at two in the morning, after I answered, talked, and ended the call, I put the phone on the table and met Sabina's curious eyes.

"You heard. I said I'd go over."

"James *Meade*? Calling you at this hour?"

"We have this . . . this *thing* going on. We kept bumping into each other mornings when we were out running, and I had no idea it would amount to anything, because you remember the mess I got into with Aidan, but then it just kind of . . . became . . . something."

"*Sex*?" she asked.

"I'm not a virgin, Sabina. It was only when I was growing up here that I was a total nonwoman."

She was smiling. "Sex with James Meade? That's amazing. He doesn't do it with anybody. I mean, there's a reason he had to adopt a child."

I might have enlightened her on that score if I felt it were my place to do it, but it wasn't. As much as I liked this new honesty between Sabina and me, I couldn't betray James. So I put the cell phone back in my bag and took my car keys from the basket of keys on the counter. "I don't know as we'll be doing that tonight. He's upset about the fire, and he's nervous about tomorrow. I think he just doesn't want to be alone."

Sabina remained amazed. "James Meade is *always* alone. Either that image is bogus, or you've done something to him."

"The image is bogus," I said on my way to the door. I paused with a hand on the knob and looked at her reflection in the mullioned glass. "I may not be back until morning. I

mean, it won't be about sex, but if he wants
me to stay, it might be in all of our best interests. I'm going to be lobbying for your job."

There was enough teasing in Sabina's
smile to make it even through the glass. "I'll
just bet you are."

James saw my headlights and was waiting
at the side door. He wore nothing but jeans,
and much as I might tell my sister that this
visit wasn't about sex, I was ready when he
pressed my back to the wall.

So was it only about sex after all? Had he
called with that in mind, needing physical release in a time of stress, as so many men
did? Was I fooling myself, as so many
women did?

No. Absolutely not. It was just that we
were newcomers at communicating, and in
this way we did it very well. And then, just to
show you how really new we were at understanding each other, he asked a question
that took me by surprise. Actually, it wasn't
even the question that took me by surprise,
as much as the confrontational tone of his
voice.

"How was Greg?"

I was tingling inside. He had just left me, but my legs were still wrapped around his waist, with his hands holding them there. My back was still to the wall—my bare back, now, because our clothes were strewn about—and my arms were around his neck.

Freeing one, I swatted at his head. "You jerk!" I cried. I would have separated from him completely, if his torso hadn't been so fixed. "Is *that* what this is about—you marking your territory?"

Had he smiled, I would have hit him again. But he didn't. His face was as sober as I'd seen it. "It's about my feeling too much for you and wondering when the hurt's gonna come."

My heart melted. "You *are* a jerk," I said, but more gently. Then I told him about Greg. It was not a betrayal of Greg's privacy, because James and I had crossed some kind of line into a relationship that included trust. He trusted me with the truth about Mia; I trusted him with the truth about Greg. I didn't make him promise not to tell anyone about Greg, any more than he had made me promise not to tell anyone about Mia—any more than Tom had made me promise not to tell anyone what I'd learned about him. Trust was implicit in each of these instances. It went with the territory of being a true friend.

Tom was that. And yes, James was that, too.

We talked more. We spent most of the night at it. The light of day had already risen beyond his bedroom window before we finally fell asleep, but Mia woke us shortly thereafter. Since neither of us felt comfortable having her see us in bed, I let James go to her while I showered and slipped back into my clothes.

James caught me at the door just as I was about to leave. Yes, he was holding Mia, but she didn't see anything remotely untoward. All he did was put a hand on my cheek.

"Thank you," he said. "I needed that."

I nodded. "Today is important."

He raised his brows in wry agreement. "See you at four?"

I nodded again. Touching the tip of Mia's tiny nose, I said, "Bye, Mia."

She put a finger in her mouth and smiled.

You love him, Grace said, but it wasn't an accusation. After my blasting her on the plane, she had mellowed.

I don't know, I said.

I think you do. Will you marry him?

Aren't we getting ahead of ourselves here?

Are we? she asked. *Isn't this what girls do? We dream, and then we picture how things will be in those dreams. When I was in high school, I'd doodle "Mrs. George Metalious" in my notebook*.

Well, I'm not in high school. There are weighty decisions to be made, and I don't have the facts. James may or may not take over his company, he may or may not be leaving Middle River, and he may or may not love me. He has never used those words.

Do you want him to?

I don't know.

Do you want him to take over the mill?

I don't know.

What about leaving Middle River? Do you want him to do that?

I don't *know*. Why are you asking me all this? I've told you I don't have the facts.

Facts? Facts don't matter. It's the heart that counts.

Thank you, Dear Abby.

Now wait just a minute. Aren't you the one who spent half a plane ride giving me advice? You said I moved around too much and that maybe if I'd stayed in one place and put down roots, I'd have been happier. So was that about facts? It was not. It was about heart. That's what happiness is. It's about heart.

I couldn't argue with that. Nor was I surprised that she said it. Grace had lived life with her heart on her sleeve. It made sense that she would see it. And me, I was prevaricating. That was all.

I may have botched the execution of it, she went on, *but I knew what I wanted. I wanted success. I wanted the freedom to live life on my own terms, and I wanted a man to do it with.*

You wanted a man to dote on you.

Okay. Yes. Fine. And I wanted children, lots of children, only my body gave up after three, so my books became my children, and then they failed me, too, but at least I tried, because those were the things I wanted. So maybe you don't want those things. But do you know what you want? Do you know?

I did not. And I couldn't spend time right now thinking about it. A revived Phoebe—not perfect, but improved after a day of therapy, three days of rest, and a huge dose of optimism—insisted on being at the store, and if the crowds were big when she had been at the clinic, they were even bigger now. Sales were strong. But affection? Through the roof. It was one large community show of love for our sister, and Sabina and I were impressed.

That said, we were also distracted. Mindful of a four o'clock deadline, we made phone call after phone call trying to find one more person who would testify to the tie between exposure to mercury at Northwood and chronic illness. We had a significant list by this time, though, of course, most of the people on it were being taken care of by the Meades and had no desire to rock that boat. I ran down the street, caught Emily and Tom McCreedy at their store, and made my arguments yet again. I drove across town and did the same thing with Susannah Alban, but none of the three would commit. Sabina even managed to dig deep enough into mill records to identify the chef who had prepared dinner for the Women in Business that night in March so many years ago. He had cooked; his wife had served. They had left Middle River soon after the fire to work at a resort in Vermont, but when we called there, the manager said that the pair hadn't been reliable enough to keep on for more than a year, and he had no idea where they had gone.

Very possibly their "unreliability" had to do with health problems stemming from the leak of those underground drums. But if we couldn't locate the two, the point was moot.

Badly discouraged and needing respite, we took to filling the time between calls by reading Mom's journals. There were a remarkable number of them, going all the way back to the time when she gave up writing for shopkeeping, and it wasn't that they held any major surprises, just that they were so . . . so *Mom*. There's reason why she had wanted to be a writer—she was good. That came through when she wrote about her feelings for Daddy even years after his death, when she wrote about all that she wanted for us that she feared she couldn't provide, when she wrote about Phoebe's divorce (which pained her) and Sabina's absorption in computers (which confounded her) and my leaving Middle River and never returning (which hurt her deeply).

More than once, Sabina and I would sit

back with tears in our eyes after one or the other of us read a passage aloud. I'm not sure I would have wanted to do this alone. It had more meaning being with her. There was certainly more comfort.

But four o'clock approached. Setting the notebooks aside, we made a final round of calls, reminding all we spoke with of the meeting at the mill, and then Sabina went out on foot again to lobby for help, while I returned home to shower and dress. I put on a skirt, the only one I had brought with me. It was white and worked well with a red blouse and red sandals. I was definitely going for power. To that end, rather than letting my hair shout *WOMAN,* I swept it back and anchored it with a clasp. I put on makeup with care, then lined my lips with a pencil and applied lipstick with a brush. I straightened before the mirror and checked to make sure I approved.

Feeling reasonably attractive and suitably professional, I set off—and only then did it occur to me that not once since I had been back in Middle River had I seen Sandy Meade in person, which means, of course, that you don't know what he looks like.

Picture a lion with a large head and a full

mane of silver hair. Erect, he stands at the same height as Aidan, several inches shorter than James. Picture a barrel chest above slim hips, and legs that are strong and agile enough to allow for stalking, which he had always done, and which, I'm told, hadn't slowed with age. Picture strong hands that like tapping a table with a pen, a mouth that is turned down, and eyes that drill whatever they see.

Intimidating? Definitely. By the time I arrived at the executive offices, I was wondering if I was up to the task. It didn't help that the last time I had faced Sandy Meade in a confrontational situation, I had lost badly.

But I was here. And I wasn't turning back. The lingering scent of the fire stiffened my resolve. Last time, the Meades had succeeded in sweeping truth under the carpet. This time, I had maturity, evidence, and James on my side.

Since there were more cars than usual, I had to park a short distance down the drive. I was walking toward the redbrick Cape with its tall pediment, dormers, and white shutters, when a man emerged from a car immediately ahead. He was very much like James in appearance, though I realized afterward that it was only the height, the neatness, and the air of authority that corresponded.

He put out his hand. "Ben Birmingham. I'm James's friend."

In that instant, I recalled what TrueBlue had said. Adding that to what Sabina had learned hacking into the system e-mail, I said, "His college roommate, the lawyer from Des Moines."

Ben smiled. "That's it. He described you well, too. I couldn't have missed you."

"Cute. I'm the only woman around. This is still an all-male board."

"There's Sandy's secretary."

"She's sixty."

"True." He hitched his head toward the building and said more seriously, "The board members are all inside. They'll be conducting business on their own for a while. While they do that, James wants us to wait outside the conference room. I'd be honored if you'd let me walk you there."

"Are you my lawyer?" I asked, only half in jest.

"If you're on James's side, I believe I am."

Chapter 28

THE CONFERENCE room was large and paneled in dark wood, with thick Persian rugs underfoot and weighty oil portraits on the walls. The oils were of Sandy Meade's parents and grandparents, painted from timeworn photographs but depicting each subject doing the kind of aristocratic activity that, even James knew, was more fiction than fact. His great-great-grandfather hadn't hunted for fox, any more than his great-great-grandmother had been a socialite. James believed that the dishonesty detracted from their memory, that it somehow said that who they really were wasn't good enough. He had always felt that the truest things in this room were the open meadow, the scattered trees, and the river, pictured in all their glory in the wide wall of windows.

He didn't see those today; his back was to the glass. He did see the long mahogany conference table, with a glass of iced water for each attendee and, evenly spaced, a pair of paper cubes. Sandy was at the far end, with Aidan on his immediate right, and the non-Meade members of the board, five in all, randomly seated between those two and James. They included Lowell Bunker, the mill's lawyer; Sandy's longtime friend Cyrus Towle, president of the country club and another longtime friend; and Harry Montaine, vice president of a private college in nearby Plymouth and the token intellect. Brad Miller, a state senator and the newest member of the board, was a friend of Aidan's and had been brought on the year before for that reason alone. And finally, at the far end of the table, was Sam Winchell. A sometime friend, sometime adversary of Sandy's, he was on the board to show that Sandy wanted the town to know the truth about the state of the mill. Unfortunately, since meetings were carefully scripted, Sam had nothing to report to the town that Sandy didn't want known.

James wore a navy suit, not his everyday attire, but that didn't keep him from resting easily in his seat with his elbows on the upholstered arms, while Sandy gave his opening. It always followed the same chatty format, more a personal update than business, and it seemed all the more outlandish to James today, given the fire the night before. Not that the lingering smell of it penetrated the conference room. The air here was as carefully controlled as the agenda of the meeting. If the board members hadn't already learned of the fire, they might not have known it had happened at all.

But that was how Sandy worked. A master at manipulating the opinions of others, he was saving the real purpose of the meeting for last, solely to minimize its importance.

Setting a pair of slim reading glasses halfway down his nose, he began reading excerpts of papers pertaining to the financial health of the mill. He went on to discuss the newest paper being produced by the mill for digital-imaging processes, put due emphasis on the customization capability of its coatings and the resultant demand by hospitals, praised it as being the wave of the future—and he did all of this without once mentioning James, whose brainchild it was. He talked of other directions the mill hoped to take and presented more figures and charts, and by the time he was done, more than one pair of eyes had begun to glaze.

But not James's. He knew what his father was doing and had to work harder to swallow his fury as the minutes passed, had to work harder to keep his mind on his father's words. And then, finally, there they were.

". . . really too bad," Sandy was saying about the fire at last, "because the Children's Center is a vital part of this community. But Aidan has already found temporary quarters for the day care center until it can be rebuilt here on our grounds. Aidan, please speak about that."

Aidan came to. He had been sitting in something of a trance with his fingers steepled and his eyes out the window. It was his rendition of the thinking man's look.

Thinking man's look? James knew damn well what he was thinking about, and it wasn't business. It was his Executive Assistant, who was giving him some kind of trouble, to judge from Aidan's ranting that morning.

More galling to James, though, *Aidan* hadn't found a thing with regard to the Children's Center. Sandy was the one who had called the Catholic church and convinced Father William that it was in Our Lady's best interest to shift CCD classes to the evening and offer the space to the Children's Center during the day. Sandy knew about Father William and his housekeeper, and would have had no qualms in using that as leverage.

No, Aidan hadn't *found* a thing. Sandy had laid it all out for him an hour before the start of this meeting. Presenting it as Aidan's idea was the typical spin. Aidan had to be established as being competent, active, and concerned.

James might have laughed at that ruse, if it hadn't been so pathetic. Putting Aidan in charge would be the beginning of the end for the mill. Aidan was a puppet. He neither knew nor cared about the nuts and bolts of the mill. Sandy wanted a tool for promulgating secrecy, bribery, and fraud even long after he was gone. Aidan would be that tool.

James was angry, but not out of envy. He was angry because he cared about the mill, and because he cared about Middle River. Had either not been true, he would have left town and tried to salvage his relationship with April for the baby's sake. Likewise, had either not been true, he would be thinking of leaving town to follow Annie Barnes. But that was a whole *other* can of worms.

"So we're on top of this," Sandy assured the board, taking over again that quickly, because Aidan hadn't had a whole lot to say. "Now," setting one piece of paper aside, he cleared his throat and pulled up another, "there's the matter of mercury. Rumors have been going 'round town that the mill has problems, so if you move on in your folders, you'll see copies of the latest certification we received from the state."

James knew how this would go, too. By way of damage control, Sandy launched into a presentation of scientific facts and figures that were far too complex for the board to understand, but understanding wasn't the point. Sandy didn't understand them himself. He had admitted that to James more than once, back in the days when they talked, when he thought he was grooming James for his job and wanted to teach his son how to lead. The goal, he claimed, was simply to have the board *believe* that he knew what he was talking about, in this case that the state had truly declared that Northwood was mercury free.

"Hold it," James said now. They were the first words he had spoken, and they brought all eyes his way. "That's only half the truth."

Sandy shot him an ingratiating smile. "Well, it's the half we need to hear," he said and turned to the others, prepared to go on.

"No," James insisted, straightening in his seat and thereby drawing attention. "That's wrong. We need the whole truth."

"What's the whole truth?" Sam Winchell asked.

"Christ, Sam, don't encourage him," Sandy snapped.

James held up a hand to appease Sam. The man had a natural curiosity, not to mention a few philosophical differences with Sandy, hence the question. But James knew what had to be said. "The whole truth is that there are potential hot spots scattered around the mill campus, where toxic waste was buried in drums. If those drums start to leak, the potential for causing harm to human health will spawn the kind of legal action against Northwood that will run us out of business and take the town down with us."

"James," Sandy said with an exaggerated

sigh. "You're being dramatic. There is no such risk."

"It's already happened twice," James said.

Apologetically, Sandy looked at his board. "I'm sorry. He simply doesn't know the facts."

James was satisfied. Sandy was playing right into his hands. The more he said to discredit James, the more he cooked his own goose.

Meanwhile, Aidan wasn't saying anything. He wasn't good on his feet. To his credit, though, he did look alarmed.

James took his briefcase from where it had rested so innocuously on the floor by his chair. He stood, put the case on the table, undid the strap, pulled out papers of his own, and began passing them around—and still Sandy tried to act as though James were nothing more than a misguided son who didn't know any better.

He gave a long-suffering sigh. "What are you doing, James?" To the others, he said, "I wouldn't bother yourselves with whatever it is that's on those sheets. It's misinformation."

James ignored him. He addressed the board members. "These are lists of people whom Northwood has been supporting through bad times. You'll see that each has been unable to work because of the ailments listed, and you'll see the dates when the problems began. You'll also see the dates of the last two fires here at the mill. What you won't see is that those fires, like the one last night, were deliberately set after mercury leaked from those underground drums and made people sick."

Sandy stood. "That's *enough.*"

James went on. "Mercury poisoning is typically mistaken for a dozen other diseases. That was what happened. Except we knew the truth."

"Bull*shit,*" Sandy roared, but James wasn't done, not by a long shot. He was actually just getting going, just finding his rhythm. He half wished Annie were in the room, but it wasn't time yet. She would be pleased, though. The board members were listening. If they didn't believe yet, they would soon.

"If you turn to page three," he said, "you'll find copies of the first of the internal memos sent at the time of each leak. They're coded, so that there was no direct mention of words like *mercury* or *illness.* But if you turn to page six," he waited through the rustle of pages, "you'll see that Aidan says it all in one memo."

Sandy turned to Aidan. "What the hell?"

Aidan seemed startled. "Not me. I didn't send this. It's a forgery."

James eyed his brother. "I have the originals in a safe place under lock and key. Your initials are right there in ink."

"I didn't send any memo."

"Your initials, Aidan. In your hand, in ink," James said with quiet confidence, and the tone served him well. Brad, who had held back, was paying close attention now. Same with Harry. Lowell, Sandy's lawyer, had his glasses on and was closely reading the pages. And Sam was sitting back, hanging on James's every word.

So James gave him more. "After Aidan's memos, you'll find ones that deal with the men who helped with the cleanup. There are invoices detailing the gear that was purchased so that they wouldn't be exposed themselves. They weren't told about any leak, by the way. They were told that wearing the gear was a precaution in case something happened while they worked. They were told that since the Gazebo, in this case, had to be rebuilt, it just made sense to get rid of the drums."

"*That* was the truth," Sandy charged. "There were no leaks."

James went on calmly, "The last two pages detail Sandy's relationship with certain people at both the local and the state level, people who may have been in a position to question the toxicity at Northwood. As you can see, he's treated them very nicely over the years. It's not surprising that they looked the other way."

"This is slander," Sandy growled in the direction of his lawyer. He took his seat, ceding responsibility for the fight.

Lowell looked over his glasses at James and said with a Brahmin air, "Your father may be right, if all you have is paper." He shook the papers, as though they weren't worth a cent. "These things can be fabricated."

Before he had finished, James was on his way to the door. He hadn't ever doubted that Ben and Annie would be there, but the sight of them warmed a cold spot in his gut. They were two of the people in his life that he most admired, albeit one for years and the other for only days. He gestured them in.

Sandy didn't like that. "This is a board meeting. They don't belong here."

"We often have guests," James said. He introduced Ben as his longtime friend and personal attorney, and Annie as the voice of the people.

"Voice of the people?" Aidan cried, sounding outraged. "Oh, come on. She's a woman wanting vengeance for something that happened years ago."

James turned on him. "And she deserves it," he said tightly, "only that's not why she's here. She's here because she has proof—listen and hear—proof that her sister Phoebe had mercury poisoning." His eyes returned to the group. "And don't anyone try to suggest that she was exposed to mercury elsewhere in town, because the very last pages of the printout you have detail the years when the mill used mercury in its chlor-alkali plant, and acknowledgment of that by the state. We used mercury, and we produced mercury waste. Those facts are a matter of public record." He drew up chairs for Annie and Ben.

Sandy was on his feet, addressing the board. "What my son hasn't told you is that

Ben Birmingham is his buddy. They were roommates in college. Makes you wonder why James couldn't go to a real lawyer, and as for Anne Barnes, she's emotionally unstable. You'll get people all over town testifying to what she did here years ago."

James was undaunted. He knew just how stable Annie was. He guessed that she had as level a head on her shoulders as anyone else here.

Smiling, he looked around the room. He was his father's son. He could match the old man for grit. "Tell you what," he said with ironclad good nature, "if any of you feels that my guests have nothing valid to say and you want to leave now, go ahead. You'll miss history being made in this town." He extended a hand toward the door in invitation, raised his brows, looked from face to face.

"Well, I just might," Sandy said. "Too much hot air in here. I'd be better outside," he added with a glance at the window, but the glance stuck. He focused, frowned, craned his neck. "What is this—'take a walk in the meadow day'? It isn't five yet. What in the hell are they doing out there?"

James looked outside, as did everyone else in the room. At first glance, it did appear that people were just walking across the grass. Second glance showed them not to be walking once they reached the point directly in front of the windows. They were turning and staring up.

James exchanged questioning looks with Annie, but she shook her head, as mystified as he.

Sandy grinned. "There. You see? Don't know how they found out about this meeting, but they're not buying what James is trying to sell." He leaned forward, hands on the table, piercing eyes going from face to face. "Because that's what he's doing—trying to sell you a bill of goods. My eldest son sees the writing on the wall. He knows his time here is limited."

James didn't reply. Nor did he back down. "Annie, please. You first." He sat down and watched her rise, and for a minute, he was startled all over again. The Annie Barnes they all remembered from the past was worlds away from the attractive and sophisticated woman who stood here now. Her clothes were perfect for the occasion, her hair was perfect, her face was perfect. Even her tone of voice, with its tentative start, was perfect. Had she come on too strong too soon, people would have called her arrogant. Starting this way, she seemed more committed than seasoned and therein genuine.

He loved her honesty. He truly did.

She began with mention of her mother's death, but gained strength as she mentioned the people in town with whom she had talked, the physical problems they faced, their reluctance to blame the mill, and their reasons for it. She brought a personal touch to the less personal lists he had produced, and grew even more impassioned when she began speaking of Phoebe. And that was the thing that worked best—the passion. James knew everything she was going to say, but found himself rapt. He loved that passion. He couldn't doubt that she cared for the people in town or that she valued her family. She told of the Middle River Women in Business, ran through dates of their meetings at the Clubhouse prior to the fire, and gave an update of the subsequent decline in health of each attendee. She produced her mother's notebook, read the entry telling of Phoebe's presence at that one fateful meeting, spoke of miscarriages and recent illness. By the time she finished telling of Phoebe's tests in New York, the first of the treatments here in Middle River, and the latest lab results, the room was silent.

By then, the crowd outside had swelled. James didn't know if this was good or bad.

Sandy didn't seem to notice the crowd. He was too busy being smug. "Well, we certainly know why she sells books. She tells a good tale."

James came to his feet. "No tale," he said, impassioned himself now. "It's the truth, which brings us full circle." He shot a thumb toward the window. "We can't do a goddamned thing about the people from th'other side who eat fish from the river after they've been told not to, but when we let people send their children daily to a building that we know sits on a powder keg, that's immoral. When we let people suffer without the medical care they could have if only they had a clue to what's wrong, *that's* immoral. Some of it can't be fixed. We can't do anything about the McCreedys' autistic son. That kind of *in utero* damage is permanent, but we can help with his education and set up a fund to help support him—"

"We're doing that!" Sandy said.

"—and we can let Emily McCreedy know that her asthma's from chronic mercury poisoning, and that maybe, just maybe, there's a treatment for it."

"Ah, *Christ,* James," Sandy cried, throwing a hand up in exasperation. "What're you going to do—tell the world the truth and have panic here and downriver and everywhere else that people want something to blame for their being sick?"

"If I have to, I will," James vowed. "We can either do this thoroughly but quietly, or we can open it up and make a big noise. Your choice."

"*My* choice? Forget mercury. The *truth* can kill people. Tell them the truth in this case, and that'll be the end of the mill. Don't you care about that? Doesn't it matter? Or are you just as happy to pack up that little girl, who isn't going to fit in here anyway, and move on?"

James was livid. "That little girl is my daughter."

"Your adopted daughter," Sandy clarified coolly.

"No," James stated and looked slowly away from his father and toward the others in the room. "Mia is my daughter. Her mother is a woman I was with for six years—"

Sandy cut in. "Did any of us ever *see* this

woman he was with for six years?" He rolled his eyes.

"Oh, she was here," James said, still speaking to the board. "She was here on three different occasions, and on each she was made to feel like such a non-person by my good father that she swore she would never return." His gaze locked with Sam Winchell's. "I'm Mia's biological father. I have sole custody, because April wouldn't live here and I couldn't leave, which tells you what I feel for this town. So go ahead. Print any or all of the above in the *Times*. I'm tired of playing Sandy's game. And while you're at it—"

"James . . ." Sandy warned.

But James was beyond restraint. "—while you're at it, you can print the fact that my mother is alive and well and happily married to a far better man than the one at the head of this table. And *that*," he said, oblivious to the reaction of those listening, "brings me to the reason why my college roommate is here. Ben happens to be one of the best legal specialists in the country when it comes to family businesses. His work involves psychology as much as it does law, and he's been talking to me for a while, counseling me on how to try to get my father to understand how bad this whole mercury thing is, but Sandy refuses to see. It's my way or the highway, he says, and that would be fine, if hundreds of people in this town weren't in need of help. So we're going to help them."

"Like hell we are," Sandy said, though with less force than before. He was looking out the window, seeming confused.

James remained focused on the board. "We're going down through our lists, making sure that the people who can be helped will be helped, that the others are compensated, and that every last bit of mercury waste is removed."

"You'll have lawsuits," Lowell warned.

"Not if we do this right," James said, "and we have the money for it, with plenty to spare for future growth of the mill. Don't you see," he asked now of the others, "that it's the right thing to do? The mill has a moral obligation to come clean."

"Over my dead body," Sandy vowed, but he sounded tired. He was still looking out the window.

Now James turned to look. He must have reacted somehow, because suddenly everyone else was looking, too, including those on the far side of the table, now rising to see what was out on the lawn.

The crowd had grown. It covered most of the immediate area, stretching from side to side, but even more striking was the sign they had raised. There was only one. It, too, stretched from side to side, with letters large enough to be read from the boardroom with ease.

WE WANT A CLEAN MIDDLE RIVER, it said.

Annie was clutching James's arm. "Do you see who's there?" she asked softly, excitedly.

James nodded. Front and center, he saw Sabina and Phoebe. But he also saw the McCreedys. He saw the Albans and the Dahills. He saw Ian Bourque, John DeVoux, and Caleb Keene. He saw people from both sides of the river, sick ones, well ones, ones who worked at the mill and ones who did not. The implication, of course, was that if push came to shove, these people would fight the mill.

Looking back at his father, he saw that Sandy understood it, too. In that handful of minutes, he had grown smaller and paler. And suddenly James was saddened. He disagreed with Sandy on most everything important, but the man was still his father.

"I don't want this done over your dead body," he said as though he and Sandy were alone in the room. "What I want is a clear-headed transition, with everyone's needs respected." He reached over to take the folder Ben held. Opening it, he pushed it to the center of the table. "These papers specify that you'll stay on as chairman of the board, but that I'll be named president and CEO."

Sandy looked stunned. "Christ," he murmured, "that gives you all the power."

James remained silent.

Aidan had been looking at his father, clearly waiting for him to say more. When he didn't, he turned on James. "What about *me*?"

"What about you?" James replied.

"I'm in line for what you're taking."

"You'll stay where you are," James said. "You're a good front man."

"But you're taking what's mine."

James might have pointed out that he was only taking *back* what had been *his*. But it was beside the point. "This is for the good of the mill. We need a change in leadership, if for credibility alone. Those people out there? Look at them. They have power by dint of sheer numbers, and they're right—we have to clean up the mess we made. This isn't about you or me, Aidan, and it isn't about Sandy. It's about Middle River. As the mill goes, so goes the town. It's as simple as that."

no

Chapter 29

MY HEART WAS chock-full. I don't know how else to describe what I felt when I saw all those people with their sign. Not only had I *not* been expecting them, but, sitting in that anteroom and wondering if Phoebe's case alone could do it, I had been thinking that Grace was right, that small-town people were small-*minded* people who were just stubborn enough to keep their ugly little secrets even when it went against their own self-interest.

Yet there they were out there on the lawn, people who had no idea what James had planned but who were putting their jobs at stake nonetheless. In all the running around I had done Saturday and Sunday, visiting dear friends and favorite haunts in Washington, I hadn't felt the sheer joy I felt now.

So there was the sign on the lawn—literally and figuratively, the writing on the wall for Sandy Meade. And here on the table were the agreements James would insist Sandy sign. And in their elegant upholstered chairs were the board members, alternately looking subdued, cautious, or, in Sam's case, barely restrained. And Aidan was clearly dying.

I wanted to stay. I wanted to see Sandy sign those agreements with my own two eyes. I wanted to watch Aidan swallow the bitter pill he deserved. I wanted to throw my arms around James, give him a hug (maybe even a kiss, if he was amenable to that in front of others), and tell him how *good* this all felt.

I did none of the above, because James asked me to leave. He did it politely—"Annie, we need privacy now"—and I understood. The tone he set in his first few moments at the helm was crucial. There was no place here for a nonlawyer, non–board member, non–Middle Riverite. There was no place here for a novelist, much less the protégée of Grace Metalious.

That was the bad news. The good news was that I didn't have to wait in the anteroom like a good little girl. I got to go outside.

That was where things got really weird. I had been calm through my presentation to the board. I had no illusions about the men in that room. Other than James and Sam (and Ben, whom I'd only just met), I had no friends here. Nor did I wish to. I could present the facts as I saw them, with the emotion that I felt, and that was that. It was like when I stood at a podium, a successful author talking to hundreds of people at a charity dinner, I could keep my cool because my audience was largely nameless.

Out here was something else. I went out the front door of the handsome redbrick Cape and walked around to the back on a paved path. But when the people came into sight—actually, when I came into *their* sight—I felt a qualm. This was their victory, far more than mine. I had come here simply to find out why my mother had been sick.

I had already stopped. Now I took several steps back. That was when Sabina spotted me. She was with Phoebe—had an arm around her waist in support, because Phoebe was still far from well. Moving as a pair under Sabina's guidance, they broke from the crowd and came toward me. Their faces held a look of victory, even defiance, but it went only so far. Only then did it occur to me that they didn't have the foggiest idea of what had gone on upstairs.

Taking a deep breath, I raised my brows, pressed my lips together, and nodded.

Two sisters, two relieved sighs, one grand whoop of delight (this, from Sabina). Then it seemed they were running, though I know Phoebe barely had the strength, but in the next instant we were hugging, all three of us, sharing victory for what I do believe was the very first time in our lives.

"Sharing" was the operative word here, of course. We'd each had victorious moments in our lives—weddings for my sisters, child-birth for Sabina, best-sellerdom for me, to name a few. But we had never "shared" a victory, in the sense of feeling it equally with each other—in fact, feeling it all the *more* because we were *with* each other.

To this day, we've never talked about it. As close as we became in the months following the takeover of the mill, some things remained unsaid. I think we simply wanted to enjoy the closeness without analyzing it to death.

But back to that afternoon. The three of us had barely had our moment when others were beside us, asking about the meeting, giving joyous shouts of their own. I was hugged by more people in that short time, many of whose names I didn't even know, than I had been hugged by even during the last breast cancer walk that I did. Have you ever done one of those? I can't begin to describe the feeling of solidarity. Hugging is its outward show.

Well, solidarity was what we had there in Middle River that afternoon. We felt the heat of the air, smelled the warmth of the grass, the dampness of the river, the sweetness of the leaves on the trees on its banks. The smell of the fire lingered, but it had fallen behind those other, more potent and heavenly smells.

And the neat thing was that the men in that conference room didn't know any of it. All they knew was that the crowd had dispersed. They were embroiled in signing their papers and agreeing to change. By the time the meeting adjourned, we were gone.

Well. I say *we* in the communal sense. I

didn't leave. I couldn't. I just sat in my car while the last of the Middle Riverites returned to their lives, just sat there with the top down and the sun, no longer oppressive, listing toward the treetops to the west of my car. I pulled the pins from my hair and let my fingers be a brush.

Annie.

Shhhh.

Putting on sunglasses, I waited and watched and wondered who would be first to leave and what his mood would be.

Ben was the first. He had a plane to catch, but he saw me in the car behind his, trotted back, and gave me a peck on the cheek. He was pleased about the meeting's outcome. No doubts about that.

Annie.

Not now, I insisted.

Sam was next. One step out the door and he lit a cigar. I suspected he was as relieved by that as by what had happened inside. He was in his car, backing out of his parking place, before he spotted me. Then he drove right over and stopped, door to door.

"That was something," he said around the cigar.

"Good, don't you think?" I asked.

"Good, I do think." He looked at me with affection. "If your mother could have seen you today, she'd have been proud."

My throat closed up. I couldn't answer. Glassy-eyed, I smiled my thanks. His remark meant more than he could know.

"I have copy to write," he said on a lighter note. "Want to help?"

I shook my head.

"Didn't think so," he said. He winked at me, put his car in gear, and stepped on the gas.

Yes, my mother would have been proud. I was a minute in recovering from that realization. It helped that Aidan and Sandy were the next to leave the building. They kept me from having to hear Grace. Aidan strode out in obvious anger, slammed into that bold black SUV, and left rubber on the road. It was juvenile, but not at all surprising. Aidan was used to getting his way. For perhaps the first time in his life, he had not.

Sandy was more restrained, but his disappointment was no less marked. His shoulders were slack, his steps were slower, his movements—hand on the sedan door, body folding into the seat, head bowing once he was inside—were tired. I knew that had things gone just a hair differently, he would have been inside at this very moment, as determined as ever to keep Middle River in the dark about what ailed it.

I could never feel for this man, as I knew James did. But I could acknowledge the passing of an era.

Sandy drove off. Eyes on the door, I waited.

Annie.

I tried to ignore her.

Why won't you let me speak? We're friends.

You're my past, maybe even my present, but this is about the future.

I just want you to know . . . just want you to know . . .

I was distracted then, because Cyrus and Harry left. Looking somber, they exchanged several words before each disappeared in his own car and drove away. It was another five minutes before Brad left, and then another five before Lowell left. Not long after that, Marshall Greenwood came from the back of the building. Deep in thought, he eased himself into the cruiser and drove off without ever seeing me there.

And still I waited—waited and wondered—until James finally emerged.

Chapter 30

NICOLE HAD taken a sick day. She was still trying to process what had happened between Aidan and her, was trying to figure out how she felt about it and what it meant for her job. Aidan had been calling all morning, alternately angry and apologetic, but she had no desire to see him. She did go to a friend's house for lunch, though, and once word spread that she wasn't working, her home phone rang off the hook. So she knew exactly what was about to happen at Northwood, and even *then* she had no intention of going anywhere near the place.

But she kept looking at her watch through

the afternoon, and when four o'clock neared, she was in her car and on her way. She didn't make a conscious decision to take part in the protest. But something caught her up and kept her there.

It wasn't until the whole front of the meadow was filled and the sign had been raised—wasn't until they had all stood there in silence looking up at the windows behind which the fate of Middle River was being decided—wasn't until Annie Barnes came around the corner of the building and was mobbed first by her sisters, then by everyone else—that Nicole knew what the something was that had caught her up and kept her there, and it had nothing to do with mercury.

It was her daughter, likely not even aware Nicole was there but looking totally comfortable—and totally *right*—in that group surrounding Annie. It was her daughter, looking totally *attractive* with these people, certainly accepted by them and clearly happy. Her daughter. *Her* daughter. Was this Kaitlin? Grown? Independent? Perhaps acting more on what she believed than Nicole did herself?

Nicole couldn't stop staring at her. Inevitably, Kaitlin glanced her way, did a startled double-take, and stood looking back.

Nicole tried to smile, but failed miserably. She was too confused.

That must have come across, because when the others around Kaitlin started to head for their cars, the girl held back. She seemed as confused as her mother.

At last Nicole approached her. "I didn't know you'd be here."

Even in confusion, there was a touch of defiance. "We closed the store early. I thought you were sick."

Nicole might have said that she had been sick that morning but was feeling better now, only that wasn't true, and suddenly she was tired of making excuses. "I needed a break from the office. When I found out about this, it just seemed like the best place to be."

"Does he know you're here?"

Nicole knew she meant Aidan. She thought of the possibilities—his getting a call from the guard at the gate, seeing her out the window, hearing it through the grapevine after the fact. Did he know at that moment?

She shrugged. She had no idea.

"What about your job?" Kaitlin asked. "He fired Sabina for daring to mention the word *mercury* to someone, and, like, here you are, taking part in a protest? He'll fire you if he finds out."

"If he does, he does."

Kaitlin seemed shocked. "I can't believe you're so calm. Your job means the world to you."

"It did. Maybe it doesn't anymore. I just don't know."

The girl was suddenly nervous. Her expression changed in the space of a breath. "Has something happened? You and Dad . . . ?"

Nicole shook her head. She smiled sadly. "No. Everything's the same. I'm not even sure your father knows anything about this." It was pathetic, truly. Anton was a nonplayer in her life. They passed in the house like ships in the night. It wasn't much fun. And was it reparable? She just didn't know. But Kaitlin was waiting for her to say something. "So," she said, "you like working at the store."

"Yes."

"They're your friends, the Barneses?"

"They like me," Kaitlin said, defiant again.

"So do I," Nicole argued. "You're my *daughter,* for God's sake."

"You think I'm a loser. They don't."

"I *never* said you were a loser."

"You don't have to say it to make the point, Mom." Her chin came up in an ominous way. "I have a boyfriend. You didn't even know that, did you?"

No. She hadn't. A boyfriend? What did that mean?

"His name is Kevin Stark," Kaitlin hurried on, "and he's from th'other side, and before you get all hot and bothered and tell me that you'll have Aidan fire his dad, you need to know that I *know* what I'm doing. I am not marrying Kevin. He is my boyfriend right now. That's all. I'm going to college, so I can get out of Middle River, and I'll get married after I have a career, so that it'll never matter again whether I'm ugly or fat, because people will see me for who I am," she thumped her chest, "*me,* first."

Nicole didn't know where to begin. She wanted to know about the boyfriend part, but this was paving new ground. So she said meekly, "You're not ugly or fat."

"The thing is, Kevin doesn't care. And my friends at school think it's really cool that I'm with someone."

"*With.* What does *with* mean?"

"*With.* You know, Mom."

"Are you sleeping with him?" Nicole asked.

Kaitlin didn't blink. "Are you sleeping with Aidan?"

"Hold on, there. I'm an adult. I know how to take care of myself. You're a *child*."

"I am not!" Kaitlin cried. Her body had gone rigid, but there were tears in her eyes. "That's the problem with you and me. I am not a child. Why can't you see that? I'm more than able to *have* a child, which is what *three* of my classmates in school are doing, but you probably don't know that, because you'd look at the loose clothes they wear and just think they're fat. I know how to protect myself, Mom. And I know what I want. Kevin treats me like a someone. Same with the people at Miss Lissy's Closet. I *want* to be with people who treat me like a someone."

Abruptly she stopped. It was as though she had reached the heart of the matter and had nothing more to say.

The ball was in Nicole's court. The problem was, she had never been good at sports. So she tried to clear her head and distance herself from the emotional tangle that had always tripped up Kaitlin and her. She tried to think of what she would say if Kaitlin were the daughter of a friend.

"I would like to treat you like a someone," she said quietly.

"So *do* it," Kaitlin begged.

And Nicole wanted to. "But I can't do anything if I don't know what you're thinking and feeling. Maybe I still see you as a child because you don't share your feelings and thoughts."

"It's my fault?" Kaitlin asked in dismay.

Quickly, Nicole said, "No. Mine. I don't ask. So maybe I can change that." She was the one who stopped this time. She had hit on the heart of the matter.

Kaitlin must have sensed it, too, because the chin came down and there was suddenly such a look of wanting in her eyes that Nicole was touched. Without another word, she closed the space between them and wrapped her arms around her daughter, and it felt good, felt *really* good. She didn't know what to do about her husband and didn't have a clue what to do about Aidan. But Kaitlin was different. Here was something to save.

Sabina felt victorious as she left Northwood. She was proud to be a Barnes and the sister of Annie—was proud to be a Middle

Riverite and a friend to all those who had come out to the mill. She was even humbled to be the mother of Lisa and Timmy, who had seen all there was to like in Annie long before she had.

Victory. Pride. Humility. Beyond it all, though, was a hollowness. It didn't take a name or a face until she pulled in at the pale blue Cape on Randolph Road and the embodiment was there in the flesh, leaning against the trunk of his car with his arms folded and his ankles crossed, looking for all the world like he was waiting for her.

Ron.

He didn't look angry, exactly. She couldn't quite gauge his mood.

She parked and went forward, stopping just beyond arm's reach. "Hi," she said tentatively.

"Hi," he said in kind, then, "I guess there was quite a show?"

She nodded. "Did you hear about James?"

"Yes. Word spread through shipping in no time. People are pleased."

"Are you?"

He nodded. "The change will be good for the mill. Sandy deserves a rest, and Aidan, well, maybe James will rein him in." He nodded again, more slowly this time, seeming to not know what to say.

Sabina knew what he needed to say, certainly knew what she needed to hear. What he proceeded to ask wasn't it.

"Do you think you'll get your job back?"

"I have no idea!" she burst out. "That wasn't the point of the protest. The people who came to the mill today did it knowing that they were risking retaliation by the Meades, and still they came. This was about something bigger than my job or your job, Ron. It was about right and wrong. It's about the world we leave for our kids."

"I know," he said, sounding duly chastised.

"I'm not sorry for anything I did, Ron, certainly not for talking to Toni even if she *did* go to Aidan. What happened today would have happened whether I'd lost my job or not, because one person in this town took on the cause. Annie has more courage than the whole of us on this block combined!"

"I know."

"I sold her short. I'm ashamed of that."

"I sold you short. I'm sorry for that."

Hearing the words, it struck her that what she had seen on his face earlier was remorse. She hadn't recognized it, because it was new.

Ron had never before had cause for remorse. This was the first real blow-out they'd ever had. That was really quite remarkable.

He reached out with arms that were longer than hers, snagged her wrists, and pulled her close. "You've been a better person than me in all this. Can you live with me, knowing that?"

The hollowness in Sabina dissolved. She smiled. "I think I can."

James didn't breathe freely until the last of the board had left. That last member was Lowell. His father's lawyer, longtime friend and confidant, he held tremendous sway when it came to helping Sandy accept defeat.

"Not defeat," James had suggested to Lowell. "Can't we call it semi-retirement?"

But Lowell knew Sandy well. "He'll call it defeat. You definitely trumped him, James."

"That wasn't my bottom line. If he'd agreed to clean things up, I'd have stayed where I was."

"This is fine. It's time. Your papers are duly signed and witnessed. I'll add my resignation in the morning."

"No, Lowell. There's no need for that."

"I'm old-school. You'll want your own team."

"But this team is good. A change in quarterback may be enough. I want you to stay. I'll need your help. Let's not throw out the baby with the bathwater."

"Speaking of which—"

James held up a hand. "Mia's for another time." He extended the hand. Lowell shook it, lifted his briefcase, and left—which was when James went to the window and took a long, relieved breath. Triumph, smugness, and certainly satisfaction were mixed in. Plus excitement. That was there, too. Mostly, though, it was about relief. He had been fighting his father for too long. Finally it was done.

A movement off by one of the trees caught his eye. It was Marshall Greenwood, in his denim shirt, jeans, and brown boots, looking up at the window.

James gestured for him to come inside and pointed toward the back door. He was waiting at the top of the stairs when Marshall labored up.

"You were waiting for me," James surmised.

One step short of the top, the chief of police put a supporting hand on the banister. His breath was short and his voice raspy. "We know your father's out. I'm wondering if I am, too."

"Not unless that's what you want."

"No matter what I want. You're the boss now."

"Only of the mill. I won't be controlling the town manager like my father did. I have enough work to do without that."

Marshall seemed dubious. "You won't fire me just because . . . because . . ."

"Because you were hard on Annie? You backed off. That's done. But what you have to do next is to make it so that you aren't afraid of people finding you out."

Marshall stared at him. "What do you mean?"

James didn't say. He simply stared right back.

Marshall was the first to blink. "And how do you suggest I do that?"

James went to the table, took a piece of paper from the cube in the middle, and jotted down the name of a treatment center. He handed the paper to Marshall. "It's in Massachusetts. You and Edna can take a 'vacation.' No one'll even know you're there."

"Is this a requirement if I want to keep my job?"

James shook his head. "Just a suggestion from a caring party."

Marshall stared at him a minute longer, then turned and plodded down the stairs. James had no idea whether the man would take his advice, but he had given it. That was all he could do.

Besides, he was suddenly eager to leave. Back in the conference room, he gathered up his things. He didn't know where Annie had gone, but he needed to find her. They had to talk.

Chapter 31

I WAS INSTANTLY alert when he appeared, and I have to say, he was something. In a business suit, he was as dignified as any Washington luminary I had seen. The slim build, chiseled features, and salt-and-pepper hair only augmented the image. Eyes on the SUV parked in the lot, he was striding forward when he glanced to the side and saw my car. Ours were the only two in sight.

His features lifted—and, oh my, what that did to me. I actually put a hand to my chest to calm my traitorous heart lest it take off and run on its own.

As he strode my way now, he loosened

his tie and took off his jacket. He was grinning by the time he reached me. Leaving the jacket on the hood of the car, he opened the passenger door and slid in. He sat at an angle to face me, then put a hand on the back of my seat.

I saw triumph on his features and was fully expecting him to remark on the victory that had just been won for the mill and the town, when he said, "You were remarkable."

I think I blushed—I think, because I hadn't had much cause to do that when I was growing up, and when I was an adult, well, blushing wasn't particularly cool. "I just talked about what I knew," I said.

His smile vanished. "I'm sorry I had to be secretive beforehand, but I've been wanting to do this for a while. When the opening came, I knew I had one shot. If something went wrong, there wouldn't be a second chance. So I was paranoid. I didn't trust that they wouldn't find out and sabotage our efforts. My father is very good at that. Money talks. I was afraid."

"I understand."

"But I was wrong. Relationships, capital *R*, imply trust."

Capital R? I was suddenly feeling equal

amounts of wanting and fear. The wanting was pure heart, the fear the better part of the rest, and that had to take precedence, didn't it? My fear involved practicality. It involved reality. It involved fact.

I shrugged. It was my best shot at being casual. "Ours is so new. We barely know each other."

James's eyes were midnight brown and intense. "Then if I asked you to stay here and move in with me, you wouldn't?"

Forget casual. My breath totally left me for a minute. That hand went back to my chest. "James," I protested, "don't *say* things like that."

"I mean every word."

"But we *do* barely know each other. You haven't a clue what happens to a writer when she's in the throes of writing a book."

"Is it like being in the throes of orgasm?" he asked, straight-faced.

"No," I replied. "Absolutely not. It's like you're living with someone who isn't there."

"Greg has managed."

"Greg isn't—he doesn't—it's just *different,* James. But there's one basic fact. I live in Washington. I own a condo and I have a life. For another, I'm a Barnes and you're a

Meade. Barneses and Meades don't—don't cohabitate. For a third, there's Mia. If I were to move in, find that you and I hated each other, and move out, Mia would be hurt."

"I don't think that will happen."

"What, Mia being hurt?"

"Your moving out. I think this is good, Annie. I haven't felt this way about any woman. Ever."

"But I'm *Annie Barnes*!" I cried, saying the name with a teenager's pique.

"Yes, you're Annie Barnes," he repeated, but instead of the pique, there was reverence.

Reverence? For real? What woman didn't want to be revered by a man? What woman didn't *dream* of it? But dreaming didn't make it so.

"What about April?" I asked, grasping at straws. "You were with her for six years before you realized it wouldn't work."

James was shaking his head even before I finished. "I was with April that long *because* I realized it wouldn't work. At least, deep down I realized it. I never asked her to marry me."

I held up both hands. "Don't *say* that word. It's terrifying right now." Prior to dropping my hands, I used one to swat at a fly that buzzed around my head. "Truly, James," I said in earnest. "Look what you have ahead of you. However you look at it, you have just taken on a major headache. You have the whole mill to head now, not just product development, and then there's the work involved in finding out who all was affected by those spills and making arrangements for their care. This is not the time for you to make a major change at home."

"I disagree. It's the best time," he reasoned. "I'm going to want to come home to someone."

"Mia."

His look said it wasn't the same.

"Okay," I said, "but the three arguments I made before still fit. We're adults. We have to be sensible." The fly was buzzing around his head now. I watched him swat at it. "I think that's Grace, by the way."

James snorted. "It isn't grace. It's annoyance."

"Grace Metalious. She takes various forms. A cat purring. A fly buzzing."

He seemed both curious and amused. "Grace Metalious has been dead for years."

I looked him in the eye. "I have conversations with her."

"Do you?" he asked indulgently.

I nodded, daring him to laugh at me. "We started talking when I was a kid."

He was suddenly as serious as I was. "If you're trying to scare me away by suggesting you're nuts, it won't work. You're a novelist. Having an imagination goes with that. I'd also guess that when you were a kid you needed a friend. I can't imagine it was easy for you growing up here. So tell me what she said."

He had pegged the situation so well that I couldn't help but answer. "She told me I was a good writer and that I'd be someone someday. She was like an older sister."

"An imaginary friend."

"She gave me encouragement."

"And now?"

I had to think about that. The answer wasn't as simple. "We argue a lot now," I said. "It's like she's pushing me to do things for the wrong reasons. She wants me to be angry."

"Angry?"

"About every little thing that's wrong with Middle River. Omie said Grace was using me as a vehicle for fulfillment, but if fulfillment means revenge, I can't *do* that, because there are things wrong with *every* town, if you dig deep enough. And there's another thing: being angry is *exhausting*. I don't want to be that way for the next twenty years. Nothing is perfect. No *man* is perfect." I paused, realizing I was arguing with Grace. "By the way, she thinks you're gorgeous. She was telling me that way back when I first saw you running."

"She was, was she?" James asked, pleased in a very male sort of way. "What else did she say about me?"

"She asked questions—where did you live, were you married, that kind of thing. She kept telling me to run faster to catch up."

"Good advice."

"She also called you Archenemy Number One. She liked the drama of that. She still thinks I should be writing a book about what's happened here. That's mostly what we argue about."

James was silent. A frown crossed his brow. Finally he said, "It would make a good book."

"I told you I wouldn't," I said in a fierce burst. "I told her I wouldn't. I told my sisters I wouldn't."

"But you've earned the right—"

"I don't *want* to write a book about this,

James. I've *lived* it. Why would I want to *re-live* it?"

"Isn't that what authors do?"

"Some. But not me. And certainly not in this case. Besides, the story isn't *over*. It remains to be seen what finally happens to the mill, once word about this is fully out."

"I know that," James said with a return of gravity. "But that brings us full circle. Whatever does happen, I want you with me."

Exasperated, I tossed a hand in the air. "How can you tell?"

"I can tell. I know. You're different from any woman I've known."

"Uh-huh. Different, bizarre, prickly—oh, and what was it your dad said before—unbalanced?"

"He was dead wrong."

"But how do you know?" I asked, suddenly distracted. He was unbuttoning that pale blue shirt. "What are you doing?"

"I want you to feel something."

"James," I said in a half whisper and with a glance around. The day was definitely winding down, but it still had a ways to go. "*Here*?"

Catching my hand, he slipped it inside his shirt. "Feel that?"

I did. Omigod. I did. I felt the roughness of chest hair on warm, firm skin, but that wasn't what he meant. It was his heart, beating strongly against my palm.

"That is what happens to me when I'm with you. It's like I'm more alive than I was the minute before."

I wanted to cry and say, *Yes! Yes! That's how it is!* But I was frightened. Too much was happening too soon. Even aside from the three very logical reasons I had listed earlier, there was the one about love.

Love—yikes! I hadn't come to Middle River for love. Moreover, had I known that first night that I would find love *here,* I might have turned right around and driven back to Washington.

Did I love James? How could I possibly know that? I hadn't seen him in the life situations that a couple faced.

My parents had loved each other. I wanted what they had. And yes, I wanted the perfect Adam that Grace had sought. Maybe James was it. But could I tell now, when we were riding an adrenaline rush after beating Sandy and Aidan? How many people do *you* know who came together in unusual circumstances and thought they were madly in love, only to find that when they got into the nitty-gritty of living together, they were incompatible?

Besides, James hadn't said the word love. Had he. I don't ask that. I *say* it. It is fact.

Leaving that hand on his heart, I touched his face with my free one, put my soul in my voice, and begged, "I need time, James. Can you give me that?"

It took four months, during which I vacillated between falling more and more deeply in love and fighting what I felt. It was scary. I had liked people before. I had even loved people. But nothing compared with my feelings for James. They crept into every aspect of my life, from running to eating to reading to sleeping to working to being with friends. And to sex. Can't forget that. It kept getting better. Can you believe that?

Oh, we discovered differences. I like my coffee black, he likes his light. I like mine in a ceramic mug, he likes his in a travel mug. I like Starbucks, he prefers Dunkin' Donuts. And that's only coffee. There were other differences—like his taking jelly beans over chocolate any day—but they were trivial when it came to the Big Picture.

First, Mia. By early fall she was walking, and let me tell you, there was *nothing* better than returning to Middle River after a week or two in D.C. and having Mia make a bee-line for me with her little arms raised for a lift and a hug. I suspect that she and I bonded over James, but bond we did—and if you're thinking how convenient it was now for James to have someone to take care of Mia when he was late at work or preoccupied or tired, don't. I happily took care of her, because I loved the child, but he rarely missed time with her. He genuinely enjoyed feeding and playing and bathing, and if there was a bad diaper—I mean, a really *bad* diaper—he never asked me to do it, he just did it himself. And I let him. Hey, I'm no glutton for punishment. But my point here is that we shared the chores. Never once did I feel used.

Second, work. To this day I'm in awe of the way James addressed the mercury problem. He did it—really did it—used profits from the mill to provide for the people who had been hurt, and he did it in a way that made lawsuits unnecessary. He was generous to a fault. Naturally, neither Sandy nor Aidan was pleased. They accused him of stealing the guts from the mill, but James stuck to his guns. He brought in experts to handle the legal and economic angles, and they got it right.

As for my work, it didn't matter where I lived. I actually did all of the proofing for my book in James's home office because even with Mia and the nanny around, it was quieter there than in my city condo.

Third, family. James's was difficult, the proverbial thorn in his side, because neither Sandy nor Aidan took defeat lightly. Once they got their second wind after the takeover, they pushed and prodded, strong-armed Cyrus and Harry to their side, and generally did their best to sabotage what James had planned. In theory, it would have worked. With eight on the board, there was no tie-breaker. One had never been needed before. Now, though, James outfoxed his father by naming a representative from Middle River as the ninth member, and he did it in headlines on the front page of the *Middle River Times.* Once that happened, it was too late. Middle River would have raised holy hell if Sandy had nullified the appointment.

And my family? Unbelievable. Sabina got her job back. And Phoebe very slowly recovered, though that process continues as I write this. How did they take to James as my significant other? Easily. Like they weren't at all surprised. Like the new light in which they viewed me was perfectly compatible with how they viewed James. Like they loved the idea of adding a Meade to the family.

The last took me most by surprise. I had assumed that they felt the Meade-Barnes antipathy as strongly as I did. I had assumed that the whole town felt it. But it was me. My mind. My anger. When I let go of that anger, I could see **TRUTH #10: What's in a name? Not a helluva lot. It isn't the name that matters. It's the person.**

Speaking of persons, want to know what my sister Sabina did? She convinced James to let her take Mia for a weekend (this, in mid October) so that he could surprise me in Washington. And was I ever surprised! James was deep into the mercury settlements at the time, and might have used the weekend I was gone to crash. Yet there he was at the door of the condo, replete with chocolate pennies and the warmest brown eyes. He wanted to meet Greg, he said. He wanted to meet Berri, Jocelyn, and Amanda. He wanted to see where I slept when I wasn't with him, wanted to see what it was that I loved about Washington.

Naturally, everything in Washington was better with James there.

I want to tell you about that, but first, let me add one more thing about Sabina. What she and I had found in each other, in those few days between her firing and James's coup, continued to grow. She actually became one of my closest friends, which made it easier for me to make the break.

So. Washington.

I love Washington. I always will. But even back in August, when I returned to be with Greg and his broken leg that one weekend, something had changed. Run as I might all over the city, as I had done that weekend and continued to do during the time I spent there in the fall, trying to remind myself of all that I loved and convince myself that there was absolutely no other place on earth where I could live, it didn't work. Yes, I had friends in D.C. But they had frequent-flyer numbers, too. And New Hampshire welcomed visitors. After he met James (and likely saw where my heart was leaning), Greg came. So did Berri. And each week that I spent in Middle River, I made another friend.

So the good news was that **now I had**

two homes, rather than one. Do we count this as **TRUTH #11?**

Why not? Because I was wrong in this instance, too. When James asked me to live with him, I assumed it meant giving up my Washington life. The truth is that between phone, fax, FedEx, digital cameras, and the Internet, geography has been redefined. Isolationists can babble all they want about the evils of globalism, but the world has become a smaller place. My Washington life goes on, though I have now sold my share of the condo to Greg and moved north.

Which brings me to Middle River. It's a fabulous town. But you're not surprised to hear me say that, are you? And I'll be honest. It helps to be in love with one of the leaders of the town and know that even if I never wrote another book, I wouldn't have a financial worry in the world. Folks on th'other side may not be as bullish on Middle River as I am, but they would agree with me about one thing: Middle River is home.

And that was what I told Grace. It was a flurrying day in December. I was home for the holidays (yes, home—it rolls off the tongue like honey), and I had been thinking about Grace a lot. We hadn't talked since that day in August when I had shut her out. I was feeling guilty for that, was feeling that there was unfinished business between us. Since she wouldn't come to me, I went to her.

Grace Metalious is buried in Gilmanton, the small town where she lived for longer stretches of her adult life than anywhere else. The town is south of the Lakes Region, which makes it a good drive from Middle River. James knew I was planning to go. I had been talking about it, had printed directions from MapQuest, had Xed out the date on the calendar. When the weather that day dawned iffy, he insisted on driving me there himself.

It was a bleak winter day. The trees were bare, their once colorful leaves now faded and dry at their feet, covered with a blanket of snow that was thin but growing. Once we passed under the iron arch that marked the cemetery's entrance, we had no trouble finding her stone. It stood alone in an open patch, but there were evergreens nearby and a gentle slope with trees that would be green again come spring, and beyond the trees the peaceful water of Meetinghouse Pond, near frozen now and still.

The setting was one Grace might have described in a book to be read by ten million people. But she was here all alone, her name the only one on the stone.

James parked. He stayed in the car while I pulled up my hood and walked over that frail covering of snow. I had brought gerbera daisies; of all the flowers I had considered at the McCreedys' store, these seemed to best suit Grace. They were wildly vivid, a mix of reds, oranges, yellows, and pinks, but they were simple in form. A contradiction? No more so than Grace.

I propped the flowers in the snow so that they rested against the protruding base of the large granite stone. The name Metalious was printed in block letters high on that stone and, beneath the surname, a smaller Grace. Beneath that were the years of her birth and death, nothing else.

"I'm sorry," I said softly. "I was not nice last time we talked. I cut you off. But you deserved better. You were a good friend to me when I needed one."

I paused. A red squirrel cheeped away somewhere in the trees, and though the lightly falling snow muted the woods, the breeze caused a creak now and again. My parka rustled when I drew it tighter around me. But I didn't hear Grace.

"I've learned so much," I went on in that same quiet voice. "Mostly it's about me and how bullheaded I've been about certain things. I'm good at writing books. So are you. But we trip up when it comes to personal gripes."

I paused again. Still Grace didn't reply.

"Remember when we had that last big argument? It was after the fire, after I'd spent the night with James. You said I loved him, and I denied it, and you kept badgering, goading me to say what it was I wanted in life, and I just didn't know. I know now. A lot of what I want is what you had. I want books, and I want children, and I want James. But there's something else I want, and maybe it's what you never had. You wrote about Peyton Place, and I lived Peyton Place, and the two of us kept coming *back* to Peyton Place, even though we swore we hated it. So what is its appeal? It's home, Grace, *home.*"

The breeze whipped a handful of snowflakes at my face, and for a minute I wondered if that was Grace's way of telling me to pack up my brilliant insights and leave.

But there was something soothing in the aftermath of that brush with cold. I felt cleansed.

"You had started to say something to me that day. You know, the last time we talked. It was after the meeting at the mill. I was sitting in my car waiting for James to come out, and you kept saying, *I just want you to know . . . just want you to know . . .* What did you want me to know?"

I waited, but she didn't answer.

So I whispered, "Can you hear me, Grace?"

After another minute, I smiled sadly and let out a misty breath. Grace was dead.

And still I stood looking at her stone, though the cold seeped in and I started to shiver. In time, I looked at the trees, then the pond, then the stone again. This time my eyes were drawn to the daisies I had brought. I was glad I had chosen these. They were so bright, they fairly glowed in the snow.

Releasing a final misty breath, I turned and looked back. There at the end of the boxy track in the snow left by my Uggs, like so many crumbs leading me home, was the big, warm SUV with James inside.

"I just want you to know . . ." I whispered a final time, because my boots wouldn't move yet, "just want you to know . . ." I paused. What to say? That I would miss her? That she would forever be part of me? That I wouldn't have been who I am today without her, and that I loved her for that?

I didn't say any of it. I didn't need to. If the spirit of Grace was anywhere near this place, she would know.

With that realization, my boots came unglued and I started back toward the car. The walking was easier than it should have been, because I had been relieved of a weight. I had needed to come here. Now I could move on.

Lighthearted and more eager with each step, I was halfway to James when it came to me. *He's the one.* That was what Grace had wanted to say. I could hear it now. *He's the perfect one.*

I don't know about perfect. There was the thing about drinking coffee from a travel mug in the comfort of his own kitchen. But I could forgive him that.

Smiling at the thought, I jogged over the snow. When the door opened, I slipped inside.

"I saw the starry tree Eternity, put forth the blossom Time, she thought, and remembered Matthew Swain and the many, many friends who were part of Peyton Place. I lose my sense of proportion too easily, she admitted to herself. I let everything get too big, too important and world shaking. Only here do I realize the littleness of the things that can touch me."

From *Peyton Place,* by
Grace Metalious

A NOTE FROM THE PUBLISHER
Afterword

When we hear the words "Peyton Place," certain images come to mind: debauchery in small-town America, the pent-up passions that lurk behind white-picket fences, and sensational scandal. Banned across the globe and denounced as "moral filth," coveted by curious teenagers and hidden under the pillows of desperate housewives, Grace Metalious's *Peyton Place* was nothing less than a bombshell of a book at the time of its publication in 1956. Its provocative energy has since permeated our whole culture, touching everything from art and politics to gender relations and the role of women in society. Its enduring impact can still be felt today. As film director John Waters put it, *Peyton Place* was "the first dirty book the baby-boom generation ever read; the 'shocker' they never got over."

But in the beginning, the bombshell had trouble getting off the ground. Turned down by five publishers, the book finally found a champion in Kitty Messner, a straight-shooting feminist and one of the only women to serve as president of a major publishing house. Messner knew a good thing when she saw it.

With an unprecedented first printing of a million copies, the book went on to sell well over 10 million, making *Peyton Place* the biggest bestseller of its time—and still one of the top bestsellers of the twentieth century. No matter how hard the obscenity police tried to block its path, the book just bulldozed on through, delighting and titillating its count-less devoted readers. But *Peyton Place* was much more than just its notorious reputation as a "dirty" book. It was a revolution.

"Sinclair Lewis would no doubt have hailed Grace Metalious as a sister-in-arms against the false fronts and bour-

**geois pretensions of allegedly re-
spectable communities."**

—Carlos Baker,
The New York Times Book Review

The book's author, Grace Metalious, was a thirty-two-year-old stay-at-home mother of three. Truly a desperate housewife herself, she wrote *Peyton Place* in the hopes that it would lift her out of the "cage of poverty and mediocrity" in which she felt trapped. Driven and passionate, she neglected her wifely duties to write ten hours a day, leaving dishes to pile up in the sink and dust to accumulate throughout her ramshackle home. Independent and single-minded, she was not a typical woman of her time.

In the famous 1956 photo of the author, Metalious is shown dressed in rolled-up blue jeans, a man's flannel shirt, and sneakers. Her hair is unceremoniously pulled back in a practical ponytail, her cigarette is burning in a nearby ashtray, and her typewriter sits before her, waiting to pound out the next steamy scene to emerge from her uncensored mind. The photo is entitled *Pandora in Blue Jeans*. It is an appropriate title. Like her mythological namesake, Metalious lifted the lid off the box of repressive 1950s society, unleashing the demons of adultery, abortion, incest, rape, and the unbridled yearnings of female sexuality. A controversial figure, to say the least, she became an instant celebrity, both applauded and reviled, as some readers deluged her with fan letters while others hurled stones and obscenities. She was, for a while, the most talked about woman in America. Soon the pressure of fame became too much for Metalious, and she turned to alcohol, losing her fortune and eventually her life before she would reach her fortieth birthday.

Today, though the talk about Metalious has long since quieted down, she is often hailed as one of the first liberated modern women and a courageous predecessor of the feminist movement. However she is remembered, as popular heroine or hell-bound pornographer, one fact is undeniable: It took tremendous courage for a wife and mother, in the 1950s, to write and publish a book like *Peyton Place*.

"A vivid, vigorous story of a small town and an expert examination of the lives of its people—their drives and vices, their ambitions and defeats, their pas-

sivity or violence, secret hopes and kindnesses, their cohesiveness and rigidity, their struggles, and oftentimes their courage."   ***—Boston Herald***

The story begins in a picturesque New England town, where Indian summer has come to heat up the chilly autumn landscape. "Indian summer," wrote Metalious in her famous first line, "is like a woman. Ripe, hotly passionate, but fickle, she comes and goes as she pleases so that one is never sure whether she will come at all, nor for how long she will stay." The author walks us down the leafy streets of this seemingly peaceful suburb, introducing us to the players in her drama. On one side of the tracks, there are the wealthy Leslie Harrington and his spoiled son Rodney, as well as the good-hearted doctor Matthew Swain. On the other side, living in a tar-paper shack, is young Selena Cross and her wretched family. And in the middle class stand the book's two central characters, single mother Constance McKenzie and her teenage daughter, Allison.

On its surface, Peyton Place looks picture-perfect, but, as Metalious herself once put it, "if you go beneath that picture it's like turning over a rock with your foot. All kinds of strange things crawl out." And that's exactly what Metalious does next; she turns over that rock and lets the dark truths of the American small town come crawling.

Overprotected Allison McKenzie, desperate for a friend, grows close to Selena Cross, who is just as desperate to escape poverty and the clutches of her violent and sexually abusive stepfather, Lucas. Selena works in Constance McKenzie's dress shop, seeking maternal love at a time when Allison pushes her own mother away. Constance, rigid and cool, forbids her daughter to run around with boys, especially boys like Rodney Harrington, who knocks up the town's bad girl. Constance, once a bad girl herself, is terrified that Allison will end up like her—a single woman with a child born out of wedlock, forever hiding from scandal. This truth about her daughter's birth is a carefully guarded secret, until Constance begins to thaw in the arms of the new school principal, Tom Makris. Arguing with Allison, she blurts out the secret, wounding her daughter, who flees the small town, running away to be a writer in New York City.

After four years pass, smarting from a disastrous affair with her literary agent, Allison returns to Peyton Place to attend the murder trial of her old friend, Selena Cross. The girl admits to killing her stepfather and burying him in the sheep pen, claiming self-defense. But she doesn't specify what she was defending herself against, ashamed and afraid of losing her fiancé, Ted Carter. The trial turns around when Doc Swain testifies that he performed an illegal abortion for Selena, who was raped by her stepfather. This shocking admission blows Peyton Place wide open. Unable to hide from their secrets anymore, the townspeople must stand in the harsh light of truth. For Constance and Allison, two fiercely independent women fighting to make it in a man's world, this means reconciliation and a sense of peace.

"Captures a real sense of the tempo, texture and tensions in the social anatomy of a small town." *—Time*

Sensational and unstoppable, after publication *Peyton Place* expanded into an Academy Award–nominated film, a bestselling sequel, and a wildly popular television series. Starring Ryan O'Neal and Mia Farrow, the 1960s TV show was the first primetime soap opera, paving the way for future hits like *Dallas, Twin Peaks,* and, most recently, *Desperate Housewives*—where Wisteria Lane is truly an extension of Elm Street and Maple, the thoroughfares that crisscrossed the town of Peyton Place.

When asked in a television interview if she thought her creation would be remembered, Grace Metalious, without a moment's hesitation, responded, "I doubt it very much." Luckily for us, Grace was wrong. Half a century after its publication, *Peyton Place* lives on, still influencing popular culture. As we approach its fiftieth anniversary in 2006, there is renewed interest in the book and its groundbreaking author. Actress Sandra Bullock is planning a major motion picture based on Metalious's biography, and writers like Barbara Delinsky are finding new inspiration from the pioneering woman writer. Metalious, says Delinsky, "was a free-thinker who was way ahead of her times where the plight of women—indeed women's rights—was concerned." In *Looking for Peyton Place,* Delinsky has crafted a creative homage to this almost-forgotten heroine, the Pandora who helped to liberate women's hidden desires and domestic sorrows, to free them from the darkness and bring them out into the light, where shame has no place. For Grace Metalious, it was just something that had to be done. "I don't know what all the screaming is about," she said in an interview soon after publication. "*Peyton Place* isn't sexy at all. Sex is something everybody lives with—why make such a big deal about it?"

A Reading Group Guide for *Looking for Peyton Place* that explores the similarities between Barbara Delinsky's novel and Grace Metalious's book is available at www. barbara delinsky.com and at www. simon says authors.com/barbaradelinsky.